Computational Music Science

Series editors

Guerino Mazzola
Moreno Andreatta

More information about this series at http://www.springer.com/series/8349

Guerino Mazzola

The Topos of Music II: Performance

Theory, Software, and Case Studies

Second Edition

Springer

Guerino Mazzola
School of Music
University of Minnesota
Minneapolis, MN, USA

This is the second of four volumes in the set "The Topos of Music" in the series Computational Music Science. Volume I is ISBN 978-3-319-64363-2, Volume III is ISBN 978-3-319-64479-0, Volume IV is ISBN 978-3-319-64494-3. The first edition of The Topos of Music appeared in one volume in the Birkhäuser program ISBN 978-3-7643-5731-3.

ISSN 1868-0305 ISSN 1868-0313 (electronic)
Computational Music Science
ISBN 978-3-030-09719-6 ISBN 978-3-319-64444-8 (eBook)
https://doi.org/10.1007/978-3-319-64444-8

Mathematics Subject Classification (2010): 14K99, 53A45, 68T35, 54H99

Printed on acid-free paper

This Springer imprint is published by the registered company Springer International Publishing AG part of Springer Nature.
The registered company address is: Gewerbestrasse 11, 6330 Cham, Switzerland

Preface to the Second Edition

Comprendre, c'est
attraper le geste
et pouvoir continuer
Jean Cavaillès [181, p. 186]

A major reason for a second edition of *The Topos of Music*—besides the simple fact that the first edition is now sold out—goes back to spring 2002, when I was completing its first edition, published in fall 2002. I was asked to give a talk in the MaMuX seminar of the IRCAM in Paris, to explain how I applied the mathematics of *The Topos of Music* to my free jazz improvisations.

While preparing my talk I realized that despite the presence of mathematical music theory the decisive generator of my instant compositions was the gestural deployment of formulas, the "action painting" of musical thoughts, not the abstract formulas in their static facticity. First and foremost this was a shocking insight in view of the forthcoming publication of the formulaic setup in *The Topos of Music*.

Fortunately, I knew from Hermann Hesse that "every end is a beginning"[1], which meant in my case that the end of a scientific development as traced in the book's first edition initiated the next step: a music theory of gestures. It goes without saying that this new phase would not destroy the previous research, but incorporate it as the stratum of facticity in an extended ontology of embodiment, where facts are the output of processes and their gestural generators.

In the sequel, I discovered that I was far from being the first scholar and artist to discover the crucial role of gestures in music. For instance, free jazz pianist Cecil Tayor, music philosopher Theodor Wiesengrund Adorno, or lateral thinker Paul Valéry had clearly stressed the dancing essence of art, an insight that I had embodied in my own pianist's art, but never understood on an intellectual level.

Of course, I could not be satisfied by the very existence of gesture philosophy or gestural practice, just as I could not accept traditional music and performance theory when I started my enterprise of mathematical music theory in 1978. The gesturally colored thoughts and actions needed a rigorous conceptualization in the same vein as my efforts before the first edition of *The Topos of Music*.

In 2002, I was in the privileged position to work in the multimedia division of Peter Stucki at the Institut für Informatik of the University of Zurich. I had excellent PhD students, and we could, with one of them, Stefan Müller, realize a first experimental software for the gestural representation of a pianist's hand, a work presented at the ICMC conference in 2003 [772].

This experimental preliminary work was then taken as a point of departure for a mathematical theory of musical gestures. I presented this theory in a course in spring 2005 at École normale supérieure in Paris, a course that was later in 2007 taken as the material basis of my French book *La vérité de beau dans la musique* [718]. The first publication of a formally more evolved mathematical theory of musical gestures was

[1] Actually, he says that "Jeder Anfang ist ein Ende," but the reverse is immediate.

written with co-author Moreno Andreatta in 2007 [720]. This date could be called the birthday of a valid mathematical theory of musical gestures.

Until the publication of this second edition of *The Topos of Music*, several important conceptual extensions of the mathematical theory of musical gestures, models of musical gestural processes, as well as a number of theorems have been published. The decade from date of birth to the presence proved that the mathematical theory of musical gestures is an important added value to the theory described in the first edition of *The Topos of Music*.

We can however not state that this theory of gestures is in a complete state, quite the opposite is true: The coming years will reveal important news of theoretical as well as practical nature. So why did we make the decision to publish the present state of the art? The first argument is that the present state is rich enough to define concrete new directions, be it in music theory, such as harmony or counterpoint, be it in performance theory, or be it in the understanding of embodiment in the making of music. The second argument is that the present material, roughly 500 pages of new material, is ample enough to present a book's stature. And the third and very important argument is that we would like to communicate the state of the art in the spirit of Cavaillès: *Understanding is catching the gesture and being able to continue*. The co-authors of the gesture theory part, René Guitart, Jocelyn Ho, Alex Lubet, Maria Mannone, Matt Rahaim, and Florian Thalmann, are a wonderful confirmation of this philosophy. So let us continue!

Here is a summary of the new material and its authorship. Whenever I don't mention the author, it is my own contribution, all others are mentioned explicitly.

Until Part XIV, nearly everything is as in the book's first edition, refer to the preface of that edition (also included in this edition) for detailed summaries. The only new content—besides errata corrections—is Chapter 45 in Part XI, which is a shortened version of a paper [110] on a statistical analysis of Chopin's Prélude op. 28, No. 4, written with Jan Beran, Robert Goswitz, and Patrizio Mazzola.

Gesture theory starts with Part XV: Gesture Philosophy for Music. Chapter 57 gives an overview of philosophical aspects of gestures, including works by Jean-Claude Schmitt, Vilém Flusser, Michel Guérin, Adam Kendon, David McNeill, Juhani Pallasmaa, André Chastel, Émile Benveniste, Marie-Dominique Popelard, and Anthony Wall. In Chapter 58, we discuss the presemiotic approch to gestures in the French perspective of Maurice Merleau-Ponty, Gilles Deleuze, Jean Cavaillès, Charles Alunni, and Gilles Châtelet. Paul Valéry is also referenced in Section 59.4.

Chapter 59 deals with gestural aspects in cognitive science. After a review of Embodied AI and anthropology, Alex Lubet in Section 59.5 introduces us to gestural disability studies, focusing on two famous disabled pianists: Horace Parlan and Oscar Peterson (in his last years). Then in Section 59.6 Lubet reflects on perception of musical gesture as being inherently synaesthetic.

Chapter 60 concludes this part with a review of musical models of gesturality as proposed by Wolfgang Graeser, Theodor W. Adorno, Neil P. McAngus Todd, David Lewin, Robert Hatten, Marcelo Wanderley, Claude Cadoz, and Marc Leman.

Part XVI introduces the mathematics of gestures. Chapter 61 presents the mathematical concept of a gesture in a topological space and states the Diamond Conjecture, which deals with a hypothetical big space that unites algebraic and topological categories. Chapter 62 extends the theory from gestures in topological spaces to gestures in topological categories and introduces functorial gestures, i.e., functors on topological categories with values in the category of gestures, similar to functorial compositions in the previous theory.

Chapter 63 presents a generalized singular homology, where cubes are replaced by general hypergestures. Hypergesture homology applies to a gestural model of counterpoint and to a gestural refinement of performance stemma theory.

Chapter 64 presents—similar to Chapter 63—Stokes' Theorem for hypergestures. This theorem applies to problems in gestural modulation theory.

Chapters 65 and 66 discuss categories of local and global compositions, processes/networks, and gestures, together with their functorial relationships. This triple typology composition/process/gesture corresponds to the ontological dimension of embodiment with its three coordinates facts/processes/gestures.

In Sections 67.1–67.7 of Chapter 67, René Guitart develops a fascinating and demanding mathematical model of mathematical creativity, where thought is viewed as an algebra of gestures.

In Section 61.14, we, Maria Mannone and Guerino Mazzola, present a group-theoretical model of Georg Wilhelm Friedrich Hegel's initial discourse in his *Wissenschaft der Logik*, a model that applies to the Yoneda concept of creativity [726, Chapter 19.2]. We illustrate the method with an experimental composition by Mannone. This discussion extends over Sections 67.8–67.15 and completes Chapter 67.

Part XVII deals with concept architectures and software for musical gesture theory. Chapter 68 explains the denotator formalism for gestures over topological categories. Chapter 69 is a summary of the Java-based RUBATO® Composer software [739], written to sketch the framework of Chapters 70–74, where Florian Thalmann presents his gesture-oriented software component, the BigBang rubette. This discourse again follows the coordinates of the dimension of embodiment: facts, processes, gestures, which in this situation specify to: visualization and sonification of denotators (Chapter 71), BigBang's operation graph (Chapter 72), and gestural interaction and gesturalization (Chapter 73). In the final Chapter 74 of this part, Thalmann discusses musical examples.

Part XVIII is entitled *The Multiverse Perspective* because it opens up the relationship of gesture theory with string theory in theoretical physics. After a critical review of Hermann Hesse's *Glasperlenspiel* with regard to its gestural deficiencies, we, Mazzola and Mannone, develop the Euler-Lagrange formalism of world-sheets for musical gestures. This theory extends to functorial global gestures over global topological categories.

Part XIX is dedicated to applications of gesture theory to a number of musical themes.

Chapter 79 deals with singular gesture homology being applied to counterpoint.

Chapter 80 introduces a gestural restatement of modulation theory, applying in particular Stokes' Theorem for hypergestures.

Chapter 81 applies gesture theory to a gestural performance stemma theory.

Chapter 82 is written by Jocelyn Ho as a creative presentation of composition and analysis as embodied gestures in an inter-corporeal world. She presents two compositions, Toru Takemitsu's *Rain Tree Sketch II* and her own composition *Sheng* for piano, smartphones, and fixed playback.

Chapter 83 is Mannone's analysis and classification of a conductor's movements from the viewpoint of gesture theory.

Chapter 84 is a review of gestural aspects that were developed in *Flow, Gesture, and Spaces in Free Jazz* [721].

Chapter 85 is written by Matt Rahaim and presents the gestural approach to understanding Hindustani music in its vocal gesturality.

Chapter 86 is a first approach, written by Mannone, to a future theory of vocal gestures. The short addendum was written by Mazzola.

The Appendix has been enriched by additional complements on mathematics (Chapter J) plus complements on physics (Chapter K).

The Leitfaden III has been added to the original Leitfaden I & II for the gestural chapters.

The ToM_CD has been updated, containing now the present book's pdf ToposOfMusic.pdf. However the original CD is no longer added to the book, instead the ToM_CD can be downloaded from

`www.encyclospace.org/special/ToM_CD.zip`.

Concerning the division of the now very large book into parts, this is the split:

- Volume I: *Theory*, Prefaces and Table of Contents, Parts I to VII
- Volume II: *Performance*, Parts VIII to XIV
- Volume III: *Gestures*, Parts XV to XIX
- Volume IV: *Roots*, Appendices

My sincere acknowledgments go to my co-authors and to Springer's Ronan Nugent and Frank Holzwarth as well as to Birkhäuser's Thomas Hempfling.

Minneapolis, October 2017 Guerino Mazzola

Preface

Man kann
einen jeden Begriff,
einen jeden Titel,
darunter viele Erkenntnisse gehören,
einen logischen Ort nennen.
Immanuel Kant [519, p. B 324]

This book's title subject, *The Topos of Music*, has been chosen to communicate a double message: First, the Greek word "topos" ($\tau \acute{o} \pi o \varsigma$ = location, site) alludes to the logical and transcendental location of the concept of music in the sense of Aristotle's [40, 1154] and Kant's [519, p. B 324] topic. This view deals with the question of *where music is situated* ***as*** *a concept*—and hence with the underlying ontological problem: *What is the type of being and existence of music?* The second message is a more technical understanding insofar as the system of musical signs can be associated with the mathematical theory of *topoi*, which realizes a powerful synthesis of geometric and logical theories. It laid the foundation of a thorough geometrization of logic and has been successful in central issues of algebraic geometry (Grothendieck, Deligne), independence proofs and intuitionistic logic (Cohen, Lawvere, Kripke).

But this second message is intimately entwined with the first since the present concept framework of the musical sign system is technically based on topos theory, so the topos of music receives its topos-theoretic foundation. In this perspective, the double message of the book's title in fact condenses to a unified intention: to unite philosophical insight with mathematical explicitness.

According to Birkhäuser's initial plan in 1996, this book was first conceived as an English translation of my former book *Geometrie der Töne* [682], since the German original had suffered from its restricted access to the international public. However, the scientific progress since 1989, when it was written, has been considerable in theory and technology. We have known new subjects, such as the denotator concept framework, performance theory, and new software platforms for composition, analysis, and performance, such as RUBATO® or OpenMusic. Modeling concepts via the denotator approach in fact results from an intense collaboration of mathematicians and computer scientists in the object-oriented programming paradigm and supported by several international research grants.

Also, the scientific acceptance of mathematical music theory has grown since its beginnings in the late 1970s. As the first acceptance of mathematical music theory was testified to by von Karajan's legendary Ostersymposium "Musik und Mathematik" in 1984 in Salzburg [383], so is the significantly improved present status of acceptance testified to by the Fourth Diderot Forum on Mathematics and Music [711] in Paris, Vienna, and Lisbon 1999, which was organized by the European Mathematical Society. The corresponding extension of collaborative efforts in particular entail the inclusion of works by other research groups in this book, such as the "American Set Theory", the Swedish school of performance research at Stockholm's KTH, or the research on computer-aided composition at the IRCAM in Paris.

Therefore, as a result of these revised conditions, *The Topos of Music* appears as a vastly extended English update of the original work. The extension is visibly traced in the following parts which are new with respect to [682]: Part II exposes the theory of denotators and forms, part V introduces the topological theories of rhythms and motives, part VIII introduces the structure theory of performance, part IX deals with the expressive semantics of performance in the language of performance operators and stemmata (genealogical trees of successively refined performance), part X is devoted to the description of the RUBATO® software platform for representation, analysis, composition, and performance, part XI presents a statistical analysis of musical analysis, part XII concludes the subject of performance with an inverse performance theory, in fact a first formalization of the problem of music criticism.

This does however not mean that the other parts are just translations of the German text. Considerable progress has been made in most fields, except the last part XIV which reproduces the status quo in [682]. In particular, the local and global theories have been thoroughly functorialized and thereby introduce an ontological depth and variability of concepts, techniques, and results, which by far transcend the semiotically naive geometric approach in [682]. The present theory is as different from the traditional geometric conceptualization as is Grothendieck's topos theoretic algebraic geometry from classical algebraic geometry in the spirit of Segre, van der Waerden, or Zariski.

Beyond this topos-theoretic generalization, the denotator language also introduces a fairly exceptional technique of circular concept constructions. This more precisely is rooted in Finsler's pioneering work in foundations of set theory [322], a thread which has been rediscovered in modern theoretical computer sciences [5]. The present state of denotator theory rightly could be termed a Galois theory of concepts in the sense that circular definitions of concepts play the role of conceptual equations (corresponding to algebraic equations in algebraic Galois theory), the solutions of which are concepts instead of algebraic numbers.

Accordingly, the mathematical apparatus has been vastly extended, not only in the field of topos theory and its intuitionistic logic, but also with regard to general and algebraic topology, ordinary and partial differential equations, Pólya theory, statistics, multiaffine algebra and functorial algebraic geometry. It is mandatory that these technicalities had to be placed in a more elaborate semiotic perspective. However, this book does not cover the full range of music semiotics, for which the reader is referred to [703]. Of course, such an extension on the technical level has consequences for the readability of the theory. In view of the present volume of over 1300 pages, we could however not even make the attempt to approach a non-technical presentation. This subject is left to subsequent efforts. The critical reader may put the question whether music is really that complex. The answer is yes, and the reason is straightforward: We cannot pretend that Bach, Haydn, Mozart, or Beethoven, just to name some of the most prominent composers, are outstanding geniuses and have elaborated masterworks of eternal value, without trying to understand such singular creations with adequate tools, and this means: of adequate depth and power. After all, understanding God's 'composition', the material universe, cannot be approached without the most sophisticated tools as they have been elaborated in physics, chemistry, and molecular biology.

So who is recommended to read this book? A first category of readers is evidently the working scientist in the fields of mathematical music theory, the soft- and hardware engineer in music informatics, but also the mathematician who is interested in new applications from the above fields of pure mathematics. A second category are those theoretical mathematicians or computer scientists interested in the Galois theory of concepts; they may discover interesting unsolved problems. A third category of potential readers are all those who really want to get an idea of what music is about, of how one may conceptualize and turn into language the "ineffable" in music for the common language. Those who insist on the dogma that precision and beauty contradict each other, and that mathematics only produces tautologies and therefore must fail when aiming at substantial knowledge, should not read such a book.

Despite the technical character of *The Topos of Music*, there are at least four different approaches to its reading. To begin with, one may read it as a philosophical text, concentrating on the qualitative passages, surfing over technical portions and leaving those paragraphs to others. One may also take the book as a dictionary for computational musicology, including its concept framework and the lists of musical objects and processes (such as modulation degrees, contrapuntal steps) in the appendices. Observe however, that not all existing important lists have been included. For example, the list of all-interval series and the list of self-addressed chords are omitted, the reader may find these lists in other publications. Thirdly, the working scientist will have to read the full-fledged technicalities. And last, but not least, one may take the book as a source for ideas of how to go on with the whole subject of music. The GPL (General Public License[2]) software sources in the appended CD-ROM may support further development.

The prerequisites to a more in-depth reading of this book are these. Generally speaking, a good acquaintance with formal reasoning as mathematics (including formal logic) preconizes, is a conditio sine qua non. As to musicology and music theory, the familiarity with elementary concepts, like chords, motives, rhythm, and also musical notation, as well as a real interest in understanding music and not simply (ab)using it, are recommended. For the more computer-oriented passages, familiarity with the paradigm of object-oriented programming is profitable. We have not included the appendix on mathematical basics because it should help the reader get familiar with mathematics, but as an orientation in fields where the specialized mathematician possibly needs a specification of concepts and notation. The appendix was also included to expose the spectrum of mathematics which is needed to tackle the formal problems of computational musicology. It is by no means an overkill of mathematization: We have even omitted some non-trivial fields, such as statistics or Lambda calculus, for which we have to apologize.

There are different supporting instances to facilitate orientation in this book. To begin with, the table of contents and an extensive subject and name index may help find one's key-words. Further, following the list of contents, a leitfaden (on page xlv) is included for a generic navigation. Each chapter and section is headed by a summary that offers a first orientation about specific contents. Finally, the book is also available as a file `ToposOfMusic.pdf` with bookmarks and active cross-references in the appended CD-ROM (see page xlvii for its contents). This version is also attractive because the figures' colors are visible only in this version.

In order to obtain a consistent first reading, we recommend chapters 1 to 5, and then appendix A: Common Parameter Spaces (appendix B is not mandatory here, though it gives a good and not so technical overview of auditory physiology). After that, the reader may go on with chapter 6 on denotators and then follow the outline of the leitfaden (see page xlv).

This book could not have been realized without the engaged support of nineteen collaborators and contributors. Above all, my PhD students Stefan Göller and Stefan Müller at the MultiMedia Laboratory of the Department of Information Technology at the University of Zurich have collaborated in the production of this book on the levels of the LaTeX installation, the final production of hundreds of figures, and the contributions sections 20.2 through 20.5 (Göller) and sections 47.3 through 47.3.6.2 (Müller). My special gratitude goes to their truly collaborative spirit.

Contributions to this book have been delivered by (in alphabetic order): By Carlos Agon, and Gérard Assayag (both IRCAM) with their precious Lambda-calculus-oriented presentation of the object-oriented programming principles in the composition software OpenMusic described in chapter 52, Moreno Andreatta (IRCAM) with an elucidating discourse on the American Set Theory in section 11.5.2 and section 16.3, Jan Beran (Universität Konstanz) with his contribution to the compositional strategies in his original composition [103] in section 11.5.1.1, as well as with his inspiring work on statistics as reported in chapters 43 and 44, Chantal Buteau (Universität and ETH Zürich) with her detailed review of chapter 22, Roberto Ferretti (ETH Zürich) with his progressive contributions to the algebraic geometry of inverse performance theory in sections 39.8 and 47.2, Anja Fleischer (Technische Universität Berlin) with her short but critical preliminaries in chapter 23, Harald Fripertinger (Universität Graz) with his 'killer' formulas concerning enumeration of finite local and global compositions in sections 11.4, 16.2.2 and appendix C.3.6, Jörg Garbers (Technische Universität Berlin) with his portation of the RUBATO® application to Mac OS X, as documented in the screenshots in chapters 40, 41, Werner Hemmert (Infineon) with a very up-to-date presentation of room acoustics in section A.1.1.1 and auditory physiology in appendix B.1 (we would have loved to include more of his knowledge), Michael Leyton (DIMACS, Rutgers University) with a formidable cover figure entitled "Dark Theory", a beautiful subtitle to this book, as well as with innumerable discussions around time and its reduction to symmetries as presented in chapter 48, Emilio Lluis Puebla (UNAM, Mexico City)

[2] A legal matter file is contained in the book's CD-ROM, see page xlvii.

with his unique and engaged promotion and dissipation of mathematical music theory on the American continent, especially also in the preparation and critical review of this book, Mariana Montiel Hernandez (UNAM, Mexico City) with her critical review of the theory of circular forms and denotators in section 6.5 and appendix G.2.2.1, Thomas Noll (Technische Universität Berlin) with his substantial contributions to the functorial theory of compositions, and for his revolutionary rebuilding of Riemann's harmony and its relations to counterpoint, Joachim Stange-Elbe (Universität Osnabrück) with a very clear and innovative description of his outstanding RUBATO® performance of Bach's contrapunctus III in the *Art of Fugue* in sections 42.2 through 42.4.3, Hans Straub with his adventurous extensions of classical cadence theory in section 26.2.2 and his classification of four-element motives in appendix O.4, and, last but not least, Oliver Zahorka (Out Media Design), my former collaborator and chief programmer of the NeXT RUBATO® application, which has contributed so much to the success of the Zürich school of performance theory. To all of them, I owe my deepest gratitude and recognition for their sweat and tears.

My sincere acknowledgments go to Alexander Grothendieck, whose encouraging letters and, no doubt, awe inspiring revolution in mathematical thinking has given me so much in isolated phases of this enterprise. My acknowledgments also go to my engaged mentor Peter Stucki, director of the MultiMedia Laboratory of the Department of Information Technology at the University of Zurich; without his support, this book would have seen its birthday years later, if ever. My thanks also go to my brother Silvio, who once again (he did it already for my first book [670]) supported the final review efforts by an ideal environment in his villa in Vulpera. My thanks also go to the unbureaucratic management of the book's production by Birkhäuser's lector Thomas Hempfling and the very patient copy editor Edwin Beschler. All these beautiful supports would have failed without my wife Christina's infinite understanding and vital environment—if this book is a trace of humanity, it is also, and strongly, hers.

Vulpera, June 2002 Guerino Mazzola

Volume II Contents

Part VIII Structure Theory of Performance

32 Local and Global Performance Transformations ... 547
32.1 Performance as a Reality Switch ... 548
32.2 Why Do We Need Infinite Performance of the Same Piece? ... 549
32.3 Local Structure ... 550
32.3.1 The Coherence of Local Performance Transformations ... 550
32.3.2 Differential Morphisms of Local Compositions ... 551
32.4 Global Structure ... 554
32.4.1 Modeling Performance Syntax ... 556
32.4.2 The Formal Setup ... 557
32.4.3 Performance qua Interpretation of Interpretation ... 560

33 Performance Fields ... 561
33.1 Classics: Tempo, Intonation, and Dynamics ... 561
33.1.1 Tempo ... 561
33.1.2 Intonation ... 563
33.1.3 Dynamics ... 564
33.2 Genesis of the General Formalism ... 565
33.2.1 The Question of Articulation ... 565
33.2.2 The Formalism of Performance Fields ... 568
33.3 What Performance Fields Signify ... 568
33.3.1 Th.W. Adorno, W. Benjamin, and D. Raffman ... 569
33.3.2 Towards Composition of Performance ... 571

34 Initial Sets and Initial Performances ... 573
34.1 Taking Off with a Shifter ... 573
34.2 Anchoring Onset ... 574
34.3 The Concert Pitch ... 576
34.4 Dynamical Anchors ... 578
34.5 Initializing Articulation ... 578
34.6 Hit Point Theory ... 580
34.6.1 Distances ... 580
34.6.2 Flow Interpolation ... 582

35 Hierarchies and Performance Scores ... 585
35.1 Performance Cells ... 585
35.2 The Category of Performance Cells ... 586
35.3 Hierarchies ... 588
35.3.1 Operations on Hierarchies ... 590
35.3.2 Classification Issues ... 591
35.3.3 Example: The Piano and Violin Hierarchies ... 593
35.4 Local Performance Scores ... 594
35.5 Global Performance Scores ... 598
35.5.1 Instrumental Fibers ... 599

Part IX Expressive Semantics

36 Taxonomy of Expressive Performance ... 603
36.1 Feelings: Emotional Semantics ... 604
36.2 Motion: Gestural Semantics ... 606
36.3 Understanding: Rational Semantics ... 609
36.4 Cross-semantical Relations ... 612

37 Performance Grammars ... 615
37.1 Rule-Based Grammars ... 615
37.1.1 The KTH School ... 617
37.1.2 Neil P. McAngus Todd ... 618
37.1.3 The Zurich School ... 619
37.2 Remarks on Learning Grammars ... 619

38 Stemma Theory ... 621
38.1 Motivation from Practising and Rehearsing ... 622
38.1.1 Does Reproducibility of Performances Help Understanding? ... 622
38.2 Tempo Curves Are Inadequate ... 623
38.3 The Stemma Concept ... 626
38.3.1 The General Setup of Matrilineal Sexual Propagation ... 628
38.3.2 The Primary Mother—Taking Off ... 629
38.3.3 Mono- and Polygamy—Local and Global Actions ... 632
38.3.4 Family Life—Cross-correlations ... 634

39 Operator Theory ... 637
39.1 Why Weights? ... 638
39.1.1 Discrete and Continuous Weights ... 638
39.1.2 Weight Recombination ... 639
39.2 Primavista Weights ... 640
39.2.1 Dynamics ... 640
39.2.2 Agogics ... 643
39.2.3 Tuning and Intonation ... 644
39.2.4 Articulation ... 645
39.2.5 Ornaments ... 645
39.3 Analytical Weights ... 646
39.4 Taxonomy of Operators ... 648
39.4.1 Splitting Operators ... 649
39.4.2 Symbolic Operators ... 650

39.4.3 Physical Operators 651
39.4.4 Field Operators 652
39.5 Tempo Operator 653
39.6 Scalar Operator 654
39.7 The Theory of Basis—Pianola Operators 655
39.7.1 Basis Specialization 656
39.7.2 Pianola Specialization 659
39.8 Locally Linear Grammars 659

Part X RUBATO®

40 Architecture 665
40.1 The Overall Modularity 666
40.2 Frame and Modules 668
40.3 Postscriptum: The Rubato Composer Environment 668

41 The RUBETTE® Family 669
41.1 MetroRUBETTE® 669
41.2 MeloRUBETTE® 672
41.3 HarmoRUBETTE® 674
41.3.1 A Set of New Harmonic Analysis Rubettes on RUBATO® Composer 678
41.4 PerformanceRUBETTE® 679
41.5 PrimavistaRUBETTE® 685

42 Performance Experiments 687
42.1 A Preliminary Experiment: Robert Schumann's "Kuriose Geschichte" 687
42.2 Full Experiment: J.S. Bach's "Kunst der Fuge" 688
42.3 Analysis 688
42.3.1 Metric Analysis 688
42.3.2 Motif Analysis 692
42.3.3 Omission of Harmonic Analysis 693
42.4 Stemma Constructions 694
42.4.1 Performance Setup 694
42.4.2 Instrumental Setup 700
42.4.3 Global Discussion 701

Part XI Statistics of Analysis and Performance

43 Analysis of Analysis 707
43.1 Hierarchical Decomposition 707
43.1.1 General Motivation 707
43.1.2 Hierarchical Smoothing 708
43.1.3 Hierarchical Decomposition 710
43.2 Comparing Analyses of Bach, Schumann, and Webern 711

44 Differential Operators and Regression 719
44.0.1 Analytical Data 720
44.1 The Beran Operator 721
44.1.1 The Concept 721
44.1.2 The Formalism 723

44.2 The Method of Regression Analysis 726
44.2.1 The Full Model 726
44.2.2 Step Forward Selection 727
44.3 The Results of Regression Analysis 727
44.3.1 Relations Between Tempo and Analysis 727
44.3.2 Complex Relationships 729
44.3.3 Commonalities and Diversities 729
44.3.4 Overview of Statistical Results 738

45 Relating Tempo to Metric, Melodic and Harmonic Analyses in Chopin's Prélude op. 28, No. 4 743
45.1 Introduction 743
45.2 Data 745
45.2.1 Analytical Data 745
45.2.2 Tempo Data 745
45.3 Short Summary of the Results 747
45.4 Some Philosophical Comments 748

Part XII Inverse Performance Theory

46 Principles of Music Critique 751
46.1 Boiling Down Infinity—Is Feuilletonism Inevitable? 751
46.2 "Political Correctness" in Performance—Reviewing Gould 752
46.3 Transversal Ethnomusicology 754

47 Critical Fibers 755
47.1 The Stemma Model of Critique 755
47.2 Fibers for Locally Linear Grammars 756
47.3 Algorithmic Extraction of Performance Fields 759
47.3.1 The Infinitesimal View on Expression 759
47.3.2 Real-Time Processing of Expressive Performance 760
47.3.3 Score-Performance Matching 761
47.3.4 Performance Field Calculation 762
47.3.5 Visualization 763
47.3.6 The EspressoRUBETTE®: An Interactive Tool for Expression Extraction 764
47.4 Local Sections 766
47.4.1 Comparing Argerich and Horowitz 768

Part XIII Operationalization of Poiesis

48 Unfolding Geometry and Logic in Time 773
48.1 Performance of Logic and Geometry 774
48.2 Constructing Time from Geometry 775
48.3 Discourse and Insight 776

49 Local and Global Strategies in Composition 779
49.1 Local Paradigmatic Instances 780
49.1.1 Transformations 780
49.1.2 Variations 780

49.2 Global Poetical Syntax 781
49.2.1 Roman Jakobson's Horizontal Function 781
49.2.2 Roland Posner's Vertical Function 781
49.3 Structure and Process 782

50 The Paradigmatic Discourse on *presto*® 783
50.1 The *presto*® Functional Scheme 783
50.2 Modular Affine Transformations 786
50.3 Ornaments and Variations 786
50.4 Problems of Abstraction 789

51 Case Study I: "Synthesis" by Guerino Mazzola 791
51.1 The Overall Organization 791
51.1.1 The Material: 26 Classes of Three-Element Motives 792
51.1.2 Principles of the Four Movements and Instrumentation 793
51.2 1st Movement: Sonata Form 794
51.3 2nd Movement: Variations 794
51.4 3rd Movement: Scherzo 798
51.5 4th Movement: Fractal Syntax 799

52 Object-Oriented Programming in OpenMusic 801
52.1 Object-Oriented Language 802
52.1.1 Patches 802
52.1.2 Objects 803
52.1.3 Classes 803
52.1.4 Methods 803
52.1.5 Generic Functions 804
52.1.6 Message Passing 804
52.1.7 Inheritance 805
52.1.8 Boxes and Evaluation 805
52.1.9 Instantiation 806
52.2 Musical Object Framework 806
52.2.1 Internal Representation 806
52.2.2 Interface 808
52.3 Maquettes: Objects in Time 811
52.4 Meta-object Protocol 813
52.4.1 Reification of Temporal Boxes 815
52.5 A Musical Example 817

Part XIV String Quartet Theory

53 Historical and Theoretical Prerequisites 825
53.1 History 825
53.2 Theory of the String Quartet Following Ludwig Finscher 826
53.2.1 Four Part Texture 826
53.2.2 The Topos of Conversation Among Four Humanists 827
53.2.3 The Family of Violins 828

54 Estimation of Resolution Parameters 831
54.1 Parameter Spaces for Violins 831
54.2 Estimation 833

55 The Case of Counterpoint and Harmony 839
55.1 Counterpoint 839
55.2 Harmony 840
55.3 Effective Selection 840

Part XXIV References and Index

References R.1

Index R.33

Book Set Contents

Part I Introduction and Orientation

1 **What Is Music About?** ... 3
1.1 Fundamental Activities ... 3
1.2 Fundamental Scientific Domains ... 5

2 **Topography** ... 9
2.1 Layers of Reality ... 10
2.1.1 Physical Reality ... 10
2.1.2 Mental Reality ... 11
2.1.3 Psychological Reality ... 11
2.2 Molino's Communication Stream ... 11
2.2.1 Creator and Poietic Level ... 12
2.2.2 Work and Neutral Level ... 13
2.2.3 Listener and Esthesic Level ... 13
2.3 Semiosis ... 14
2.3.1 Expressions ... 15
2.3.2 Content ... 15
2.3.3 The Process of Signification ... 15
2.3.4 A Short Overview of Music Semiotics ... 15
2.4 The Cube of Local Topography ... 17
2.5 Topographical Navigation ... 19

3 **Musical Ontology** ... 21
3.1 Where Is Music? ... 21
3.2 Depth and Complexity ... 23

4 **Models and Experiments in Musicology** ... 27
4.1 Interior and Exterior Nature ... 29
4.2 What Is a Musicological Experiment? ... 30
4.3 Questions—Experiments of the Mind ... 31
4.4 New Scientific Paradigms and Collaboratories ... 32

Part II Navigation on Concept Spaces

5 Navigation ... 35
5.1 Music in the EncycloSpace ... 36
5.2 Receptive Navigation ... 39
5.3 Productive Navigation ... 39

6 Denotators ... 41
6.1 Universal Concept Formats ... 42
6.1.1 First Naive Approach to Denotators ... 43
6.1.2 Interpretations and Comments ... 48
6.1.3 Ordering Denotators and 'Concept Leafing' ... 50
6.2 Forms ... 52
6.2.1 Variable Addresses ... 53
6.2.2 Formal Definition ... 54
6.2.3 Discussion of the Form Typology ... 56
6.3 Denotators ... 57
6.3.1 Formal Definition of a Denotator ... 57
6.4 Anchoring Forms in Modules ... 59
6.4.1 First Examples and Comments on Modules in Music ... 60
6.5 Regular and Circular Forms ... 64
6.6 Regular Denotators ... 66
6.7 Circular Denotators ... 72
6.8 Ordering on Forms and Denotators ... 75
6.8.1 Concretizations and Applications ... 78
6.9 Concept Surgery and Denotator Semantics ... 83

Part III Local Theory

7 Local Compositions ... 89
7.1 The Objects of Local Theory ... 89
7.2 First Local Music Objects ... 92
7.2.1 Chords and Scales ... 92
7.2.2 Local Meters and Local Rhythms ... 96
7.2.3 Motives ... 99
7.3 Functorial Local Compositions ... 101
7.4 First Elements of Local Theory ... 103
7.5 Alterations Are Tangents ... 107
7.5.1 The Theorem of Mason-Mazzola ... 108

8 Symmetries and Morphisms ... 113
8.1 Symmetries in Music ... 114
8.1.1 Elementary Examples ... 116
8.2 Morphisms of Local Compositions ... 128
8.3 Categories of Local Compositions ... 132
8.3.1 Commenting on the Concatenation Principle ... 134
8.3.2 Embedding and Addressed Adjointness ... 136
8.3.3 Universal Constructions on Local Compositions ... 138
8.3.4 The Address Question ... 140
8.3.5 Categories of Commutative Local Compositions ... 142

9 Yoneda Perspectives ... 145
9.1 Morphisms Are Points ... 147
9.2 Yoneda's Fundamental Lemma ... 150
9.3 The Yoneda Philosophy ... 152
9.4 Understanding Fine and Other Arts ... 153
9.4.1 Painting and Music ... 153
9.4.2 The Art of Object-Oriented Programming ... 155

10 Paradigmatic Classification ... 157
10.1 Paradigmata in Musicology, Linguistics, and Mathematics ... 158
10.2 Transformation ... 162
10.3 Similarity ... 163
10.4 Fuzzy Concepts in the Humanities ... 164

11 Orbits ... 167
11.1 Gestalt and Symmetry Groups ... 167
11.2 The Framework for Local Classification ... 168
11.3 Orbits of Elementary Structures ... 168
11.3.1 Classification Techniques ... 169
11.3.2 The Local Classification Theorem ... 170
11.3.3 The Finite Case ... 177
11.3.4 Dimension ... 178
11.3.5 Chords ... 180
11.3.6 Empirical Harmonic Vocabularies ... 181
11.3.7 Self-addressed Chords ... 185
11.3.8 Motives ... 187
11.4 Enumeration Theory ... 190
11.4.1 Pólya and de Bruijn Theory ... 190
11.4.2 Big Science for Big Numbers ... 196
11.5 Group-Theoretical Methods in Composition and Theory ... 198
11.5.1 Aspects of Serialism ... 199
11.5.2 The American Tradition ... 202
11.6 Esthetic Implications of Classification ... 211
11.6.1 Jakobson's Poetic Function ... 212
11.6.2 Motivic Analysis: Schubert/Stolberg "Lied auf dem Wasser zu singen..." ... 214
11.6.3 Composition: Mazzola/Baudelaire "La mort des artistes" ... 218
11.7 Mathematical Reflections on Historicity in Music ... 220
11.7.1 Jean-Jacques Nattiez' Paradigmatic Theme ... 221
11.7.2 Groups as a Parameter of Historicity ... 223

12 Topological Specialization ... 225
12.1 What Ehrenfels Neglected ... 225
12.2 Topology ... 226
12.2.1 Metrical Comparison ... 228
12.2.2 Specialization Morphisms of Local Compositions ... 230
12.3 The Problem of Sound Classification ... 232
12.3.1 Topographic Determinants of Sound Descriptions ... 232
12.3.2 Varieties of Sounds ... 238
12.3.3 Semiotics of Sound Classification ... 240
12.4 Making the Vague Precise ... 241

Part IV Global Theory

13 Global Compositions ... 245
13.1 The Local-Global Dichotomy in Music ... 246
13.1.1 Musical and Mathematical Manifolds ... 251
13.2 What Are Global Compositions? ... 252
13.2.1 The Nerve of an Objective Global Composition ... 253
13.3 Functorial Global Compositions ... 256
13.4 Interpretations and the Vocabulary of Global Concepts ... 258
13.4.1 Iterated Interpretations ... 258
13.4.2 The Pitch Domain: Chains of Thirds, Ecclesiastical Modes, Triadic and Quaternary Degrees ... 259
13.4.3 Interpreting Time: Global Meters and Rhythms ... 266
13.4.4 Motivic Interpretations: Melodies and Themes ... 270

14 Global Perspectives ... 273
14.1 Musical Motivation ... 273
14.2 Global Morphisms ... 274
14.3 Local Domains ... 280
14.4 Nerves ... 281
14.5 Simplicial Weights ... 283
14.6 Categories of Commutative Global Compositions ... 285

15 Global Classification ... 287
15.1 Module Complexes ... 287
15.1.1 Global Affine Functions ... 288
15.1.2 Bilinear and Exterior Forms ... 290
15.1.3 Deviation: Compositions vs. "Molecules" ... 291
15.2 The Resolution of a Global Composition ... 292
15.2.1 Global Standard Compositions ... 293
15.2.2 Compositions from Module Complexes ... 294
15.3 Orbits of Module Complexes Are Classifying ... 298
15.3.1 Combinatorial Group Actions ... 299
15.3.2 Classifying Spaces ... 300

16 Classifying Interpretations ... 303
16.1 Characterization of Interpretable Compositions ... 303
16.1.1 Automorphism Groups of Interpretable Compositions ... 306
16.1.2 A Cohomological Criterion ... 307
16.2 Global Enumeration Theory ... 309
16.2.1 Tesselation ... 309
16.2.2 Mosaics ... 310
16.2.3 Classifying Rational Rhythms and Canons ... 312
16.3 Global American Set Theory ... 314
16.4 Interpretable "Molecules" ... 316

17 Esthetics and Classification ... 319
17.1 Understanding by Resolution: An Illustrative Example ... 319
17.2 Varèse's Program and Yoneda's Lemma ... 323

18 Predicates ... 327
18.1 What Is the Case: The Existence Problem ... 327
18.1.1 Merging Systematic and Historical Musicology ... 328
18.2 Textual and Paratextual Semiosis ... 329
18.2.1 Textual and Paratextual Signification ... 330
18.3 Textuality ... 331
18.3.1 The Category of Denotators ... 331
18.3.2 Textual Semiosis ... 334
18.3.3 Atomic Predicates ... 339
18.3.4 Logical and Geometric Motivation ... 345
18.4 Paratextuality ... 349

19 Topoi of Music ... 351
19.1 The Grothendieck Topology ... 351
19.1.1 Cohomology ... 354
19.1.2 Marginalia on Presheaves ... 356
19.2 The Topos of Music: An Overview ... 357

20 Visualization Principles ... 361
20.1 Problems ... 361
20.2 Folding Dimensions ... 363
20.2.1 $\mathbb{R}^2 \to \mathbb{R}$ ... 363
20.2.2 $\mathbb{R}^n \to \mathbb{R}$ ... 364
20.2.3 An Explicit Construction of μ with Special Values. ... 365
20.3 Folding Denotators ... 366
20.3.1 Folding Limits ... 366
20.3.2 Folding Colimits ... 367
20.3.3 Folding Powersets ... 368
20.3.4 Folding Circular Denotators ... 369
20.4 Compound Parametrized Objects ... 370
20.5 Examples ... 371

Part V Topologies for Rhythm and Motives

---

21 Metrics and Rhythmics ... 375
21.1 Review of Riemann and Jackendoff-Lerdahl Theories ... 375
21.1.1 Riemann's Weights ... 375
21.1.2 Jackendoff-Lerdahl: Intrinsic Versus Extrinsic Time Structures ... 376
21.2 Topologies of Global Meters and Associated Weights ... 378
21.3 Macro-events in the Time Domain ... 380

22 Motif Gestalts ... 383
22.1 Motivic Interpretation ... 384
22.2 Shape Types ... 385
22.2.1 Examples of Shape Types ... 386
22.3 Metrical Similarity ... 388
22.3.1 Examples of Distance Functions ... 389
22.4 Paradigmatic Groups ... 390
22.4.1 Examples of Paradigmatic Groups ... 391
22.5 Pseudo-metrics on Orbits ... 393
22.6 Topologies on Gestalts ... 394

22.6.1 The Inheritance Property ... 395
22.6.2 Cognitive Aspects of Inheritance ... 396
22.6.3 Epsilon Topologies ... 397
22.7 First Properties of the Epsilon Topologies ... 399
22.7.1 Toroidal Topologies ... 401
22.8 Rudolph Reti's Motivic Analysis Revisited ... 404
22.8.1 Review of Concepts ... 404
22.8.2 Reconstruction ... 406
22.9 Motivic Weights ... 408

Part VI Harmony

23 Critical Preliminaries ... 413
23.1 Hugo Riemann ... 414
23.2 Paul Hindemith ... 414
23.3 Heinrich Schenker and Friedrich Salzer ... 414

24 Harmonic Topology ... 417
24.1 Chord Perspectives ... 418
24.1.1 Euler Perspectives ... 418
24.1.2 12-Tempered Perspectives ... 422
24.1.3 Enharmonic Projection ... 425
24.2 Chord Topologies ... 427
24.2.1 Extension and Intension ... 427
24.2.2 Extension and Intension Topologies ... 429
24.2.3 Faithful Addresses ... 431
24.2.4 The Saturation Sheaf ... 434

25 Harmonic Semantics ... 435
25.1 Harmonic Signs—Overview ... 436
25.2 Degree Theory ... 437
25.2.1 Chains of Thirds ... 437
25.2.2 American Jazz Theory ... 439
25.2.3 Hans Straub: General Degrees in General Scales ... 442
25.3 Function Theory ... 442
25.3.1 Canonical Morphemes for European Harmony ... 444
25.3.2 Riemann Matrices ... 447
25.3.3 Chains of Thirds ... 447
25.3.4 Tonal Functions from Absorbing Addresses ... 449

26 Cadence ... 453
26.1 Making the Concept Precise ... 454
26.2 Classical Cadences Relating to 12-Tempered Intonation ... 454
26.2.1 Cadences in Triadic Interpretations of Diatonic Scales ... 455
26.2.2 Cadences in More General Interpretations ... 456
26.3 Cadences in Self-addressed Tonalities of Morphology ... 457
26.4 Self-addressed Cadences by Symmetries and Morphisms ... 459
26.5 Cadences for Just Intonation ... 460
26.5.1 Tonalities in Third-Fifth Intonation ... 460
26.5.2 Tonalities in Pythagorean Intonation ... 461

27 Modulation . . . 463
27.1 Modeling Modulation by Particle Interaction . . . 464
27.1.1 Models and the Anthropic Principle . . . 464
27.1.2 Classical Motivation and Heuristics . . . 465
27.1.3 The General Background . . . 467
27.1.4 The Well-Tempered Case . . . 469
27.1.5 Reconstructing the Diatonic Scale from Modulation . . . 471
27.1.6 The Case of Just Tuning . . . 473
27.1.7 Quantized Modulations and Modulation Domains for Selected Scales . . . 477
27.2 Harmonic Tension . . . 481
27.2.1 The Riemann Algebra . . . 481
27.2.2 Weights on the Riemann Algebra . . . 482
27.2.3 Harmonic Tensions from Classical Harmony? . . . 484
27.2.4 Optimizing Harmonic Paths . . . 485

28 Applications . . . 487
28.1 First Examples . . . 488
28.1.1 Johann Sebastian Bach: Choral from "Himmelfahrtsoratorium" . . . 488
28.1.2 Wolfgang Amadeus Mozart: "Zauberflöte", Choir of Priests . . . 490
28.1.3 Claude Debussy: "Préludes", Livre 1, No.4 . . . 492
28.2 Modulation in Beethoven's Sonata op.106, 1st Movement . . . 495
28.2.1 Introduction . . . 495
28.2.2 The Fundamental Theses of Erwin Ratz and Jürgen Uhde . . . 497
28.2.3 Overview of the Modulation Structure . . . 498
28.2.4 Modulation $B_\flat \rightsquigarrow G$ via e^{-3} in W . . . 499
28.2.5 Modulation $G \rightsquigarrow E_\flat$ via U_g in W . . . 499
28.2.6 Modulation $E_\flat \rightsquigarrow D/b$ from W to W^* . . . 499
28.2.7 Modulation $D/b \rightsquigarrow B$ via $U_{d/d_\sharp} = U_{g_\sharp/a}$ within W^* . . . 500
28.2.8 Modulation $B \rightsquigarrow B_\flat$ from W^* to W . . . 500
28.2.9 Modulation $B_\flat \rightsquigarrow G_\flat$ via $U_{b_\flat}$ within W . . . 501
28.2.10 Modulation $G_\flat \rightsquigarrow G$ via $U_{a_\flat/a}$ within W . . . 501
28.3 Rhythmical Modulation in "Synthesis" . . . 501
28.3.1 Rhythmic Modes . . . 502
28.3.2 Composition for Percussion Ensemble . . . 503

Part VII Counterpoint

29 Melodic Variation by Arrows . . . 507
29.1 Arrows and Alterations . . . 507
29.2 The Contrapuntal Interval Concept . . . 508
29.3 The Algebra of Intervals . . . 509
29.3.1 The Third Torus . . . 510
29.4 Musical Interpretation of the Interval Ring . . . 511
29.5 Self-addressed Arrows . . . 514
29.6 Change of Orientation . . . 515

30 Interval Dichotomies as a Contrast ... 517
30.1 Dichotomies and Polarity ... 517
30.2 The Consonance and Dissonance Dichotomy ... 520
30.2.1 Fux and Riemann Consonances Are Isomorphic ... 521
30.2.2 Induced Polarities ... 523
30.2.3 Empirical Evidence for the Polarity Function ... 523
30.2.4 Music and the Hippocampal Gate Function ... 527

31 Modeling Counterpoint by Local Symmetries ... 531
31.1 Deformations of the Strong Dichotomies ... 531
31.2 Contrapuntal Symmetries Are Local ... 533
31.3 The Counterpoint Theorem ... 534
31.3.1 Some Preliminary Calculations ... 534
31.3.2 Two Lemmata on Cardinalities of Intersections ... 536
31.3.3 An Algorithm for Exhibiting the Contrapuntal Symmetries ... 536
31.3.4 Transfer of the Counterpoint Rules to General Representatives of Strong Dichotomies ... 540
31.4 The Classical Case: Consonances and Dissonances ... 540
31.4.1 Discussion of the Counterpoint Theorem in the Light of Reduced Strict Style ... 541
31.4.2 The Major Dichotomy—A Cultural Antipode? ... 542
31.4.3 Software for Counterpoint and Theoretical Extentions ... 543

Part VIII Structure Theory of Performance

32 Local and Global Performance Transformations ... 547
32.1 Performance as a Reality Switch ... 548
32.2 Why Do We Need Infinite Performance of the Same Piece? ... 549
32.3 Local Structure ... 550
32.3.1 The Coherence of Local Performance Transformations ... 550
32.3.2 Differential Morphisms of Local Compositions ... 551
32.4 Global Structure ... 554
32.4.1 Modeling Performance Syntax ... 556
32.4.2 The Formal Setup ... 557
32.4.3 Performance qua Interpretation of Interpretation ... 560

33 Performance Fields ... 561
33.1 Classics: Tempo, Intonation, and Dynamics ... 561
33.1.1 Tempo ... 561
33.1.2 Intonation ... 563
33.1.3 Dynamics ... 564
33.2 Genesis of the General Formalism ... 565
33.2.1 The Question of Articulation ... 565
33.2.2 The Formalism of Performance Fields ... 568
33.3 What Performance Fields Signify ... 568
33.3.1 Th.W. Adorno, W. Benjamin, and D. Raffman ... 569
33.3.2 Towards Composition of Performance ... 571

34 Initial Sets and Initial Performances ... 573
34.1 Taking Off with a Shifter ... 573
34.2 Anchoring Onset ... 574
34.3 The Concert Pitch ... 576
34.4 Dynamical Anchors ... 578

34.5 Initializing Articulation ... 578
34.6 Hit Point Theory ... 580
34.6.1 Distances ... 580
34.6.2 Flow Interpolation ... 582

35 Hierarchies and Performance Scores ... 585
35.1 Performance Cells ... 585
35.2 The Category of Performance Cells ... 586
35.3 Hierarchies ... 588
35.3.1 Operations on Hierarchies ... 590
35.3.2 Classification Issues ... 591
35.3.3 Example: The Piano and Violin Hierarchies ... 593
35.4 Local Performance Scores ... 594
35.5 Global Performance Scores ... 598
35.5.1 Instrumental Fibers ... 599

Part IX Expressive Semantics

36 Taxonomy of Expressive Performance ... 603
36.1 Feelings: Emotional Semantics ... 604
36.2 Motion: Gestural Semantics ... 606
36.3 Understanding: Rational Semantics ... 609
36.4 Cross-semantical Relations ... 612

37 Performance Grammars ... 615
37.1 Rule-Based Grammars ... 615
37.1.1 The KTH School ... 617
37.1.2 Neil P. McAngus Todd ... 618
37.1.3 The Zurich School ... 619
37.2 Remarks on Learning Grammars ... 619

38 Stemma Theory ... 621
38.1 Motivation from Practising and Rehearsing ... 622
38.1.1 Does Reproducibility of Performances Help Understanding? ... 622
38.2 Tempo Curves Are Inadequate ... 623
38.3 The Stemma Concept ... 626
38.3.1 The General Setup of Matrilineal Sexual Propagation ... 628
38.3.2 The Primary Mother—Taking Off ... 629
38.3.3 Mono- and Polygamy—Local and Global Actions ... 632
38.3.4 Family Life—Cross-correlations ... 634

39 Operator Theory ... 637
39.1 Why Weights? ... 638
39.1.1 Discrete and Continuous Weights ... 638
39.1.2 Weight Recombination ... 639
39.2 Primavista Weights ... 640
39.2.1 Dynamics ... 640
39.2.2 Agogics ... 643
39.2.3 Tuning and Intonation ... 644
39.2.4 Articulation ... 645
39.2.5 Ornaments ... 645

39.3 Analytical Weights ... 646
39.4 Taxonomy of Operators ... 648
39.4.1 Splitting Operators ... 649
39.4.2 Symbolic Operators ... 650
39.4.3 Physical Operators ... 651
39.4.4 Field Operators ... 652
39.5 Tempo Operator ... 653
39.6 Scalar Operator ... 654
39.7 The Theory of Basis—Pianola Operators ... 655
39.7.1 Basis Specialization ... 656
39.7.2 Pianola Specialization ... 659
39.8 Locally Linear Grammars ... 659

Part X RUBATO®

40 Architecture ... 665
40.1 The Overall Modularity ... 666
40.2 Frame and Modules ... 668
40.3 Postscriptum: The Rubato Composer Environment ... 668

41 The RUBETTE® Family ... 669
41.1 MetroRUBETTE® ... 669
41.2 MeloRUBETTE® ... 672
41.3 HarmoRUBETTE® ... 674
41.3.1 A Set of New Harmonic Analysis Rubettes on RUBATO® Composer ... 678
41.4 PerformanceRUBETTE® ... 679
41.5 PrimavistaRUBETTE® ... 685

42 Performance Experiments ... 687
42.1 A Preliminary Experiment: Robert Schumann's "Kuriose Geschichte" ... 687
42.2 Full Experiment: J.S. Bach's "Kunst der Fuge" ... 688
42.3 Analysis ... 688
42.3.1 Metric Analysis ... 688
42.3.2 Motif Analysis ... 692
42.3.3 Omission of Harmonic Analysis ... 693
42.4 Stemma Constructions ... 694
42.4.1 Performance Setup ... 694
42.4.2 Instrumental Setup ... 700
42.4.3 Global Discussion ... 701

Part XI Statistics of Analysis and Performance

43 Analysis of Analysis ... 707
43.1 Hierarchical Decomposition ... 707
43.1.1 General Motivation ... 707
43.1.2 Hierarchical Smoothing ... 708
43.1.3 Hierarchical Decomposition ... 710
43.2 Comparing Analyses of Bach, Schumann, and Webern ... 711

44 **Differential Operators and Regression** ... 719
44.0.1 Analytical Data ... 720
44.1 The Beran Operator ... 721
44.1.1 The Concept ... 721
44.1.2 The Formalism ... 723
44.2 The Method of Regression Analysis ... 726
44.2.1 The Full Model ... 726
44.2.2 Step Forward Selection ... 727
44.3 The Results of Regression Analysis ... 727
44.3.1 Relations Between Tempo and Analysis ... 727
44.3.2 Complex Relationships ... 729
44.3.3 Commonalities and Diversities ... 729
44.3.4 Overview of Statistical Results ... 738

45 **Relating Tempo to Metric, Melodic and Harmonic Analyses in Chopin's Prélude op. 28, No. 4** ... 743
45.1 Introduction ... 743
45.2 Data ... 745
45.2.1 Analytical Data ... 745
45.2.2 Tempo Data ... 745
45.3 Short Summary of the Results ... 747
45.4 Some Philosophical Comments ... 748

Part XII Inverse Performance Theory

46 **Principles of Music Critique** ... 751
46.1 Boiling Down Infinity—Is Feuilletonism Inevitable? ... 751
46.2 "Political Correctness" in Performance—Reviewing Gould ... 752
46.3 Transversal Ethnomusicology ... 754

47 **Critical Fibers** ... 755
47.1 The Stemma Model of Critique ... 755
47.2 Fibers for Locally Linear Grammars ... 756
47.3 Algorithmic Extraction of Performance Fields ... 759
47.3.1 The Infinitesimal View on Expression ... 759
47.3.2 Real-Time Processing of Expressive Performance ... 760
47.3.3 Score-Performance Matching ... 761
47.3.4 Performance Field Calculation ... 762
47.3.5 Visualization ... 763
47.3.6 The EspressoRUBETTE®: An Interactive Tool for Expression Extraction ... 764
47.4 Local Sections ... 766
47.4.1 Comparing Argerich and Horowitz ... 768

Part XIII Operationalization of Poiesis

48 **Unfolding Geometry and Logic in Time** ... 773
48.1 Performance of Logic and Geometry ... 774
48.2 Constructing Time from Geometry ... 775
48.3 Discourse and Insight ... 776

49 Local and Global Strategies in Composition 779
49.1 Local Paradigmatic Instances 780
49.1.1 Transformations 780
49.1.2 Variations 780
49.2 Global Poetical Syntax 781
49.2.1 Roman Jakobson's Horizontal Function 781
49.2.2 Roland Posner's Vertical Function 781
49.3 Structure and Process 782

50 The Paradigmatic Discourse on *presto*® 783
50.1 The *presto*® Functional Scheme 783
50.2 Modular Affine Transformations 786
50.3 Ornaments and Variations 786
50.4 Problems of Abstraction 789

51 Case Study I: "Synthesis" by Guerino Mazzola 791
51.1 The Overall Organization 791
51.1.1 The Material: 26 Classes of Three-Element Motives 792
51.1.2 Principles of the Four Movements and Instrumentation 793
51.2 1st Movement: Sonata Form 794
51.3 2nd Movement: Variations 794
51.4 3rd Movement: Scherzo 798
51.5 4th Movement: Fractal Syntax 799

52 Object-Oriented Programming in OpenMusic 801
52.1 Object-Oriented Language 802
52.1.1 Patches 802
52.1.2 Objects 803
52.1.3 Classes 803
52.1.4 Methods 803
52.1.5 Generic Functions 804
52.1.6 Message Passing 804
52.1.7 Inheritance 805
52.1.8 Boxes and Evaluation 805
52.1.9 Instantiation 806
52.2 Musical Object Framework 806
52.2.1 Internal Representation 806
52.2.2 Interface 808
52.3 Maquettes: Objects in Time 811
52.4 Meta-object Protocol 813
52.4.1 Reification of Temporal Boxes 815
52.5 A Musical Example 817

Part XIV String Quartet Theory

53 Historical and Theoretical Prerequisites 825
53.1 History 825
53.2 Theory of the String Quartet Following Ludwig Finscher 826
53.2.1 Four Part Texture 826
53.2.2 The Topos of Conversation Among Four Humanists 827
53.2.3 The Family of Violins 828

54 Estimation of Resolution Parameters ... 831
54.1 Parameter Spaces for Violins ... 831
54.2 Estimation ... 833

55 The Case of Counterpoint and Harmony ... 839
55.1 Counterpoint ... 839
55.2 Harmony ... 840
55.3 Effective Selection ... 840

Part XV Gesture Philosophy for Music

56 The Topos of Gestures ... 843

57 Gesture Philosophy: Phenomenology, Ontology, and Semiotics ... 845
57.1 A Short Recapitulation of Musical Ontology ... 845
57.1.1 Ontology: Where, Why, and How ... 845
57.1.2 Oniontology: Facts, Processes, and Gestures ... 846
57.2 Jean-Claude Schmitt's Historiographic and Philosophical Treatise "La raison des gestes dans l'Occident médiéval" ... 846
57.2.1 Comments ... 847
57.3 Vilém Flusser's *Gesten: Versuch einer Phänomenologie* ... 848
57.3.1 A Short Introduction to Flusser's Essay ... 848
57.3.2 The Semiotic Neurosis ... 848
57.4 Michel Guérin's *philosophie des gestes* ... 850
57.4.1 The Essay's Structure ... 850
57.4.2 Gestural Ontology and Four Elementary Gestures ... 850
57.5 Flusser and Guérin: Some Consequences ... 851
57.6 A Program ... 852
57.6.1 Circularity ... 852
57.7 The Semiotic Gesture Concept of Adam Kendon and David McNeill ... 853
57.7.1 Comments ... 855
57.8 Juhani Pallasmaa and André Chastel: The Thinking Hand in Architecture and the Arts ... 855
57.9 Benveniste, Popelard, Wall ... 856

58 The French Presemiotic Approach ... 859
58.1 Maurice Merleau-Ponty ... 860
58.2 Francis Bacon and Gilles Deleuze ... 860
58.3 Jean Cavaillès and Charles Alunni ... 861
58.4 Gilles Châtelet ... 862

59 Cognitive Science ... 867
59.1 Embodiment ... 867
59.1.1 Embodiment Science ... 868
59.2 Neuroscience ... 871
59.2.1 Embodied AI ... 873
59.3 Anthropology ... 874
59.4 Dance ... 874
59.5 Disabled Gestures Versus Gestures Disabled: Parlan's Versus Peterson's Pianism ... 875
59.5.1 Performative Gestures: Disabled Jazz Pianists ... 876
59.5.2 Horace Parlan: Disabled Gestures ... 877
59.5.3 Parlan with Bass (and Drums) ... 877

59.5.4 Parlan with Rhythm Section ... 878
59.5.5 Parlan as Soloist ... 879
59.5.6 Parlan's Duets with Archie Shepp ... 880
59.5.7 Disabled Gestures ... 881
59.5.8 Gestures Disabled: Oscar Peterson ... 882
59.5.9 Conclusion ... 884
59.6 Aristotle, Blind Lemon Jefferson, and Vilayanur S. Ramachandran Walk into a Bar ... 885
59.6.1 Introduction ... 885
59.6.2 Division by (Almost) Zero: Many Blind Bluesmen but Few Blind Blues ... 885
59.6.3 Seeing Blind Blues: Gesture, Flow, Circuitry, and Amplification ... 886
59.6.4 Epilogue: Puns as Gestures ... 887

60 Models from Music ... 889
60.1 Wolfgang Graeser ... 890
60.2 Adorno, Wieland, Sessions, Clynes ... 890
60.2.1 Theodor Wiesengrund Adorno ... 891
60.2.2 Renate Wieland ... 893
60.2.3 Roger Sessions ... 894
60.2.4 Manfred Clynes ... 895
60.3 Johan Sundberg and Neil P. McAngus Todd ... 896
60.4 David Lewin and Robert S. Hatten ... 898
60.5 Marcelo Wanderley and Claude Cadoz, Rolf Inge Godøy and Marc Leman ... 901

Part XVI Mathematics of Gestures

61 Fundamental Concepts and Associated Categories ... 907
61.1 Introduction ... 907
61.2 Towards a Musical String Theory ... 909
61.3 Initial Investigations: Diagrams of Curves ... 910
61.4 Modeling a Pianist's Hand ... 912
61.4.1 The Hand's Model ... 912
61.4.2 Transforming Abstract Note Symbols into Symbolic Gestures ... 912
61.4.3 From Symbolic Hand Gestures to Physical Gestures ... 913
61.5 The Mathematical Definition of Gestures ... 914
61.6 Hypergestures ... 915
61.6.1 Spatial Hypergestures ... 917
61.7 Categorically Natural Gestures ... 918
61.8 Connecting to Algebraic Topology: Hypergestures Generalize Homotopy ... 919
61.9 Gestoids ... 922
61.9.1 The Fundamental Group, Klumpenhouwer Networks, and Fourier Representation ... 924
61.10 Gabriel's Spectroids and Natural Formulas ... 925
61.10.1 Solutions of Representations of Natural Formulas by Local Networks ... 927
61.11 The Tangent Category ... 927
61.12 The Diamond Conjecture ... 929
61.13 Topos Logic for Gestures ... 930
61.14 The Escher Theorem for Hypergestures ... 931
61.14.1 Hypergestures and the Escher Theorem for Fux Counterpoint ... 931
61.14.2 Rebecca Lazier's Vanish: Lawvere, Escher, Schoenberg ... 933

62 Categories of Gestures over Topological Categories ... 937
62.1 Gestures over Topological Categories ... 939
62.1.1 The Categorical Digraph of a Topological Category ... 940
62.1.2 Gestures with Body in a Topological Category ... 940
62.1.3 Varying the Underlying Topological Category ... 942
62.2 From Morphisms to Gestures ... 942
62.2.1 Diagrams as Gestures ... 944
62.2.2 Gestures in Factorization Categories ... 944
62.2.3 Extensions from Homological Algebra Are Gestures ... 945
62.2.4 The Bicategory of Gestures ... 945
62.2.5 Entering the Diamond Space ... 946
62.3 Diagrams in Topological Groups for Gestures ... 947
62.4 Gestural Interpretation of Modulations in Beethoven's op.106/Allegro ... 950
62.4.1 Recapitulation of the Results from Section 28.2 ... 951
62.4.2 The Modulation $B_\flat$-major $\rightsquigarrow$ G-major Between Measure 31 and Measure 44 ... 952
62.4.3 Lewin's Characteristic Gestures Identified? ... 955
62.4.4 Modulation $E_\flat$-major $\rightsquigarrow$ D-major/B-minor from W to W^* ... 957
62.4.5 The Fanfare ... 958
62.5 Conclusion for the Categorical Gesture Approach ... 960
62.6 Functorial Gestures: General Addresses ... 961
62.7 Yoneda's Lemma for Gestures ... 962
62.8 Examples from Music ... 963
62.8.1 Collections of Acoustical Waves ... 963
62.8.2 Collections of Spectral Music Data ... 964
62.8.3 MIDI-Type ON-OFF Transformations ... 964

63 Singular Homology of Hypergestures ... 965
63.1 An Introductory Example ... 965
63.2 Chain Modules for Singular Hypergestural Homology ... 967
63.3 The Boundary Homomorphism ... 968

64 Stokes' Theorem for Hypergestures ... 973
64.1 The Need for Stokes' Theorem for Hypergestures ... 973
64.2 Almost Regular Manifolds, Differential Forms, and Integration for Hypergestures ... 973
64.2.1 Locally Almost Regular Manifolds ... 974
64.2.2 Differential Forms ... 975
64.2.3 Integration ... 975
64.3 Stokes' Theorem ... 976

65 Local Facts, Processes, and Gestures ... 979
65.1 Categories of Local Compositions ... 979
65.2 Categories of Local Networks ... 980
65.3 Categories of Local Gestures ... 982
65.3.1 Local Gestures on Topological Categories of Points ... 982
65.4 Connecting Functors ... 984
65.5 Hypernetworks and Hypergestures ... 985
65.5.1 Escher Theorems ... 985
65.6 Singular Homology of Hypernetworks and Hypergestures ... 985

66 Global Categories ... 987
66.1 Categories of Global Compositions ... 987
66.1.1 Simplicial Methods ... 988
66.2 Classification of Global Compositions ... 989
66.3 Non-interpretable Global Compositions ... 990
66.4 Categories of Global Networks ... 990
66.4.1 Non-interpretable Global Networks ... 991
66.5 Categories of Global Gestures ... 993
66.6 Globalizing Topological Categories: Categorical Manifolds ... 993
66.7 Globalizing Skeleta ... 996
66.8 Functorial Global Gestures ... 998

67 Mathematical Models of Creativity ... 1001
67.1 Forewarning: Invention of Gestures in Mathematics ... 1001
67.1.1 Thinking Exactness, Like a Rolling Mind ... 1001
67.1.2 Thought as an Algebra of Gestures ... 1002
67.2 Method and Objects, Summarily Explained: I—Preamble ... 1003
67.2.1 Prelude to a Discourse of a Method: "Caminos", "Aletheia", Irreverence ... 1003
67.2.2 Our Posture ... 1015
67.3 Method and Objects, Summarily Explained: II—Data ... 1021
67.3.1 Simple Objects, Structures and Invariants in Mathematics ... 1021
67.3.2 Complete Frameworks, Computations and Representations ... 1029
67.4 Creativity in Mathematics: Gestures in Historical Contexts ... 1031
67.4.1 Creativity: Phenomenology, Psychology and Skills, and Life ... 1031
67.4.2 Determination of Mathematics as a History of Its Gestures ... 1036
67.4.3 Invention in the Art of Mathematics ... 1044
67.5 On the Mathematical Invention of Coordinations ... 1046
67.5.1 Emergence of Coordinations ... 1047
67.5.2 Arrows ... 1050
67.5.3 Bodies, Implicit Surfaces, Abstract Relations ... 1055
67.5.4 Sketches ... 1056
67.6 Pulsation in the Living Process of Invention Among Shapes ... 1058
67.6.1 Productions: Objects and Relations, Problems, Pulsation ... 1058
67.6.2 Creativity in the Mathematical World Seen as a Living System of Shapes, in a Categorical Framework ... 1060
67.7 Conclusion: Categorical Presentation of Pulsations ... 1064
67.8 The Hegel Group Action on a Critical Concept's Walls ... 1066
67.9 Introduction ... 1066
67.10 The Hegel Concept Group $\mathcal{G}$ ... 1067
67.10.1 Hegel's Initial Thought Movement in *Wissenschaft der Logik* ... 1067
67.10.2 The Implicit Group Structure ... 1070
67.10.3 The Conceptual Box Structure ... 1072
67.11 The $\mathcal{G}$ Action on the Yoneda Model of Creativity ... 1073
67.12 The Hegel Body $\mathcal{B}$ in the Concept Architecture of Forms and Denotators ... 1073
67.13 The Usage of $\mathcal{G}$ for the Dynamics of Creativity ... 1074
67.13.1 Two Preliminary Examples ... 1074
67.13.2 The Challenge: Creating a Spectrum of Conceptual Extensions ... 1075
67.13.3 Escher's Theorem for Beethoven's Fanfare in the *"Hammerklavier" Sonata op. 106* 1075
67.13.4 The Rotation $S@N$ as a Driving Creative Force in the Incipit of Liszt's *Mephisto Walzer* No.1 ... 1076
67.14 An Experimental Composition ... 1078
67.15 Still More Symmetries? Future Developments ... 1080

Part XVII Concept Architectures and Software for Gesture Theory

68 Forms and Denotators over Topological Categories ... 1085
68.1 The General Topos—Theoretical Framework ... 1085
68.1.1 The category **TopCat** of Small Topological Categories ... 1085
68.2 Forms and Denotators ... 1086
68.3 Mathematics of Objects, Structures, and Concepts ... 1087
68.4 Galois Theory of Concepts ... 1087
68.4.1 Introduction ... 1088
68.4.2 Form Semiotics ... 1089
68.4.3 The Category of Form Semiotics ... 1092
68.4.4 Galois Correspondence of Form Semiotics ... 1093

69 The Rubato Composer Architecture ... 1095
69.1 The Software Architecture ... 1096
69.2 The Rubette World ... 1097
69.2.1 Rubettes for Counterpoint ... 1097
69.2.2 Rubettes for Harmony ... 1098
69.2.3 MetroRubettes ... 1099

70 The BigBang Rubette and the Ontological Dimension of Embodiment ... 1101

71 Facts: Denotators and Their Visualization and Sonification ... 1103
71.1 Some Earlier Visualizations of Denotators ... 1103
71.1.1 Göller's PrimaVista Browser ... 1103
71.1.2 Milmeister's ScorePlay and Select2D Rubettes ... 1105
71.2 An Early Score-Based Version of BigBang ... 1106
71.2.1 The Early BigBang Rubette's View Configurations ... 1107
71.2.2 Navigating Denotators ... 1111
71.2.3 Sonifying Score-Based Denotators ... 1111
71.3 BigBangObjects and Visualization of Arbitrary $Mod^{@}$ Denotators ... 1111
71.3.1 A Look at Potential Visual Characteristics of Form Types ... 1112
71.3.2 From a General View Concept to BigBang Objects ... 1114
71.3.3 New Visual Dimensions ... 1115
71.4 The Sonification of BigBangObjects ... 1116
71.5 Examples of Forms and the Visualization of Their Denotators ... 1117
71.5.1 Some Set-Theoretical Structures ... 1117
71.5.2 Tonal and Transformational Theory ... 1119
71.5.3 Synthesizers and Sound Design ... 1121

72 Processes: BigBang's Operation Graph ... 1127
72.1 Temporal BigBangObjects, Object Selection, and Layers ... 1128
72.1.1 Selecting None and Lewin's Transformation Graphs ... 1128
72.1.2 The Temporal Existence of BigBangObjects ... 1129
72.1.3 BigBang Layers ... 1131
72.2 Operations and Transformations in BigBang ... 1132
72.2.1 Non-transformational Operations ... 1132
72.2.2 Transformations ... 1136
72.3 BigBang's Process View ... 1138
72.3.1 Visualization of Processes ... 1138
72.3.2 Selecting States and Modifying Operations ... 1139

72.3.3 Alternative and Parallel Processes ... 1139
72.3.4 Structurally Modifying the Graph ... 1141
72.3.5 Undo/Redo ... 1141

73 Gestures: Gestural Interaction and Gesturalization ... 1143
73.1 Formalizing: From Gestures to Operations ... 1143
73.1.1 Modes, Gestural Operations, and the Mouse ... 1144
73.1.2 Affine Transformations and Multi-touch ... 1148
73.1.3 Dynamic Motives, Sound Synthesis, and Leap Motion ... 1149
73.1.4 Recording, Modifying Operations and MIDI Controllers ... 1151
73.2 Gesturalizing and the Real BigBang: Animated Composition History ... 1152
73.2.1 Gesturalizing Transformations ... 1152
73.2.2 Gesturalizing Other Operations ... 1154
73.2.3 Using Gesturalization as a Compositional Tool ... 1154

74 Musical Examples ... 1157
74.1 Some Example Compositions ... 1157
74.1.1 Transforming an Existing Composition ... 1157
74.1.2 Gesturalizating and Looping with a Simple Graph ... 1158
74.1.3 Drawing UPIC-like Motives and Transforming ... 1159
74.1.4 Drawing Time-Slices ... 1160
74.1.5 Converting Forms, Tricks for Gesturalizing ... 1160
74.1.6 Gesturalizing a Spectrum ... 1162
74.1.7 Using Wallpapers to Create Rhythmical Structures ... 1163
74.2 Improvisation and Performance with BigBang ... 1163
74.2.1 Improvising by Selecting States and Modifying Transformations ... 1164
74.2.2 Playing Sounds with a MIDI Keyboard and Modifying Them ... 1164
74.2.3 Playing a MIDI Grand Piano with Leap Motion ... 1165
74.2.4 Playing a MIDI Grand Piano with the Ableton Push ... 1166
74.2.5 Improvising with 12-Tone Rows ... 1167

Part XVIII The Multiverse Perspective

75 Gesture Theory and String Theory ... 1173

76 Physical and Musical Multiverses ... 1175

77 Hesse's Melting Beads: A Multiverse Game with Strings and Gestures ... 1177
77.1 Review of Hesse's Glass Bead Game ... 1177
77.2 Frozen Glass Beads of Facticity ... 1178
77.3 The Revolution of Functors ... 1178
77.4 Gestures in Philosophy and Science ... 1180
77.5 Gesture Theory in Music ... 1181
77.6 A Remark on Gestural Creativity ... 1183
77.7 Gestures and Strings ... 1183
77.8 Playing the Multiversed Game in a Pre-semiotic Ontology ... 1184

78 Euler-Lagrange Equations for Hypergestures 1185
78.1 The Problem in Performance Theory with the Physical Nambu-Goto Lagrangian 1185
78.1.1 Complex Time and Descartes' Dualistic Ontology 1186
78.2 Lagrangian Density for Complex Time 1187
78.2.1 The Lagrangian Action for Performance 1188
78.2.2 The World-Sheet of Complex Time 1190
78.2.3 The Space for a Hand's Gestures 1192
78.2.4 The World-Sheet for a Simple Case 1192
78.2.5 The Elementary Gesture of a Pianist 1192
78.2.6 The Overarching Framework Between Note Performance and Gesture Performance 1195
78.2.7 Examples of Functional Relations Between Potential and Physical Gesture 1196
78.2.8 Calculus of Variations for the Physical Gesture 1202
78.2.9 A First Solution. World-Sheet Potentials Determine a Pianist's Gesture: Calculus of Variations and Fourier Analysis 1203
78.2.10 The Calculus with Vanishing Potential 1204
78.2.11 The Calculus with General Potential 1209
78.2.12 Solution of the Differential Equation Using 2D Fourier Series 1212
78.2.13 Parallels Between Performance Operators for Scores and for Gestures 1216
78.2.14 Complex Time and the Artistic Effort 1218
78.2.15 Opening the Aesthetic Question that Is Quantified in Lagrange Potentials 1218
78.2.16 A Musical Composition by Maria Mannone Realized Using These Ideas 1219
78.3 Global Performance Hypergestures 1223
78.3.1 The Musical Situation: An Intuitive Introduction 1223
78.4 Categorical Gestures and Global Performance Hypergestures 1224
78.4.1 Categorical Gestures: The Case of Potentials 1224
78.4.2 The Mathematics of Global Performance Hypergestures 1226
78.5 World-Sheet Hypergestures for General Skeleta 1227
78.6 A Global Variational Principle for the Lagrange Formalism 1231

Part XIX Gestures in Music and Performance Theory, and in Ethnomusicology

79 Gesture Homology for Counterpoint 1235
79.1 Summary of Mathematical Theory of Counterpoint: What It Is About and What Is Missing ... 1235
79.2 Hypergestural Singular Homology 1236
79.3 A Classical Example of a Topological Category from Counterpoint 1237
79.3.1 Generators of $H_1({}^G X)$ for a Groupoid ${}^G X$ Defined by a Group Action 1238
79.4 The Meaning of H_1 for Counterpoint 1240
79.5 Concluding Comments 1241

80 Modulation Theory and Lie Brackets of Vector Fields 1243
80.1 Introduction 1243
80.1.1 Short Recapitulation of the Classical Model's Structure 1244
80.2 Hypergestures Between Triadic Degrees That Are Parallel to Vector Fields 1245
80.3 Lie Brackets Generate Vector Fields That Connect Symmetry-Related Degrees 1245
80.4 Selecting Parallel Hypergestures That Are Admissible for Modulation 1247
80.5 The Other Direct Modulations 1249

81 Hypergestures for Performance Stemmata 1253
81.1 Motivation, Terminology, and Previous Results 1253
81.1.1 Performance Stemmata and Performance Gestures of Locally Compact Points 1254
81.2 Gestures with Lie Operators in Stemma Theory 1255
81.3 Connecting Stemmatic Gestures for Weights and Performance Fields 1256
81.4 Homology of Weight Parameter Stemmata 1258
81.5 A Concrete Example 1260
81.6 A Final Comment 1261

82 Composing and Analyzing with the Performing Body 1263
82.1 Gesture: A Sign or a Totality? 1264
82.2 A Gesture-Based Structural Reading in *Rain Tree Sketch II* by Toru Takemitsu 1266
82.2.1 Process I: Synergy of Mirroring and Parallel Gestures 1267
82.2.2 Process II: Towards Relaxation, Balance, and Weightfulness 1270
82.3 The Last Leg of a Bodily Journey 1274
82.3.1 *Sheng* for Piano, Smartphones, and Fixed Playback 1278
82.3.2 Cross-modality of Gestures 1279
82.3.3 Learning the Smartphone Instrument 1279
82.3.4 Kinesthetic Awareness and Modes of Listening 1281
82.4 Conclusion: Foregrounding the Performer's Body 1283

83 Gestural Analysis and Classification of a Conductor's Movements 1285
83.1 Gestures and Communication in Orchestral Conducting: A Case Study 1285
83.1.1 Problematics and Solving Methods 1286
83.1.2 Results, Consequences, Applications 1288
83.1.3 Some Remarks 1289
83.2 Hints for a Mathematical Description 1289
83.3 Data Analysis 1290
83.4 Conclusion 1292
83.5 Addendum 1293

84 Reviewing Flow, Gesture, and Spaces in Free Jazz 1295
84.1 Improvisation: Defining Time 1295
84.2 Flow, Gestures, Imaginary Time and Spaces in the Music Movie *Imaginary Time* 1295
84.2.1 The Compositional Character of the Pieces 1297
84.2.2 Large Forms 1298
84.2.3 Precision of Attacks 1298
84.2.4 Co-presence of Different Time Layers 1299
84.2.5 The Reality of Imaginary Time 1300
84.2.6 Measuring Flow 1300
84.2.7 Explicit Perception of Gestures 1300

85 Gesture and Vocalization 1301
85.1 Vocal Gesture 1301
85.2 Vocal and Manual Motion 1303
85.3 Gait 1304
85.4 Hindustani Vocal Music 1305
85.5 Notic Models and Kinetic Models 1306
85.6 The Realist Pitfall 1307
85.7 The Subjectivist Pitfall 1308
85.8 Speech Gesture 1311

86 Elements of a Future Vocal Gesture Theory 1313
86.1 Why a Theory of Vocal Gestures? 1313
86.1.1 Studying the Voice Without the Singer? 1314
86.1.2 Parts of the Phonatory System and Their Functions 1314
86.1.3 Imaginary Gestures in Real Time? 1315
86.1.4 Space of Voice Parameters Gestures 1315
86.1.5 About the Importance of Breathing and of Laryngeal Movements 1316
86.1.6 Mathematical Description of Vocal Gestures 1317
86.1.7 Gestures Thought by Singers 1321
86.2 A Powerful Tool from the Past 1322
86.2.1 Gestures in Gregorian Chant Didactics 1323
86.2.2 Concept of Rhythm and Time 1323
86.2.3 The Neumes 1327
86.3 Connecting Physiology, Gestures and Notation. Toward New Neumes? 1329
86.3.1 New Neumes 1330

Part XX Appendix: Sound

A Common Parameter Spaces 1335
A.1 Physical Spaces 1335
A.1.1 Neutral Data 1336
A.1.2 Sound Analysis and Synthesis 1338
A.2 Mathematical and Symbolic Spaces 1347
A.2.1 Onset and Duration 1347
A.2.2 Amplitude and Crescendo 1348
A.2.3 Frequency and Glissando 1349

B Auditory Physiology and Psychology 1353
B.1 Physiology: From the Auricle to Heschl's Gyri 1353
B.1.1 Outer Ear 1354
B.1.2 Middle Ear 1354
B.1.3 Inner Ear (Cochlea) 1355
B.1.4 Cochlear Hydrodynamics: The Travelling Wave 1357
B.1.5 Active Amplification of the Traveling Wave Motion 1358
B.1.6 Neural Processing 1360
B.2 Discriminating Tones: Werner Meyer-Eppler's Valence Theory 1362
B.3 Aspects of Consonance and Dissonance 1364
B.3.1 Euler's Gradus Function 1364
B.3.2 von Helmholtz' Beat Model 1365
B.3.3 Psychometric Investigations by Plomp and Levelt 1367
B.3.4 Counterpoint 1367
B.3.5 Consonance and Dissonance: A Conceptual Field 1367

Part XXI Appendix: Mathematical Basics

C Sets, Relations, Monoids, Groups 1371
C.1 Sets 1371
C.1.1 Examples of Sets 1371

C.2 Relations ... 1372
C.2.1 Universal Constructions ... 1374
C.2.2 Graphs and Quivers ... 1375
C.2.3 Monoids ... 1376
C.3 Groups ... 1378
C.3.1 Homomorphisms of Groups ... 1378
C.3.2 Direct, Semi-direct, and Wreath Products ... 1380
C.3.3 Sylow Theorems on p-groups ... 1380
C.3.4 Classification of Groups ... 1381
C.3.5 General Affine Groups ... 1382
C.3.6 Permutation Groups ... 1382

D Rings and Algebras ... 1385
D.1 Basic Definitions and Constructions ... 1385
D.1.1 Universal Constructions ... 1386
D.2 Prime Factorization ... 1389
D.3 Euclidean Algorithm ... 1389
D.4 Approximation of Real Numbers by Fractions ... 1389
D.5 Some Special Issues ... 1390
D.5.1 Integers, Rationals, and Real Numbers ... 1390

E Modules, Linear, and Affine Transformations ... 1391
E.1 Modules and Linear Transformations ... 1391
E.1.1 Examples ... 1391
E.2 Module Classification ... 1392
E.2.1 Dimension ... 1392
E.2.2 Endomorphisms on Dual Numbers ... 1394
E.2.3 Semi-simple Modules ... 1394
E.2.4 Jacobson Radical and Socle ... 1395
E.2.5 Theorem of Krull-Remak-Schmidt ... 1396
E.3 Categories of Modules and Affine Transformations ... 1396
E.3.1 Direct Sums ... 1397
E.3.2 Affine Forms and Tensors ... 1397
E.3.3 Biaffine Maps ... 1399
E.3.4 Symmetries of the Affine Plane ... 1402
E.3.5 Symmetries on $\mathbb{Z}^2$ ... 1402
E.3.6 Symmetries on $\mathbb{Z}^n$ ... 1403
E.3.7 Complements on the Module of a Local Composition ... 1403
E.3.8 Fiber Products and Fiber Sums in **Mod** ... 1404
E.4 Complements of Commutative Algebra ... 1405
E.4.1 Localization ... 1406
E.4.2 Projective Modules ... 1406
E.4.3 Injective Modules ... 1407
E.4.4 Lie Algebras ... 1408

F Algebraic Geometry ... 1411
F.1 Locally Ringed Spaces ... 1411
F.2 Spectra of Commutative Rings ... 1412
F.2.1 Sober Spaces ... 1413

F.3 Schemes and Functors 1414
F.4 Algebraic and Geometric Structures on Schemes 1415
F.4.1 The Zariski Tangent Space 1415
F.5 Grassmannians 1416
F.6 Quotients 1417

G Categories, Topoi, and Logic 1419
G.1 Categories Instead of Sets 1419
G.1.1 Examples 1420
G.1.2 Functors 1421
G.1.3 Natural Transformations 1422
G.2 The Yoneda Lemma 1423
G.2.1 Universal Constructions: Adjoints, Limits, and Colimits 1423
G.2.2 Limit and Colimit Characterizations 1425
G.3 Topoi 1427
G.3.1 Subobject Classifiers 1428
G.3.2 Exponentiation 1429
G.3.3 Definition of Topoi 1429
G.4 Grothendieck Topologies 1430
G.4.1 Sheaves 1431
G.5 Formal Logic 1432
G.5.1 Propositional Calculus 1432
G.5.2 Predicate Logic 1435
G.5.3 A Formal Setup for Consistent Domains of Forms 1437

H Complements on General and Algebraic Topology 1443
H.1 Topology 1443
H.1.1 General 1443
H.1.2 The Category of Topological Spaces 1444
H.1.3 Uniform Spaces 1444
H.1.4 Special Issues 1445
H.2 Algebraic Topology 1445
H.2.1 Simplicial Complexes 1445
H.2.2 Geometric Realization of a Simplicial Complex 1446
H.2.3 Contiguity 1447
H.3 Simplicial Coefficient Systems 1447
H.3.1 Cohomology 1447

I Complements on Calculus 1449
I.1 Abstract on Calculus 1449
I.1.1 Norms and Metrics 1449
I.1.2 Completeness 1450
I.1.3 Differentiation 1451
I.2 Ordinary Differential Equations (ODEs) 1451
I.2.1 The Fundamental Theorem: Local Case 1452
I.2.2 The Fundamental Theorem: Global Case 1453
I.2.3 Flows and Differential Equations 1455
I.2.4 Vector Fields and Derivations 1455
I.3 Partial Differential Equations 1455

J More Complements on Mathematics 1457
J.1 Directed Graphs 1457
J.1.1 The Category of Directed Graphs (Digraphs) 1457
J.1.2 Two Standard Constructions in Graph Theory 1459
J.1.3 The Topos of Digraphs 1459
J.2 Galois Theory 1460
J.3 Splines 1461
J.3.1 Some Simplex Constructions for Splines 1461
J.3.2 Definition of General Splines 1462
J.4 Topology and Topological Categories 1463
J.4.1 Topology 1463
J.4.2 Topological Categories 1464
J.5 Complex Analysis 1465
J.6 Differentiable Manifolds 1466
J.6.1 Manifolds with Boundary 1467
J.6.2 The Tangent Manifold 1467
J.7 Tensor Fields 1468
J.7.1 Alternating Tensors 1468
J.7.2 Tangent Tensors 1468
J.8 Stokes' Theorem 1469
J.9 Calculus of Variations 1471
J.10 Partial Differential Equations 1471
J.10.1 Explicit Calculation 1472
J.11 Algebraic Topology 1474
J.11.1 Homotopy Theory 1474
J.11.2 The Fundamental Group(oid) 1475
J.12 Homology 1475
J.12.1 Singular Homology 1476
J.13 Cohomology 1477

Part XXII Appendix: Complements in Physics

K Complements on Physics 1481
K.1 Hamilton's Variational Principle 1481
K.1.1 Euler-Lagrange Equations for a Non-relativistic Particle 1482
K.2 String Theory 1482
K.3 Duality and Supersymmetry 1484
K.4 Quantum Mechanics 1485
K.4.1 Banach and Hilbert Spaces 1486
K.4.2 Geometry on Hilbert Spaces 1490
K.4.3 Axioms for Quantum Mechanics 1492
K.4.4 The Spectral Theorem 1493

Part XXIII Appendix: Tables

L Euler's Gradus Function 1497

M Just and Well-Tempered Tuning 1499

N Chord and Third Chain Classes ... 1501
N.1 Chord Classes ... 1501
N.2 Third Chain Classes ... 1506

O Two, Three, and Four Tone Motif Classes ... 1513
O.1 Two Tone Motifs in $OnPiMod_{12,12}$ ... 1513
O.2 Two Tone Motifs in $OnPiMod_{5,12}$ ... 1513
O.3 Three Tone Motifs in $OnPiMod_{12,12}$ ... 1514
O.4 Four Tone Motifs in $OnPiMod_{12,12}$ ... 1517
O.5 Three Tone Motifs in $OnPiMod_{5,12}$ ... 1523

P Well-Tempered and Just Modulation Steps ... 1525
P.1 12-Tempered Modulation Steps ... 1525
P.1.1 Scale Orbits and Number of Quantized Modulations ... 1525
P.1.2 Quanta and Pivots for the Modulations Between Diatonic Major Scales (No.38.1) . 1527
P.1.3 Quanta and Pivots for the Modulations Between Melodic Minor Scales (No.47.1) . . 1528
P.1.4 Quanta and Pivots for the Modulations Between Harmonic Minor Scales (No.54.1) 1530
P.1.5 Examples of 12-Tempered Modulations for All Fourth Relations ... 1530
P.2 2-3-5-Just Modulation Steps ... 1531
P.2.1 Modulation Steps Between Just Major Scales ... 1531
P.2.2 Modulation Steps Between Natural Minor Scales ... 1532
P.2.3 Modulation Steps from Natural Minor to Major Scales ... 1532
P.2.4 Modulation Steps from Major to Natural Minor Scales ... 1533
P.2.5 Modulation Steps Between Harmonic Minor Scales ... 1533
P.2.6 Modulation Steps Between Melodic Minor Scales ... 1534
P.2.7 General Modulation Behaviour for 32 Alterated Scales ... 1535

Q Counterpoint Steps ... 1537
Q.1 Contrapuntal Symmetries ... 1537
Q.1.1 Class No. 64 ... 1537
Q.1.2 Class No. 68 ... 1538
Q.1.3 Class No. 71 ... 1539
Q.1.4 Class No. 75 ... 1540
Q.1.5 Class No. 78 ... 1541
Q.1.6 Class No. 82 ... 1542
Q.2 Permitted Successors for the Major Scale ... 1543

Part XXIV References and Index

References ... R.1

Index ... R.33

Leitfaden

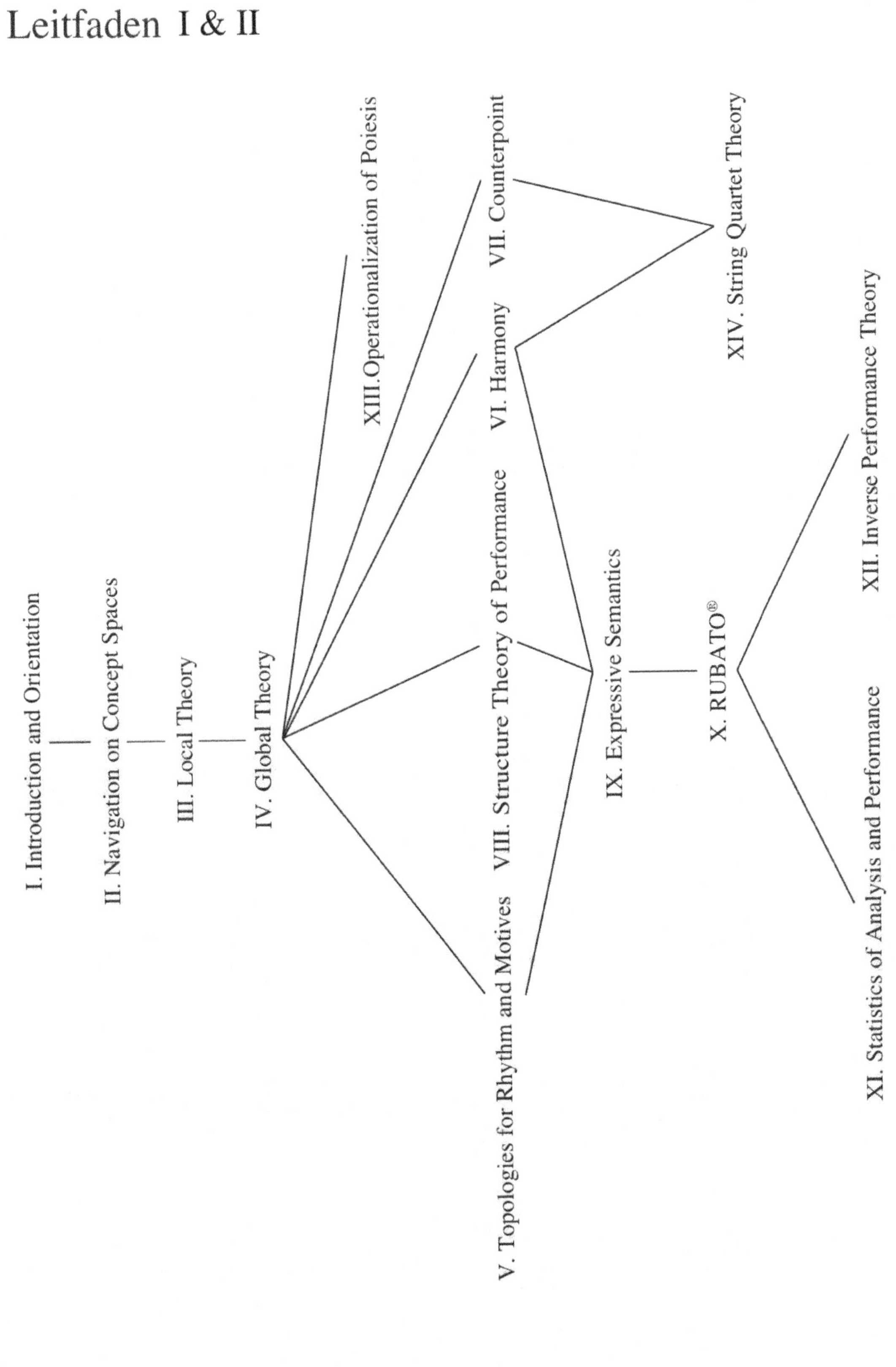
Leitfaden I & II
I. Introduction and Orientation
II. Navigation on Concept Spaces
III. Local Theory
IV. Global Theory
V. Topologies for Rhythm and Motives
VIII. Structure Theory of Performance
VI. Harmony
VII. Counterpoint
XIII. Operationalization of Poiesis
IX. Expressive Semantics
X. RUBATO®
XIV. String Quartet Theory
XI. Statistics of Analysis and Performance
XII. Inverse Performance Theory

Leitfaden III

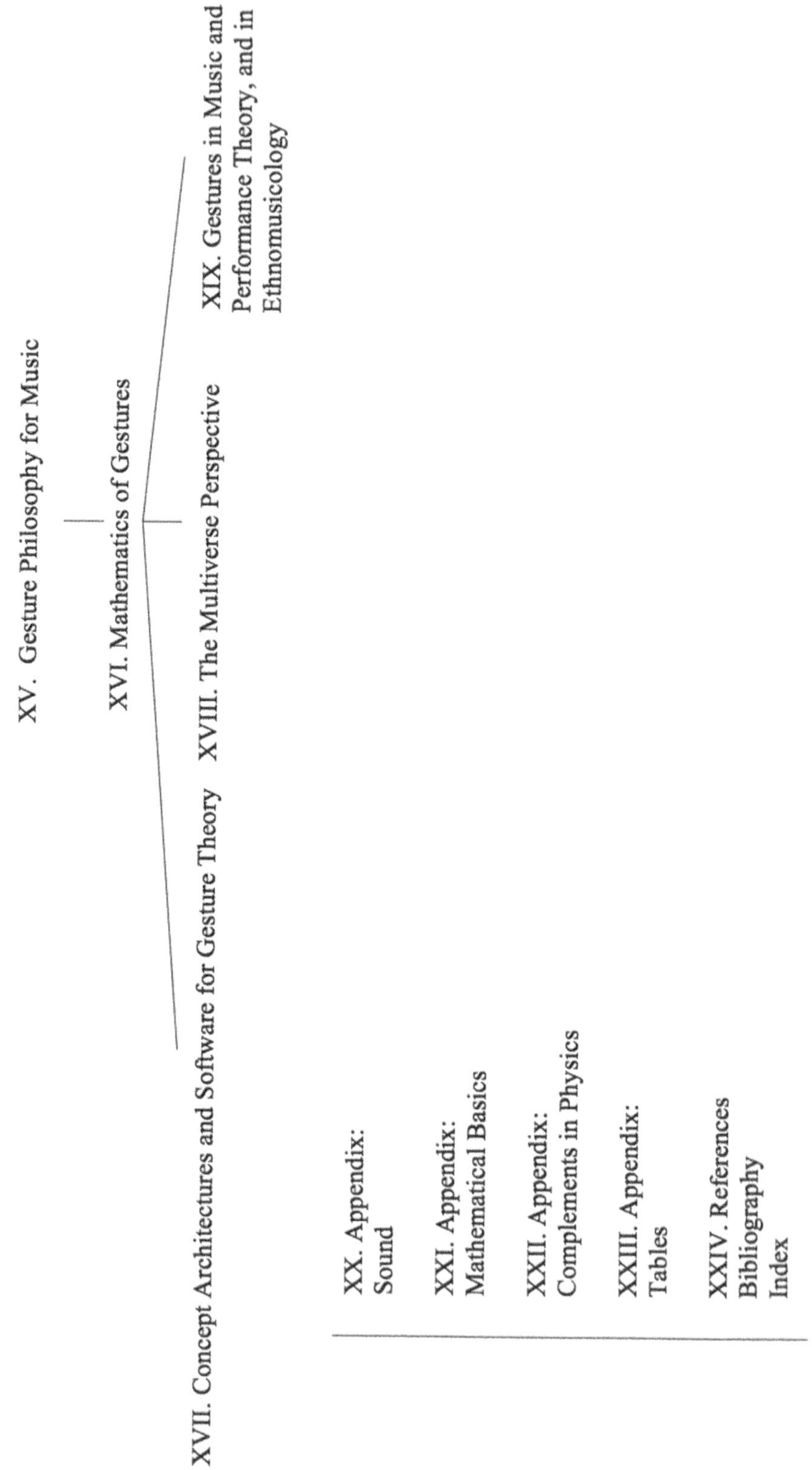

ToM_CD

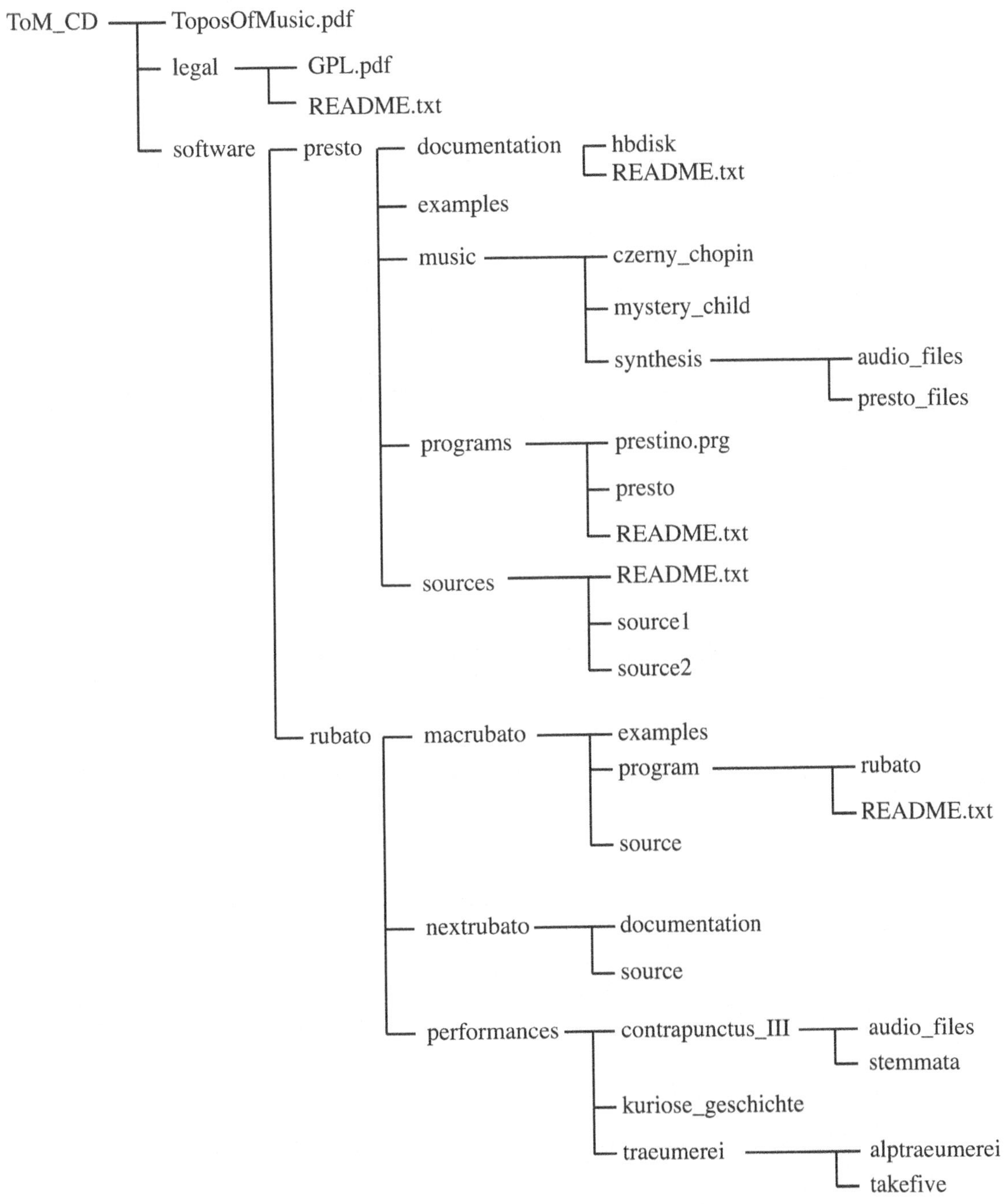

ToM_CD
ToposOfMusic.pdf
legal
GPL.pdf
README.txt
software
presto
documentation
hbdisk
README.txt
examples
music
czerny_chopin
mystery_child
synthesis
audio_files
presto_files
programs
prestino.prg
presto
README.txt
sources
README.txt
source1
source2
rubato
macrubato
examples
program
rubato
README.txt
source
nextrubato
documentation
source
performances
contrapunctus_III
audio_files
stemmata
kuriose_geschichte
traeumerei
alptraeumerei
takefive

Part VIII

Structure Theory of Performance

32

Local and Global Performance Transformations

C'est l'exécution du poème qui est le poème.
Paul Valéry

Summary. Performance includes a non-trivial transformation from the mental reality of a score to the physical reality of acoustic and, in the limit, gestural realization. We discuss a model of local and global transformation structures and present a preliminary discourse on the need to "shape mental reality" in performance. Local performance transformation structures, together with their syntactic combination to global structures, are formally developed. They involve extrapolation from discrete to continuous or differentiable data; the latter are induced by use of spline techniques. We give a justification of such a procedure from the musical and mathematical points of view, in particular with stress on expressive coherence.

– Σ –

This is the second big part of the book (in its second edition): It deals with performance as opposed to the traditional structural theory of rhythm, melody, and harmony.

Structure theory of performance is a turning point in the entire theory of the topos of music. In fact, the preceding parts dealt with general structure theory and then, on a more musicological and music-theoretic focus, with mental perspectives of rhythm, motives, harmony, and counterpoint. In contrast, performance is concerned with the transformation of mental structures into physical ones. This is what traditionally happens in a concert where human artists are performing on physical instruments, including all the richness of human expression on the gestural, emotional, or structural level of physiological, social, and physical parameters.

Evidently, a comprehensive theory of performance is out of reach as long as major constituents have not even been attacked on a scientific level. For example, there is no deeper understanding of the emotional function of music, one knows some extremely elementary facts, for example those concerning the emotional impact of contrapuntal intervals on the emotional brain, as described in chapters 29, 30, and 31. Also on the level of instrumental factors in the quality of a performance, very little is known, for example, concerning the role of instrumental parameters in the communication of musical contents. Worse than that: There is not even a commonly accepted structure theory of performance, i.e., a theory which deals with the precise and general description of what is a performance in its most elementary shape. For example, the discussion of tempo has not yet been carried to a point of general acceptance of what tempo can be, including a fundamental disagreement on its hierarchical ramifications. One of these discussions could germinate around the distinction between mental (symbolic, logical, call it as you wish) time and physical time. Naively, mental time, as it is encountered on the score notation, looks like something discrete, to be encoded on integers, or some isomorphic submodule $\delta.\mathbb{Z}$ of the real numbers, whereas physical time is parametrized by the full line of real numbers. This is also the point of view of Desain and Honing in [261]: They reduce mental time to

discrete time intervals of a metrical structure,

leaving the smooth part to the

continuous time scales of tempo changes and expressive timing.

G. Mazzola, *The Topos of Music II: Performance*, Computational Music Science,
https://doi.org/10.1007/978-3-319-64444-8_1

This procedure is mathematically incorrect because

- the metrical time is infinitely divisible in itself: No positive lower limit for mental durations has ever been envisaged, metrical time is a topologically dense, not a discrete set in the field of real numbers. Hence, any reasonable (more precisely: uniformly continuous) time function from mental time E to physical time e can uniquely be extended to a time function on the reals (see [527]). There is no conceptual reason to restrict metrical time to a discrete subdomain of the reals.
- Tempo does not deal with something more continuous than metrical time. It is another concept[1]: the inverse differential quotient of a function $E \mapsto e(E)$ between two copies of the real number axis with irreducibly different ontological specifications, namely the musical mental status of the score and the physical status of performed music.

Other misunderstandings floating around in musicological environments [382] maintain that tempo is a locally constant function, i.e., a step function, much like the medieval theories on velocity in the spirit of Oresme (see also our discussion in chapter 4). He would decompose accelerated movements into a succession of uniform movements [971, contribution of Isabelle Stengers on Galileo Galilei].

In what follows, we have not tried to downsize the complexity of performance, it is indeed the most complex subject in musicology. It involves all kinds of considerations concerning the three basic realities of physics, mentality, and psychology. But beyond this ontological diversity, it is of a sophisticated structural nature, involving differential geometry, ODEs, and PDEs, and evidently all the music theory which is presupposed in any reasonable performance theory (though: not in everybody's performance, but blurred gesticulations of musicians "playing their ass off"[2] are not the subject of any performance theory).

We shall however not discuss proper psychological aspects, but restrict ourselves to the mental and physical perspectives. For psychological concerns, refer to [546]. It may appear that we thereby omit an essential and basic point of view, and that such a restriction would hamper the entire discourse. This is true insofar as we omit an essential perspective. But it is false that this hampers the discourse. The performance discourse has to deal with structural descriptions of performance: What happens if a score is played in acoustical and gestural reality? This is beyond psychology. Also do the rationales of a performance not uniquely rely on the psychological reality (of emotions as stressed in the naive romantic approach), but, in the spirit of Theodor Wiesengrund Adorno or Walter Benjamin, on analytical facts or, in another approach, on gestural paradigms, as investigated by Johan Sundberg, for example.

Performance research is only in its initial phase, in particular with respect to the high level of performing artists. But this is no reason to abbreviate the scientific path with the risk of misunderstanding the beauty and complexity of performance, and to fall into some crevasse of oversimplification. Hopefully, this should help the reader to follow chapters on performance theory with due patience.

32.1 Performance as a Reality Switch

Summary. There is a sharp dichotomy of realities, communication, and semantical levels between a score and its performances. We expose these facts and their consequences for a theory of performance *versus* mental music theory.

– Σ –

Performance is more than simply playing an instrument "all improviso"; even in free jazz (if it merits that name), musicians always refer to an inner score. The concept of a score is used here in its generic meaning, i.e., a *generic score* is any written, imagined or conceived scheme for the execution of a musical composition (see [703] for details). In European or Japanese classical music for example, a score is realized as a denotator structure of more or less complex form (see [760], for an explicit score form englobing classical Viennese composition). In improvised music, the scheme is less of a structural nature than of a processual one.

[1] We shall introduce and discuss such concepts in Section 33.1.1.

[2] Answer of free jazz saxophonist Werner Lüdi to Cecil Taylor's question: "What's your concept?"

The jazz musician, for example, follows action and reaction patterns and rules according to blues schemes and individual dictionaries of motivic, rhythmic, and harmonic elements. We propose this informal but reliable definition:

Definition 97 *Performance is defined as the physical realization of an interpretation of a generic score.*

We leave the meaning of "interpretation" open in its common understanding, but we do include, and this is a core issue of the subsequent discourse, interpretation in the technical sense of interpretations of local compositions. We also stress the generic attribute "physical" which does include the acoustic realization, but does not exclude further performance parameters, such as the gestural dynamics of a performer. But we do, as already mentioned, exclude psychic parameters and therefore mean strictly physical performance. Notice that this subsumes intermediate technological strata as a special subsystem of physical realization.

32.2 Why Do We Need Infinite Performance of the Same Piece?

Summary. This section deals with the *a priori* necessity of infinite performance. The argumentation relates to infinite analysis—due to the Yoneda philosophy—and to its communication on the rhetoric level of expressive performance.

– Σ –

The basic problem here is a very common one: Why do we need infinite performance? We are talking about real performances in real concerts. Do we, and why do we need new performances, again and again? Couldn't it arrive that all possible or—at least—all relevant performances are definitively played at a given moment, and that all successive performances are doomed to the existence of superfluous variants of the arsenal of core performances.

Seemingly, a musical composition, as it is fixed in a score and an associated denotator, is a finite object. So it would be a logical consequence of this finiteness that infinity is not inscripted in a musical composition and its performance. We therefore have two questions: Is there a substantial infinity in the interpretative variety of a given score? And, if this is the case, is there an associated infinity of performances, and why should such varieties have a sense for the listener or for the artist?

The first question is easily answered: Yes, there is an infinity of interpretations of a given finite score denotator. This was already shown in the discussion of iterated interpretations in Section 13.4.1.

This is an affirmative answer on the mental level, and one may easily add an infinity of evaluations of any such finite or infinite interpretation, for example on the level of rhythmic, motivic, or harmonic analyses as described previously in the respective chapters. One may also enrich the possibilities of viewing a given interpretation by a variation of the address, a technique which evokes Yoneda's Lemma, of course, but which is also very concretely developed in the context of harmonic topologies.

We shall see later in chapter 44.7 on performance operators that the analytical propaedeutics to performance also includes an important class of analyses: the analytical weights, i.e., numerical functions associated with more abstract analyses (such as harmonic topologies). This type of numerical evaluation which was already introduced in the power series λ_{w_0} of Section 13.4.2 is not just a "boiled-down" version of 'serious analysis', it is just another way of looking at complex configurations. The point is that the categories of 'understanding' a (finite) musical composition are of an infinite character in many respects, and that there is no reason to claim complete understanding by any finitistic argument. So the expression of mental content relies upon an infinite arsenal:

Fact 18 *Performance conveys an infinite message.*

But infinity in performance also comes in from a second point of view: Performance is inevitably an experiment in understanding or advancing comprehension. The realization of a piece of music in physical spaces creates a view, a flight through a virtual landscape which may reveal new insights to the involved actors, insights which are not anticipated but emerge from a particular visual angle of a determined 'performance flight'. Such an experience is a *mental experiment* as it has been described in chapter 4, and as such it is a creative investigation, not only a reproductive activity for the sake of social coherence and psycho-hygiene.

In this sense, performance need not be agreeable or pleasant—recognizing new aspects within an infinite repertory of analytical structures can be painful, but healthy. Glenn Gould's work is a brilliant piece of history in this investigative field of musical performance. Even his most awkward performances of Beethoven's piano music (such as the strange, cranky performance of the "Hammerklavier" sonata op. 106 or the funny and blasphemic review of the "Appassionata" sonata op. 57) are masterpieces in revealing new aspects of compositions, be it solely to learn how *not* to play them...

32.3 Local Structure

Summary. Just as with the analytical work, the performance task is also composed of local subtasks which constitute the morphemes of performance transformations. This section describes the emergence of such units and their structure.

– Σ –

In a first approximation, performance p could be seen as a set map which associates with every element X of a local composition (in a score-related form) a physical event, encoded by an element $x = p(X)$ in a second local composition whose parameters pertain to a form of physical signification.

The coherence of such a performance map is however very rarely a global one. For example, the tempi may change instantaneously, by the indication *istesso tempo*, forcing the artist to restart from the initial tempo after a deformation via a sequence of agogical indications. Or else, left and right hands may follow their separate tempi—except for some meeting points where the onsets should coincide (in Chopin rubato, for example). In orchestral music, the tuning is a function of the instrument, although everything is written in the same orchestra score. Or it may happen that the local performance of an ornament (a complicated trill, say) is an autonomous shaping map, operating independently of the map on the events which neighbor the ornament. We therefore observe a global structure, a patchwork of local chart maps, much like morphisms on global compositions.

32.3.1 The Coherence of Local Performance Transformations

Summary. Arguments for building local units—based on local compositions—of performance transformations are given. The idea of making performance on local units *coherent* is explicated: The mapping on the unit's domain is defined by a coherent rule, typically described by continuously differentiable transformations on connected neighborhoods of the local unit.

– Σ –

In the theory of local and global compositions, the definition of morphisms was roughly this: You are given a set map between two sets of music objects and you want to express that this map is not just any set-theoretic map, but shares a type of coherence. This is done by the extensibility of such a set map to the ambient space by an affine space map. This is one way of expressing coherence between the images of different points of the local composition domain: they are mapped under a formula of affine type. This kind of coherence is also required in classical algebraic geometry when defining morphisms between algebraic sets.

For performance maps, the narrow affine formalism and also the formalism of polynomial maps are both too algebraic in nature. Let us give some arguments to make this clear. If we want to model a fermata tempo shape, we can evidently not restrain to linear expressions since the fermata tempo must decrease in the beginning and increase towards the end of the fermata duration. Moreover, the cognitive rules of logarithmic perception of differences[3] would enforce changes in tempo in the sense that the change of physical onset

[3] See also Appendix A.2.2.

as a function of mental onset, i.e., de/dE at onset E proportional to the actual value of physical onset e, i.e., $de/dE \sim e$ which means that $dE/de \sim 1/e$ and therefore $E \sim \ln(e)$, i.e., $e \sim exp(E)$ so we cannot use polynomials for such a shaping of physical onset. The general shape of physical parameters against mental ones is also strongly documented in glissando and crescendo effects, i.e., changes of pitch h or l as functions of mental time E. These two effects are also a strong argument for continuous mental time: Glissando and crescendo must be defined at each moment of a continuous time parameter, otherwise they cannot be realized, not more than the continuous (even differentiable) movement of the conductor's baton.

Of course, such general function requirements could be worked around by gluing of polynomial functions to polynomial spline functions. But this is exactly what we do not want here: We want a unique class of functional expressions for the local coherence. It is this principle: What we can achieve by gluing together local pieces should be left to the global theory instead of hiding the gluing technique on the level of function classes!

32.3.2 Differential Morphisms of Local Compositions

Summary. This section describes the categories of local compositions with *differential morphisms* on real vector spaces. This formalism captures unreflected prerequisites of musicological approaches on performance instances, such as tempo or tuning.

– Σ –

Before constructing the categories of differential morphisms, we should specify their objects. As the existing performance theory is only developed for the zero address and for objective compositions, we shall only look at such objects. In general, score denotators live in complex form spaces, including components which are far from being performable, such as bar-lines or pauses. We will not consider these full-fledged structures, but only local compositions which consist of objects that are candidates for events to be performed in physical spaces.

Such local compositions live in space forms which are limits (products are sufficient here), whose factor spaces are submodules of real vector spaces. For example, pitch can be encoded by integers (such as is the case for MIDI key numbers), or by tuples of rationals (as is the case with the Euler spaces built from (just) octaves, fifths, thirds, sevenths etc.), whereas onset is encode by a module $\mathbb{Z}.x$ or $\mathbb{Z}[1/2]$ or $\mathbb{Q}$. Dynamics is usually encoded by integers (such as suggested by the MIDI key velocity numbers $0, 1, 2, \ldots 127$. With these usual values, one may consider local compositions (K, M) with M a submodule of $\mathbb{R}^n$. The same observations are valid when representing local compositions which are related to physical parameter space forms. We shall therefore, without loss of generality, assume that we are given local compositions (K, S) in simple space forms $S \underset{Id}{\longrightarrow} \mathbf{Simple}(\mathbb{R}^n)$ of modules[4] $\mathbb{R}^n, n \in \mathbb{N}$. Later, we shall specify the form names when more concrete parameters will be discussed.

Let us first explain the general construction of tangent compositions and their morphisms:

Definition 98 *Given an objective, zero-addressed commutative local composition (K, S) in a simple ambient space $S \underset{Id}{\longrightarrow}$* **Simple**$(M)$ *with R-module M, then the* tangent composition *TK of K is the composition $(K \times R.K, S^2)$, where $S^t \underset{Id}{\longrightarrow}$* **Limit**$(S, \ldots S)$ *is the t-fold product. The local composition K is called* the basis of TK. *For a natural number t, the* t-fold tangent composition *$T^t K$ is the tangent composition $T(T^{t-1}K) = K \times R.K \times (R.K \times R.K)^{t-1}) \subset S^{2t}$, with special value $T^0 K = K$.*

The tangent space $T_k K$ *of K at point $k \in K$ is the subset $\{k\} \times R.K$, identified with $R.K$ for its module structure.*

Given two tangent compositions TK of K and TL of L (both compositions over the same ring R). A tangent morphism *$Tf : TK \to TL$ is a set map $Tf : TK \to TL$ between local compositions TK, TL which factors through the canonical projections $p_K : TK \to K, p_L : TL \to L$ and a (necessarily uniquely*

[4] The case $n = \infty$, the countable direct sum of copies of $\mathbb{R}$ for possible Fourier coefficients or similar parameters in sound color spaces will not be considered except for some explicitly described special considerations

determined) map $f : K \to L$, and such that all the fiber maps $Tf_k : T_kK \to T_{f(k)}L$ are linear. For $1 < t$, a t-fold tangent morphism $T^tf : T^tK \to T^tL$ is a tangent morphism $T(T^{t-1}K) \to T(T^{t-1}L)$ whose basis map is a $(t-1)$-fold tangent morphism.

The obvious category of t-fold tangent compositions and t-fold tangent morphisms is denoted by Tan^t_R.

Lemma 49 *Let ComLoc^0_R be the category of commutative objective local compositions over the commutative ring R. Let $f : (K, S) \to (L, U)$ be a morphism in ${}_1\mathrm{ComLoc}^0_R$. Then the map*

$$f \mapsto Tf = f \times R.f : TK \to TL$$

defines an injective natural transformation $\mathrm{ComLoc}^0_R \to \mathrm{Tan}^1_R$ onto the subcategory of those tangent morphisms $Tf : TK \to TL$ between tangent compositions such that the fiber maps Tf_k are all the same linear map T, and we have $f(k_1) - f(k_2) = T(k_1 - k_2)$ for all couples $k_1, k_2 \in K$, i.e., the map f is defined via T and the value on one single point.

The proof of this lemma is left to the reader. For the real number field $R = \mathbb{R}$, we are not only interested in the tangent categories, but in those morphisms which extend to differentiable maps on the underlying vector space, or, more generally in maps which extend to any maps of a specific category *Cat*:

Definition 99 *A* t-fold tangent morphism $T^tf : T^tK \to T^tL$ *of t-fold tangent compositions with positive t in $\mathrm{Tan}^t_{\mathbb{R}}$ is said to be t-fold differentiable iff there is a t-fold differentiable map F between the underlying vector spaces such that $T^tF|T^tK = T^rf$. For $0 \leqslant t$, such a map is said to be C^t iff it may be extended to a C^t map F between the underlying vector spaces.*

Clearly, for positive t, the t-fold differentiable and the C^t morphisms, respectively, define subcategories $\mathrm{LocDiff}^t$ and LocC^t of $\mathrm{Tan}^t_{\mathbb{R}}$, respectively.

More generally, if Cat is a category of morphisms on real vector spaces, we denote by $\mathrm{Tan}^t_{\mathrm{Cat}}$ the category of t-fold tangent morphisms which extend to morphisms from Cat. In practice, this mathematical catechism is not very practical, we shall rather use the wording of "an extension of a given t-fold tangent morphism to a morphism of the respective category Cat".

Example 49 In traditional performance research, tempo is an important feature. It is usually described via Mälzel's metronomic tempo indication (M.M.) of the type "x quarters per minute". This means that we compare mental time E (quarters) to physical time e (minutes) via the quotient $\Delta E/\Delta e$ as a function of E. Forgetting about the other parameters for the sake of simplicity, we have a (not necessarily finite) local composition of (real-valued) mental onsets $MyOnsets : 0 \rightsquigarrow OS(\{E_1, \ldots E_i, \ldots\})$ with form $OS \underset{Id}{\longrightarrow} \mathbf{Power}(Onset)$ (see list of mental forms in formula 6.69 in Section 6.6). Suppose that we have a performance set map p associating each E_i with a physical onset $p(E_i) = e_i$ of a denotator $MyPhysOnsets : 0 \rightsquigarrow PhysOS(\{e_1, \ldots e_i, \ldots\})$ in the form

$$PhysOS \underset{Id}{\longrightarrow} \mathbf{Power}(PhysOnset)$$

over the physical onset form $PhysOnset \underset{Id}{\longrightarrow} \mathbf{Simple}(\mathbb{R})$ (see 6.6). Suppose that the tempo indications are given on $MyOnsets$ by the values T_i at the onsets E_i. This data defines a tangent morphism $Tp : MyOnsets \to MyPhysOnsets$ whose linear fibers over E_i are the linear maps $E \mapsto T_i.E$

For the other onsets, tempo is not declared. This is the usual situation in music scores. However, it is not clear what should be the tempo in between the indicated onsets. The point is that one would like to turn the tangent morphism into a differential morphism, at least into a piecewise differential morphism, ie., one that extends to a piecewise differentiable function having the derivatives T_i as required at the arguments E_i.

Some musicologists even do not agree with this extensibility requirement, but maintain that tempo is not defined except at the given arguments. We have already countered that position, however, if tempo has to be extended in the sense that we ask for a differentiable morphism giving rise to the present tangent morphism, there are many ways to do so. For example, tempo could be set to a piecewise constant function, giving rise to a step tempo curve which integrates to a piecewise linear map P that extends p.

As to the mental contents of a score, this question could be viewed as secondary, but if we agree with Paul Valéry in saying that *c'est l'exécution du poème qui est le poème* (see the catchword of this chapter), then the *total* content of a composition must include the extension of p to P.

It should also be stressed that the European understanding of musical time is a type of negative account, time is only interesting if it is over, i.e., the time between two successive notes is non-existent, or, at least, of no existential relevance. Tempo does not exist between two notes since there is no time feeling, performed music has no time except when a note onset intervenes. This is a severe lack of understanding of what musically happens. Such a music understanding is poor and proves a negative, immature relation to time. Positively stated, European time is more a kind of trigger, a Turing machine unit slot transporting the logical processes (typically in harmony).

It is the merit of performance research, in particular the Swedish school of Johan Sundberg and his collaborators, to have pointed at this delicate time question. We shall come back to the extension type problem for tangent morphisms in chapter 36 on expressive performance.

Exercise 71 *Let $Tf : TK \to TL$ be a tangent morphism of $\mathrm{Tan}^1_{\mathbb{R}}$ over a basis $K = \{x, y\}$ of cardinality 2 in $\mathbb{R}$. Then Tf is extensible by exactly one 1-fold differential morphism whose coordinate functions $P_i(t)$ are polynomials in t of degree three. Moreover, these polynomials are rational functions in $x, y, f_i(x), f_i(y), t$ which are polynomials for fixed x, y. If $Tf_x = Tf_y = 0$, we have $min(f_i(x), f_i(y)) \leqslant P_i(t) \leqslant max(f_i(x), f_i(y))$ for $x \leqslant t \leqslant y$.*

Exercise 72 *Let $p : K \to L$ be a set map of local compositions with $card(K) = m + 1$, then there is a polynomial morphism extension P of p whose coordinate functions have degree $\leqslant m$. (Hint: Find a straight line through the origin and not parallel to any of the finitely many lines connecting pairs of points of K. Take the projection q of K into the orthogonal space to this line. This is a bijection onto the image $q(K)$. Within this orthogonal space, repeat the same procedure until you have projected K into a one-dimensional subspace, call K^* the image of K in this subspace. Then, looking at the map $p^* : K^* \to L$ induced by p on this one-dimensional projection of K, the claim is a classical result of polynomial interpolation.)*

Exercise 73 *If the points of K in exercise 72 are in general position, then p is a morphism of local compositions and therefore automatically t-fold differentiable for every t.*

32.3.2.1 A Recursive Interpolation Algorithm

The next result is related to what we shall call analytical weights, a key structure in the theory of performance operators. But it pertains to the theory of differential morphisms, so we present it right here. Since the essential statement is a recursive algorithm which is of essential use in programming performance theory, we do not present the result as a lemma but as a construction.

The situation is this: We are given a finite local composition in a simple ambient space $S \underset{Id}{\longrightarrow}$ **Simple**$(\mathbb{R}^n)$, consisting of points $K = \{x_1, \ldots x_s\}$, and contained in the open n-cube $C^n =]u_1, v_1[\times \ldots]u_n, v_n[$ of $\mathbb{R}^n$. For $m = 1, \ldots n$, we denote by $C^m =]u_1, v_1[\times \ldots]u_m, v_m[$ the projection onto the first m coordinates, and by $C_m =]u_m, v_m[$ the projection onto the m-th coordinate.

On K, we are given a tangent morphism $Tf : TK \to TL$, where the codomain tangent composition TL sits over a local composition L in an ambient space $W \underset{Id}{\longrightarrow}$ **Simple**$(\mathbb{R})$ of "weights", and where the linear components Tf_k vanish for every $k \in K$. We want to find a C^1 extension of Tf which adds no extra extrema to the extrema required per definition on the K points, and which evaluates to the constant value 1 on the complement of the frame cube C^n. The meaning of this requirement is that we want to extend the "discrete weight" f to a function F on the entire space S such that the extension is C^1, normalizes to the constant value 1 when the arguments tend to infinity, and has values very close to the given values in a small neighborhood of K without producing new extremal values not defined on K.

For the following construction, we concentrate on the module $\mathbb{R}^n$ and forget about the underlying space forms. The construction is a recursive one, and we start with $n = 1$. In this case, we suppose that the s points of K are ordered by size, i.e., we have these $n+2$ points on the real axis: $u_1 < x_1 < \ldots x_s < v_1$. Extending f to the frame arguments u_1, v_1, these arguments are mapped to the values $1 = f(u_1), f(x_1), \ldots f(x_s), 1 = f(v_1)$. From exercise 71 we have a (unique) cubic polynomial extension of the tangent map on every pair of successive points $\{u_1, x_1\}, \{x_1, x_2\}, \ldots \{x_{s-1}, x_s\}, \{x_s, v_1\}$ with zero fibers $Tf_{u_1} = Tf_{x_1} = \ldots Tf_{x_s} = Tf_{v_1} = 0$. We now consider the spline function defined by gluing all the polynomial extensions at the common points of the pairs, i.e., at the elements of K. Outside the frame, we extend this spline function by the constant function of value 1. This function is C^1 and adds no extra extremal values to the values on K inside the frame.

Suppose now that we have succeeded by induction to construct the following extension: We are given the decomposition $x_i = (z_i, w_i), z_i \in \mathbb{R}^{n-1}, w_i \in \mathbb{R}$. Let $p_n(K) = \{w_{i_1} < w_{i_2} < \ldots w_{i_r}\}$ be the n-th projection of K with the different values in increasing order. For every hyperplane $H_x = p_n^{-1}(x), x \in \{u_n, w_{i_1}, w_{i_2}, \ldots w_{i_r}, v_n\}$, we are given either the recursively defined C^1-functions $P_x : H_x \to \mathbb{R}$, or the constant function 1 for the frame points $x = u_n, v_n$. For any point $v = (z, w) \in \mathbb{R}^{n-1} \times \mathbb{R}$, the value $F(v)$ is defined as follows. If $w \notin [u_n, v_n[$, we set $F(v) = 1$. Else, there is exactly one interval $[a, b]$ from the successor pairs $(u_n, w_{i_1}), (w_{i_1}, w_{i_2}), \ldots (w_{i_{r-1}}, w_{i_r}), (w_{i_r}, v_n)$ such that $a \leqslant w < b$. We then evaluate to the value

$$F(v) = P_{a,b,P_a(z),P_b(z)}(w)$$

of the rational function P defined in exercise 71. Since a, b are fixed here, the function is a polynomial in the arguments $P_a(z), P_b(z)$, and w. So the function is a C^1-function in $v = (z, w)$. On each hyperplane H_x, the derivatives of neighboring functions coincide, and the entire function is C^1. Finally, since the values are constant 1 outside the frames on the hyperplanes, the constant 1 value is guaranteed outside the n-cube C^n and we are done.

This construction has the disadvantage of depending on the order of the coordinates used in the recursion. For every permutation π of the n coordinates, we have such a function, call it F_π. Each such function extends one and the same original tangent function and is 1 outside the cube C^n. So we can symmetrize the construction by the weighted sum $F_{sym} = \frac{1}{n!}\sum_\pi F_\pi$. However, for programming tasks, such a symmetrization is very time-consuming.

32.4 Global Structure

Summary. The global structure of a performance transformation is a patchwork of local performance transformations. The combination and gluing data of local units expresses the syntax of performance. In turn, this performance syntax defines an interpretation of what has been recognized on the analytical level. This "interpretation of the interpretation" obeys its own rules and constitutes a relatively autonomous *rhetorical* shaping of the given "text" and its analytical comprehension.

– Σ –

Evidently, the local performance structures described above are far from sufficient to grasp realistic performance situations. Basically, there are four reasons for this insufficiency:

- *Instrumental variety*
The events of a score are not always in one and the same parameter space, especially if we deal with compositions for different instruments. We typically work in a colimit space

$$ScoreOrchestra \xrightarrow[Id]{} \mathbf{Colimit}(ScoreInstr_1, \ldots ScoreInstr_k)$$

of instrumental spaces

$$ScoreInstr_i \xrightarrow[Id]{} \mathbf{Simple}(\mathbb{R}^{n_i})$$

with individual coordinator modules. For example, the piano notes have a four-dimensional space associated with pitch, onset, loudness, and duration, whereas the violin space adds crescendo and glissando to the piano space. Ditto for the physical codomain spaces

$$PhysInstr_i \xrightarrow[Id]{} \mathbf{Simple}(\mathbb{R}^{m_i})$$

where the mental instrumental events are mapped. But here, the codomains may be very different in dimension according to the real physical instrument which is addressed on the mental score. The typical colimit space over the physical instruments is

$$PhysOrchestra \xrightarrow[Id]{} \mathbf{Colimit}(PhysInstr_1, \ldots PhysInstr_k).$$

This means that performance starts from a local composition K in the ambient space *ScoreOrchestra* and maps into the local composition of physical events L in the ambient space *PhysOrchestra*. The local compositions K, L are disjoint unions of the subcompositions $K_i, L_i, i = 1, \ldots k$ corresponding to the instrumental cofactors in both, K, L, and the performance splits into a coproduct of k individual local instrumental performances $p_i : K_i \to L_i$. Even if the mental score level composition K is not a disjoint union of instrumental subcomposition, i.e., a proper colimit, it can be lifted to the coproduct of its cofactors to obtain a disjoint family of performances, so our hypothesis is not restrictive. On the other side, the proper colimits on the physical level are superfluous since it is always (technically) possible to realize the disjoint union of instrumental voices if necessary. It follows that the above splitting is necessary and sufficient for the *orchestral* globalization of local performances.
- *Gluing of local extension strategies*
The local approach is insufficient when sudden, non-continuous changes of the performance map happen. For example, if starting with tempo M.M. quarter = 120 per minute, and then performing a chain of accelerandi and rallentandi, it may be asked by the composer that we reset the tempo after this tempo variations, by the command "istesso tempo", meaning to return to the original M.M. quarter = 120 per minute. It would be artificial to construct a continuous tempo curve of transition to the reset value. In this case, it is natural to split the performance map into two contiguous domains with their own tempi. Or else, it may happen that the extension category *Cat* is rather strict, allowing only polynomial maps, say. Such a restriction may intervene for reasons of cognitive nature, or because there are hypotheses about the dynamics of performance, such as mechanical models (see the work of McAngus Todd [1060], for example). Then it is better to glue the performance map from parts which are conformal with *Cat* and coincide on the intersection of charts of the atlas patchwork. This latter strategy is nothing else than the well-known spline approach to the extension of discrete map data, see [994, 995] for a typical reference.
- *Special roles of selected parameters of a given local performance map (hierarchy)*
This type of globalization effects is perhaps the most interesting. We stated that in performance theory, we are only interested in the performance map on the mental event which will effectively be played, and not in abstract objects such as bar-lines or pauses. But abstract objects may emerge from effectively played event in the following sense: Suppose that a local composition K in some space S has to be performed, for example a space parametrizing onset E, duration D, and pitch H and suppose that a space S' of selected parameters of S, for example onset and duration. We then have the projection $\pi_{S'} : S \to S'$ and the associated projection $K \to \pi_{S'}(K)$. In a great number of performances, it happens

that the performance of K factorizes through $\pi_{S'}$, which means this: We look at the performance map $p : K \to L$, and if L lives in the corresponding physical space PhS with the corresponding projection $\pi_{PhS'} : PhS \to PhS'$ and the induced projection $L \to \pi_{PhS'}(L)$. Factorization means that we have a performance map $p_{S'} : \pi_{S'}(K) \to \pi_{PhS'}(L)$ which commutes with the projections, i.e., $\pi_{PhS'} \circ p = p_{S'} \circ \pi_{S'}$. Here, the time events in S' are not really played, but their performance determines the time performance of the really played events. Hence, it may be reasonable to add the projected mental event set $\pi_{S'}(K)$ to the real set K (plus some projection and commutativity conditions as shown above) in order to describe the overall situation. This idea is the basis of the so-called performance hierarchies which control the special roles of parameters.

- *Stemmatic deployment of performance*
  Performance is never realized on the spot or by means of a unit process which grasps the analytical data and presents their "rhetorical" shaping at once. Humans have to rehearse again and again, continuously refining their results until they reach the final (or provisionally final) performance.
  This stemmatic deployment process from the sight reading (primavista) performance to the artistically well-devised presentation could be just a human learning process which a machine can achieve at once if it is able to learn the performative substance in abstracto. But this is erroneous, since the logic of performance, the anatomy of the shaping process is a stemmatic one, a multilayered unfolding of deformations of the mental score symbols.
  Rather than a learning process rehearsal is a meditation on the refinement of understanding. So performance is also global in the sense of a multilayered time-dynamic process. Each stemmatic layer is a logical step in the understanding of the rhetorical expressivity.

32.4.1 Modeling Performance Syntax

Summary. We review the syntactical mechanism of global performance transformation: orchestration, contiguity, hierarchy, and stemmatic layers.

– Σ –

As was shown above, performance syntax is a bundle of four streams of completely different nature. Contiguity is perhaps the most obvious stream. It reminds us of the syntactic juxtaposition of units within the stream of ordinary language.

The orchestration stream is much more difficult to understand since it conveys a layering of the performative mapping which creates the interplay of voices. The structure of this interplay must be viewed in a multidimensional space of geometric parameters, such as onset, duration, pitch, loudness, of sound color parameters for envelopes, Fourier, FM, or wavelet coefficients and the like, and of gestural coefficients for curves of body movements.

The hierarchical stream is of high cognitive relevance since it describes the leading and slave parameters of performance. Such a hierarchy is likely to produce orientation in the perception of a performance. But it is also of a purely structural relevance since hierarchies are the turning point and key to the stemmatic layering logic. In fact, the successive refinement process of performance is often managed via deformations of hierarchies, i.e., adjustments of relations and functional dependencies among the members within a given hierarchy.

Whereas the three preceding streams are "in praesentia" as semioticians would say, the fourth: stemmatic layering, is of completely different nature. It is not, and this is decisive, of paradigmatic nature, since it describes a development of logical enchainment, i.e., of a juxtaposition of logical stages. But this type of syntagm is not unfolded in time, it is a kind of encapsulated history whose presence cannot be unveiled without a huge amount of ambiguity and uncertainty (see chapter 48 on inverse performance theory), its presence is virtual.

The complexity of global performance is not only due to the four-fold stream of global structures, it is enforced by the completely divergent extension morphisms underlying any two different local charts within a particular stream, and even more dramatically when distributed over different streams.

32.4.2 The Formal Setup

Summary. This section introduces the formal performance transformation setup via the category $GlTan^t_R$ of global tangent compositions with *t-fold tangent morphisms* and, for $R = \mathbb{R}$, with the associated subcategories $GlDiff^t$ and GlC^t of $GlTan^t_{\mathbb{R}}$.

– Σ –

From the four types of globalization phenomena, instrumental variety, gluing of local parts, hierarchy of parameters, and stemmatic deployment, the first three will be covered by the following categorical description, while the last one—being of more processual nature—will be treated in chapter 38. Here is the definition which is coined on the definition 36 of a global objective composition.

Definition 100 *For a positive integer t, a* global t-fold tangent composition *over a commutative ring R is defined by the following data:*

(*i*) *A set G and a finite, non-empty covering I of G,*
(*ii*) *a family* $(T^t K_u, S_u)_{u \in U}$ *of local t-fold tangent compositions* $T^t K_u$ *over R, with bases* K_u,
(*iii*) *a surjection* $I_? : U \to I : u \mapsto I_u$,
(*iv*) *a bijection* $\phi_u : T^t K_u \xrightarrow{\sim} I_u$ *for each* $u \in U$,
(*v*) *for each couple* $u, v \in U$, *the induced subcomposition*

$$T^t K_{u,v} = \phi_u^{-1}(I_u \cap I_v)$$

equals the tangent subcomposition of $T^t K_u$ *induced on the basis*

$$K_{u,v} = p_{K_u}(T^t K_{u,v}),$$

(*vi*) *for each couple* $u, v \in U$, *the induced bijection*

$$\phi_{u,v} = \phi_u^{-1} \circ \phi_v : T^t K_{u,v} \to T^t K_{u,v}$$

is an isomorphism of local tangent compositions.

The data (ii) - (vi) *are called an* atlas Φ for the covering I of G. *Two atlases* Φ, Ψ *for the covering I of G are called equivalent iff their disjoint union is also an atlas for the covering I of G. A* global tangent composition over R *is a covering I of G, together with an equivalence class of atlases, a fact which we abbreviate by the symbol* G^I *or even by G if the atlases or the covering, respectively, are clear from the context.*

If two global tangent compositions G^I, H^J *over R are given, a* t-fold tangent morphism from G^I to H^J *is a couple* (f, ι) *where*

1. $f : G \to H$ *is a set map,*
2. $\iota : I \to J$ *is a set map such that* $f(i) \subset \iota(i)$ *for all covering sets* $i \in I$,
3. *for any atlases* $(T^t K_u, S_u)_U$ *for G and* $(L_v, M_v)_V$ *for H, if we take the chart isomorphisms* $\phi_u : K_u \xrightarrow{\sim} I_u$ *and* $\psi_v : L_v \xrightarrow{\sim} J_v$ *for some pair u, v of indexes which correspond under the map* ι *(i.e.,* $\iota(I_u) = J_v$*), then the induced maps* $f_u : T^t K_u \to T^t L_v$ *define a morphism of local tangent compositions.*
The global t-fold tangent compositions and their morphisms define the category $GlTan^t_R$.

For the real number, $R = \mathbb{R}$, we may consider those global t-fold tangent compositions and morphisms which are t-fold differentiable or C^t, respectively, on the local data, and thereby define the subcategories $GlDiff^t$ and GlC^t of $GlTan^t_{\mathbb{R}}$.

Example 50 Let G^I be zero-addressed objective global composition which is interpretable by a simple ambient space S of module M over the commutative ring R. WLOG we have $G \subset M$, and the atlas is $I = \{G_1, \dots G_n\}$. The transition morphisms $\phi_{i,j}$ between intersections $G_{i,j} = G_i \cap G_j$ are the identities. We then have the inductive system $G_{..} = (G_i, G_{i,j}, \rho_{i,j}, \phi_{i,j})$ with the inclusions $\rho_{i,j} : G_{i,j} \subset G_i$, and $G = colim(G_{..})$. Using the functor from Lemma 49, we have an induced inductive system $TG_{..} = (TG_i, TG_{i,j}, T\rho_{i,j}, T\phi_{i,j})$, together with a morphism of inductive systems $p_{..} : TG_{..} \to G_{..}$. This gives the colimit map $p : colim(TG_{..}) \to G$. The domain is $TG^I = colim(TG_{..})$ is covered by the images of the charts TG_i. Using Theorem 12, we see that the canonical maps $\phi_i : TG_i \to TG^I$ are injective. In fact, the transition morphisms are the identities, therefore the relation $\sim$ of that theorem is an equivalence relation, and the theorem can be applied. Call TG^I the tangent interpretation associated with G^I. This is an interpretation with ambient space $S \times S$, and the projection $p : TG^I \to G^I$ is a morphism of global compositions.

Since the transition morphisms are the identities here, one may also refine the tangent construction to one of differentiable nature.

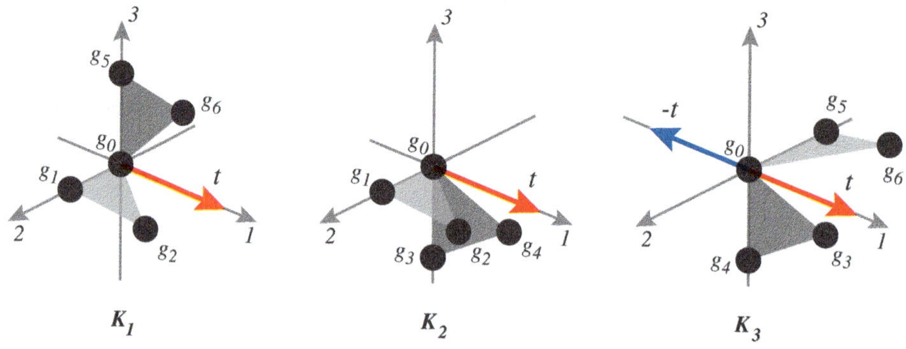

Fig. 32.1. The seven-element global composition of which the three charts are shown here is a Möbius bottle. It may not be extended to a global tangent composition by the obvious colimit of its tangent charts and their intersections since, for example, a tangent vector t is equivalent to its negative $-t$ under the colimit construction.

Exercise 74 *For a non-interpretable composition the colimit does not yield a reasonable global tangent composition. The following example illustrates this fact, see also Figure 32.1. We consider a global composition* G^I *whose support*

$$G = \{g_0, g_1, g_2, g_3, g_4, g_5, g_6\}$$

consists of seven different points, and whose covering consists of three 5-element charts G_i *having the common point* g_0*:*

$$G_1 = \{g_0, g_1, g_2, g_5, g_6\}, G_2 = \{g_0, g_1, g_2, g_3, g_4\}, G_3 = \{g_0, g_3, g_4, g_5, g_6\}.$$

They are in bijection with these local compositions K_i *in* $\mathbb{R}^3$*:*

$$\phi_1 : G_1 \to K_1 :$$
$$g_0 \mapsto (0,0,0), g_1 \mapsto (0,1,0), g_2 \mapsto (1,1,0), g_5 \mapsto (0,0,1), g_6 \mapsto (1,0,1)$$

$$\phi_2 : G_2 \to K_2 :$$
$$g_0 \mapsto (0,0,0), g_1 \mapsto (0,1,0), g_2 \mapsto (1,1,0), g_3 \mapsto (0,0,-1), g_4 \mapsto (1,0,-1)$$

$$\phi_3 : G_3 \to K_3 :$$
$$g_0 \mapsto (0,0,0), g_3 \mapsto (1,0,-1), g_4 \mapsto (0,0,-1), g_5 \mapsto (0,-1,0), g_6 \mapsto (1,-1,0)$$

The transition isomorphisms are these:

$$\phi_{12} = Id$$

$$\phi_{23}:$$
$$g_0 \mapsto g_0, (0,0,-1) \mapsto (1,0,-1), (1,0,-1) \mapsto (0,0,-1)$$

$$\phi_{13}:$$
$$g_0 \mapsto g_0, (0,0,1) \mapsto (0,-1,0), (1,0,1) \mapsto (1,-1,0)$$

In the colimit of TK_i, TK_{ij}, the tangent $t = (g_0, (1,0,0)) \in TK_1$ identifies to $T\phi_{12}(t) = t$, and then to $T\phi_{23}(t) = -t \in TK_{32}$, whereas $T\phi_{13}(t) = t \in TK_{32}$, which means $t \sim -t$, i.e., the canonical maps $TK_i \to colim(TK_i, TK_{ij})$ are not injective—gluing tangent spaces is not feasible. Clearly, this phenomenon is due to the non-interpretability of G^I, in fact, this is a kind of Möbius bottle on which any global affine function must identify the values of the pairs $(g_1, g_2), (g_3, g_4), (g_5, g_6)$.

Exercise 75 *However, if any global composition is such that its nerve is one-dimensional, the colimit construction yields a global tangent composition with the expected chart injections. Give a proof of this fact.*

Let us discuss the reason why the three globalization perspectives are included in the above definitions. Clearly, the disjoint union of the interpretation morphisms on different instrumental parts is represented by a disjoint union of morphisms on local (or global, if they exist) tangent compositions on each of the instrumental parameter spaces.

Let us give an example of a global gluing operation between local tangent compositions. To keep the discussion simple, let us consider the performance map on the onset axis. Suppose that $G = \{E_0, E_1, E_2\}$ is a sequence of three increasing onsets $E_0 < E_1 < E_2$ which are mapped onto the set of the three physical onsets $g = \{e_0, e_1, e_2\}$, $e_i = P(E_i)$, under the performance map $P : G \to g$. Suppose that the musical conditions are such that one starts with an initial velocity $de/dE|_{E_0} = V_0, de/dE|_{E_2} = V_2$, whereas the velocity is $de/dE|_{E_1} = V_1$ when we arrive at onset E_1. But the successive path should be started with velocity $de/dE|_{E_1} = V_1^*$, and end on velocity $de/dE|_{E_2} = V_2$.

This apparent incompatibility at onset E_1 can be resolved by the coverings $I = \{G_1 = \{E_0, E_1\}, G_2 = \{E_1, E_2\}\}$, $i = \{g_1 = \{e_0, e_1\}, g_2 = \{e_1, e_2\}\}$ of G, g, respectively. Consider the tangent compositions TG^I, Tg^i of the two interpretable compositions G^I, g^i, respectively, see Figure 32.2.

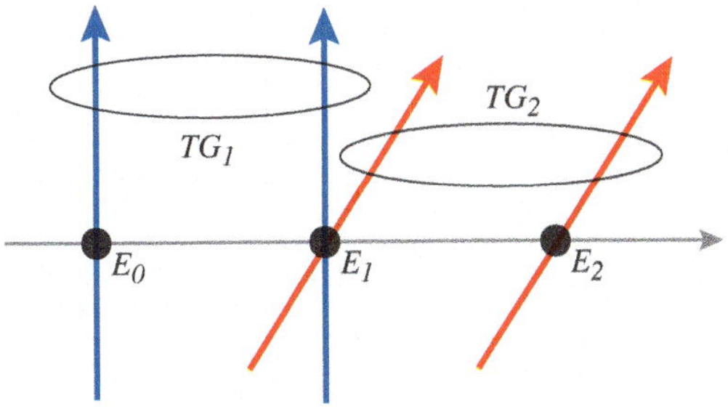

Fig. 32.2. The global tangent composition built from two charts TG_1, TG_2 of successive tempo regions.

Consider an extension of the global morphism $TP : TG^I \to Tg^i$ associated with the map P and the above velocity conditions as linear maps on the four tangent spaces

$$TG_{1,E_0}, TG_{1,E_1}, TG_{2,E_1}, TG_{2,E_2}$$

to a given category of maps on the real axis, C^1, say. Then we have a "gluing" of two velocities, one from below, one from above, at onset E_1.

32.4.3 Performance qua Interpretation of Interpretation

Summary. We finally compare the analytical interpretation in the framework of categories *Glob* of global compositions with associated global performance transformations.

$- \Sigma -$

Recalling the analytical interpretation of a given score denotator, this one a priori regards a variety of involved object-types, such as bar-lines, notes, macros, rests—whatever is needed. Such an interpretation is guided by the analytical approach, such as the topological interpretation using local meters, or motives, or chords, as exposed in the respective chapters.

In contrast, the performance-guided interpretation of the given score is firstly restricted to special event types. Usually, for example, bar-lines are not subjected to performance transformations. Secondly, the interpretation within a performable event-type is not primarily guided by analytical considerations, rather it is related to considerations of coherence in the performative process. For example, the agogical variations in a tempo curve, say, can be shaped by complex weight functions as defined via the maximal meter topology or the nerve topology on the composition's nerve. Such a tempo curve will not refer to the complex covering by maximal meters, but define a unique tangent morphism on a single tangent chart for performance.

So the interpretations of analytical and performative nature can and usually will be very different. Nonetheless, the latter interpretations are related in a complex way to the analytical background interpretations. We shall devote the following chapters to the explication of these relations which belong, to our belief, to the most fascinating challenge in mathematical music theory.

33

Performance Fields

La musique mathématiquement discontinue
peut donner les sensations les plus continues.
Paul Valéry [1073, I]

Summary. Performance fields are the core of an in-depth theory of performance structure. They are a distinguished type of vector fields which give an infinitely precise, i.e., infinitesimal, account of the 'shaping forces' of a given local performance transformation. Although performance fields are not recognized as such in musicology and traditional performance research, they arise in a completely natural way in the traditional context of *tempo*, *intonation*, and *dynamics*. We give a careful account of this basic fact. A closer look at *articulation* and further sound parameters (apart from onset, pitch and loudness which are used for tempo, intonation, and dynamics) reveals that performance fields should be viewed within a fairly general approach. We define the formal setup. In order to provide a deeper understanding of the semiotic signification process of performance fields, we review the performance philosophies of Theodor W. Adorno, Benjamin, and Diana Raffman.

– Σ –

33.1 Classics: Tempo, Intonation, and Dynamics

Summary. This section recapitulates and analyzes the concepts of tempo, intonation, and dynamics. We take the opportunity to make the point of blurred concepts in musicology, in this case in the sense of a fascinating quest for expressive precision, paired with denial of formal explicitness.

– Σ –

33.1.1 Tempo

Summary. Tempo is one of the best 'known' features of performance. However, its concept is blurred and far from standardized among musicians and performance scientists. We analyze the state of the art and its deficiencies. It is shown that tempo is a local concept which is charged with a large amount of semantics. We proceed in filtering out the semantics from the structural data. After this, a precise definition of tempo as a one-dimensional performance field in the onset axis is given.

– Σ –

IN music, Tempo is the Italian word for "velocity" (originating from "time"). In musical notation and performance, tempo refers to the pace at which the mental events, typically written on a score or imagined by the performer, are projected into physical reality. This is done either to describe the relative velocity or the velocity change with respect to a given absolute tempo situation; typically indicated by notation such

G. Mazzola, *The Topos of Music II: Performance*, Computational Music Science,
https://doi.org/10.1007/978-3-319-64444-8_2

as verbal annotation "accelerando", "rallentando", or by corresponding pictorial signs. Or else, the tempo is described with an absolute meaning. The prototype of this second situation is Mälzel's Metronome (M.M.), quantified by a quotient such as "quarter = 120", i.e., a quarter note is given the physical duration of 0.5 second, resulting in 120 quarters per minute. In contrast to this precise indication, a relative tempo sign within a Mälzel context does not define precisely how much the tempo should change, and in which way this should happen., e.g., by a linear curve or in a quadratic way, etc. The relative tempo signs are massively ambiguous. We shall have to spend quite some time in order to set up a reliable handling of such blurred signs in algorithmic contexts.

Absolute tempo is also indicated by more poetic indications, such as "andantino", "prestissimo", "maestoso". Clearly, such wordings are loaded with a good portion of connotation which exceeds the mere tempo and targets emotional refinement, expressed via other parameters, such as articulation (legati, staccati), dynamics (loudness variations) or intonational deformations (on violins or for the human voice).

Whatever the specific expression, *tempo relates to the transformation of mental onsets to physical onsets.* Note that this does not imply that such a transformation is independent of the other event parameters. We shall learn that the performance of onsets may involve all other parameters. A simple example of such a relation is given by the so-called "Chopin rubato" which lets the left hand perform a "mother" tempo while the right hand performs local deviations of the left-hand tempo in order to generate the effect of temporal tension ("daughter" tempi). In this case, the common onset of left and right hand notes may lead to different images in physical time, as a function of the interpretation of the total score by the left-hand and right-hand charts.

Formally, local tempo is related to a tangent morphism $Tf : TK \to Tk$ of local tangent compositions TK, Tk derived from compositions $K \subset Onset, k \subset PhysOnset$ in onset spaces. Tempo is then defined iff the linear fiber maps $Tf_X, X \in K$ are invertible and there, we define $Tempo_X = (Tf_X)^{-1}$, the inverse slope at X. Moreover, tempo is always supposed to be positive, while negative tempi are without evident musical meaning.

In such a situation, tangent morphisms are extended to differentiable morphisms with positive, continuous derivatives at every point. The *tempo curves* associated with such differentiable morphisms are the continuous, positive functions whose values are the inverse derivatives at all points in an open interval containing the points of K. Usually, the tangent morphisms are not of immediate interest in performance theory, this means that one is given differentiable morphisms, ie., the morphisms F inducing tempo curves as their inverse derivatives $Tempo_X = (TF(X)^{-1})$. However, it may then also be required that these functions be differentiable extensions of tangent morphisms, but this is not the mandatory situation.

Nonetheless, the music(ologic)al point of view which negates differentiable extensions is not correct: tempo is also present in between the points of K, and its structure is an essential information of the performance maps. Formally speaking, we are dealing with a local composition $K \subset Onset$, a local composition $k \subset PhysOnset$, two closed intervals $OFrame = [A, B] \subset Onset, POframe = [a, b] \subset PhysOnset$ such that $K \subset OFrame, k \subset POframe$, and a C^1 diffeomorphism $F : OFrame \xrightarrow{\sim} POframe$ with positive derivative $TF(X)$ at each point $X \in OFrame$, and such that $F(K) = k$ (and possibly F extending a given tangent morphism $Tf : TK \xrightarrow{\sim} Tk$ if that is required). The *tempo field of F on $OFrame$* is the continuous field *Tempo* with values $Tempo_X = (TF(X)^{-1}), X \in OFrame$.

By construction, if the performance map $x_0 = F(X_0)$ is defined on any point $X_0 \in OFrame$, the performance $F(X)$ on $X \in K$ is defined by the integral[1]

$$F(X) = x_0 + \int_{X_0}^{X} \frac{1}{Tempo} \tag{33.1}$$

of the inverse tempo function. Musically, the "initial value" $x_0 = F(X_0)$ means that the "conductor" defines a starting time x_0 of performance from which the remaining performance onsets can be deduced by means of the given tempo curve.

[1] We follow the Douady notation and do not write the old-fashioned infinitesimal "dY" in the integrand if this one is clear.

Exercise 76 *With the above notation, calculate the onset function F for these tempo types: (1) $Tempo_X = q_0 + q_1X$, (2) $Tempo_X = q_0 + q_1X + q_2X^2$, (3) $Tempo_X = e^X$, (4) $Tempo_X = 2 + \sin(X)$. Discuss the possibilities of coping with these tempo types with tangent conditions.*

In general, the tempo must also cope with more than one given performance value, $x_0 = F(X_0), x_1 = F(X_1), \ldots x_t = F(X_t)$, and then, we have to ask for the conditions

$$x_j - x_i = \int_{X_i}^{X_j} \frac{1}{Tempo} \tag{33.2}$$

for $i, j = 0, 1, \ldots t$ to guarantee compatibility of the tempo integral with the given values.

33.1.2 Intonation

Summary. Intonation deals with pitch just as tempo deals with onset time. The definition of intonation as a one-dimensional performance field in the pitch axis is given.

– Σ –

Intonation, including the specialization of tuning, deals with the relation between mental pitch and physical pitch. It is analogous to tempo insofar as it suggests a map from the pitch space *Pitch* to the pitch space *PhysPitch*. But the situation is not so easy, if we observe the common pitch spaces in music theory. Recall that symbolic pitch is usually represented in the Euler module space *EulerModule*, and that we have a space morphism $E2M : EulerModule \to MathPitch$ defined in equation (6.28). This is an injective $\mathbb{Q}$-linear map on the supporting modules $\mathbb{Q}^3$, and $\mathbb{R}$, respectively. If we want to apply calculus to this situation, we have to extend the scalars to real numbers and then consider the induced map $E2M \otimes \mathbb{R} : Euler\mathbb{R}Module \to Pitch$ with underlying vector spaces $\mathbb{R}^3$, and $\mathbb{R}$, respectively. This latter morphism is no longer injective (in fact: surjective with two-dimensional kernel $pv^{\perp}$, see formula (6.27)). If we consider the performance map composed from the given performance of pitch, $p : Pitch \to PhysPitch$, and the above $E2M \otimes \mathbb{R}$, we are confronted with a performance map $p \circ E2M \otimes \mathbb{R}$ from Euler space to physical pitch which is far from being a diffeomorphism. This cannot be repaired since there is no physical space corresponding to the Euler construction, at least not in the simple physical pitch dimension. The solution could consist in a construction of a direct performance map from *Euler*$\mathbb{R}$*Module* to a physical "pitch" space of higher dimension. This one would evidently not restrict to mere physical pitch (logarithm of frequency, see Appendix A.2.3), but include other parameters, such as sound color or the like. This is however an open problem in performance research: How should we perform Euler space pitch?

We will stick to the state of the art and build pitch performance upon the maps of type $F : Pitch \to PhysPitch$. Such maps should be the C^1-extensions of tangent maps $Tf : TS \to Ts$ over bijective basis maps $f : S \to s$. Here, $S \subset Pitch, s \subset PhysPitch$ are a number of pitches and corresponding physical pitches. This situation is formally equivalent to the tempo situation. We may indeed take over mutatis mutandis the statements and formulas established for tempo as follows. We are given a mental pitch frame $PFrame = [U, V]$, a physical pitch frame $PPframe = [u, v]$, inclusions $S \subset PFrame, s \subset PPframe$, and a C^1-diffeomorphism $G : PFrame \xrightarrow{\sim} PPframe$, the *intonation curve*, with positive derivative $TG(X)$ for every $X \in PFrame$ (and possibly extending a given tangent morphism $Tg : TS \xrightarrow{\sim} Ts$ if that is required). The *intonation field of G on $PFrame$* is the continuous field *Intonation* with values $Intonation_X = TG(X)^{-1}, X \in PFrame$.

By construction, if the performance map $x_0 = G(X_0)$ is defined on any point $X_0 \in PFrame$, the performance $G(X)$ on $X \in S$ is defined by the integral

$$G(X) = x_0 + \int_{X_0}^{X} \frac{1}{Intonation} \tag{33.3}$$

of the inverse intonation function. Musically, the "initial value" $x_0 = G(X_0)$ means that the piece is played from a starting pitch, i.e., the "chamber pitch" x_0 of performance from which the remaining performance pitches can be deduced by means of the given intonation curve.

In general, the intonation must also cope with more than one given performance value, $x_0 = G(X_0), x_1 = G(X_1), \ldots x_t = G(X_t)$, and then, we have to ask for the conditions

$$x_j - x_i = \int_{X_i}^{X_j} \frac{1}{Intonation} \tag{33.4}$$

for $i, j = 0, 1, \ldots t$ to guarantee compatibility of the intonation integral with the given values. This is, in particular, the case when we are given a specific intonation between semitones in a fixed tuning mode, such as well-tempered or just tempered tunings.

Again, musicological approaches only deal with intonation values on the given tangent composition TS, while the intermediate values of the intonation curve are neglected (or even negated—in the worst case). However, if glissando effects are present, discrete intonation fails to give information about the intermediate values, much like the intermediate tempo is necessary to perform glissandi in their development along the time axis!

33.1.3 Dynamics

Summary. Dynamics deals with loudness just as tempo deals with onset time, and intonation deals with pitch. The definition of dynamics as a one-dimensional performance field in the loudness axis is given.

– Σ –

Dynamics is the physical shaping of loudness symbols such as *ff, mf, ppp, mp, sf* and words such as *sforzato, meno forte, diminuendo, crescendo.* In this classical setup of score notation, the scope is much less quantified than with tempo or intonation. There is no such norm as Mälzel's metronome or the chamber pitch in dynamics. Also, the shaping of dynamics is dramatically different, i.e., refined, with respect to the written prescriptions. It may happen that a section is written in *mezzo forte*, but within this section, the performance of *mezzo forte* is quite variable, within a certain tolerance bandwidth of what *mezzo forte* can be felt. However, just for these reasons, it is much more accepted by musicologists that dynamics is a continuous phenomenon when compared to tempo and intonation.

Sticking again to the state of the art, we build loudness performance upon the maps of type $F : Loudness \to PhysLoudness$. Such maps should be the C^1-extensions of tangent maps $Tf : TL \to Tl$ over bijective basis maps $f : L \to l$. Here, $L \subset Loudness, l \subset PhysLoudness$ are a number of loudness values and corresponding physical loudness (logarithms of pressure amplitude, see Appendix A.2.2). This situation is also formally equivalent to the tempo situation. We may indeed take over mutatis mutandis the statements and formulas established for tempo as follows. We are given a mental loudness frame $LFrame = [U, V]$, a physical loudness frame $PLframe = [u, v]$, inclusions $L \subset LFrame, l \subset PLframe$, and a C^1-diffeomorphism $H : LFrame \xrightarrow{\sim} PLframe$, the *dynamics curve*, with positive derivative $TH(X)$ for every $X \in LFrame$ (and possibly extending to a given tangent morphism $Th : TL \xrightarrow{\sim} Tl$ if that is required). The *dynamics field of* H *on* $LFrame$ is the continuous field $Dynamics$ with values $Dynamics_X = TH(X)^{-1}, X \in LFrame.$

By construction, if the performance map $x_0 = H(X_0)$ is defined on any point $X_0 \in LFrame$, the performance $H(X)$ on $X \in L$ is defined by the integral

$$H(X) = x_0 + \int_{X_0}^{X} \frac{1}{Dynamics} \tag{33.5}$$

of the inverse intonation function. Musically, the "initial value" $x_0 = H(X_0)$ means that the piece is played from a starting dynamics, i.e., the "mezzo forte" x_0 of performance from which the remaining performance dynamics can be deduced by means of the given dynamics curve.

In general, the dynamics must also cope with more than one given performance value, $x_0 = H(X_0), x_1 = H(X_1), \ldots x_t = H(X_t)$, and then, we have to ask for the conditions

$$x_j - x_i = \int_{X_i}^{X_j} \frac{1}{Dynamics} \tag{33.6}$$

for $i, j = 0, 1, \ldots t$ to guarantee compatibility of the dynamics integral with the given values. This is, in particular, the case when a specific dynamics range between successive symbols, such as *ppp, mpp, pp, mp, p, mf, f, mff, fff*, is required for acoustical reasons. In the precise technical sense this is a rare situation which occurs more in studio situations than in live performances. But also in common (what is that in the age of complex man-machine interfaces?) human performance, it may occur that one musician or instrument dictates the dynamical ranges by individual approaches, and that other musicians in the orchestra have to cope with these constraints. For example, the enhanced dynamics of a strong beating drummer can trigger dramatically the dynamical range of a whole jazz band.

33.2 Genesis of the General Formalism

Summary. Tempo, intonation and dynamic fields are one-dimensional special cases of performance fields. We discuss the a priori two-dimensional performance field of tempo and articulation. From this special case, a general type of performance fields is deduced and motivated by a set of representative examples. We conclude the section with a rigorous definition of the concept of a performance field.

– Σ –

If we consider a local composition $L \subset Onset \oplus Pitch$ (see Section 6.4.1 for this space), the performance of such a composition is a bijection $h : L \xrightarrow{\sim} l$ with codomain $l \subset PhysOnset \oplus PhysPitch$. Suppose that this bijection is induced by bijections on the onset and pitch axes as described above. With the above notation, we consider the projections $p_{Onset} : Onset \oplus Pitch \to Onset, p_{Pitch} : Onset \oplus Pitch \to Pitch$. Consider the two projections $L \to K = p_{Onset}(L)$, and $L \to S = p_{Pitch}(L)$. Suppose that these projections are performed conforming to the above rules, i.e., there are frames $OFrame, PFrame$ with $K \subset OFrame$, and $S \subset PFrame$. Suppose also that we have diffeomorphisms F, G as above. Let our performance h be defined by the projection morphisms, i.e., for $(X, Y) \in L$, $h(X, Y) = (F(X), G(Y))$. This means that we are given a product C^1-diffeomorphism $H = F \times G : OFrame \times PFrame \xrightarrow{\sim} POframe \times PPframe$ which restricts to h on L. Evidently, the two one-dimensional vector fields $Tempo, Intonation$ induce a two-dimensional vector field $Tempo \times Intonation$ on the product frame $OFrame \times PFrame$, see also Figure 33.1. Whereas the factor fields are derived from the tangent morphisms TF, TG, the product vector field is evidently not derived from the tangent morphism TH. In fact, the latter is a two-dimensional linear transformation

$$TH_{X,Y} = TF_X \times TG_Y$$

and not a vector. We recognize however that the product field verifies this equation:

$$Tempo_X \times Intonation_Y = (TH_{X,Y})^{-1}.\Delta \tag{33.7}$$

where $\Delta = (1, 1)$ is the diagonal unit vector in the tangent space at the H-image (x, y) of (X, Y). This means that *the tempo-intonation field is derived from the diffeomorphism H which extends the map h on the underlying composition L via the inverse image of the diagonal field Δ on the image product frame $POframe \times PPframe$.*

33.2.1 The Question of Articulation

Summary. Special attention is given to the delicate relation between tempo and articulation. It reveals that one-dimensional performance fields are a priori insufficient to describe local performance transformations. Articulation requires a priori two-dimensional fields.

– Σ –

It may appear as if the two-dimensional tempo-intonation field were an artificial generalization of an essentially one-dimensional situation. This is however not the case as we now show for the two-dimensional performance on the plane of onset and duration.

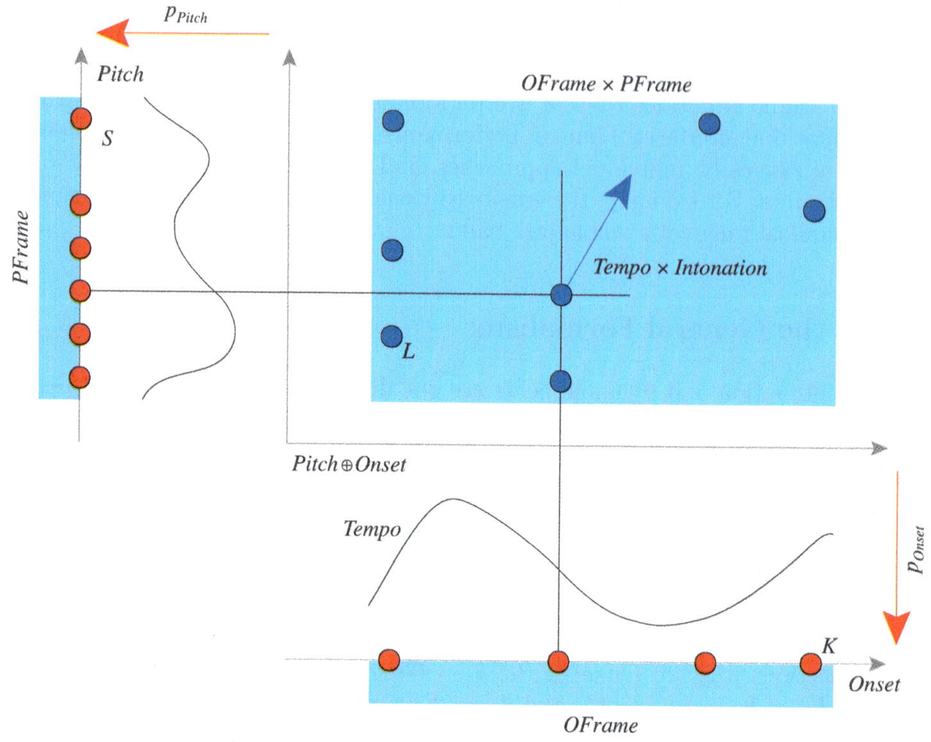

Fig. 33.1. The two one-dimensional vector fields $Tempo, Intonation$ induce a two-dimensional vector field $Tempo \times Intonation$ on the product frame $OFrame \times PFrame$.

This situation is as follows. Suppose that we have a local composition $K \subset Onset \oplus Duration$ which should be performed with respect to a tempo performance of the onset projection. More precisely, call such a performance $F(E, D), (E, D) \in K$, and recall the sweeping alterator already used in counterpoint (and introduced in Section 7.5):

$$\alpha_+ : Onset \oplus Duration \to Onset : (E, D) \mapsto E + D$$

with the omission of the dual number formalism here because we do not need this algebraic enrichment of structure in performance theory. Suppose that the onset projection $K_O = p_{Onset}(K)$ of K is performed according a performance map $f : K_O \to k_O$ onto an image set k_O. Where should we map the duration components? The canonical recipe is to consider the "offsets" of our events, i.e., for $(E, D) \in K$, we take its alteration $Off = \alpha_+(E, D) = E + D$ and look for its image $f(Off)$. We may then calculate the duration d in the image $(e, d) = F(E, D)$ by the formula $d = f(Off) - f(E)$.

This is however not well defined if we do not know what is the image of onsets outside K_O! This is another strong argument for the existence of tempo outside a composition's onsets. So we have to embed the onset performance f in a C^1-differentiable extension, to be defined on an onset frame $Frame$ as above. Call this extension again f. Observe that $Frame$ must now also contain the local composition of alterates $\alpha_+(K)$, i.e.,

$$K_O \cup \alpha_+(K) \subset Frame. \tag{33.8}$$

With these conventions, we may calculate the tangent morphism TF associated with the map $F(E, D) = (e(E, D), d(E, D)) = (f(E), f(E + D) - f(E))$. The tangent map at (E, D) is described by the Jacobian

$$JF_{(E,D)} = \begin{pmatrix} \partial_E e & \partial_D e \\ \partial_E d & \partial_D d \end{pmatrix} = \begin{pmatrix} 1/Tempo_E & 0 \\ 1/Tempo_{E+D} - 1/Tempo_E & 1/Tempo_{E+D} \end{pmatrix}$$

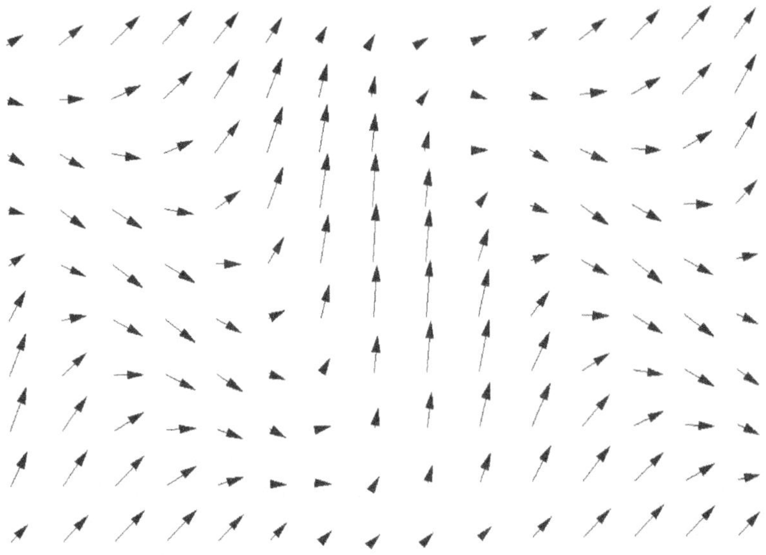

Fig. 33.2. This parallel articulation field $\partial Tempo$ is derived from the tempo curve $Tempo_E = 1 + 0.4\sin(E)$. The horizontal axis is onset, the vertical axis is duration.

which yields the inverse value

$$(JF_{(E,D)})^{-1} = \begin{pmatrix} Tempo_E & 0 \\ Tempo_{E+D} - Tempo_E & Tempo_{E+D} \end{pmatrix}$$

whence the inverse image vector field of the constant diagonal field Δ on the physical plane:

$$(JF_{(E,D)})^{-1}.\Delta = \begin{pmatrix} Tempo_E \\ 2Tempo_{E+D} - Tempo_E \end{pmatrix} = \partial Tempo_{E,D} \tag{33.9}$$

which we call the *parallel field* ∂*Tempo of articulation* since the duration component is calculated in parallel to the onset by use of offset values. Figure 33.2 shows such a parallel field. We see that the direction of such an articulation field is from $-\pi/4$ to $\pi/2$.

From this example, we learn that the performance field of the parallel articulation map on the onset-duration plane is not a product of one-dimensional fields, even under the completely innocuous assumption of durations being induced by offset data.

Exercise 77 *Calculate the performance field of articulation if the performance map has a dilatation by* $0 < \lambda$ *in the duration component, i.e.,* $F(E, D) = (f(E), \lambda(f(E + D) - f(E)))$.

The example of a parallel field in articulation can be taken over to the pitch domain: Instead of duration we have to think of glissando. This is the proportion of pitch shift with respect to the pitch coordinate if the pitch at the end of an event must have a different pitch with respect to the onset pitch. This generates the parallel glissando performance field $\partial Intonation$ with exactly the same formalism as for articulation. The same method can be applied to generate a parallel crescendo field $\partial Dynamics$ if *Dynamics* is the field associated with loudness performance. The details are left to the reader. Whenever we deal with basis and pianola spaces, we use this notation: If $B = B_1 \oplus \ldots B_k$ is a product space of basis parameters (such as onset, pitch, loudness), and if $P = P_1 \oplus \ldots P_k$ is the product space of corresponding pianola parameter spaces, we denote by α_+ the alteration map $B \oplus P \to B$ defined by $(b, p) \mapsto b + p$ on each basis-pianola component, whereas p_B is the first projection. So parallel performance maps and the corresponding parallel performance fields are defined by use of the alteration α_+. If the basis field at the basis point X is Z_X, the parallel field at the basis-pianola point Q is

$$\partial Z_Q = (Z_{p_B(Q)}, 2.Z_{\alpha_+(Q)} - Z_{p_B(Q)}), \tag{33.10}$$

a linear operator in Z. We shall use this operator not only for basis-pianola couples, but in general situations of a direct product of isomorphic simple spaces B, P with the associated alteration map.

33.2.2 The Formalism of Performance Fields

Summary. This section puts the previous considerations and special cases into a generic formalism, the concept of a performance field in general musical parameter spaces.

– Σ –

The general situation is this. We suppose that a local composition $K \subset S$ is contained in a closed rectangle $R = [a_1, b_1] \times \ldots [a_n, b_n]$ of the underlying real n-space. We also suppose that a bijective performance map $\wp : K \to \wp(K)$ onto a local composition $\wp(K) \subset PS$ in a simple physical space PS of same dimension n is given, and that this map can be extended to a C^1-diffeomorphism $\wp : R \to \wp(R)$, i.e., a C^1-diffeomorphism which is defined on a neighborhood of R. The *performance field* of this diffeomorphism is a continuous vector field[2] $\mathfrak{T}$ on R which is defined by the inverse Jacobian

$$\mathfrak{T}^{\wp}_X = (J\wp_X)^{-1}.\Delta \tag{33.11}$$

applied to the diagonal unit vector $\Delta = (1, \ldots 1)$, i.e., $\mathfrak{T}^{\wp} = T\wp^{-1}.\Delta.\wp$.

If x_0 is a point in PS, the integral curve[3] $\int_{x_0} \Delta$ of Δ through x_0 evaluates to $\int_{x_0} \Delta(t) = x_0 + t.\Delta$. For the existence and uniqueness of maximal integral curves of $\mathfrak{T}^{\wp}$, we suppose that the performance field $\mathfrak{T}^{\wp}$ is locally Lipschitz[4]. Then, evidently, integral curves of $\mathfrak{T}^{\wp}$ are transformed into integral curves of the diagonal field under $\wp$. Therefore, if $X \in R$, if $X_0 = \int_X \mathfrak{T}^{\wp}(t)$, and if the image $x_o = \wp(X_0)$ is known, then we have

$$\wp(X) = x_0 - t.\Delta. \tag{33.12}$$

This means that the performance map $\wp(X)$ can be calculated via the integral curve $\int_X \mathfrak{T}^{\wp}$ of the performance field if there is at least one point on such a curve for which the performance map $x_0 = \wp(X_0)$ is known. This generalizes what we have already learned for tempo curves and initial performances on selected onsets.

Therefore, performance can be calculated from a Lipschitz-continuous performance field on all points X whose integral curves hit a set $Initial \subset R$ of points whose performance is known in advance. We may now forget about the performance map $\wp$ and start from the performance field and a 'good' initial set, *defining* the performance map via the integral curves and equation (33.12). This performance map will usually only be calculated on the local composition $K \subset R$, or at least for a selected set of points of that frame, and not for any point in R, but this is exactly what we want, see Figure 33.3.

33.3 What Performance Fields Signify

Summary. The very complexity of performance fields parallels a strong impact on semantic layers of performance. We want to lay bare this crucial relation. In a first approach, we discuss the contributions of Adorno, Benjamin, and Raffman to the very ineffability of performance nuances. We then deduce the adequacy of performance fields to deal with this "ineffability" and to control it on the level of the sophisticated language of calculus. Finally, we deal with the tension between structural and performative parameters in music. We expose and discuss Helga de la Motte's thesis that, historically, there is an increasing number of performance parameters being transformed into structural score parameters.

– Σ –

[2] $\mathfrak{T}$ stands for German "Tempo-Stimmung" and is symbolized by the Hebrew letter "tsadeh".
[3] For integral curves of vector fields, see Appendix I.2.3.
[4] See Appendix I.2.1.

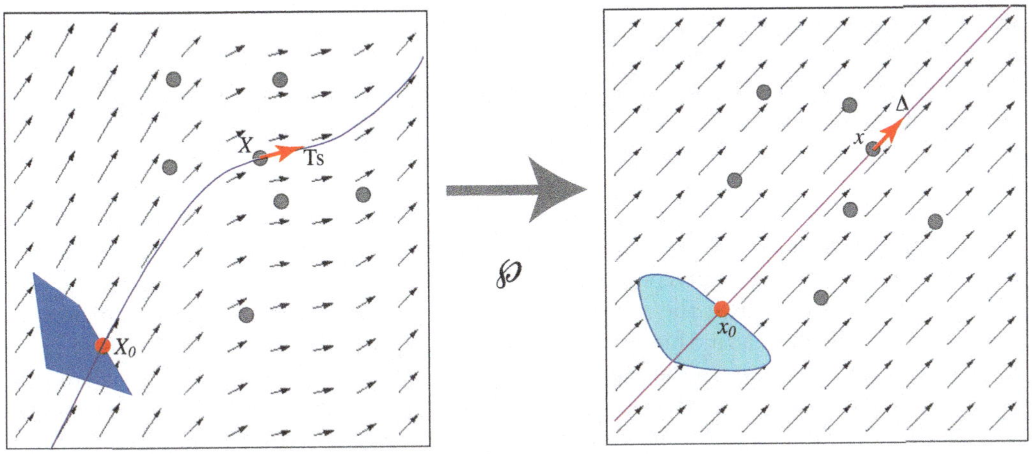

Fig. 33.3. Performance $x = \wp(X)$ can be defined upon the performance field $\mathfrak{Z}$ and on an initial set (left polygon) where the performance is known in advance.

33.3.1 Th.W. Adorno, W. Benjamin, and D. Raffman

Summary. Adorno and Benjamin [243] have associated performative adequacy with an activity of "infinitesimal precision". We make plausible that their language suggests the language of vector fields—though not explicitly stated by these authors. Diana Raffman's argument for ineffability of musical nuances in performance [875] is discussed. We relate this admittance of ineffability to the search for a powerful language as an extension of the powerless common language.

– Σ –

The previous discussion of musicological approaches to performance might have given the impression that in general, musicologists share the tendency to oversimplify the complexity of performance, be it in discrete tempo concepts, be it in correspondingly simplistic understanding of intonation. In fact, the usual understanding of articulation is not better than that. It is defective to the point of not realizing that performance of duration is related to the onset performance plus some deformation of the duration according to articulation rules. We have not encountered any such structural description to date—worse: discussions about these phenomena were dominated by a complete ignorance of this kind of effects.

However, on a non-quantitative level, very intelligent observations have been advanced by the most sophisticated theorists of performance: Theodor W. Adorno, and Walter Benjamin in [9]. Here is their basic text which introduces the "micrological procedure":

> *(Dieses mikrologische Verfahren) darf nicht als ein dem künstlerisch produktiven Entgegengesetztes verstanden werden. Walter Benjamin hat 'das Vermögen der Phantasie' als 'die Gabe, im unendlich Kleinen zu interpolieren' definiert. Das beleuchtet blitzhaft die wahre Interpretation. Der Forderung, Phantasie, als Medium des Lebens der Werke, und Genauigkeit als das ihrer Dauer, zu vereinen, der Grundfrage, welcher der verantwortliche Interpret sich gegenüber sieht, wird genügt nur durch den gebannten und bannenden Blick auf den Notentext der Werke. In seinem dicht gewobenen Zusammenhang sind die minimalen Hohlräume zu entdecken, in denen sinnverleihende Interpretation ihre Zuflucht findet. (...) Das Medium künstlerischer Phantasie ist nicht ein Weniger an Genauigkeit sondern das noch Genauere.*

As the wording is chosen, micrologic is a logic in the smallest dimensions of a composition and its performance. The text suggests that this procedure could be misunderstood as opposed to artistic fantasy. Adorno evokes Benjamin's observation that fantasy is involved in the infinitely small ("im unendlich Kleinen"), more precisely in the interpolation towards the infinitely small. For Adorno, this is a revealing insight into the ultimate, true performance. Adorno asks for the discovery and inspection of the innermost interspaces ("sind die minimalen Hohlräume zu entdecken"). Infinite interpolation is the tool to do so. And this is not contrary to artistic fantasy, it is, so to speak, the strongest microscopic instrument we have, and should use. Artistic fantasy is not the pseudo-romantic blurredness, but a maximum of precision, of intensity and interplay of minimal movements and forces.

This absolutely central insight of Adorno and Benjamin is not only astonishing in the musicological environment (though not as a category of Adorno's and Benjamin's discourse), it is also a very problematic approach insofar as the humanities—where their text belongs—do not have any means of making such allusions precise. The text is a kind of schizophrenic claim of non-mathematical experts in the words of mathematical concept frameworks: Interpolation, infinitely small, etc.

To the mathematically trained, the allusion to calculus is straightforward. No doubt, the language of the infinitely small is calculus. Is it this kind of language which Adorno and Benjamin were aiming at? What is intriguing is that they are talking of infinite interpolation. Between what? The score is a radically discrete symbolism. The infinite interpolation is not a priori inscripted into the score structure. And, what are those cavities ("Hohlräume") in terms of music parameters and processes? In between the discrete score events, there must be some infinitely divisible space which encompasses the cavities, Adorno and Benjamin are zooming in and penetrating. A solution of this conceptual approach could in fact be the continuous and differentiable interpolation suggested in the previous considerations of extensions of local compositions and maps. We claim that our theory is the mathematically adequate concept framework to the Adorno-Benjamin approach.

More precisely, performance fields which include infinitesimal information about the performance process "between the score units" seem to conceptualize this world of infinitely refined reading of what is happening in the cavities of time, pitch, loudness, articulation, glissandi, and crescendi.

Without anticipating the expressive power of performance fields, it appears that performance fields in their very rich structure could englobe a deep semantic richness towards human expression of all the intentions which human performance commits. We have to imagine that given a score with its discrete event set, the layer of a performance field which is superposed to that score adds an infinitely fine interpolation in the sense of Adorno–Benjamin. It is like an optical lens system which deforms the "mechanical, rigid" score data into a rhetorical expression of the interpreter's understanding.

Performance fields implement a powerful language of performative rhetorics which transcend the discrete vocabulary of common scores. This opens a substantial discussion about ineffability in music. Recall that musical performance is still a strong argument for the ineffability of musical reality: In [875], Diana Raffman has argued that ineffability is a characteristic feature of musical expression, and that this is related to the quale objects as defined by Clarence Irving Lewis in [606]. Quales are those qualities of immediate human experience which cannot be conceptualized and are of private, individual, irreproducible and antilexical nature, such as colors, sounds, hunger, anger, sadness, or happiness. Raffman argues that musical experience is strongly related to quales and therefore shares strongly ineffable characteristics.

However, ineffability means escaping the power of language. And this is the critical point: which language is escaped, transcended, what is the boundary to effability? Clearly, the infinitely small, the infinite interpolation are such ineffabilities to the common language. But they are not to the mathematical language of calculus. Ineffability is a relative concept and not a static verdict. This means that ineffability is a challenge for language extension: Are we able to find a richer language which captures phenomena which were—hitherto—ineffable?

Principle 24 *We argue that performance fields are precisely such an extension of the music description language which turns ineffable instances of musical expressivity into regions which may be controlled by such a language enrichment.*

After all, vector fields are a very romantic subject: The experience of wind and weather, of stormy rains, of water streams and lava breakouts is a valid metaphor of the forces and processes of our souls. It is not miraculous that performance fields are the exact counterpart of musical expressivity in its most refined appearance as preconized by Adorno and Benjamin in their visionary text.

33.3.2 Towards Composition of Performance

Summary. Helga de la Motte has hypothesized [257] that, historically, there is an increasing number of performance parameters being transformed into structural score parameters. We discuss this argument and make a picture of its consequences for the composition of performance fields; a language reflecting such a refined performance data must be radically different from the known "digital" language as is common in western score notation.

– Σ –

It is undeniable that from the early days of neumes to the present, or at least, to the classical European notation, an increasing number of music parameters can be observed in the score notation. For example, the bar-lines were only introduced around 1420, whereas the dynamic signs or the instrumental specification were not present in Bach's *Art of Fugue*, and the metronomic indications became standard only after Mälzel's invention (though sometimes in a problematic way, such as in Beethoven's "Hammerklavier" sonata).

This fact must be seen in the context of the very concept of a score. The concept of a score is that of a mediator between musical ideas and their physical execution. As such, the score of a music piece cannot be narrowed to the level of a notated sign system. The score begins in the mental layers of the composer and musician and is only supported by, not identified with, the schematic and digitized material form. The distribution between exterior and interior[5] score depends on the specific music culture. In jazz, the major part of the score is present on its interior level, the rudimentary lead-sheet notation sketches only the most elementary ingredients such as harmonic progression and melodic core structures.

This has deep implications on the concept of a composition. If one agrees that a composition is all that is fixed on the "neutral level" of the score, the stratification into inner and exterior score makes the composition concept a bit difficult: Where does the composition begin, where does it end? In fact, the interior part of the score may include mental sketches as well as schemes of performance or construction, such as realized in jazz improvisation or processual schemes in the sense of Cage. If we agree that the composition is everything that can be traced either on the exterior or the interior score, then composition may very well include performative instances if the interior score shapes the mapping from mental to physical reality by a well-defined concept.

In this sense, Helga de la Motte's observation of an increasing shift of performative instances to the score level can be made more precise: It is a shift from inner score instances to those of the exterior score. Now, such a shift is a function of two factors: First, as in jazz, and even more in free jazz, it may not be the objective of a musical approach to aggregate exterior score signs, and then, no such shift is needed. Nonetheless, transcriptions of jazz improvisations may be desired (such as the famous transcriptions of John Coltrane soli by Andrew White), and also to a very refined degree. Second, the composer may want to give a precise description of instances of hitherto interior scores, such as happened with the performance signs of *accelerando*, *crescendo*, etc.

In both cases, the shift can only be achieved by a refined language since the interior score may comprise ineffable spots, regions that cannot be conceptualized on the verbal level: Non-verbal concepts are a widespread phenomenon among musicians! Performance fields may be an extension of the exterior score language which helps shift non-verbal concepts to the verbal level. But we do not insist on the verbal character of a vector field, at least not in the common sense of verbalization. Mathematical concepts are beyond common language. And they are also beyond quales, they are an effective extension of language which can help think things which in the past were completely out of reach to the human intellectual power. Evidently, the score concept which includes such sophisticated objects as vector fields will look completely different from the traditional discrete system. Perhaps it will also per default and as a mandatory condition be related to computer-aided representation and editing. In this sense, the medium computer may help profile a message which musicians of all cultures have always dealt with and tried to communicate so desperately.

[5] To my knowledge, the term "interior score" was introduced by the Jazz theorist and musician Jacques Siron in his remarkable book "La partition intérieure" [979] on jazz theory.

Check for updates

34

Initial Sets and Initial Performances

Jeder Anfang ist ein Ende.
Hermann Hesse (1877–1962)

Summary. Performance has to start somewhere. The theory of this initialization deals with initial sets and initial performance. Naively speaking, initial sets are the first notes of a performance. Semiotically, initial theory describes a turning point from lexicality to reality in music which is supported by shifters; we explain this rationale. Since music deals with many parameters, initialization has to be specified in all dimensions. We first comment on the classical initial sets in onset, pitch, and duration. On an initial set, the performance cannot be calculated from previous performance, an initial performance has to be defined. We discuss ways to do so. On a more technical level, we introduce the hit point theory, a mathematical account for the control of performance field flows as their curves approach initial sets. Strategies of guessing best approaches to initial sets are presented.

– Σ –

We now situate our investigations in the framework developed in Section 33.2.2. This means that we are given

1. a frame $R = [a_1, b_1] \times \ldots [a_n, b_n]$ in the n-space $\mathbb{R}^n$,
2. a local composition[1] $K \subset R$,
3. a performance field $\mathfrak{Z}$, i.e., a locally Lipschitz vector field defined[2] on R,
4. and an initial set $I \subset R$.

According to the previous theory, we shall suppose that the points X_0 of the initial set I are given an *initial performance* $\wp_I(X_0)$, and that we may define the performance $\wp(X)$ for $X \in K$ by means of the integral curve $\int_X \mathfrak{Z}$ if this one hits a point $X_0 \in I$ via formula (33.12).

In this chapter we shall first discuss the meaning of the initial set approach and then techniques to find initial points for polyhedral initial sets.

34.1 Taking Off with a Shifter

Summary. When music performance takes off, a magic moment takes place. This well-known effect relates to shifter signs in the semiotic system of music. We make a point of this magic, in fact a moment with deep

[1] We are sloppy about the underlying forms here and just consider the coordinates and modules since the forms are not of primary interest here. It is however subtended that everything happens within well-defined forms that may be evoked if necessary, e.g., if the mathematical methods need a justification via parameter forms beyond numerical coordinates.

[2] By definition, this means that the field is defined in a neighborhood of R. But the portions of integral curves of this field contained within the frame R are only a function of the field values within the frame.

G. Mazzola, *The Topos of Music II: Performance*, Computational Music Science,
https://doi.org/10.1007/978-3-319-64444-8_3

consequences for performance as a whole which is fully appreciated by the auditory on an emotional level. This "juncture of fiction and reality" is singular but does in fact happen in several places of a composition's performance. The magic is distributed among the entire performance, we introduce this subject as an interface between semiotics and psychology.

– Σ –

Even before discussing the different parameter-specific initial sets, it is important to understand the deep meaning of the fact of initial performance. We have known performance as a transitional process (formally described by the performance map $\wp$) from mental to physical reality. On the level of documentation this is a transition from the score to the acoustical realization, to be archived on sound media such as a CD. Following the valid doctrine—as preconized by Valéry[3] and Adorno[4], for example—the performance is an integral part of the work of art, and this means that, in the sense of communication theory of art as described by Jean Molino (see our introduction in Section 2.2), performance is part of the semiosis of the work, its meaning is not complete except when it is performed. Put it the other way round: The mental score (interior as well as exterior) conveys a part but not the whole significate, and only via performance can we complete the work's semiosis. Performance englobes a kind of usage of the mental score sign by a performer, much like the usage of a sign in the pragmatic dimension is part of its semiosis. More specifically, those signs whose significate are not only instantiated but substantially depend on the user, are the well-known shifters. Performance of a mental score is such a shift from lexicality to full-fledged meaning, since the pure score is *essentially* less than the work of art. In other words, performance is a shifter characteristic of the score semantics. Production of full-fledged meaning is only possible by means of performance, and this adds a semantic aspect to the sign which is a non-trivial function of the performer(s).

So the anchorage of a score in the physical performance is not only a transformation but also a completion of the score's meaning. It is a completion of shifter type, i.e., adding a new, user-dependent value of the score sign, a value which turns the abstract score structure into a concrete, existential entity.

Within this dramatic transformation process which the conductor Sergiu Selibidache has so violently defended against the musical reproduction industry [944], there is a particularly dramatic moment of initiation of the existential kernel. This is what happens in the beginning of a performance.

This beginning is when the conductor (or the soloist in the case of a solo performance) appears on stage, steps to the conductor's desk, takes the baton and freezes every movement in order to get off with the first onset gesture. The moment, when the conductor lowers the baton and unfreezes the time process, is the real magic of performance: Some moments ago, the work was still in its lexical potential state, everything could have happened. But now, we are getting off into reality. The first note puts an end to the potentiality, the shifter level has come into life. We do not know whether it is this switch of existentialities which so incredibly fascinates the auditory (and the orchestra), but it is an objectively dramatic event. It might be compared to the reading of the article on the first person singular pronoun "I" in an encyclopedia, and pronouncing the word "I" as a living person with all the shifted meaning pointed at when you say and mean "I".

The question of what really initiates when the performance gets off is not simple. It was possibly suggested in the previous discussion that it is only a time initialization, but this is as wrong as it would be wrong to claim that music reduces to onset time. In the initial moment of a musical performance, many different settings are instantiated, in fact as many as we have parameters to describe the sound events (and the gestural parameters in an extended performance theory) in our piece of music. In what follows, we want to investigate these initializations in more detail.

34.2 Anchoring Onset

Summary. The most elementary and important initialization is that of onset. We give the overview of its structure and function, including the shifter nature of onset initialization. The problem of multiple onset initialization is discussed.

[3] "C'est l'execution du poème qui est le poème." [1073]

[4] "Die Idee der Interpretation gehört zur Musik selber und ist ihr nicht akzidentiell." [7]

– Σ –

This section uniquely deals with the *Onset* space form and performance fields on this space, i.e., tempo curves. Musically speaking, this approach is a naive one, in fact more involved performance of time does not happen independently of other parameters and therefore cannot be described by tempo curves. For example, if we are involved in a performance where onset is a function of pitch, a situation which may happen in an arpeggio. This is viewed as a temporary onset distortion as a function of pitch. A typical such performance map $\wp$ is as follows:

$$\wp(E, H) = (E - 4e^{-(H-5)^2}, H).$$

The corresponding performance field $\mathfrak{Z}^{\wp}$ is shown in Figure 34.1. Observe that onset components of the field may also be negative, according to the retard of onset in the middle arpeggio region of pitch around $H = 5$ in this generic example.

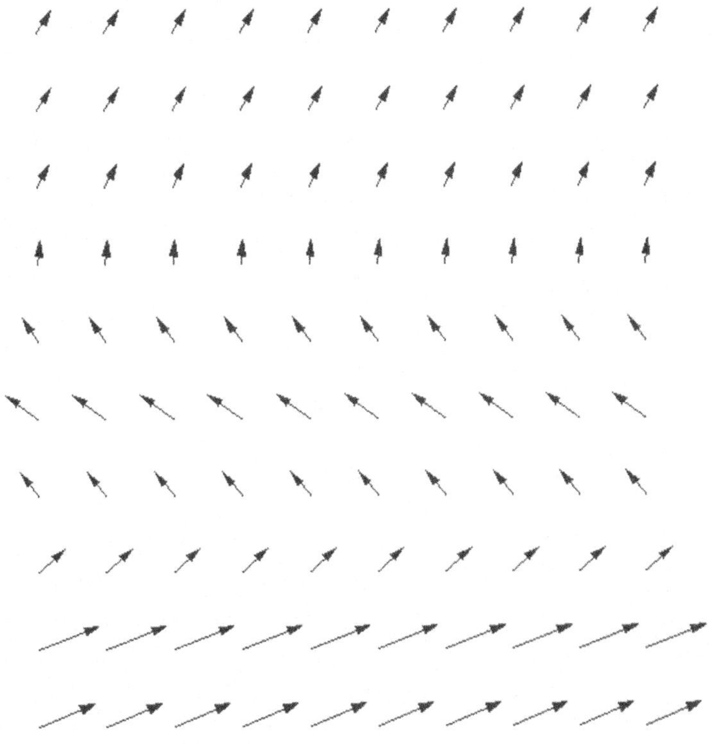

Fig. 34.1. The performance field on the plane of onset (horizontal axis) and pitch (vertical axis), corresponding to a prototypical arpeggio. This field cannot be built upon a tempo curve.

The data for our pure onset performance is as follows: We are given a frame interval $R = [a, b] \subset Onset$, a tempo "field", i.e., a continuous positive tempo curve $\mathfrak{Z}(E) = T(E)$ defined on R, and a finite initial set $I = \{E_0 < E_1 < \ldots E_f\} \subset R$ within the frame. On this set, we are given performance data $\wp(E_i) = e_i, i = 0, \ldots f$ with $e_0 < e_1 < \ldots e_f$. The meaning of this data is that the composition K's onset set $K_E = p_E(K)$ is also a subset of the frame, $K_E \subset R$, and that its performance is calculated by the performance formula (33.1). More precisely, in the common situation, the initial set is the singleton $I = \{E_0\}$ consisting of the composition's first referential onset. This could be the first element in K_E or else the first bar-line's onset. In this case, no further discussion is required, the integral exists, and we can calculate all required onset performances.

However, realistic situations are more involved. It may happen that one is given several initial points, for example if the left hand plays in constant tempo, whereas the right hand is allowed to vary locally against the left hand in a "Chopin rubato", i.e., in such a way that the onset performances coincide on each bar-line, but it may differ locally. In this case, the initial points for the left hand would have a fixed performance whereas the right hand tempo curve would have to fit the left hand onset values. This will in fact happen in our discussion of tempo hierarchies, see Section 38.2.

This means that we have additional conditions, the integrals of inverse tempo must coincide with the given initial onset performances, i.e., for $0 \leqslant i < j \leqslant f$, we must have

$$e_j - e_i = \int_{E_i}^{E_j} \frac{1}{T}. \tag{34.1}$$

Under these conditions, the calculation of any onset performance $\wp(E)$ is clearly independent of the reference initial point $E_i \in I$. In this case we say that *the tempo curve is adapted to the initial performance* $\wp_I$ *on the initial set* I.

Proposition 56 *With the above notation and hypotheses, there is always a unique continuous tempo curve* $T(E)$ *which is linear on each interval* $[E_i, E_{i+1}], i = 0, \ldots f-1$*, constant outside* $[E_0, E_f]$*, and prescribes an arbitrary positive tempo* $T(E_i)$ *for one index* $0 \leqslant i \leqslant f$.

Proof. We prove the case of a fixed start tempo, the general case works in complete analogy. By induction on f, we may restrict the proof to $f = 1$. It is sufficient to show that we may find a final positive tempo value $x = T(E_1)$ such that the linear tempo $T_x(E) = (x - T(E_0))(E - E_0)(E_1 - E_0) + T(E_0)$ fulfills $d = e_1 - e_0 = q(x) = \int_{E_0}^{E_1} \frac{1}{T_x}$ for any positive value d. But $q(x) = (E_1 - E_0)\frac{\log(x) - \log(T(E_0))}{x - T(E_0)}$, which is a continuous function of x, converging to $\frac{E_1 - E_0}{T(E_0)}$ as $x \to T(E_0)$, and ranging from ∞ to 0 for positive x. Therefore there is exactly one positive x_1 such that $d = q(x_1)$, and $T(E_1) = x_1$ solves our problem, QED.

Corollary 21 *Suppose that for the increasing sequence* $E_0 < E_1 < \ldots E_f$ *of symbolic onsets, we are given positive tempi* $T_i = T(E_i), i = 0, \ldots f$*, and that the tempo curve* $T(E)$ *on the interval* $[E_0, E_f]$ *is the polygon through these values. Let* $\Delta = \int_{E_0}^{E_f} \frac{1}{T}$*, and select a positive real scalar* σ*. Then there is a positive scalar* τ *such that the polygonal tempo curve* T_τ *for the vertex values* $T_\tau(E_0) = T_0, T_\tau(E_f) = T_f, T_\tau(E_i) = \tau T_i, 0 < i < f$*, has* $\int_{E_0}^{E_f} \frac{1}{T_\tau} = \sigma.\Delta$.

Proof. The integral $\int_{E_1}^{E_{f-1}} \frac{1}{T_\tau} = \frac{1}{\tau}\int_{E_1}^{E_{f-1}} \frac{1}{T}$ assumes any positive value if τ varies. Further, Proposition 56 guarantees that the initial and terminal integrals $\int_{E_1}^{E_0} \frac{1}{T_\tau}, \int_{E_f}^{E_{f-1}} \frac{1}{T_\tau}$ can also be adapted to any positive value with varying τ, so we are done.

This means that given a tangent morphism Tg on the initial values E_0, E_f, and a polygonal tempo curve T which extends Tg, we can "deform this curve" to a new polygonal curve T_τ without altering the tangent data Tg_{E_0}, Tg_{E_f}, but such that the total duration is stretched by any positive value σ. In practical situations—such as the composition software *presto*®'s Agologic module (see also Chapter 50)—this has the following application: We are given a polygonal tempo curve as in the Corollary, with duration $\Delta = \int_{E_0}^{E_f} \frac{1}{T}$. By graphically interactive editing, the curve may be altered in that the vertical position (the tempo coordinate) of one inner vertex of the polygon is augmented or diminished. This changes the duration from Δ to $\rho.\Delta$, but we do want to conserve duration. This can be achieved by a deformation of the graphically altered polygon as in the lemma, setting $\sigma = 1/\rho$, and we recover the original duration. So the new shape of the polygon can be conserved as far as possible.

34.3 The Concert Pitch

Summary. In the pitch dimension, initialization deals with concert pitch, i.e., initial intonation or tuning. We compare this initialization with the onset's take-off and discuss the specific difference in deicticity: Concert pitch is dominated by a lexical dimension.

– Σ –

In this first approach, we shall view pitch performance as a parallel situation to onset performance as discussed in the previous section. This means that pitch performance is viewed independently of other parameters, in particular, independent of time. As with tempo curves, this is a musically naive, but basic approach. In fact, pitch performance may be a function of onset, according to the tonal system of the given composition, for example in modulatory parts.

So we are given a frame $R = [a, b] \subset Pitch$, an intonation field, i.e., a continuous intonation curve[5] $\mathfrak{I}(H) = I(H)$ on R, and a finite initial set $I = \{H_0 < H_1 < \ldots H_f\} \subset R$. On this set, we are given performance data $\wp(H_i) = h_i, i = 0, \ldots f$ with $h_0 < h_1 < \ldots h_f$. In the standard case, this initial set is not as variable as for tempo. In fact, there are several constraints of tuning which are above all given by the orchestral instrumentation. For the piano, to begin with a common reference for orchestral tuning, we are given all semitone pitches $I = \{H_0 < H_1 < \ldots H_f\}$ of the chromatic 88 keys[6], together with their rigid performance pitches which are $\wp(H_i) = h_i = h_0 + i.\log(2)/12$ in the common 12-tempered tuning. The underlying intonation curve must cope with all 88 values. We have the following result which replicates Proposition 56 for intonation:

Proposition 57 *With the above notation and hypotheses, there is always a unique continuous intonation curve $I(H)$ which is linear on each interval $[H_i, H_{i+1}], i = 0, \ldots f-1$, constant outside $[H_0, H_f]$, and prescribes an arbitrary positive intonation slope $I(H_i)$ for one index $0 \leqslant i \leqslant f$.*

The less rigid case is the tuning of the (continental European) chamber pitch $a' \sim 440\ Hz$, together with the octave periodicity condition on the integral, i.e.,

$$\int_H^{H+12} \frac{1}{I} = log(2)$$

for all pitches H and the semitone encoding of pitch. Such a tuning is independent of local variations of intonation due to intonation specificities for just fifths and the like which violinists and singers may prefer. It is also open to glissandi which include all pitch values in a determined real interval.

There is, however, a qualitative difference between intonation initialization and onset initialization. The latter is a "magic shifter" which by its very construction instantiates the fictitious music score time in real time. The former correspondingly instantiates fictitious score pitch in physical pitch, but its value is not a question of individual construction, it is lexicalized on the standards of music culture and tradition. It is also lexicalized by the absolute pitch perception of a number of musicians, for example Herbert von Karajan, which makes it virtually impossible to alter this conventional initialization on a shifter basis of individual, spontaneous usage. The usage of intonation and its initialization is therefore much more restricted than the usage of onset initialization. One may distinguish the local initialization within a fixed tonal context, between two successive modulations in just tuning of octaves and selected fifths (tonic, dominant for example), say, and the more individual shaping of the individual intonation of less prominent intervals within one such tonal context.

It is clear that such a one-dimensional intonation field component is as artificial as the one-dimensional tempo field. It is not artificial in the sense of a useless academic exercise, but in the sense of first approximation to artistic performance—something like a zero-state of performance—which needs refinement. This is an important observation insofar as it suggests an investigation of the problem of refining performance, of seeking for paradigms of unfolding performance shaping. The theory of performance stemmata in Chapter 38 will deal with this approach.

---

[5] We are sorry for the homonymous symbol I for initial sets and intonation fields, a confusion is however very unlikely.

[6] For the added keys on Boesendorfer's Imperial model, extend the key numbers by $-1, -2, \ldots$.

34.4 Dynamical Anchors

Summary. In the loudness dimension, initialization deals with reference dynamics. This is a dramatic and completely shifting phenomenon which stays in contrast to the onset's take-off and the lexical pitch initialization.

– Σ –

Instead of repeating the propositions presented in the previous sections, which are, of course still valid mutatis mutandis, we should rather focus on the specific character of dynamical initialization. Above all, the performance of loudness in its physical expression is a complete shifter: There is no lexical normalization since every concert determines its dynamical initialization as a function of the concert hall, the orchestra, its disposition, and the public. Whereas the intonation curve may remain more or less the same over different performances of a determined orchestra, the initial anchoring will vary considerably, though not as fundamentally as with onset, because onset will be an existentially different one in each performance.

There is another shifting character of dynamics initialization: the reproduction of a performance or the live broadcast from an electronic media, such as a radio, TV or internet concert broadcast or the simple playing of a CD on the private sound equipment. In these cases, the individual user's preferences may initialize a very specific dynamic anchorage. And, more dramatically: During the ongoing performance, the initialization may be reset according to hearing dispositions and temporary irritations from disturbing environmental noise, including the abrupt lowering of dynamics while a phone-call or a verbal intervention of another person happens.

Similar to onset initialization, dynamics initialization is quite strong and shifting, in contrast to pitch initialization. But it is also more existential, together with onset initialization: It reflects the human condition of when and how and why music is enjoyed. In this shifter process, evidently, the lexical musical content seems not to suffer, it is more the anchorage in human life which is profiled. Whereas absolute pitch seems to play a certain role for the understanding of the musical message—even to the vast majority of non-absolute pitch listeners—absolute onset and dynamics play the contrary role: They give the listener his/her coordinates of existence where they want to meet this particular music.

34.5 Initializing Articulation

Summary. The initial theory becomes less trivial when applied to articulation. Initial articulation reveals the complex recursive structure within initial data, i.e., sets and performance. We explicate different approaches to initial articulation, as based upon initial onsets.

– Σ –

Whereas initialization on one-dimensional performance fields is a question of selecting finite point sets, initialization on higher-dimensional frames is dramatically more complex. Let us have a look at the most elementary two-dimensional situation: the articulation field, see Figure 34.2.

In this example, we are given a frame R defined by its low left vertex (E_a, D_a) and the high right vertex (E_b, D_b). (In this figure, we even suppose that $D_a = 0$, but this is not the general case.) Let us first look at the right integral curve (terminating at the point on its arrow-head to be performed) which hits the frame boundary on $X_h = (E_h, D_a)$. Initialization can be decomposed in two partial problems: Initialization of onset can be reduced to the one-dimensional case since we have a tempo curve here. This means that we may consider the projection of the articulation field onto the tempo curve and accordingly the projection of the integral curve onto its onset component. Suppose further for simplicity that the onset initialization is defined on the frame boundary value E_a. Then we are done with the onset performance, and we may proceed with duration. Let us look at initial duration in X_h. Here, we have the onset performance and should define performance of duration D_a. If this value vanishes, it seems natural to perform it to physical duration zero, too. But if the lower bound D_a is positive, it is not clear what should be done. The canonical idea would

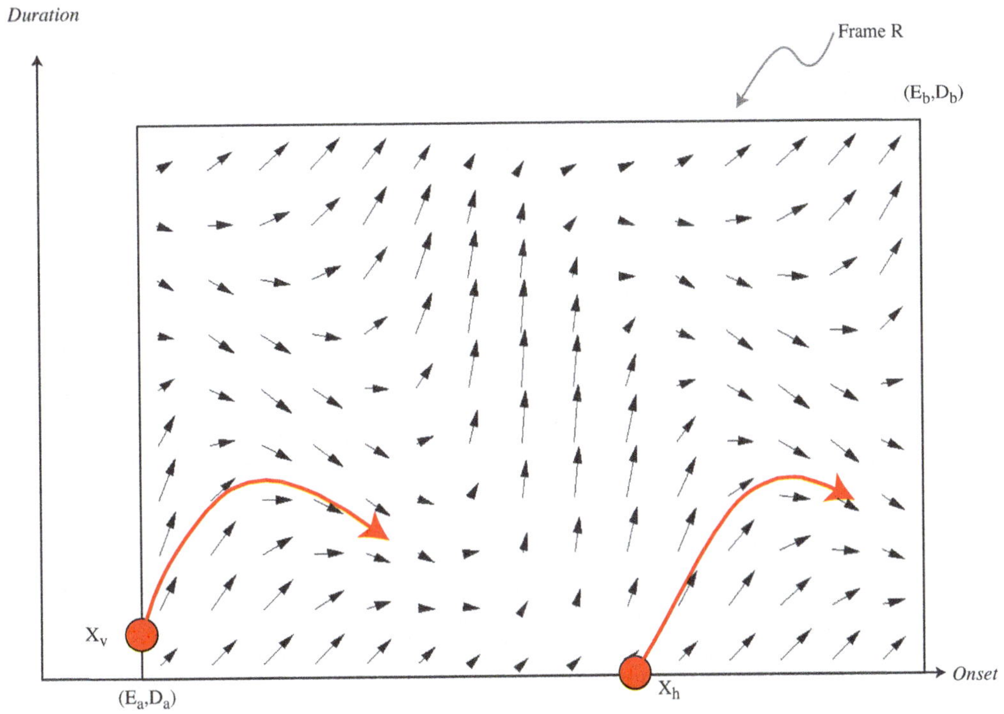

Fig. 34.2. The integral curves of an articulation field may typically hit boundary points of the given frame on different positions: the horizontal lower boundary line or the vertical left one. This entails dramatically more complex initialization data than for the one-dimensional case of tempo and intonation.

be to identify D_a as a difference $D_a = (E_h + D_a) - E_h = \alpha_+(X_h) - p_E(X_h)$. We could then apply the well-known construction of duration performance as a difference of onset performances, i.e.,

$$\wp(X_h) = (\wp_E(p_E(X_h)), \wp_E(\alpha_+(X_h)) - \wp_E(p_E(X_h))) \tag{34.2}$$

which is built upon the onset performance $\wp_E$ via the tempo curve and the initial performance on E_a. This is not mandatory though: Initial duration performance need not be without articulation, be it legato, be it staccato. Whatever: We have to be sure that not only the onset component of X_h is on the frame, the alteration component $\alpha_+(X_h)$ must also be within the frame's onset bounds! This implies that score points to be performed on this basis must stay to the left of the descending diagonal through the lower right corner of the frame. Other points must be given a different initial performance data.

If, on the other hand, the frame is hit in a point $X_v = (E_a, D_v)$ on the left boundary line, its (initial) performance can be settled by the same formula as above

$$\wp(X_v) = (\wp_E(p_E(X_v)), \wp_E(\alpha_+(X_v)) - \wp_E(p_E(X_v))), \tag{34.3}$$

but now, the initial duration can be any long duration, not just the lowest admitted duration of our frame. In this setup, the initial performance of the left boundary line is a function of the onset performance of the whole onset interval $[E_a, E_a + (D_b - D_a)]$, and not only of the initial onset performance at E_a. This is quite dramatic as a contrast to the formula (34.2) which is local on the onset of X_h and the slightly shifted onset $\alpha_+(X_h)$. Here, we have to know a lot about the future onset performance for the initial values at the beginning of the frame.

An even more dramatic effect happens if the frame has an upper boundary line which enforces integral curves to hit the upper boundary when reaching the rectangle's boundary ∂R. In our example, this could happen if the frame ended at the half height (with the same field). There, the initial performance of duration can fail to be controllable by onset performance alone, and we have to design new initial performance strategies.

Difficult initial performance problems can also happen if the field doesn't have a positive onset component, as in Figure 34.1. Here, initial performance must be defined on the right boundary hyperplane of the frame, i.e., for the last, not for the first events. And if the performance configuration is such that initial values happen to be positioned anywhere within the frame, initial performance has to be defined by use of general strategies which work for any initial set configuration. We shall review this topic in Chapter 38 after the discussion of the basic problem: How can we effectively know where an integral curve hits the initial set?

34.6 Hit Point Theory

Summary. Hit point theory deals with the control of access to initial sets along the integral curves of performance fields. These curves describe the flow associated with performance fields. In general, they can be quite "wild" so that a generic strategy of seeking hit points of such curves with initial sets is required.

– Σ –

The general setup for initial sets is a (locally Lipschitz) performance field $\mathfrak{Z}$ defined on (a neighborhood of) a (closed) frame R in a parameter space which—up to isomorphism of functors—is defined by $\mathbb{R}^n$. In many situations, the initial set for the performance $\wp$ is not any set, but will be a polyhedron defined as a union $I = \bigcup_i S_i$ of a finite family $(S_i)_i$ of simplexes $S_i \subset R$. Each simplex S_i is given by a sequence $s^i. = (s^i_0, s^i_1, \ldots s^i_{n(i)})$ of (pairwise different) points in R. The points need not be in general position, i.e., $dim(\mathbb{R}.s^i.) < n(i)$ is admitted[7]. If we view this data as being the vertexes of the simplex $S_i = \{\Sigma_j \lambda_j s^i_j | 0 \leqslant \lambda_j \leqslant 1, \Sigma_j \lambda_j = 1\}$, this means that we admit degenerate simplexes as constituents of initial sets. Also is it not required that the simplexes build a simplicial complex, their intersections can be arbitrary. We also do not require that these simplexes have $n+1$ vertexes: it is allowed to have a number of isolated points and straight lines in three-space, for example. Clearly, this type of initial sets allows virtually every shape—up to approximations by triangulations.

The previous examples are: a series of zero simplexes in $\mathbb{R}$ for tempo and intonation curves, or the sequence $(\partial_E R, \partial_D R)$ of left and lower sides of R in onset-duration space. More generally, given a coordinate Y (or its index j if Y is indexed with respect to $\mathbb{R}^n$) we shall denote by $\partial_Y R$ (or $\partial_j R$) the simplex defined by the vertexes of R which have the lowest Y-coordinate (or j-th coordinate). Correspondingly, highest Y-coordinates define the simplexes denoted by $\partial^Y R$ (or $\partial^j R$). Observe that these simplexes are degenerate for $n > 2$.

The hit point problem is this:

Problem 1 *Given an integral curve, we have to decide whether and for which curve parameter value the curve will hit the initial set I. More precisely, since an integral curve $\int_X \mathfrak{Z}$ is defined starting from an event $X \in R$, we ask for the smallest parameter $|t|$ such that $\int_X \mathfrak{Z}(t) \in I$, i.e., $\int_X \mathfrak{Z}(t) \in S_i$, one of the initial set's simplexes.*

34.6.1 Distances

Summary. This section is dedicated to the calculus of distances between points on integral curves and polyhedral initial sets.

– Σ –

[7] The vector space $\mathbb{R}.s^i.$ is the module associated with the local composition defined by the vertexes $s^i.$ of S_i.

Given an polyhedral initial set I which is defined by a family $(S_i)_i$ of simplexes, and any point $X \in R$, the distance $d(X, I)$ is the minimum of the distances $d(X, S_i)$, therefore distances of points X and simplexes S must be calculated. Suppose that S is given by the sequence $(s_0, s_1, \ldots s_m)$ of points in $\mathbb{R}^n$.

If S is degenerate, it is the union of its non-degenerate sub-simplexes. This follows from the fact that it is a projection of a non-degenerate simplex, and that any point of this projection stems from a point of a side of the non-degenerate pre-image. So we are left with the calculation of the distance $d(X, S)$ for a non-degenerate S. If $dim(S) = n$, clearly, either X is in S or the distance is achieved on a point of ∂S. So in the latter case, we are in a recursive situation (the zero-dimensional simplex being trivial).

To decide upon the position of X relative to S, we consider the unique representation

$$X - s_0 = (X - s_0)^{\perp} + \sum_{j=1}^{m} \lambda_j.(s_j - s_0) \tag{34.4}$$

of the difference $X - s_0$ as a linear combination of the basis vectors $s_j - s_0$ and the vector $(X - s_0)^{\perp}$ of the (Euclidean) orthogonal space $\mathbb{R}.s.^{\perp}$. Considering the non-singular symmetric quadratic form of scalar products

$$Q = ((s_j - s_0) \cdot (s_i - s_0))_{i,j=1,\ldots m}$$

and the vectors $U = ((X - s_0) \cdot (s_i - s_0))^{T}_{i=1,\ldots m}$, $\Lambda = (\lambda_j)^{T}_{j=1,\ldots m}$, we have

$$\Lambda = Q^{-1}.U.$$

With this, the barycentric coordinates are defined by adding the coefficient $\lambda_0 = 1 - \Sigma_{i=1,\ldots m}\lambda_i$; we have

$$X = (X - s_0)^{\perp} + \sum_{i=0}^{m} \lambda_i.s_i,$$

and the component $y = \sum_{i=0}^{m} \lambda_i.s_i$ in the affine space spanned by S is in S iff $0 \leqslant \lambda_i$ for all $i = 0, \ldots m$. We now have $d(X, S) = \|(X - s_0)^{\perp}\| + d(y, S)$. If the second distance is not zero ($y \notin S$), we can proceed with the recursive calculation of $d(y, \partial S)$, and we are done. *Observe that this algorithm also gives us the coordinates of a point of S which has minimal distance to X.*

In computer programming practice, however, it is not reasonable to check for vanishing of distances $d(X, S)$ because of rounding and number representation errors. We therefore should prefer to calculate whether $d(X, S) < \epsilon$ for a selected positive neighborhood variable ϵ. The corresponding routines are obvious.

Denote the resulting point in S, which is found by this algorithm, and which has the shortest distance of points within S to X by $Min(X, S)$, whereas its barycentric coordinates are denoted by $Min(X, S)_i, i = 0, 1, \ldots m$, i.e.,

$$Min(X, S) = \sum_{i=0}^{m} Min(X, S)_i.s_i. \tag{34.5}$$

As a consequence, one can use this algorithm to calculate the distance of a straight line $L \subset \mathbb{R}^n$ to a simplex S as follows: Take the projection $p(S)$ of S on an affine hyperplane H_L orthogonal to L. Then $p(S)$ is defined by the projected points $p(s_i)$, and our algorithm applies to the singleton Y of the projection $p(L) = \{Y\}$ and to $p(S)$, giving $d(L, S) = d(Min(Y, p(S)), p(S))$. Moreover the coordinates $Min(Y, p(S))_i$ give us a point

$$Min(L, S) = \sum_{i=0}^{m} Min(Y, p(S))_i.s_i, \tag{34.6}$$

which evidently lives in S and has minimal distance to L. In our applications for integral curves, the line L is parametrized by a point X and a directional vector D, i.e., $L(t) = X + t.D$. Then the parameter λ on L such that we have $d(L(\tau), Min(L, S)) = d(L, S)$ is denoted by $\tau(X, D, S)$.

34.6.2 Flow Interpolation

Summary. This section deals with selection algorithms for searching points on an integral curve of the performance field which are successively approximated to initial sets.

– Σ –

After splitting the approximation to an initial set I to the approximation of one of its simplexes, the problem is this: We are given a positive ϵ, a point X, and a simplex S in $\mathbb{R}^n$, and we have to decide whether, and in which parameter value, the integral curve $\int_X \mathfrak{Z}$ hits the ϵ-neighborhood[8] $U_\epsilon S$ of S. Theoretically, one could just calculate the integral curve and search for a parameter that does the job—if there is any. But in computer programs, such a procedure is illusory. Numerical integration of ordinary differential equations is a time-consuming, expensive task. One cannot afford to calculate all curve points for any score event of a normal composition, which usually contains $10^4 - 10^5$ events. We shall see further in Section 39.4.4 that fields of any complexity may occur by chains of successive reshapings of given fields via arbitrary performance operators—same for the complexity of initial sets. Also the performance fields and the initial sets are of a completely arbitrary relative position.

So the situation is this: We are given a point X and a simplex S. We know in what direction—namely $\mathfrak{Z}_X$—the integral curve starts (with positive or negative curve parameter values). Nothing more. So we have to guess where the curve could approach S as well as possible. We have to evaluate the guess, and then start with another guess, etc. Eventually, we find a curve point in $U_\epsilon S$ and we are happy, or else we will have to give up the search and decide that the curve did not hit $U_\epsilon S$. This could be a wrong decision, but (calculation) time runs out and we have to resign. Intuitively, the search is best described by the following scenario: We have a paper dragon being suspended at a fixed position in the air. We also have a nervous fly, flying around in its random zig-zag (however differentiable) manner where you never know which will be the next turn. This scenario is traced on a video recorder, and the video really shows whether and when the fly hits the dragon. But we are not in the state of viewing the entire film. Rather are we given a determined sequence $VF(0) = (VideoFrame(t_{i_0}), VideoFrame(t_{i_0+1}))$ of two successive frames. This gives us the fly's position and its velocity vector at time t_{i_0} (supposing that the video is binocular...). We now have to guess, which video frame sequence could be the next best that shows the fly as near as possible to the dragon. We then forward the video to this next frame sequence $VF(1)$ and judge the new situation, and so on, until we find a hit point time or else we run out of time.

The first action is to guess a good curve time from the starting point X and the field vector $\mathfrak{Z}_X$. Denote by $x(t)$ the integral curve $\int_X \mathfrak{Z}(t)$. As nothing is known about the curve's future directions, we draw the straight line $L_X(t) = X + t.\mathfrak{Z}_X$ and look for the parameter value $t_1 = \tau(X, \mathfrak{Z}_X, S)$ defined above, which gives the nearest point to S on L_X, see Figure 34.3. We now have to compare $d(x_1, S)$, and $d(X, S)$. If the former is smaller than the latter, we may proceed, if not, we are in a bad position. Of course, this linear first approximation is not necessarily the best one to get off, since the field may be rather circular than linear in this region. We could therefore try other first approximations, such as a circle in the plane[9] spanned by the barycenter of S, X, and $\mathfrak{Z}_X$, for example the circle through the barycenter of S and X, and tangent to $\mathfrak{Z}_X$. We could follow this alternative, but we refrain from this because it does not demonstrate qualitatively new problems.

For the following interpolations, recall exercise 71 in Section 32.3.2 on cubic splines. This tells us that, given any two points $x(s), x(t)$ on our integral curve, there is exactly one cubic interpolation function $P_{s,t} : [s,t] \to \mathbb{R}^n$ with $P(s) = x(s), P(t) = x(t)$, and tangents $TP_s = \mathfrak{Z}_{x(s)}, TP_t = \mathfrak{Z}_{x(t)}$. We shall use these approximation curves to guess optimal points since we cannot calculate all curve points for an interval $[s,t]$.

We now proceed as follows. We are given the curve parameter t_1 whose point $x_1 = x(t_1)$ is nearer to S than x_0. We now repeat our linear approximation procedure with the line L_{x_1} and get a new curve parameter t^*. We also calculate the cubic interpolation point (this is a cheap calculation effort) for P_{t_0,t_1} at parameter $(t_0 + t_1)/2$. If $x(t^*)$ is nearer to S than x_1 and $P_{t_0,t_1}((t_0+t_1)/2)$, we set $t_2 = t^*$, else, if $P_{t_0,t_1}((t_0+t_1)/2)$ is

[8] This is $U_\epsilon S = \{x | \ d(x,S) < \epsilon\}$.

[9] Since the linear approximation failed, this must be a plane!

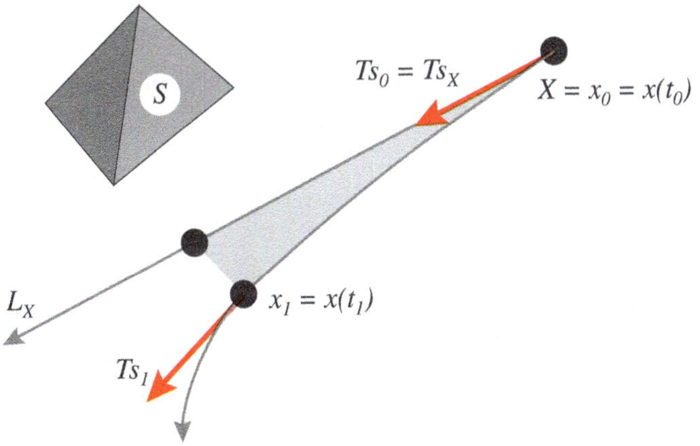

Fig. 34.3. The first approximation to a hit point on a simplex S when starting from the point X of the given composition is found by a linear approximation L_X.

nearer than x_1, we calculate $x((t_0 + t_1)/2)$, and, if this latter is still nearer than x_1, we set $t_2 = (t_0 + t_1)/2$. Else, we are stuck and have to quit.

The general situation is this: We have calculated a sequence $X = x_0, x_1, \ldots x_k$ of successively nearer points with curve parameters t_i, i.e., $x(t_i) = x_i$. Essentially we now have to check what happens to the left and to the right[10] of the best point x_k. Suppose that t_r, t_s are the right and left neighbor times to t_k. We now calculate the interpolation points $P_{t_k,t_r}((t_k+t_r)/2)$ and $P_{t_l,t_k}((t_l+t_k)/2)$. If one of them, with parameter t^*, is better than x_k, we calculate nearer one's curve point $x(t^*)$. If this is still nearer than x_k, we have found $t_{k+1} = t^*$. If not, we are stuck and give up.

This procedure will be repeated until a maximal admitted number σ of steps, and as long as the points $x_k, k \leqslant \sigma$ are not in $U_\epsilon S$. The procedure stops if we run out of steps or if we eventually hit the given neighborhood of the simplex. This algorithm has been implemented in the PerformanceRUBETTE® of the RUBATO® software, see Section 41.4.

It seems that the general situation of performing points X meets serious problems for points whose integral curve does not hit any initial point (or, at least, a point in a small neighborhood of the initial set I). This is however not so tragic since in reality, performance is not a one-step process: Stemma theory (Chapter 38) will show us that any performance is unfolded from a previous, less artistic performance, starting on the quasi-mechanical reproduction of the score. Therefore, if at a given stage, a point X cannot be performed, its performance which is already defined on a less artistic level can always be called in order to save the performability of the given score.

A final technical remark about the above algorithm: We implicitly supposed that the space where point X and simplex S live is the space of a performance field $\mathfrak{Z}$. However, we used only an integral curve through X, and not the performance field itself. *This means that we only needed a curve $x(t)$ through a point X, and no information about the curve's genealogy. So the algorithm is also valid for any such "isolated" curve.* This will be used later in our discussion of hierarchies of local performance scores in Chapter 35.

In the next chapter, we shall use this notation: If X is in the given frame R, we denote by $\int_X^- \mathfrak{Z}$ ($\int_X^+ \mathfrak{Z}$) the restriction of $\int_X \mathfrak{Z}$ to the maximal interval contained in the interval $]-\infty, 0]$ ($[0, +\infty[$) such that all values of the curve on this interval are contained in R. Given a performance field $\mathfrak{Z}$, $\epsilon \in \{+,-\}$, and an initial set I, we set:

$$I\int^{\epsilon} \mathfrak{Z} = \{X \in R |\ I \cap \int_X^{\epsilon} \mathfrak{Z} \neq \emptyset\} \qquad (34.7)$$

Here is a sorite concerning this symbol:

[10] It could happen that x_k is the right or left extremal one. Then, we have to make linear interpolation on that side, but this situation has been explained for the construction of the third point x_2 above.

Sorite 11 *In the following statements, direct products of performance fields* $\mathfrak{S}_1, \mathfrak{S}_2$*, initial sets* I_1, I_2*, and frames* R_1, R_2 *refer to the product (limit) space* SS *built from the given factors* SS_1, SS_2*.*

(i) *The operator* $I \mapsto I \int^{\epsilon} \mathfrak{S}$ *conserves inclusions, commutes with unions and is idempotent.*
(ii) *Setting* $I_1 \int^{\epsilon} I_2 = I_1 \times (I_2 \int^{\epsilon} \mathfrak{S}_2) \cup (I_1 \int^{\epsilon} \mathfrak{S}_1) \times I_2$*, we have*

$$(I_1 \int^{\epsilon} I_2) \int^{\epsilon} \mathfrak{S}_1 \times \mathfrak{S}_2 = I_1 \int^{\epsilon} \mathfrak{S}_1 \times I_2 \int^{\epsilon} \mathfrak{S}_2.$$

Proof. Statement (i) is straightforward by the uniqueness of integral curves. We show statement (ii) for $\epsilon = -$, the other case is analogous. For the inclusion "$\subset$", suppose WLOG that we have $(X, Y) \in (I_1 \times (I_2 \int^{-} \mathfrak{S}_2)) \int^{-} \mathfrak{S}_1 \times \mathfrak{S}_2$. This means that we can reach a point of $I_1 \times (I_2 \int^{-} \mathfrak{S}_2)$ from (X, Y) on the integral curve $\int^{-}_{(X,Y)} \mathfrak{S}_1 \times \mathfrak{S}_2$. Since integral curves of a product of vector fields project to integral curves of their factors, and by the idempotency of $\int^{-} \mathfrak{S}_2$, this implies $(X, Y) \in I_1 \int^{-} \mathfrak{S}_1 \times I_2 \int^{-} \mathfrak{S}_2$. As to the inclusion "$\supset$", suppose that (X, Y) is such that $\int_X \mathfrak{S}_1(s) \in I_1, V = \int_Y \mathfrak{S}_2(t) \in I_2$, and WLOG $s \leqslant t \leqslant 0$. Setting $W = \int_X \mathfrak{S}_1(t)$ we conclude $(W, V) \in I_1 \int^{-} \mathfrak{S}_1 \times I_2$. By the following lemma, we can reach (X, Y) from (W, V) on an integral curve of $\mathfrak{S}_1 \times \mathfrak{S}_2$ from t to 0, and we conclude $(X, Y) \in (I_1 \int^{-} I_2) \int^{-} \mathfrak{S}_1 \times \mathfrak{S}_2$, QED.

Lemma 50 *If we are given two locally Lipschitz vector fields* $\mathfrak{S}_1, \mathfrak{S}_2$ *on respective domains* D_1, D_2 *and two (maximal) integral curves* $x_1 : J_1(0) \to D_1, x_2 : J_2(0) \to D_2$*, the diagonal curve* $x : J_1(0) \cap J_2(0) \to D_1 \times D_2 : t \mapsto (x_1(t), x_2(t))$ *is the (maximal) integral curve of* $\mathfrak{S}_1 \times \mathfrak{S}_2$ *on* $D_1 \times D_2$ *through the couple* $x(0) = x_1(0), x_2(0))$ *of initial points.*

Proof. Clearly, the diagonal is an integral curve since differentiation goes factorwise. On the other hand, if we had a proper extension of x in the product space, its projections $p_1 \circ x, p_2 \circ x$ would yield two integral curves, one of which would have an extended domain, which contradicts the choice of x_1, x_2, QED.

35

Hierarchies and Performance Scores

On trouve toujours l'homogène à un certain degré de division.
Paul Valéry [1073, I, p.209]

Summary. As a synthesis of the structural parts described in Chapter 32 through Chapter 34 we establish the overall structure of performance. The objects of performance structure are understood as being an *additional* score type, called *performance score*, layered over the given "symbolic" score like a system of optical lenses which 'deform' the rigid configuration of note symbols. The performance score is a *global* object built from an atlas of local performance scores. A local performance score is built from a hierarchy of performance cells, the very core of performance structures. We first describe the category of performance cells. Local performance scores are defined by a hierarchical construction principle: They are particular diagrams in the category of performance cells. The conceptual and musical background of such *local hierarchies* is evidenced through a series of examples, including the piano and violin hierarchies. We end up with the definition and exemplification of the concept of a (global) performance score.

– Σ –

35.1 Performance Cells

Summary. Performance cells are the very local data of performance. They comprise the cell's *frame* (a domain of definition), the *symbolic kernel* (a set of prima vista objects), the *performance field*, the *initial set*, and the *initial performance*.

– Σ –

The innermost local structure of performance is the performance cell. We have known all of its ingredients and shall now set up the formal definition of such a cell.

Definition 101 *Given $\epsilon \in \{+,-\}$, a performance cell is a five-tuple*

$$C = (K, R, \mathfrak{Z}, I, \wp_I)$$

consisting of the following objects:

1. *a local composition $K \subset SS$ whose space is simple of underlying module $\mathbb{R}^n$, K is called the* symbolic kernel *of the cell;*
2. *a closed frame $R = [a_1, b_1] \times \ldots [a_n, b_n]$ contained in $\mathbb{R}^n$ and containing K;*
3. *a locally Lipschitz-continuous performance field $\mathfrak{Z}$ which is defined*[1] *on R;*

[1] Recall that this means that $\mathfrak{Z}$ is defined on an open neighborhood of R, but we only identify the field on R.

G. Mazzola, *The Topos of Music II: Performance*, Computational Music Science,
https://doi.org/10.1007/978-3-319-64444-8_4

4. *an initial set I such that $K \subset I\int^{\epsilon}\mathfrak{Z}$ (see equation (34.7));*
5. *an initial performance $\wp_I : I \to PS$, with codomain a physical space PS with module $\mathbb{R}^n$, and such that for any point $X \in K \cap I\int\mathfrak{Z}$ and any two points $a = \int_X^{\epsilon}\mathfrak{Z}(\alpha), b = \int_X^{\epsilon}\mathfrak{Z}(\beta)$ in I, we have $\wp_I(b) - \wp_I(a) = (\alpha - \beta).\Delta$, where $\Delta = (1, \ldots 1)$ is the diagonal vector in PS's $\mathbb{R}^n$.*

Often, I is given as a union of a finite family $(S_i)_i$ of possibly degenerate simplexes S_i, but this is not the general case. A performance cell is visualized by a tetrahedron as shown in Figure 35.1. We therefore also call $(R, \mathfrak{Z}, I, \wp_I)$ a *performance body for K*.

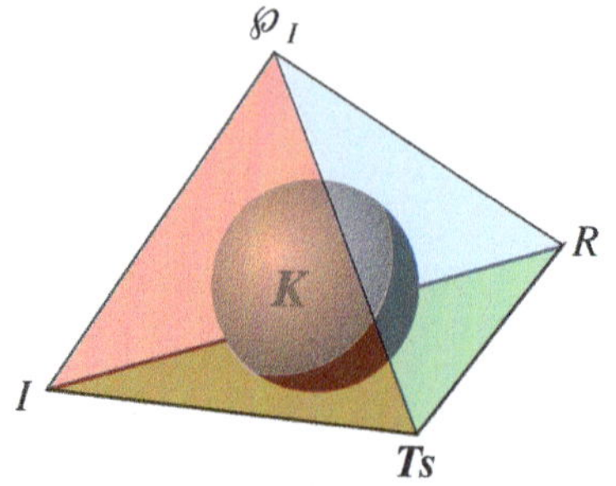

Fig. 35.1. Visualization of a performance cell by a tetrahedron, the inner ball symbolizes the cell's symbolic kernel, whereas the 4-tuple of the tetrahedron's vertexes is called the *performance body for the symbolic kernel.*

Given a performance cell C, we automatically have a well-defined performance map $\wp_C : K \to PS$ which we shall always refer to when talking about the performance $\wp$ associated with C.

Without stressing the contrary, all our cells will be $-$-cells, the theory for $\epsilon = +$ is the same. We shall only occasionally consider mixed signatures.

35.2 The Category of Performance Cells

Summary. Performance cells constitute the objects of the category *PerCell* of performance cells, the morphisms being the technical expression of relations between performance data on different parameter spaces.

– Σ –

In nuce, we have already seen phenomena of related performance cells in the discussion of parallel fields as they arise in the $Onset \oplus Duration$ space. There, we had the projection $p_{Onset} : Onset \oplus Duration \to Onset$ which was compatible with the parallel field $\partial Tempo$ and the tempo curve $Tempo$.

We shall now set up the formal statement behind those incipits. What we want is a concept of a morphism $C_1 \to C_2$ which defines the *category PerCell of performance cells* such that the associated performances are compatible with this morphism.

Definition 102 *(Recall that we have fixed the signature $\epsilon = -$ in the following discussion!) Let $C_1 = (K_1, R_1, \mathfrak{Z}_1, I_1, \wp_{I_1})$ and $C_2 = (K_2, R_2, \mathfrak{Z}_2, I_2, \wp_{I_2})$ be two performance cells living in spaces SS_1 and SS_2. Suppose further that we have a projection $p : SS_1 \to SS_2$ of the parameter space SS_1 onto the parameter space SS_2 such that the underlying projection $p : \mathbb{R}^{n_1} \to \mathbb{R}^{n_2}$ is the projection onto a subset of coordinates. Then p is a* morphism of performance cells *$p : C_1 \to C_2$ iff the following conditions are verified:*

1. *$p(K_1) \subset K_2$ (in other words: $p : K_1 \to K_2$ is a morphism of local compositions);*
2. *$p(R_1) \subset R_2$;*
3. *$Tp \circ \mathfrak{Z}_1 = \mathfrak{Z}_2 \circ p$, i.e., a morphism of vector fields $p : \mathfrak{Z}_1 \to \mathfrak{Z}_2$;*
4. *$p(I_1) \subset I_2\int\mathfrak{Z}_2$;*

5. *$p \circ \wp_{I_1} = \wp_2 \circ p|I_1$ (here, p denotes the corresponding projection on the physical spaces $p : PS_1 \to PS_2$);*

Lemma 51 *With the above notation, if $p : C_1 \to C_2$ is a morphism, then we have*

$$p \circ \wp_1 = \wp_2 \circ p. \tag{35.1}$$

with the homonymous p symbol for the mental and physical projections.

Proof. Let $X \in K_1$, and suppose that $\int_X \mathfrak{Z}_1$ hits I_1 at the point $Y = \int_X \mathfrak{Z}_1(t)$. (By axiom 5 of the definition 101 of a performance cell, it does not matter, which hit point we are selecting.) Then, $\wp_1(X) = \wp_{I_1}(Y) - t.\Delta_1$, and by linearity of p, $p \circ \wp_1(X) = p \circ \wp_{I_1}(Y) - t.\Delta_2$. But $p \circ \wp_{I_1}(Y) = \wp_2(p(Y)) = \wp_{I_2}(Z) - s.\Delta_2$, if Z is the hit point for $p(Y)$ in I_2, s is the curve parameter for this hit point. Hence, $p \circ \wp_1(X) = \wp_{I_2}(Z) - (s+t).\Delta_2$. But integral curves of $\mathfrak{Z}_1$ are projected into integral curves of $\mathfrak{Z}_2$ by axiom 3 of definition 102 and by the uniqueness of integral curves (see the fundamental theorem of ODE 86, Appendix I.2.2). Therefore, the curve parameter $s+t$ is also the parameter of the integral curve $\int_{p(X)} \mathfrak{Z}_2$ where it hits Z, and this means that $\wp_{I_2}(Z) - (s+t).\Delta_2 = \wp_2(p(X))$, QED.

This means that performances of performance cells behave in the sense of natural transformations with respect to the category *PerCell.*

Exercise 78 *Work out the example of a parallel articulation field projection and its associated performance body data, as discussed in Section 34.5, to obtain a morphism of performance cells.*

The next proposition gives us means to construct product performance bodies and product cells from two given cells.

Proposition 58 *Suppose that we are given two performance cells*

$$C_1 = (K_1, R_1, \mathfrak{Z}_1, I_1, \wp_{I_1}), C_2 = (K_2, R_2, \mathfrak{Z}_2, I_2, \wp_{I_2})$$

living in spaces SS_1 and SS_2. Let SS be the product of SS_1 and SS_2, with projections $p_i : SS \to SS_i, i = 1, 2$. Then the following data defines a performance cell C on SS:

1. *the frame is $R = R_1 \times R_2$;*
2. *the field is $\mathfrak{Z} = \mathfrak{Z}_1 \times \mathfrak{Z}_2$;*
3. *the symbolic kernel is any sub-composition $K \subset K_1 \times K_2$;*
4. *the initial set is $I = I_1 \int^{\epsilon} I_2$;*
5. *on I, the initial performance $\wp_I$ is defined as follows: if $(x, y) \in I_1 \times I_2 \int^{\epsilon} \mathfrak{Z}_2$, then we set $\wp_I(x, y) = (\wp_{I_1}(x), \wp_2(y))$, and symmetrically for the other case $(x, y) \in I_1 \int^{\epsilon} \mathfrak{Z}_1 \times I_2$.*

With his data,we have two morphisms $p_i : C \to C_i, i = 1, 2$ of performance cells.

Proof. As to the construction of C, we only have to check the relation $K \subset I \int^{\epsilon} \mathfrak{Z}_1 \times \mathfrak{Z}_2$. From sorite 11, we know that $I \int^{\epsilon} \mathfrak{Z}_1 \times \mathfrak{Z}_2 = I_1 \int^{\epsilon} \mathfrak{Z}_1 \times I_2 \int^{\epsilon} \mathfrak{Z}_2$. But the latter evidently contains $K_1 \times K_2$, and a fortiori K, whence the claim. The morphisms p_i are now straightforward by construction, QED.

The performance cell C, together with the two morphisms p_1, p_2 will be called the *product of the cells* C_1, C_2 and denoted by $C_1 \times C_2|K$, or simply $C_1 \times C_2$ if $K = K_1 \times K_2$. Although the initial set $I_1 \int^{\epsilon} \mathfrak{Z}_1 \times I_2 \int^{\epsilon} \mathfrak{Z}_2$ of the product performance cell is not a union of simplexes in general, the product formula of sorite 11 guarantees that the calculation method for hit points on simplexes discussed in Section 34.6.2 may be applied to each component of an event $X = (X_1, X_2)$ if the factor cells have initial sets which are built from simplexes.

35.3 Hierarchies

Summary. Hierarchies are space diagrams arising in local performances. They trace the functional organization among the arguments of performance fields.

– Σ –

In order to obtain more concrete results, we shall first consider a particular system of symbolic and physical spaces. We suppose that we are given a series $B_i, i = 1, 2, \ldots$ of pairwise different simple spaces, called *basis spaces*, as well as an equipollent series $P_i, i = 1, 3, \ldots$ of spaces, called *pianola spaces*, with $P_i \xrightarrow[Id]{} \mathbf{Syn}(B_i)$, while $B_i \xrightarrow[Id]{} \mathbf{Simple}(\mathbb{R}^{n_i})$. Often, we shall consider finite products $B_{i_1} \times B_{i_2} \times \ldots B_{i_k} \times P_{j_1} \times \ldots P_{j_l}$ of such spaces, and always ordered with basis spaces first, and pianola spaces second, and each space type ordered by increasing index (no repetitions!). This means that we in fact parametrize such products with finite subsets of the name set $BP = B \cup P, B = \{B_i | i = 1, 2, \ldots\}, P = \{P_i | i = 1, 2, \ldots\}$.

We shall then identify these product spaces with the simple space $B_{i_1} \oplus B_{i_2} \oplus \ldots B_{i_k} \oplus P_{j_1} \oplus \ldots P_{j_l} \xrightarrow[Id]{} \mathbf{Simple}(\mathbb{R}^N)$ associated with the direct sum $\mathbb{R}^N$ of all involved copies of real vector spaces. If the defining sequence $B_{i_1}, \ldots B_{i_k}, P_{j_1}, \ldots P_{j_l}$ is denoted by U, we shall also rename the space $B_{i_1} \oplus B_{i_2} \oplus \ldots B_{i_k} \oplus P_{j_1} \oplus \ldots P_{j_l}$ by $\oplus U$. By definition, the space associated with the empty sequence $\varnothing$ is the simple space $\oplus\varnothing \xrightarrow[Id]{} \mathbf{Simple}(\mathbb{R}^0)$ of the zero module.

This generalizes the nomenclature introduced in the description of standard spaces (see equations (6.69) ff.), such as the piano space $Onset \oplus Pitch \oplus Loudness \oplus Duration \xrightarrow{\sim} Onset \times Pitch \times Loudness \times Duration$ referring to (symbolic) onset, pitch, loudness, and duration, and represented by $\mathbb{R}^4$. For any subsequence of symbols $V = B_{u_1}, B_{u_2}, \ldots B_{u_r}, P_{v_1}, \ldots P_{v_s}$ of $U = B_{i_1}, B_{i_2}, \ldots B_{i_k}, P_{j_1}, \ldots P_{j_l}$, we have a canonical projection

$$p_{U,V} : \oplus U \to \oplus V$$

of such standard spaces, also denoted by p_V or p, if no ambiguities are possible.

For any two such sequences V, W, we denote by $V \cup W$ ($V \cap W$) the sequence defined by the union (intersection) of the basis and pianola symbols in BP. Similarly for other set-theoretic operations, such as complement in BP, differences $V - W$, etc. This means that we consider the Boolean algebra $Sub(BP)$. When considering the spaces $\oplus U$ associated with such sequences U, we often speak loosely about the sequences and mean the spaces, for example, we speak of the "union of spaces" $\oplus U, \oplus V$ and mean the space $\oplus(U \cup V) = (\oplus U) \oplus (\oplus V)$, etc., but no confusion should occur...

Definition 103 *Given a finite space collection BP, a* space hierarchy *in BP is a non-empty sub-lattice $H \subset Sub(BP)$ (closed under finite unions and finite intersections), with maximal element (the* top space*) $Top(H)$; H is viewed as a category with $U \to V$ iff $V \subset U$. The minimal non-empty elements of H are called* fundamental spaces*, their set is denoted by $Fund(H)$. For any non-empty space $U \in Sub(BP)$ which is contained in $Top(H)$, we denote by $Cl_H(U)$ the unique smallest space in H which contains U, and call it the* hierarchy closure *of U.*

A hierarchy space $U \in H$ is called indecomposable *if it is not the union of two disjoint non-empty subspaces of H.*

The motivation for the hierarchy concept is that we want to consider systems of performance cells which are related by morphisms, such as the parallel field morphisms or the product cells and their projections. Moreover, we have to group performance cells which are dominated by a root cell since all score events will be performed in one big space whose projections are however compatible with the root space.

Lemma 52 *Given a space hierarchy H, let $Fund(H) = \{F_1, \ldots F_k\}$ be the set of fundamental spaces of H. Then*

(i) *Each F_i is contained in a unique maximal indecomposable space $T(F_i)$ of H.*
(ii) *Any two different maximal indecomposable spaces in H are disjoint.*

(iii) The top space $Top(H)$ is the disjoint union of its maximal indecomposable subspaces.

Proof. The union of two indecomposable superspaces of each F_i in H is indecomposable, therefore there is a unique maximal indecomposable superspace $T(F_i)$ for each F_i. Evidently, the maximal indecomposable spaces are mutually disjoint, i.e., for $T(F_i) \neq T(F_j)$, $T(F_i) \cap T(F_j) = \oplus\emptyset$. The existence of a decomposition of $Top(H)$ as a product of maximal indecomposable subspaces follows by induction on the cardinality of $Top(H)$, QED.

The set of fundamental spaces contained in one maximal indecomposable space of H are called *the blocks* of the fundament. Hence the blocks are in bijection with the maximal indecomposable subspaces.

Definition 104 *If an indecomposable space contains a unique (proper) maximal subspace—including the zero space—it is called* irreducible.

Hence, every indecomposable space is either irreducible or it is the union of its maximal subspaces. Fundamental spaces are irreducible, the zero space is not.

With this we may now proceed to define the performance hierarchies. To this end, we consider the zero performance cell $C_\emptyset$ which is defined by a frame over the zero space $\oplus\emptyset$ and has everything trivial: unique zero field, trivial frame, one-point initial set, zero initial performance. Here is the relation of space hierarchies to performance theory (see also Figure 35.2):

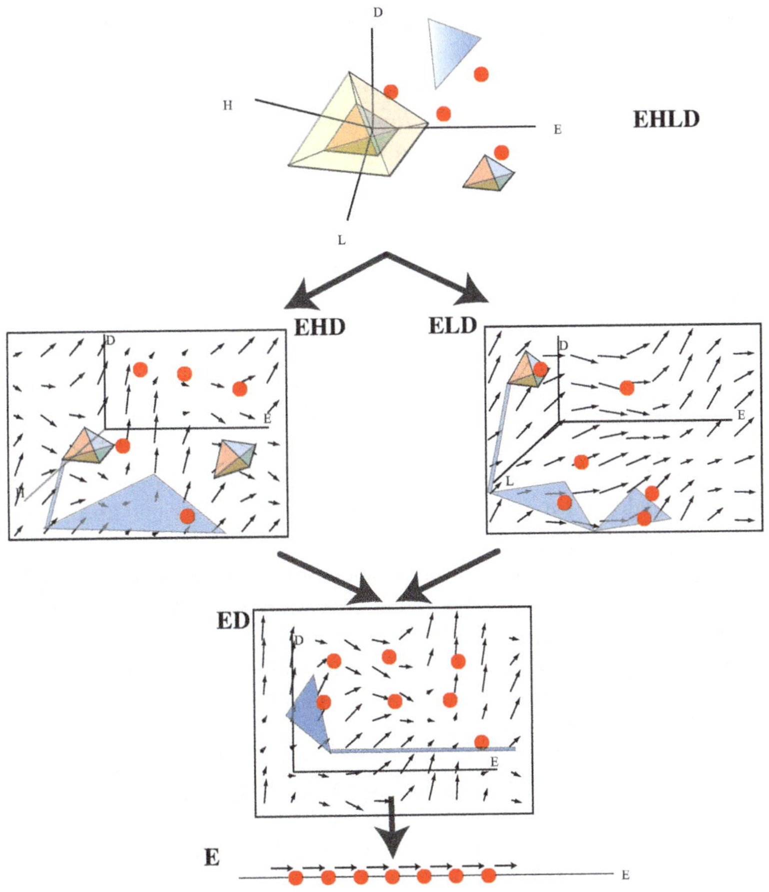

Fig. 35.2. A cellular hierarchy is shown, together with the initial sets (blue simplexes), the performance fields, the projections, and the symbolic kernels (red points).

Definition 105 *Given a space hierarchy H, a* cellular hierarchy *is a diagram (in fact, a functor on the category H) $\mathfrak{h} : H \to PerCell$ with values in the category of cells $PerCell$ such that for each $U \in H$, $\mathfrak{h}(U)$ is a cell in the space $\oplus U$, with morphisms being denoted by $\mathfrak{h}(p_{U,V})$, $p_{U,V}$, p_V, or p if no ambiguities are likely. The domain H of $\mathfrak{h}$ is called* the type *of the cellular hierarchy.*

The meaning of a cellular hierarchy is that we are given a performance on the kernel of the top cell $Top(\mathfrak{h}) = \mathfrak{h}(Top(H))$, which is compatible with the performances on all other cells $\mathfrak{h}(U)$ of $\mathfrak{h}$. So the parameters of top kernel events which are also grouped on lower hierarchy spaces can be performed independently of the other parameters. Correspondingly, the performance field components grouped on a lower hierarchy space are independent of the other space parameters, a situation already encountered for the articulation hierarchy $Onset \oplus Duration \to Duration$, for example. Moreover, if we have two cells $\mathfrak{h}(U)$, $\mathfrak{h}(V)$, their performance fields $\mathfrak{Z}_U, \mathfrak{Z}_V$ evidently define performance fields $\mathfrak{Z}_{U \cup V}, \mathfrak{Z}_{U \cap V}$ on the arguments of the union and intersection of their spaces, and this is met by the lattice structure of H.

The signification of a space being irreducible now also becomes more evident: If a hierarchy space is the union of two proper subspaces, any coordinate of the performance of an event may be calculated via projection of the event into an appropriate subspace. However, for an irreducible space, we have to make an extra effort to know how its kernel events are performed since no subspace will give us full information. The parallel field construction from Section 33.2.1 is a standard example for this: The duration component of the articulation field is not reducible to the cell of a subspace, whereas the onset component is. So the hierarchy does not tell us how to compute the duration component. The parallel construction is one possibility among an infinity of others to deduce the duration component from the fundamental tempo field by means of a special "formula" which in more generality has been explicated in formula (33.10).

If we are given a space hierarchy H with $Top(H) = \oplus BP = \oplus B \oplus P$, *we shall henceforth assume that for each pianola component P_i of P, the corresponding basis component B_i is contained in the basis B, whereas a basis component B_i may happen to live alone without its pianola counterpart in P. Such hierarchies are called* standard, *if we have to distinguish them from other, non-standard hierarchies.* For a standard space hierarchy, we therefore always have the alteration projections $\alpha_{\pm} : \oplus BP \to \oplus B$. It is the usual alteration on the couples $\oplus B_i \oplus P_i$, and the identity on the unparalleled basis components.

35.3.1 Operations on Hierarchies

Summary. Hierarchies can be altered and recombined according to a set of standard operations. Such operations intervene in practical calculations of performance transformations in software algorithms.

$- \Sigma -$

Given a cellular hierarchy $\mathfrak{h}$, if U is any subspace of the underlying space hierarchy H, we may restrict $\mathfrak{h}$ to the sublattice $H|U$of H whose top space is U. We may then restrict the cellular hierarchy $\mathfrak{h}$ to this sublattice. The restriction is denoted by $\mathfrak{h}|U$.

If we are given two cellular hierarchies $\mathfrak{h}, \mathfrak{k}$ with space hierarchies H, K, and whose spaces pertain to the given space collection BP, and such that their top spaces are disjoint, then we have the product cellular hierarchy $\mathfrak{h} \times \mathfrak{k}$ whose domain is the space lattice with spaces $U \cup V, U \in H, V \in K$. Clearly, in this product space hierarchy, a pair $U \cup V$ extends $U' \cup V'$ iff each component does so in the respective space hierarchy H, K, respectively. So for each such relation and the associated morphism $p : U \cup V \to U' \cup V'$, we have the corresponding morphism of performance cells

$$p : \mathfrak{h}(U) \times \mathfrak{k}(V) \to \mathfrak{h}(U') \times \mathfrak{k}(V').$$

The only non-trivial statement within this fact concerns property 4 of definition 102 and follows from the idempotency of the operator $?\int^{\epsilon} \mathfrak{Z}$ (sorite 11, (ii)).

Given a space U in the name space B of basis spaces, we call a *parallel space to U* and denote by ∂U the space $U \cup P|U$ consisting of U and the space $P|U$ of all the corresponding pianola space components. If a cellular hierarchy $\mathfrak{h}$ is such that for each projection $\partial U \to U$ of spaces in its space hierarchy H, the performance field over ∂U is the parallel field $\partial\mathfrak{Z}$ to the field $\mathfrak{Z}$ of U (i.e., of the cell $\mathfrak{h}(U)$) in the sense defined in formula (33.10), then we say that $\mathfrak{h}$ *is a parallel hierarchy.* Parallel hierarchies are the default hierarchies where performance is initiated.

35.3.2 Classification Issues

Summary. For small sets of parameters, frame structures of cellular hierarchies are completely classified. We describe the classification for hierarchies involving tempo, intonation, and dynamics.

– Σ –

We do not claim classification of the full-fledged cellular hierarchies, this is a much too difficult task, and it is not of primary interest; it is easier and perhaps more relevant for practical reasons to classify "frame structures" for concrete cellular hierarchies. Such a frame structure is the hierarchical organization of the hierarchy's performance fields. We shall present a complete classification of hierarchies sitting over the set $B = \{E, H, L\}$ of the three usual basis spaces $E = Onset, H = Pitch, L = Loudness$. For each hierarchy space $U = E, H, L, EH, EL, HL, EHL$, we denote the corresponding field as built from the symbols $T = Tempo, I = Intonation, D = Dynamics$. Hence, T is the field over E, I over H, D over L, whereas TD denotes the field over EL, etc., and TID the field over EHL. We write $TI \times D$ for a product field corresponding to the space EHL which is decomposable into the subspaces EH and L in the given hierarchy $\mathfrak{h}$. Figure 35.3 shows the classification Hasse diagram, with a straight line from every hierarchy to its next specializations. Here, specialization means that field components become independent from certain parameters with increasing split space hierarchies.

For a performance field $\mathfrak{Z} = \mathfrak{Z}_U, U \in H$ of a cellular hierarchy $\mathfrak{h}$ with space hierarchy H, we may ask for its functional dependence within $\mathfrak{h}$. Let $\underline{\mathfrak{Z}} = \mathfrak{Z}_{\underline{U}}$ be the field of the maximal subspace $\underline{U}$ of U in H. This is called the *territory of* $\mathfrak{Z}$, and describes the portion of $\mathfrak{Z}$ which is completely determined by lower hierarchy data. By the *extraterritorial part of* $\mathfrak{Z}$, denoted by $\mathfrak{Z}_{ex}$, we mean the projection of $\mathfrak{Z}$ to the complementary space $U - \underline{U}$, so that we have $\mathfrak{Z} = (\underline{\mathfrak{Z}}, \mathfrak{Z}_{ex})$. In general, the extraterritorial part of $\mathfrak{Z}$ is a function of its territory and of a system $Para$ of parameters which are external to the hierarchy; we write $\mathfrak{Z}_{ex} = \mathfrak{Z}_{ex}(\underline{\mathfrak{Z}}, Para)$ to indicate this dependency. So the performance fields of the fundamental spaces, the *fundamental fields*, play a primordial role in the construction of the hierarchical architecture, which is enriched by external parameters in the case of irreducible spaces (for others, the parameters are non-existent).

Example 51 *Suppose we are given a parallel hierarchy with just one parallel projection $\partial\mathfrak{Z} \to \mathfrak{Z}$ over the space projection $\partial U \to U$, according to the formula $\partial\mathfrak{Z} = (\mathfrak{Z} \circ p_U, 2.\mathfrak{Z} \circ \alpha_+ - \mathfrak{Z} \circ p_U)$ (33.10). Here, we have the empty parameter set $Para = \emptyset$, and $\underline{\partial\mathfrak{Z}} = \mathfrak{Z}, \partial\mathfrak{Z}_{ex} = 2.\mathfrak{Z} \circ \alpha_+ - \mathfrak{Z} \circ p_U$. If we generalize this configuration by the extraterritorial part, then*

$$\partial\mathfrak{Z}_{\lambda,ex} = \frac{1+\lambda}{\lambda}.\mathfrak{Z} \circ \alpha_+ - \mathfrak{Z} \circ p_U.$$

The parameter set is $Para = \{\lambda\}$, and we recover the old situation by a specialization $\lambda \to 1$. We shall see in the discussion of performance operators that this parametrization is a natural one in the context of operators for articulation (legato, staccato).

These concepts apply to the standard situations while performing special musical effects which we list here:

1. Pianola deformations[2] of parallel pianola fields of tempo:

$$\mathfrak{Z} = (T, Z_{ex}(T, Para))$$

[2] Deformation and specialization are reciprocal processes. The first is the embedding of a particular structure in a topologically dominant set of variants (usually in some Zariski topology of algebraic geometry), the second is the restriction of an irreducible set of variants to a special, closed subset, or even a single point, usually by specialization of parameters, see also Appendix F.

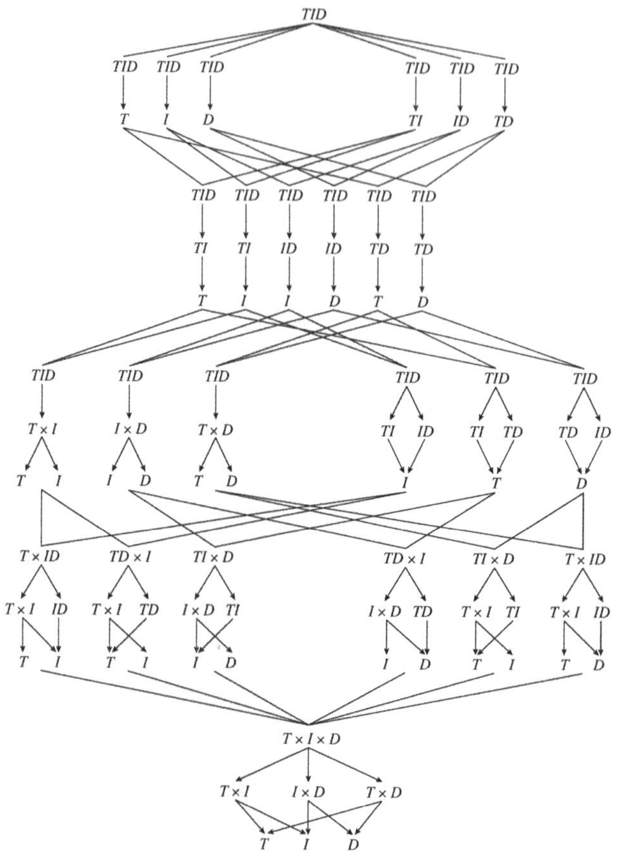

Fig. 35.3. The complete classification Hasse diagram of basis hierarchy frame structures for onset (E), pitch (H), and loudness (L) in terms of corresponding fields, including specialization (straight lines) to more split fields according to functional independency of parameters.

with specialization to ∂T. This is used for articulation as discussed above, for local performance cells of ornaments and ties.

2. Deformation of dynamics over tempo:

$$TD = (T, TD_{ex}(T, Para))$$

with specialization to $T \times D$. This is used for ondeggiando (bow vibrato) effects and prima vista dynamics accentuation following bar-lines and time signatures.

3. Deformations of tempo over intonation:

$$TI = (I, TI_{ex}(S, Para))$$

with specialization to $T \times I$. This is used for arpeggios if we agree to quit simultaneity of arpeggio events and to change their physical onsets.

The last two deformation types are shown in the lower part of Figure 35.3. More systematically, this classification shows these types of specialization (we refer to Figure 35.3 in this list):

1. Type: shown on the lowest line, no extraterritorial part, only the one-dimensional fundamental fields T, I, D.

2. Type: shown on the second lowest line, on top no extraterritorial part, in the middle of the hierarchy one two-dimensional field (e.g.,
$$ID = (D, ID_{ex}(D, Para))$$
in the right extreme hierarchy) with a one-dimensional territory (in the example: D), and two fundamental spaces (in the example: T, D).
3. Type: shown on line three from below. One dimension is extraterritorial and we have two fundamental one-dimensional fields (e.g., D in the left extreme hierarchy, with[3]
$$TID = (T, I, TID_{ex}(T \times I, Para))$$
and the fundamental fields T, I.)
4. Type: line three from above. Only one fundamental field, above which a one-dimensional extraterritorial field component is built, and the same for the top space, it is a one-dimensional extension of its two-dimensional territory (e.g., the left extreme hierarchy with $TID = (TI, TID_{ex}(TI, Para)), TI = (T, TI_{ex}(T, Para)))$.
5. Type: second line from top, right half. The fundament is two-dimensional and the top field has a one-dimensional extraterritorial part (e.g., the right extreme hierarchy with
$$TID = (TD, TID_{ex}(TD, Para)).$$
6. Type: second line from top, left half. The fundament is a one-dimensional field, and the top field has a two-dimensional extraterritorial part (e.g., the left extreme hierarchy with
$$TID = (T, TID_{ex}((T, Para)).$$
7. Type: top line. Here we are left with the single total field TID without any subspaces; it is at the same time its own irreducible territory and fundamental field.

Nonetheless, complete classification of cellular hierarchies should be a major issue of future research, although a difficult one. To this end, one notices that a morphism $\mathfrak{h} \to \mathfrak{k}$ between cellular hierarchies should be defined as a natural transformation of these functors which, by definition, starts from one and the same space hierarchy.

However, the "horizontal" natural morphisms of performance cells should be generalized beyond simple projections of parameter spaces since it is natural to say that a performance cell is isomorphic to another cell, if the second is generated by any reasonable[4] diffeomorphism between their frames. So cellular hierarchies have only projections as morphisms, while the general concept of a morphism between performance cells is a more general one. We cannot digress on this issue since nothing non-trivial is known to the date. Let us nonetheless denote the category of cellular hierarchies by $\mathfrak{H}$ although the morphisms are not made precise here, whereas the objects are.

35.3.3 Example: The Piano and Violin Hierarchies

Summary. We describe the default hierarchy associated with piano and with violin scores.

– Σ –

Figure 35.4 shows the default hierarchy for piano music. This is the hierarchy which one has to start with, when shaping performance. We see that pianola parameters are not given, except for duration, since on the piano, glissando or crescendo parameters for single notes are not feasible. The basis fields are also completely independent (no coupling), and we have no parameters for extraterritorial parts, since the parallel field ∂T is determined by the underlying territory T. Figure 35.5 shows the default hierarchy for violin music.

---
[3] We just write down the whole sequence of territorial fields in this formula.
[4] For example an affine deformation of the frames.

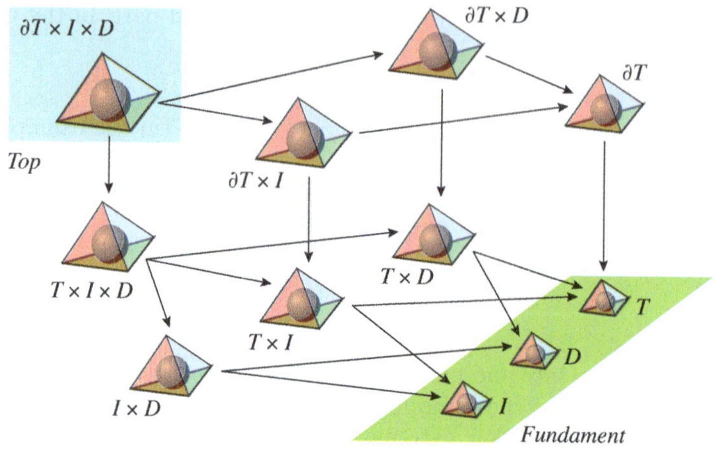

Fig. 35.4. The default piano cellular hierarchy.

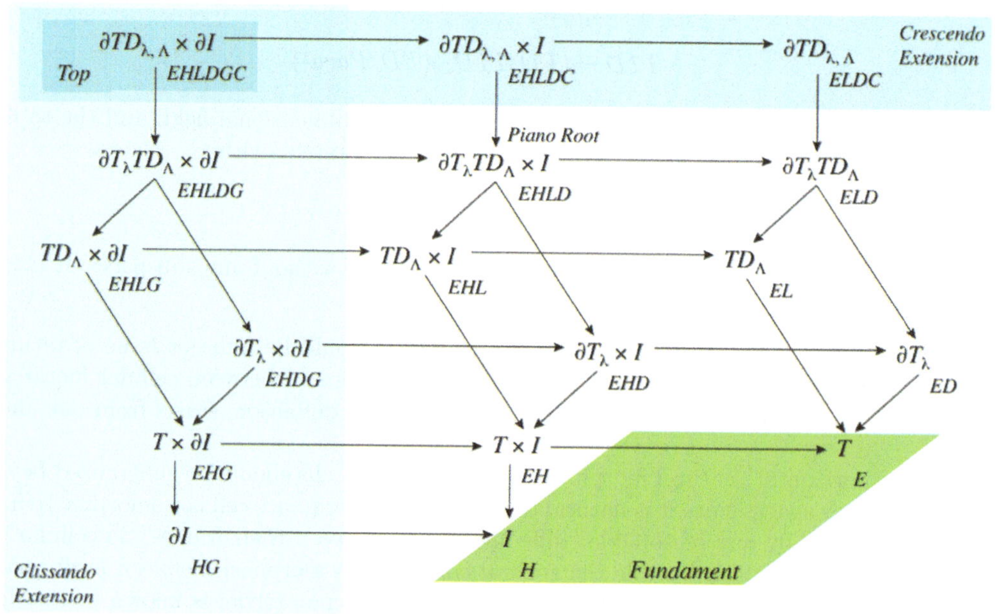

Fig. 35.5. The default violin cellular hierarchy.

It is first characterized by a double parameter extension: on one hand, we now have the crescendo extension, on the other, we have the glissando extension. Thirdly, the fundament is reduced to onset E and pitch H. This system also includes two primavista parameters λ, Λ. The point here is that there are primavista predicates for violin, such as "ondeggiando", an E-periodic change of loudness, which ask for refined parametrization already in primavista hierarchies. Same for the articulation field ∂T_λ. Details of these performance operators will be explained in Chapter 44.7.

35.4 Local Performance Scores

Summary. Cellular hierarchies are the core ingredient of the local performance units, but we are still not in state of controlling all the locally relevant parameters. Moreover, we have to get prepared for the future unfolding processes (stemma theory). Local performance scores are the complete local structures for performance, we give the technicalities in the language of the space form *LocPerScore* of local performance scores.

– Σ –

So far, a cellular hierarchy lacks several specifications which are mandatory in order to be able to perform on an instrumental basis, and in order to hook a single hierarchy into a chain of unfolding performance stages. We also do not have any reference to parameters which might contribute to the specific performance hierarchy. In what follows, we shall for several reasons set up such an environment in the language of denotators. First, we ought to englobe the language of performance theory in the general denotator concept framework, homework which we have not done to this point. This is particularly important for any software developments of performance tools since the denotator language is the lingua franca of all our theoretical perspectives when they are implemented on the software level. Second, it turns out that the very definition of a local performance score is circular and therefore cannot be expressed in usual terms of mathematical theory. It can be easily expressed on the level of object-oriented programming, and this is one reason why denotators are so useful: Their formalism fits perfectly in the object-oriented paradigm.

The following definition of the performance score space will be given top-down in the sense that it has certain spaces in its ramification tree which are not yet made fully explicit. See Figure 35.6 for its visualization by a double tetrahedron. This will be completed in subsequent chapters, but this is the right place to introduce this tree structure since the role of its components will be clearer if we have presented them in their functionality rather than in a full-size space definition.

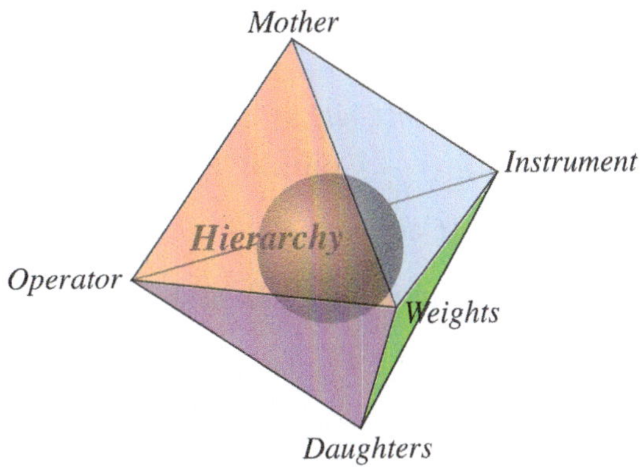

Fig. 35.6. The visualization of the six instances of a local performance score (LPS) on a double tetrahedron.

We shall henceforth abbreviate LPS = Local Performance Score. Here is the space for LPS. We thereby still suppose that we are working in a double sequence B, P of basis and pianola spaces, as introduced in Section 35.3, including this notation.

$$
\begin{aligned}
LocPerfScore_{BP} \underset{Id}{\longrightarrow} \mathbf{Limit}(\mathbb{D})& \\
\mathbb{D} = &Mother, \\
&Daughters, \\
&CellHierarchy_{BP}, \\
&Instrument, \\
&Operator, \\
&WeightList_{BP}.
\end{aligned}
\tag{35.2}
$$

Here is the meaning of these factors:

Mother. The mother form is the reference to another LPS, from which the given one may inherit a number of properties. It may also happen that there is no mother, i.e., our LPS is already a "primary mother", in which case a denotator of a score form *ScoreForm*, such as the common piano score form defined in [760], will be set instead of the referenced mother, i.e.,

$$Mother \underset{Id}{\longrightarrow} \mathbf{Colimit}(ScoreForm, LocPerfScore) \tag{35.3}$$
$$ScoreForm \underset{Id}{\longrightarrow} \ldots(\ldots) \text{ (any adequate score form).}$$

Daughters. The present LPS may be related to a finite set of other LPS which are derived from this LPS. The members of such a set are termed daughters of this LPS, the corresponding form is that of finite local compositions over *LocPerfScore*:

$$Daughters \underset{Fin(F) \rightarrowtail \Omega^F}{\longrightarrow} \mathbf{Power}(LocPerfScore) \tag{35.4}$$
$$\text{with } F = Fun(LocPerfScore).$$

CellHierarchy$_{BP}$. This space parametrizes cellular hierarchies. A cellular hierarchy may be parametrized as a set of performance cell denotators $\mathfrak{h}(U), U \in H$, H the space hierarchy of $\mathfrak{h}$, such that the space of any such denotator also identifies the space name U. Since space names are supposed to be unique for spaces, we may just retain the space names U to nominate these spaces. In this nomenclature, we shall also denote by PU the physical space associated with U within our total space system BP. So cellular hierarchies are modeled by

$$CellHierarchy_{BP} \underset{Fin(F) \rightarrowtail \Omega^F}{\longrightarrow} \mathbf{Power}(Cell_{BP}) \tag{35.5}$$
$$\text{with } F = Fun(Cell_{BP})$$
$$\text{and } Cell_{BP} \underset{Id}{\longrightarrow} \mathbf{Colimit}(\mathbb{D}) \tag{35.6}$$
$$\text{with } \mathbb{D} = (Cell_U)_{U \in BP}.$$

So we are left with the space $Cell_U$ of performance cells of space U. By definition, we have this structure:

$$Cell_U \underset{Id}{\longrightarrow} \mathbf{Limit}(InitSet_U, Frame_U, Kernel_U, Field_U, InitPerf_U). \tag{35.7}$$

The initial set space $InitSet_U$ parametrizes either sets of simplexes or just any local composition in U, i.e.,

$$InitSet_U \underset{Id}{\longrightarrow} \mathbf{Colimit}(I_U, Simplexes_U) \tag{35.8}$$
$$I_U \underset{2^{Fun(U)} \rightarrowtail \Omega^{Fun(U)}}{\longrightarrow} \mathbf{Power}(U) \tag{35.9}$$
$$Simplexes_U \underset{Fin(F) \rightarrowtail \Omega^F}{\longrightarrow} \mathbf{Power}(Simplex_U) \tag{35.10}$$
$$\text{with } F = Fun(Simplex_U)$$
$$Simplex_U \underset{Fin(G) \rightarrowtail \Omega^G}{\longrightarrow} \mathbf{Power}(U) \tag{35.11}$$
$$\text{with } G = Fun(U).$$

The *frame* is just a pair of points in U, i.e.,

$$Frame_U \underset{Id}{\longrightarrow} \mathbf{Limit}(U, U), \tag{35.12}$$

designating the lower and upper extremal points R_{min}, R_{max} of the cell frame.

The *field* is something rather mathematical which we leave in its encapsulated form as an element of the vector space[5] $Der(R)$ of vector fields over the function on the frame R of the cell, i.e., we have this space form:

$$Field_U \underset{Id}{\longrightarrow} \mathbf{Colimit}(\mathbb{D}) \tag{35.13}$$

$$\text{with } \mathbb{D} = (Field_{U,R})_{\text{frame } R \subset U}$$

$$Field_{U,R} \underset{Id}{\longrightarrow} \mathbf{Simple}(Der(R)). \tag{35.14}$$

The *kernel* is a local composition in U, i.e., we have this space for kernels:

$$Kernel_U \underset{Fin(Fun(U)) \to \Omega^{Fun(U)}}{\longrightarrow} \mathbf{Power}(U). \tag{35.15}$$

The *initial performance* is given by its graph, i.e., as a (usually infinite) local composition in the product space of U and PU:

$$InitPerf_U \underset{2^F \to \Omega^F}{\longrightarrow} \mathbf{Power}(U \oplus PU) \tag{35.16}$$

$$\text{with } F = Fun(U \oplus PU)$$

$$U \oplus PU \underset{Id}{\longrightarrow} \mathbf{Limit}(U, PU). \tag{35.17}$$

Instrument. The space for instruments is not specified in this general setup. However, if a concrete instrumental specification is needed, we shall make this component more precise. For instance, if the output could be intended to be piped to an MIDI device, a MusicN family member, a physical synthesis device, or a real physical instrument like a piano, a violin, etc.

WeightList$_{BP}$. The weight space $WeightList_{BP}$ has as denotators finite lists of weights on symbolic spaces $U \in BP$. A weight is not really something new, it is only a special kind of textual predicate which we discussed in Chapter 18. The relation is this: Given a predicate E and a denotator x, we associate a truth value x/E in $\mathbb{T}_I^A$. This truth value can be any abstract truth-oriented data, but the generality of the truth modules also includes fuzzy and similar evaluation. In particular, if we select $I = \mathbb{R}$, the truth value may be any subset of $\mathbb{R}$ (for the zero-address), and, more specifically, any open interval $]-\infty, a[$. If we identify the latter with the upper bound a, this just means that the predicate is a real-valued weighting of denotators. This is the interpretation of the weight concept here. But we do include predicates in their truth-oriented meaning. For example, one may define a set of notes K in the space U, i.e., a local composition which is the symbolic kernel of some cellular hierarchy. This can be achieved by a characteristic function $\chi_K : U \to \mathbb{T}_{\mathbb{R}}$ which takes the value $\chi_K(k) = \emptyset$ iff $k \notin K$, and $\chi_K(k) = \mathbb{R}$ else. But this also opens the path to weights which also have relevant values between the notes where they originated. Then we have

$$WeightList_{BP} \underset{Id}{\longrightarrow} \mathbf{List}(Weight_{BP}) \tag{35.18}$$

$$Weight_{BP} \underset{Id}{\longrightarrow} \mathbf{Colimit}(Weight(U), U \in BP) \tag{35.19}$$

with the space macro **List** for finite lists over a given space. The space $Weight(U)$ is this:

$$Weight(U) \underset{Id}{\longrightarrow} \mathbf{Colimit}(Weight_n(U), n = 1, 2, 3, \ldots) \tag{35.20}$$

with the indexed spaces

[5] The space of derivations on the functions R is identified with the space of vector fields, see Appendix I.2.4.

$$Weight_n(U) \underset{Id}{\longrightarrow} \mathbf{Power}(WP_n(U, \mathbb{T}_{\mathbb{R}})), \quad (35.21)$$

$$WP_n(U, \mathbb{T}_{\mathbb{R}}) \underset{Id}{\longrightarrow} \mathbf{Limit}(U_n, \mathbb{T}_{\mathbb{R}}) \quad (35.22)$$

$$U_{n+1} \underset{Id}{\longrightarrow} \mathbf{Limit}(U^n, U_n), \text{ with } U_1 = U \quad (35.23)$$

$$U^n \underset{Id}{\longrightarrow} \mathbf{Power}(U_n), \quad (35.24)$$

where values of weights are given by truth denotators on $\mathbb{T}_{\mathbb{R}}$, and the weight arguments live in the ambient space U or in one of its powers U_n. So the weight predicate is evaluated on single objects in the parameter space U, or in one of its mixed powers: local compositions in U, local compositions of local compositions, mixed with points in U, etc. Usually, weights reflect structures stemming from rhythmical, melodic, harmonic and similar music analysis, see Section 44.7 for details.

Operator. Operators are new in this setup. The point is that the cellular hierarchy which as such completely describes the performance—together with the instrumental data. But we have not dealt with the problem of generating such cellular hierarchies from system data. The operator instance has precisely this functionality: To define the cellular hierarchy. We shall deal with this very complex component in Section 44.7. Mathematically, a performance operator Ω is a map

$$\Omega : \mathfrak{H} \times \mathfrak{W} \to \mathfrak{H}, \quad (35.25)$$

where $\mathfrak{W}$ is the space of weight lists. The first argument, a cellular hierarchy, will be taken from the mother's data and inherited to the actual LPS by the operator Ω. The second argument, a selected list of weights, is usually conceived as a contribution of given musical analyses to the shaping of the present cellular hierarchy. This subject will also be dealt with in Section 44.7. But it is also conceived as a source of symbolic kernels when we need to access them via their characteristic function, see Section 38.3.2 for this approach.

35.5 Global Performance Scores

Summary. Global performance scores are atlases of local performance scores, defined for reasons of performance syntax, such as, for example, instrumentation. We describe this global approach.

– Σ –

We have already stressed in Section 32.4 that performance is a four-fold global phenomenon: There is instrumental variety, gluing of local charts, hierarchies of parameter sets, and stemmatic inheritance.

The concept of a cellular hierarchy meets the hierarchical aspect, whereas the broader LPS concept meets also the inheritance aspect by the instances of mother and daughters, and the operator—together with its weights. So we are left with the local character regarding a) instrumentation, b) local charts and strategies.

With respect to these two local aspects, a global performance score should be a finite local composition of LPS, i.e., a finite set of LPS which cover different instrumental specifications as well as local charts of a global composition which is to be performed. So we may state formally the space of global performance scores:

$$GlobPerfScore_{BP} \underset{Fin(F)\to\Omega^F}{\longrightarrow} \mathbf{Power}(LocPerfScore_{BP}) \quad (35.26)$$
$$\text{with } F = Fun(LocPerfScore_{BP}).$$

Henceforth, we abbreviate "global performance score" by "GPS". The moral of this construction is that it serves to perform all the kernel events within each top space in the respective cellular hierarchies of the LPS.

35.5.1 Instrumental Fibers

Summary. If we have the same instrument appearing for several LPS, limit constructions are necessary.

– Σ –

GPS denotators are just sets of LPS without any specific instrumental relations. If we however want to group several LPS around one and the same instrument, we have to look for limit tools. In fact, if we have k LPS with a common instrument, this is controlled by the projection $p^{BP}_{Instrument} : LocPerfScore_{BP} \to Instrument$ onto the instrumental form. We then have to take the k-fold fiber product $LPS^{BP,k}_{Instrument} = \prod^k_{p^{BP}_{Instrument}} LocPerfScore_{BP}$. The coproduct

$$LPS^{BP}_{Instrument} = \coprod_{k=1,2,\ldots} LPS^{BP,k}_{Instrument}$$

of all these k-fold fiber products gives us the possibility to build global performance scores with instrumental grouping specifications on the space type BP. If we accumulate all space types which are of interest, in a sequence $BP. = BP_1, \ldots BP_n$, say, we get the more general coproduct

$$LPS^{BP.}_{Instrument} = \coprod_{k=1,2,\ldots,\ l=1,\ldots n} LPS^{BP_i,k}_{Instrument}$$

and finally the global performance score space

$$GlobPerfScore_{BP.} \underset{Fin(F)\to\Omega^F}{\longrightarrow} \mathbf{Power}(LPS^{BP.}_{Instrument}) \tag{35.27}$$

with $F = Fun(LPS^{BP.}_{Instrument})$.

Part IX

Expressive Semantics

36

Taxonomy of Expressive Performance

This last album is not titled as a memorial album or as an album in tribute because it was titled by Coltrane himself the Friday before his death on Monday, July 17, 1967. He and Bob Thiele were considering words that might apply to the sense of this album, and finally Coltrane said, "Expression. That's what it is."
Nat Hentoff [459]

Summary. Performance structure describes a semiotic fact: expression of meaning by shaping of score data. These expressive semantics are classified according to the three layers of reality captured by the topographic cube: Psychic, physical, and mental (see Section 2.4). The first one means that performance expresses emotions, the second one deals with expression of gestural contents, and the third one—the musicologically most interesting one—aims at giving one's understanding of the musical text a rhetoric expression. In a realistic performance, all three expressive semantics will participate, however, a theory of expressive semantics must first of all deal with the "pure" types which are, each in its own way, difficult subjects of ongoing research. We will not deal with the psychological, cognitive or neurophysiological *esthesic* aspect of performance since this is part of music psychology and would exceed our subject.

– *Σ* –

To the common music lover, it is by no means clear whether and in which way performance should express contents, and what kind of contents could be addressed. The most widespread belief is that music expresses emotions, or even that "music is" emotions. The latter approach is sometimes contended by music psychologists (see 36.1). This is due to the common usage of music as a carrier for external contents. Music often just ornaments events, ceremonies, feasts, and as such it is not intended to ask music for whatever content.

Perhaps the most interesting such phenomenon is film music. It is a common saying among film music experts [253] that the best film music is the one of which the spectator does not even take notice while watching the movie. But this is a very superficial judgment, since it is in contradiction to the fact that muting the music channel in a movie virtually destroys the movie. Without music, most of its message vanishes. This is not only the case where in a criminal story of one of those cheap TV productions music has to announce that something dangerous is imminent. Omitting music evaporates the very atmosphere, the parfume of the best film! So music is quasi-absent but essential to the movie.

The point of this apparent contradiction is that the common music expert saying does not really observe what we experience while watching a movie: why are we really sitting in a movie show? First of all, we replace everyday reality by an artificial, virtual reality which is projected into the dark environment of the classical *camera obscura*. Such a reality switch needs a strong booster to work, and this booster is music. Music is well known as the exemplary environment for a counter-world, a force that lets us forget about the common things and transports us to lost, hidden, and subconscious layers of existence, see also our comments on the Depth-EEG experiments in counterpoint theory in Section 30.2.4.

In reality, *the core of a movie is its music*, not its visual process and textual story, the latter are only the pretext of what is being communicated. Pretext in its literal sense: ante textum, not irrelevant, but not

G. Mazzola, *The Topos of Music II: Performance*, Computational Music Science,
https://doi.org/10.1007/978-3-319-64444-8_5

the kernel either. One could even require from good film that *as a work of art it should transfigure the visual process and textual story into music.*

In short, this pleading for music states that music expression in performance is expression of something, not pure, self-sufficient artistry. If this is acceptable, performance as a rhetoric category should take care of how to express the contents. Therefore, on the one hand, the question must be posed about which contents can be conveyed in performance, and, on the other, about the equilibrium between contents and the medium of performance.

The following discussion deals with the basic strategies of creating performance of something. It does not claim exhaustivity, but sketches prototypical approaches to contents of expression in order to position our research approach in the field of the young science of performance. For a more extended discussion of performance research history and also psychological streams, see Reinhard Kopiez' excellent survey in [546].

36.1 Feelings: Emotional Semantics

Summary. Several authors, such as Susan Langer, Manfred Clynes, Alf Gabrielsson, Jörg Langer, and Reinhard Kopiez, have also focused their research ona relation between emotions and performance. Whereas Gabrielsson [354] contends with Langer [582] that there is an isomorphism between musical structure and emotion, Langer and Kopiez [545] develop a theory of oscillating systems that is supposed to find a physiological counterpart on the neurological level.

– Σ –

In [582, p.27], Susan Langer states that "music is a tonal analogue to emotive life", a statement which is interpreted by Alf Gabrielsson in [354, p.35] as the basic idea of "an isomorphism between the structure of music and the structure of feelings". This is also the doctrine which Gabrielsson adapts: "In summary, we may consider emotion, motion, and music as being isomorphic." This latter statement also includes motion as one of the isomorphic structures. This is related to Manfred Clynes's stress of emotion and its expression as an integrated system where motion, i.e., the gestural dimension, plays a crucial role [205]: Emotion, he calls it a "sentic state", may be expressed by "gestures, tone of voice, facial expression, a dance step, musical phrase, etc."

While this conjecture may please psychologists, it is completely useless to scientific investigation. In fact, such an isomorphism is a piece of poetic literature as long as the components, emotions, motion (gestures), and music, are not described in a way to make this claim verifiable. Presently, there is no hope for a realistic and exhaustive description of emotions. Same for gestures (see next Section 36.2), and as to music, the mathematical categories of local and global musical objects are so incredibly complicated that the claim sounds far-out, see also Figure 36.1 for an attribution of emotions to articulatory ambitus. For example, the number of isomorphism classes of 72-element motives in pitch and onset (modulo octave and onset period) is $\approx 2.23.10^{36}$ as we have seen in Section 11.4.1.3. How could the claimed isomorphism fit in this virtually infinite arsenal? Such a *terrible simplification* however does not contradict the generic insight that expression in performance may be motivated and grounded in emotional categories. But we should be careful on this point: Our present discussion is not about emotional effects of music in the listener's psyche, we are talking about emotional rationales for expressive performance, i.e., the question of how performance could be induced by emotions. This aspect is typically addressed when the score annotations require performance actions in the mood of "amoroso", "languissant", "beklemmt". Here, the performing artist has to play in and through such a mood. To be honest, to play in such a mood is not a concrete way of telling a performer what to do. The mechanism very probably works by a feedback: The performer plays and hears his/her performance so that the output may be adapted to the received impression in the artist's own ears compared to what the artist conceives as being an impression of the given type.

Evidently, such a rationale is extremely difficult to handle on a scientific level. And it is also difficult to understand a relation to the given score which transgresses the explicit annotations mentioned above. Since general feelings as a motor for performance are no good point of departure, one should (truly in the spirit of Langer, Clynes, and Gabrielsson) at least try to relate the score's contents (not the annotations,

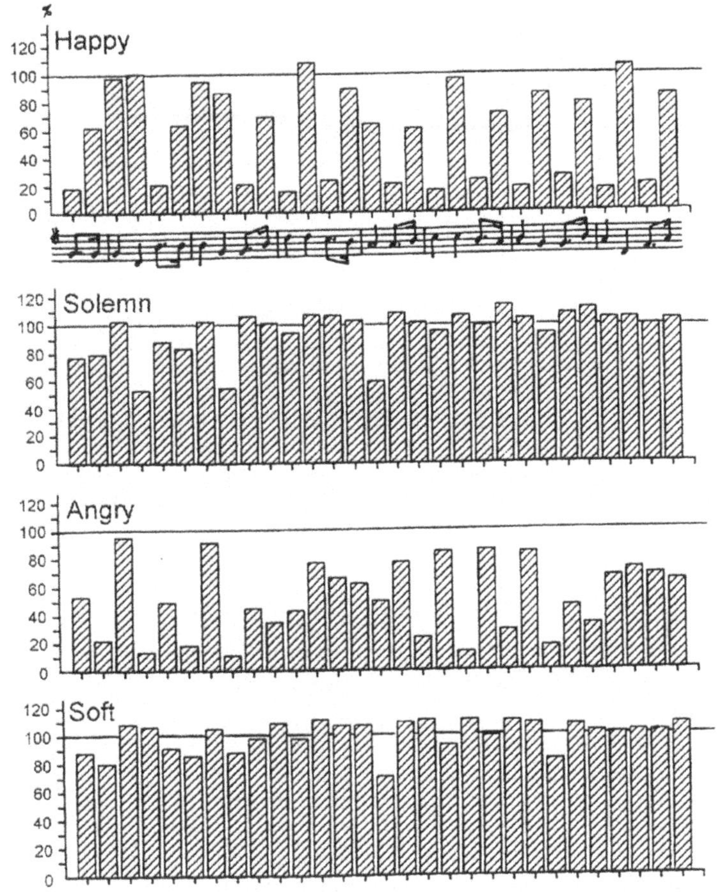

Fig. 36.1. Attribution of emotions to the articulation ambitus [354, p.42] for the song "Oh, my darling Clementine". From [354] with permission of Springer.

the structural facts) to emotions, which, in turn, could then be used to shape performance via the above feedback mechanism. But this is a very complex task. How should a determined emotion (and which?) be incited by a given structural fact, a cadence, a melody, a harmonic configuration, a rhythmic process? So the emotional rationale for performance splits into the

- association problem between emotions and performative shaping;
- and the association between non-emotive score contents and emotions.

This story has not yet been written.

A less totalitarian and more quantified approach than Langer, Clynes, and Gabrielsson to emotional rationales for performance has been proposed by Jörg Langer and Reinhard Kopiez [545, 550, 546] with what they call TOS (=Theory of Oscillating Systems). The TOS postulates a system of 120 oscillators with a determined frequency each, reaching from 8 Hz to 0.008 Hz, and distributed in logarithmic steps. In a kind of Fourier analysis (the authors have not to date published the precise formulas), the dynamical curve of a recorded piece of music is decomposed and shows the contributions of these oscillators to the given curve. The TOS is not only meant as a formal description but the authors argue that the musical progression really triggers a series of oscillators in the cognitive stratum of the human brain.

It is contended that the TOS spectrum is stored in human memory and that comparison of performances is enabled by comparison of such spectral data. The authors conclude [550, p.33] that "production and perception of expressiveness in music are essentially one entity." No reference is however made to the score

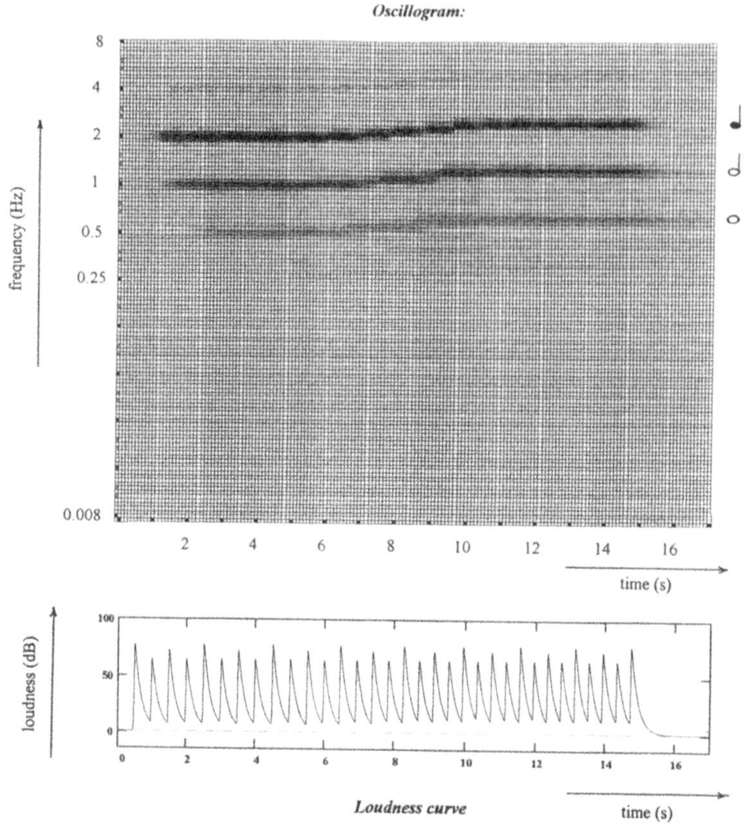

Fig. 36.2. Oscillogram (above) and loudness curve (below) belonging to quarter notes played in a 4/4-time signature by a drum computer with an additional accelerando at the end, with permission of the authors of [550, p.33]. This accelerando leads to a parallel upward movement of the dark bands in the oscillogram, which means that the activation changes to higher oscillator frequencies.

as such, i.e., TOS just measures the performance output and represents the spectral development in time by graphical means (called the oscillogram, see Figure 36.2). So TOS does not measure the performance map as such, only the image of the map. It is evidently subtended in the TOS that the spectral decomposition bears a semantics of time processes which—independently of the hidden score—transports the neural activities of the musical brain. So expressivity is correlated to a neuronal oscillator system (expressed by firing rates of neuron populations). This is a type of rationale which refers to a score-independent instance which encodes expressivity. This is a dramatic tournament since semantics, i.e., expressivity of something, is uncoupled from the text. This rationale is a kind of "pure expressivity", not a result of reflection or analysis, nor a result of gestural structures. It must therefore be an emotional rationale, although the authors are not precise on the cognitive category and the topographic origin of such neuronal oscillators (it could be that the limbic system is meant), and an experimental verification of the existence of such oscillators in human brain outstanding.

36.2 Motion: Gestural Semantics

Summary. Besides—but connected to—emotions, motion, as mediated by gestural paradigms, is a widespread rationale of expressive semantics. Neil McAngus Todd [1060], then Johan Sundberg, Violet Verillo [1022], and Ulf Kronman [560], and also David Epstein, Jacob Feldman, and Whitman Richards [312] have proposed physically motivated descriptions of retards at group endings. Richard Parncutt, Jan Sloboda, and Eric Clarke [825] have added the concrete aspect of anatomical constraints in piano performance. We describe and analyze these approaches. We shortly account for the work by Shuji Hashimoto and Hideyuki Sawada [445] concerning the Japanese kansei processing research in the musical domain of gestures.

– Σ –

The emotional argument of performance is often paralleled by the argument that gestures, "motion", are also present as a parallel phenomenon to emotion, we have already seen this in Gabrielsson's isomorphism statement in Section 36.1. It is also stressed by Clynes' sentic concept which measures the human input by a gestural device of joystick-character, and by Kopiez-Langer [550, p.32]. Citing Francois Delalande [250], they state that "body movements and gesture are in close relation to musical timing. Our neuro-psychological oscillation model implies that body movements can be triggered by musical events." And of course vice versa: Body movements trigger expressive performance. Also, in the framework of Japanese "kansei information processing"[1], Shuji Hashimoto and Hideyuki Sawada stress that "gesticulation is often employed in musical performance to express the performer's emotion." Using the dataglove, they have implemented applications which transform gestures into "MIDI control units to improve the performance in real-time" [445]. This regards expression of crescendi, vibrati, or pianissimo, for example.

Richard Parncutt has investigated the intuitively evident fact that expressive performance is strongly conditioned and induced by physical constraints from fingering [824, 825]:"good fingering is a crucial ingredient in the preparation of performances that are both technically reliable and appropriately expressive." In [823], he has also maintained the thesis that "the most important sounds conditioning the perception of rhythm may be the sounds associated with the heartbeat and walking movements of a mother, as heard by her unborn child." So good performance should been adapted to this motion trigger.

In accordance with these general observations, several performance scientists have proposed models of performance which derive performance fields from mechanical principles of accelerated motion. In particular, this has been set forth regarding tempo curves. Analyzing the first experimental studies of final retard phenomena [1022] by Sundberg and Verillo, Kronman and Sundberg [560] propose a model of final retard which is completely derived from the mechanical analogy of constant deceleration of tempo, as if it were induced by the action of a constant force upon a physical mass. However, tempo T is seen here as a kind of velocity that is a function of physical time, not symbolic time, as it is standard. So what is constant is the force as a function of physical time. We then have $T(e) = dE/de = c.e, c = const.$. The resulting formula is $T(E) = T_0\sqrt{1-E/E_0}$, where the tempo $T(E)$ at onset E is related to the starting tempo T_0, the total onset interval E_0 until total stop (which is not included in the really onging music, but has to be set as an ideal endpoint of motion). If such a formula is derived, by the same reasoning, one could also derive formulas with different, not necessarily constant force function; this has also been observed by these authors. However, there is no indication of which force should act and why so. So in principle, any tempo curve can be constructed by an adequate acceleration and therefore force function. More generally, it is not clear, why the supposed mass should be constant. As with special relativity, the mass could vary as a function of tempo. It seems that the mechanical motion model is only the construction of an intermediate layer to the real question: What are the basic forces which shape tempo? If the straightforward mechanics (constant mass, constant deceleration) are maintained, however, this explanation completely standardizes the final retard phenomenon and uncouples it from the underlying score and composition. This type of approach has also been proposed by Jacob Feldman, David Epstein, and Whitman Richards [312]. Their paper models tempo $T(E)$ as a velocity function of symbolic time E, and its derivative is meant to be determined by a quadratic force function $F(E) \sim E^2$. The Newtonian equation $F(E) = m.T(E)$ yields a cubic polynomial function $T(E) = a.E^3 + \ldots$ for the tempo. Of course, this is a completely different mechanical situation, here the force really acts on the symbolic level instead of the physical action described by Kronman and Sundberg. While that one means $T(E) \sim E^{1/2}$, this one yields $T(E) \sim E^3$. Unfortunately, the latter approach is not congruent with the examples shown in [312]: They refer to the reciprocal value $1/T(E)$ instead of $T(E)$!

[1] This is the term coined to stress a special application of information technology to implementation of emotional contents against the classical Artificial Intelligence and other, more logically and mathematically oriented approaches.

So we should have $T(E) \sim e^{1/3}$, but that requires another mechanical situation. In terms of Kronman and Sundberg, this requires a force which, as a function of physical time e, is proportional to $e^{1/2}$.

A less simplistic approach which also includes structural analysis (see also below in Section 36.3) and not only mechanical generalities is presented in Neil McAngus Todd's paper [1060]. He rightly observes that the final retard is only a very special agogical situation, and therefore models his tempo curves according to a superposition of accelerando/ritardando units which are defined by a triangular sink potential V. Accordingly, tempo is defined as a velocity v, and the total energy of the system $E = \frac{1}{2}mv^2 + V$, supposed to be constant (why?) gives the velocity formula $v = \sqrt{2(E - V)/m}$. Todd further supposes that there is an intensity variable I for loudness, with a relation $I = K.v^2$ that is common to many physical systems. This yields the relation $I = 2K(E - V)/m$ and sums up to an aggregated formula $I = \sum_l 2K(E - V_l)/m_l$ if the grouping of the piece is taken into account. Thus, the idea is that there is a physical energy and intensity parameter system that controls the "surface" of the tempo (= velocity) via classical energy formulas. So the background structure is an energetic one, i.e., the tempo curve and loudness are an expression of mechanical dynamics. The author comments on his method as follows [1060, p.3549]:

> *The model of musical dynamics presented in this paper was based on two basic principles. First, that musical expression has its origins in simple motor actions and that the performance and perception of tempo/musical dynamics is based on an internal sense of motion. Second, that this internal movement is organized in a hierarchical manner corresponding to how the grouping of phrase structure is organized in the performer's memory.*

The author also suggests a physiological correlate of this model (loc. cit.):

> *...it may be the case that expressive sounds can induce a percept of self-motion in the listener and that the internal sense of motion referred to above may have its origin in the central vestibular system. Thus, according to this theory, the reason why expression based on the equation of elementary mechanics sounds natural is that the vestibular system evolved to deal with precisely these kinds of motions.*

Todd refers to insights of neurophysiologists according to which the vestibular system is also sensitive to vibrational phenomena. So the musical expressivity is understood as an effect of transformed neurophysiological motion.

The drawback of this approach is that finer musical structures are not involved in the structuring of the energy which shapes tempo/intensity. And even if that could be done, there is an essential kernel of this shaping method which should be based upon paradigms of motion. These paradigms do however not appear clearly in the above approach. More precisely: The complex motion dynamics of the vestibular system cannot easily be mapped onto the structures of performative expressivity. What is the operator which transforms whatever structures of motion into expression parameters? If music were isomorphic to motion, no such isomorphism could be recognized from Todd's clever approach.

Beyond general motion paradigms there is the more visible level of gestural structures which can be implemented in operators for musical expressivity. Unfortunately, the existent classification of gestures is anything but detailed. Classification of gestures is only settled on a prototypical basis. But this is precisely not what music needs so urgently. For example, the extremely refined gestures of Glenn Gould's performance[2], the movements of his hand and arms, his head and thorax, this is beyond any scope of present classification.

It is the same with the incredibly refined movements of Herbert von Karajan's hands, and especially their single fingers: It is known that these gestural directives to the members of the orchestra were observed and followed with extreme attention. To measure such details is not what we can control at present.

[2] The reader should be careful on this point: In [250], Francois Delalande has recognized the expressive role of Gould's gestures (analyzing a film record of Gould's performance of Bach's *Kunst der Fuge*, see Figure 36.3) with respect to musical structure. This is an extension of the performance parameters from sound to gestural parameters, but this is not what we are discussing here. We are discussing the role of gestures as a cause for musical performance, not a media thereof. Evidently, a pianist's (and a fortiori a violinist's) performance is strongly and essentially driven by gestural shapes which are not only mediators of structural facts.

Fig. 36.3. Glenn Gould while performing. Drawing by Mitty Sheen © 2017.

To be clear: We are not contending that gestural and motion information is irrelevant to performance. On the contrary, this is an essential contribution, but it is too difficult for scientific research as long as classification of gestures is so far from being settled. And that alone does not solve the problem, since operators for shaping performance parameters must be defined from the information provided by gestural input data. All this seems to be a bit easier than the far-out emotional rationales, but still is subject of advanced research.

36.3 Understanding: Rational Semantics

Summary. As opposed to "low level" emotional and gestural expressivity, rational semantics deal with expression of rational interpretations of the score structure. This means that the text is analyzed from different points of view, such as harmony, rhythmics, motivic content. These analyses are used as an input to shape the performance structure. This aspect is dealt with in research by the group of Anders Askenfelt, Anders Friberg, Lars Frydén, and Johan Sundberg [43, 336, 337, 339], then by Neil McAngus Todd [1058], Gerhard Widmer [1117, 1118], then by Jan Beran, Guerino Mazzola, Joachim Stange-Elbe, and Oliver Zahorka [688, 689, 690, 691, 692, 699, 702].

– Σ –

Already Hugo Riemann [903] had stressed that the scope of rehearsal should be to support the communication of the motives' comprehension. And it is a fact of music psychology (see for example [200] or the excellent overview [546]) that performance as an expression of the score's structure is better understood than performance which disregards structure. Also, in Theodor W. Adorno's theory of performance [7, 9], the analytical point of view, i.e., the purpose of performance to transmit analytical insights, is prominent.

In fact, the most explicit starting point of any performance is the given score. This is a text that abounds with structure that must be shaped in a physical performance space. The reference to this structure as a rationale for performance is a straightforward logic which is completely standard in literature: Interpretation of a text is one of the most recognized and widely practised methods in text performance, especially in the actors' interpretation of dramas. Besides emotion and motion it is therefore logical to refer to the rational text analysis in order to shape performance.

Probably the first explicit and quantitatively stated contemporary approaches in this vein is the "analysis-by-synthesis" method of Sundberg and his collaborators which was first presented in [1021].

Analysis-by-synthesis means that a bunch of performance rules for the shaping of different parameters is defined in a software environment, and then applied to the production of a synthetic performance (on an MIDI-instrument, say). The result is then analyzed by an expert (in Sundberg's group this was the professional violinist Lars Frydén) who proposed alterations of the given rules and/or new ones to the programmer. In this experimental cycle, the rules are always of a general character which is based on structural data, not on direct—emotionally or gesturally driven—interventions on the performance data of the individual composition in question. This approach has been implemented on the basis of eleven rules in the computer application *Rulle*, later, renamed to *Director Musices* [339], with an extended repertory of rules.

The rules of this approach are grouped into three categories: the

- **differentiation category**, including these rules: Duration contrast, High sharp, Double duration, High loud, Accents, Melodic charge, Melodic intonation;
- **the grouping rules**, including these rules: Punctuation, Phrase arch, Leap articulation, Phrase final note, Leap tone duration, Harmonic charge, Faster uphill, Chromatic charge, Amplitude smoothing, Final retard, Inégales, Repetition articulation;
- **the ensemble rules**: Melodic synchronization, Bar synchronization, Mixed intonation, Harmonic intonation.

We shall discuss these rules in Section 37.1.1 of Chapter 37 on performance grammars. Let us just give two examples here: The rule "Faster uphill" requires in a melodic context, that if a note is followed by a higher pitch note, its physical duration is shortened by $2.k\ ms$, where k is a system variable in order to give its rules a variability in strength, default is $k = 1$. A second example is "Melodic charge", which depends on a non-negative numerical weight attribution for every pitch as a function of its position in the selected tonality. The weights are roughly proportional to the distance of the pitch from the central pitch of the tonic (its weight is zero). Proportionally to the weight, amplitude (in Decibel), duration (percents), vibrato amplitude (percents), and unevenness smoothing are effected, also multiplied with the omnipresent strength factor k.

These rules are all very elementary in their mathematical as well as in their music-theoretic approach, but they are completely concrete, and this makes this early attempt so precious. However, the present presentation and formal statement of these rules lacks a clear-cut distinction between symbolic and physical reality. For example, the above mentioned rule "Faster uphill" does not ask for the absolute duration of these notes, the change of duration is independent of the physical data, and produces either too large or too small duration changes in extremal tempi. And this is the delicate point here: A distinction between symbolic and physical reality must rely on absolute tempi. Tempi are however not mentioned in the entire machinery, although in his PhD thesis, Anders Friberg makes a comment on tempi. These rules act not in the sense that they define a map from the symbolic reality into the physical one, but they act on an already given physical image of the symbolic reality. This seems to be a kind of "prima vista" performance, but no information on this is given. *There is no performance of a score which lives outside some tempo specification*[3], *there is only performance of performance.* Moreover, the lack of agogical shaping operators is also manifest from the absence of onset in the tone parameters. The shaping of onset as such (independently of articulatory shaping) is not defined.

The example of a motion-triggered performance model by Todd which was discussed in Section 36.2 above is a realization of Todd's generic approach to rational semantics in performance [1058] which we shall now describe. The background structure of that motion paradigm in fact relies on structural data (of grouping), as we have already mentioned above. Todd's generic performance model is designed upon a bidirectional transformation pairing from a score representation Ψ to a performance P and backwards by means of:

1. a *performance procedure* Π acting on Ψ and an *encoding function* γ:
   $P = \Pi(\Psi, \gamma)$,
2. a *listening procedure* Λ acting on P and a *decoding function* δ:
   $\Psi = \Lambda(P, \delta)$.

[3] Or something of this type if simple tempo curves are not possible in complex hierarchies.

In this generality, "the theory... is sufficiently general to cover any variable of expression. At the same time, it is agnostic as to what is being communicated, be it structure, emotion, or extra-musical reference" [1058, p.407]. The generic character of Todd's approach hides an asymmetry of the transformation pairing which is due to its semiotic background; see also [703] for a modern survey on music semiotics. In fact, performance is a poietic process issued by the performer from the composer's score. In other words, a performance is *caused* by its creators and must be *understood* by the listener, not vice versa. Hence, the performance transformation has to be specified as a semiotic mechanism. This is the difficult part of the business.

Without entering into details here (see Chapter 47 for a detailed discussion) it can be said that the critical subject of performance theory—a problem which Todd thematizes in the spirit of cognitive science—is a reconstruction problem: Given a performance P, how many representations Ψ and encoding functions γ can you find such that $P = \Pi(\Psi, \gamma)$? In mathematical terms, we are looking for the fiber $\Pi^{-1}(P)$ over P. This is the so-called inverse image of P, and therefore, this branch of performance theory is called *inverse performance theory*. The listening procedure in [1058] is just a formal setup for a Section Λ to Π, i.e., the selection of an element in the fiber over P as a function of the decoding data δ.

Clearly, the fiber cannot be described in effective mathematical terms if one does not assume a well-defined transformation model. And even for very special models, the so-called locally linear performance grammars (see [694] and Section 47.2), fibers turn out to be high-dimensional algebraic varieties. Further, the encoding function must be meaningful enough to reflect the score's structure and its relations to the above categories of expressive semantics. Otherwise, performance cannot claim to interpret the selected score. In other words, the big problem of performance theory is to propose models of *adequate generality* that cope with *expressive semantics.*

In Todd's singular example to his theory, he restricts to hierarchical grouping data for the shaping of duration. Commenting on the inverse problem of listening procedure, he states that "the durations used in the calculations are from only one metrical level. Much information about tempo is given at metrical levels below the tactus and in the durations of actual notes. The representation needs to be extended downwards to include note timing, which would mean that a rubato handler would have to work in cooperation with a metrical parser, one feeding the other. Clearly, a lot of work is needed in this area." Concluding, he notes that "the known algorithms make no reference to any tonal function. Therefore, a rubato handler could be a vital component of any theory of grouping in the perception of atonal music. A complete theory must of course include dynamics, articulation and timbre."

Methodologically, this approach is tightly bound to cognitive science in that any algorithm is first of all tested upon its immediate fitting into human perception mechanisms, within real-time constraints, say. We believe this is a too narrow approach for two reasons: First of all, the investigation of general structural facts must be carried out before any relevance to human perception is taken into account. There is the general problem of getting an overview of possible models and their classification. Second, the cognitive knowledge is all but settled, more precisely: We do not know, by what processes cognition of performative expression is handled in the human brain. It could happen that a rather abstract invariant of the geometric structure of a mathematically complex fiber $\Pi^{-1}(P)$ can easily be detected by the cognitive machinery, but that this invariant would not have been detected if we were only permitting fibers which allow an immediate access by the cognitive capacities. For example, the mathematical structure of a Möbius strip shaped fiber may be too complex to be grasped by the cognitive machinery, whereas its lack of orientation may be an easy task to be tackled by a small test routine built on a neuronal basis.

Gerhard Widmer's work (e.g., [1117, 1118]) is based on the machine learning method. In this context, structural features of given scores, such as chords, or small motives, are correlated to the performance data of a given human performance. This data is then used to shape a second score. Here, the performance rule is found by an investigation of a given performance. However, there is no background theory to this, it is purely imitative with respect to human competence. We cannot see any deeper value of such an approach since nothing is really learned beyond parrot-style imitation, although such an imitation may sound quite attractive.

A completely orthogonal approach, which is more akin to Todd's intelligent setup, was undertaken by the Zurich school[4] of the author and his collaborators in [105, 106, 107, 688, 689, 690, 691, 692, 699, 702]. In our setup, the structure theory of performance in its full-fledged concept framework as developed in Chapter 35 is used as an output level for performance, whereas the LPS operators are designed to implement different output data from musical analyses, such as metrics/rhythmics, motives/melodies, harmony, counterpoint, or grouping. This approach was first sketched in [683], then presented at the SMAC in [687], at the ICMC in [689], in a paper [688] concerning tempo hierarchies, and above all in the SNSF reports "Geometry and Logic of Musical Performance I,II,III (1993–1995) [690]. This approach is completely general in the sense of Todd's scheme [1058] described above. However, we have stressed a specific approach to the communication of analytical facts to performance operators which eases many methodological questions concerning the interplay and concerted action of an entire collection of analytical results. This is the method of weights, i.e., it is required that the analysis of the music-theoretic procedure A be delivered in the form of weight functions $w_A : K \to \mathbb{R}$, where K is a local composition in some parameter space associated with the score, and more directly with the underlying cellular hierarchy of the LPS which implements a performance operator. For example, a metrical weight function $w_M : K_E \to \mathbb{R}$ typically associates a non-negative weight $w_M(E)$ for each onset E in a given local composition K_E of (symbolic) onsets. At first sight, this seems to be utterly restrictive, but the explanation is this: Ultimately, performance has to be defined via numerical indications of how to specify the instrumental parameters. So even if the analysis is of a more symbolic character, it has to be filtered or transformed into numerical values sooner or later. Of course, this could be done in the innards of a specific operator, and there is no obstruction to defining such operators. However, if we want to combine different analyses to be used as nurture for a determined operator, the problem of uniformity of such an input combination arises. But if the input is a priori a weight, the analytical arguments of an operator can be designed in a much more weight-independent format. For example, one may then feed a linear combination of a number of given weights into an operator, without being concerned about where these weights stem from.

Apart from these analytical input strategies, the approach of the Zurich school also stresses the generative nature of performance. Its differentia specifica to other approaches is that the unfolding of performance has been thoroughly formalized in the concept of a performance stemma, the genealogical tree of performance rehearsal and development history (see Chapter 38 for details). So Todd's hierarchical examples of refined grouping are not only extended to a cascade of LPS, their hierarchical nodes are also turned into autonomous agents of performance shaping processes which *trace and group the entire performance unfolding as a compound historical and logical process in the large.*

This refined input policy, extending the elementary grouping rationales as exposed by Todd, yields a more sophisticated option to test correlations between measured performances and analytical insights. In this vein, different investigations have been executed with a good success by Beran and Mazzola [105, 106, 107] on the level of statistical methods. The point of this approach is that performance is so complex that it requires a faithful representation of the analytical structure of a score. And that this analytical structure must be an essential input qua significant function of the given individual score. We argue that principles of performance shaping which are completely unspecific towards the concrete score structure cannot provide us with relevant performance directions.

36.4 Cross-semantical Relations

Summary. Since music has a communicative dimension between poiesis and esthesis, each of the three above semantical directions on the poietic level may influence the esthesic level in one or several of the other semantical directions. For example, a rational expressivity may produce a gestural understanding or vice versa. We give an account of these "cross-semantical" phenomena.

– Σ –

[4] The term was coined by Thomas Noll.

As was already observed above, it may (and will in many practical situations) very well happen that one type of semantic rationale may act only indirectly upon the shaping of performance. For example, an emotional rationale may first "shape" a gestural object which in turn will shape musical performance. There are no restrictions to that. It is only important that there be one semantical modality at least which is able to convey its contents to the effective shaping of musical parameters. The rest may be arbitrarily complicated: For instance, a rational instance (score analysis, say) may produce an emotional object as its consequence, and the emotional object may produce in turn a gestural object which may evoke a second emotional object, completely different from the first! And then it may happen that this second emotional instance acts directly upon the shaping of performance. It is not clear at present, how much and on which basis the interrelation of different types of performance rationale could be implemented as computer applications, since the cross-modal assignment procedures (e.g., emotion from ratio, ratio from motion) may be hard to realize. Undoubtedly, this question is very interesting, be it for Japanese kansei research (as discussed in Hashimoto's approach in Section 36.2), be it for the general problem of harmonizing divergent semantic directions in music, above all: harmonizing the emotional direction with the rational semantics of the text.

37

Performance Grammars

Why care for grammar as long as we are good?
Artemus Ward (Charles Farrar Browne) (1834–1867)

Summary. The idea of basing performance on rules in analogy with linguistic grammar goes back to Mathis Lussy [631]. In modern performance research, this terminology was recovered by Johan Sundberg and his school [1024]. We discuss the principles for a grammar of performance and give an overview of representative approaches to this theory.

– Σ –

To our knowledge, the term *performance grammar* was coined by Johann Sundberg on the occasion of a performance theory conference in Aarhus [1024]. The reason for such a conceptualization is that the specialists became aware that performance can be shaped in various ways, but not from an amorphous design rationale, on the contrary: It became evident that there are entire organisms for shaping performance from the given score, its different semantic approaches, and the way these approaches are transformed into concrete performance instructions.

The idea that performance should be executed along certain regular patterns that remind us of a language structure comes from the fact that performed music is viewed as a rhetoric vehicle of contents, and that these contents are, by the very nature of musical semantics, hidden, difficult and ambiguous. In other words, the way they are expressed is an essential condition for their communication. For a number of expressive methodologies, their architecture in fact resembles a language although, at present, only very elementary grammatical patterns are known. In [1023], Sundberg proposed the creation of a dictionary of expressive rules where the patterns of the performance language can be looked up. We distinguish rule based approaches from rule learning procedures (and we shall not deal with chaotic ad hoc performance being taught in the vast majority of music conservatories!). Representative research is reviewed and classified according to the semantical perspective, as discussed in Chapter 36.

But the very need for a performance "language" has also a deeper explanation. If we listen to performance, it is not just the concrete piece being performed and the concrete way of performing without further context which are perceived and judged. In fact, one cannot understand the expression of a coherent and extended text without having an access to the method of shaping contents. Understanding performance really means that we gain access to a system which is applied in this concrete expressive work, and this system is a language with its grammatical rules which shape the message. In other words, *understanding the performance of a musical composition amounts to understanding its system, the language and, in particular, the grammar which guides performance.*

37.1 Rule-Based Grammars

Summary. The generic scheme of rule-based grammars and three representative directions in rule-based grammars are presented: the KTH school, Niel Todd's approach, and the Zurich School.

G. Mazzola, *The Topos of Music II: Performance*, Computational Music Science,
https://doi.org/10.1007/978-3-319-64444-8_6

– Σ –

The known rule-based performance languages share a generic grammatical structure which can be described as follows. Basically, there are three components which shape the performance transformation $\wp : K \to \wp(K)$ of a given local composition K, i.e., the kernel of the top cell in a hierarchy, see Figure 37.1. The first component, which we call the rationale *Rat*, yields the raw material which is intended to act on $\wp$.

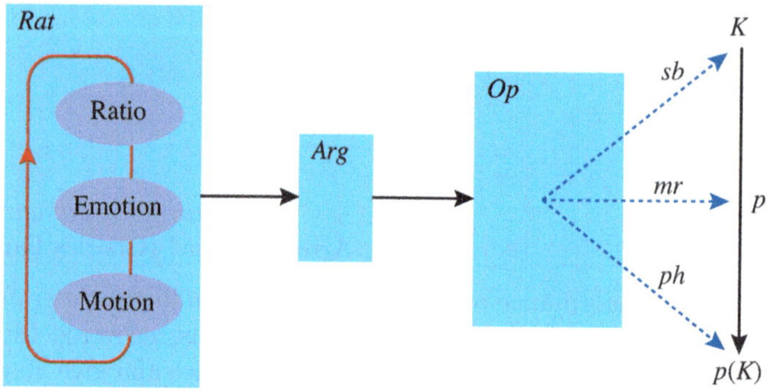

Fig. 37.1. The three components of rule-based performance grammar: the rationale *Rat*, the argument *Arg*, and the operator *Op*. They contribute to the construction of the performance transformation $\wp$ of the kernel K. The action of an operator on the performance map can be either symbolic (sb), morphic (mr), or physical (ph).

As we have seen in Chapter 36, *Rat* is a complex organism which may include emotional, motional/gestural, and rational (analytical) agents which may interact and result in a final statement to be delivered to the subsequent shaping actors. The output of these operations is the second component, call it *Arg*, the argument of performance. For example, in the analytical approach, this may be a weight function. In the motional situation, it could be an object which parametrizes a physical movement, and in the emotional setup, this may be a verbal description of a feeling, for example. The third component is an operator *Op* which "understands" the *Arg* and which, when fed with this argument, yields a determined function *Op*(*Arg*) which defines the performance transformation $\wp = \wp(Op(Arg))$. If we view this situation in the rich context of LPS theory, the argument may of course include the mother LPS and thereby determine $\wp$ not only from the LPS' proper rationale, but also from the LPS' inherited data from previous shaping activities as they are traced on its mother LPS.

The operator's action may typically be targeted to one of the three ingredients of $\wp$: its domain K, its codomain $\wp(K)$, or the map (the functional expression) as such. The first type is called a *symbolic* operator because it alters the domain. This is a very strong action since the original notes, i.e., the score's genuine structure, are changed. Symbolic operators create a new composition such that performance is defined on new input data. Whether this type is really a case of performance seems somewhat critical.

But suppose we are given the portion of a score which is written in *fortissimo*, and suppose that this dynamical attribute is a part of the kernel's specification. Then, if an additional dynamical annotation, such as *diminuendo*, is inserted in the score text, this may be seen as a performative prescription: Change the dynamics of the specified group of notes by a successive lowering of *fortissimo* dynamics in some determined range. In this case, a symbolic operator would do the job since it is an action which is required before any artistic shaping begins. We shall call this a *primavista* operator in the operator theory of Chapter 44.7.

The second operator type acts directly upon the given physical output $\wp(K)$, this is why it is termed *physical* operator. It may deform the physical data without altering the kernel K or the function $\wp$. We have to explain this seemingly contradictory argument, because it effectively changes $\wp$. We have in mind that a physical operator is a successive map $ph : \wp(K) \to ph(\wp(K))$, i.e, the original map as such remains what it is, but it is composed with the physical operator's action and yields $ph \circ \wp$.

The third operator type alters the functional description of $\wp$ into $Op(\wp)$ and therefore is termed *morphic*. This is typically the case if $\wp$ is defined via a performance field as it is implemented in the cellular hierarchy of an LPS.

Summarizing, we may restate this principle for a rule-based performance grammar:

Principle 25 *The described scheme englobes the generic framework wherein the performance language is structured. The contents which are shaped by this grammatical structure are cast in the argument Arg, whereas the grammatical structure is centered around the operator instance Op. The operator is the rhetoric element, it tells how the shaping works, whereas the argument Arg is the codified message to be conveyed after an encapsulated process of semantical elaboration in Rat.*

37.1.1 The KTH School

Summary. The KTH school's system is dominated by local rules which are based on low-level structural analysis of the text. Semantics of this analysis are rational and—to a lesser degree—gestural. They are found by the characteristic empirical "analysis-by-synthesis" method.

– Σ –

The grammatical structure of the KTH school has been described very clearly in [338], see Figure 37.2 for a reproduction of that grammatical scheme. According to this scheme, the surface level shows the transition from the symbolic score to the physical performance (horizontal arrow). This transition (in fact: the performance map $\wp$) is shaped by an analytical rationale (left lower group in the underlying level) which comprises phrase, harmony, and melody. The output of this rationale is given in the charges for melody, harmony, and chromatics. These charges are then fed into the operator unit, the transformation rules (right lower group).

The performance transformation (horizontal arrow) is visibly factorized via the transformation rules, i.e., these rules define the entire performance map. It is however not clear from this scheme, which kind of action: symbolic, morphic, or physical is taken in the concrete cases. As was already mentioned in the first discussion of the KTH system, the original transformation of symbolic data into physical data is not made explicit in these rules. It is supposed that by some underlying procedure, there is already a physical image of the score symbols before the explicit rules are activated. So the explicit rules seem to be physical operators. They act on a "primavista performance" which is implicitly assumed.

The rationales are purely analytical in their majority, but melodic rules also refer to gestural rationales. For example, "the leap tone duration modifies duration of tones in singular leaps" ([336, Rule GMI 1B]) in the sense of an expression of a gestural constraint of hand and arm movements when playing a keyboard. In fact, the duration of a high tone after a low one (or vice versa, same rule) is shortened by $\Delta DR = 4.2\sqrt{\Delta N}.k\ msec$, with the absolute pitch difference ΔN of the leap, ad k the system constant.

The structural penetration of these rules is however quite poor. For example, the harmonic analysis does not take into account longer syntagmatic units of harmony. Only values of isolated chords within a predefined tonality are evaluated. Cadences or modulations are not considered. Moreover, rhythmic structures are completely neglected as a rationale for dynamic accents, for example. Ditto for contrapuntal structures.

Further, these rules are not inductive in the sense that they are not built to shape already given performances. They just act on the physical output and do not take into consideration the special character of the already given data. And an interaction between different arguments in combined rules is not developed. A hierarchical perspective is also not envisaged in the KTH approach. But this grammar is nonetheless a very clear scientific method that can be verified/falsified upon the audible quality of its output.

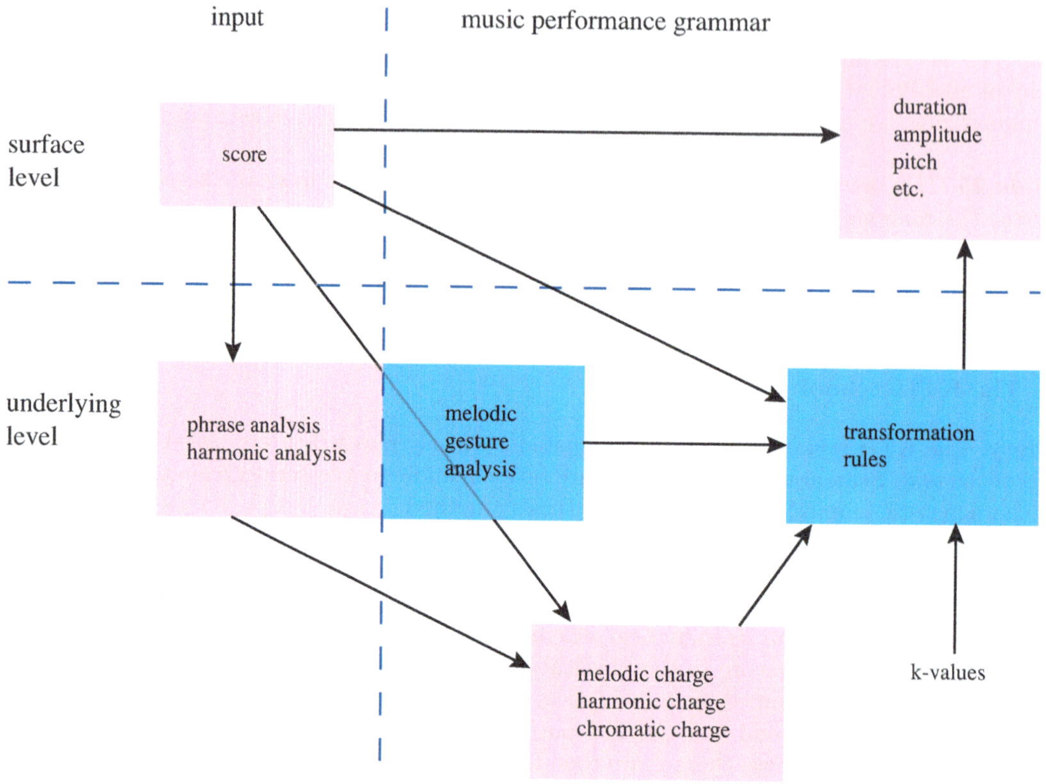

Fig. 37.2. The KTH scheme following [338] shows a clear congruence with the general scheme which was described above in 37.1.

37.1.2 Neil P. McAngus Todd

Summary. Todd's approach is backed by a systematic formalism of performance as a function of structure and specific grammatical arguments. It relates simple structural data, such as grouping boundaries, to expression by means of physically oriented transformation rules.

– Σ –

As we have seen in Section 36.3, Todd's approach to performance is a symmetric one using a performance procedure Π which acts on the input Ψ and an encoding function γ, whereas its inverse is a listening procedure Λ which acts on a performance output P and a decoding function δ. The latter is in fact meant to be a section which determines the parameters of the encoding function that lead to the given encoding values.

This distinguishes Todd's approach from the KTH approach since the requirement of a listening procedure which accompanies the performance procedure is a strong restriction to the entire theory. However, this restriction is not really needed, i.e., one may also investigate the performance procedure without knowing whether there is an inverse solution. This is also the way Todd has viewed his scheme in [1060].

As an example of an analytical rationale, Todd has described a rubato encoding formula

$$\rho_B(t, \phi.) = \phi_1 + \frac{\phi_2}{(1-\phi_6)^2}\{\frac{t}{\phi_3} - \frac{(\phi_4 - 1)}{\phi_5} - \phi_6\}^2$$

at time t, and with parameters

1. ϕ_1 = tempo,
2. ϕ_2 = (rubato) amplitude,
3. ϕ_3 = length of phrase,
4. ϕ_4 = upper limit of boundary strength,
5. ϕ_1 = offset of parabola minimum,

in [1057]. This formula is determined by a hierarchy of time span reduction segmentations in the spirit of Jackendoff's and Lerdahl's Generative Theory of Tonal Music [496], and is applied in a recursive way, between the surface at hidden levels.

This formula defines an operator which is qualitatively different from KTH operators: It includes the basic tempo variable ϕ_1 which could be interpreted as a previously defined performance information, i.e., this operator may be seen as a refinement of an already given "mother tempo ϕ_1". Although this variable is not meant as a time-dependent quantity, the formula works independently of such a restriction and therefore is a real refinement operator of the tempo field. This means that in our above scheme, the Todd operator acts in a morphic way on the already given tempo field.

Inserted in our scheme, the argument of Todd's rubato operator is $\phi_.$, and this one includes the mother tempo ϕ_1, which is in fact a reference to the mother LPS.

This approach to rubato is nevertheless poorly rooted in the composition's musical structure since it does not include the possibility to have harmonical, melodic, rhythmical or contrapuntal arguments in the rubato formula. The existence of a tempo curve is also not mandatory, as we have already seen in the discussion of arpeggio effects in Section 34.2.

37.1.3 The Zurich School

Summary. We give an overview of the approach of the Zurich school. The details will be discussed in the subsequent chapters.

– Σ –

Essentially this approach is centered around the key concept of *weights*. These are numerical functions that encode analyses and serve as an input to the core of grammatical instances: stemmata and operators. Accordingly, a performance is generated by the stemma, a genealogical tree of nodes representing local performance scores of successively refined performance quality. The generation of such node "daughters" from antecedent "mothers" involves performance operators in the role of "fathers". The latter are charged with weights and realize grammatical rules of different flavors. The nature of these rules is not further specified, and may include any of the systems proposed by other approaches as long as they are based upon weights (in particular the KTH and Todd proposals).

The qualitative difference to the KTH and Todd systems is that a clear primavista performance is defined as a starting point of successively refined performances, and therefore, the initial transition from symbolic score data to physical data is anchored. Further, the weight system is conceived in such a way that a combinations of weights may define new weights to be fed into an operator, thereby allowing a simultaneous combination of different analyses to act on performance. Further, the LPS approach is so rich that any performance situation can be dealt with: tempo hierarchies, abolition of tempo in arpeggio and rubato effects, combined deformation of parameters, and also the treatment of gestural output (beyond physical sound parameters). Finally, the stemmatic genealogy with sexual propagation from mother LPS and father operator guarantees an in-depth simulation of the process of rehearsal, where the spiritual unfolding of understanding a score may be modeled. All this has been implemented in the RUBATO® workstation, see Chapter 40.

37.2 Remarks on Learning Grammars

Summary. This section gives a very short remark about grammatical patterns generated by machine-based learning from empirical performance data.

– Σ –

For this section, we refer to [1024]. We do in fact not believe that machine-delegated statistical methods such as neural networks or proper machine learning algorithms for rule learning are of proper scientific value, since when machines learn, we do not. Of course, this is an ideological point of view, but we cannot follow methods which delegate decisions to structured ignorance: understanding cannot be delegated to engineered devices.

For example, Gerhard Widmer's approach starts with a relatively detailed structural analysis of the score, including motives, groupings etc. It then correlates these structures to empirical performance data, such as dynamics or articulation, in order to apply machine learning algorithms for extrapolation to other scores.

38

Stemma Theory

O matre pulchra filia pulchrior.
Horace (65–8 B.C.)

Summary. The stemma theory is introduced from its musicological and practical motivation. Semiotically speaking, performance is a result of a diachronic process. This is traced on the structure of a genealogical tree, the *stemma* of a performance. The stemma formalizes the diachronic process of rehearsal and practising. We describe the structure of stemmata as "family trees of performance", together with the corresponding *genetic* and *environmental* principles.

– Σ –

When we compare the performance grammars developed by the KTH school and Todd to what happens when a musician learns to perform a new composition or when a conductor rehearses a composition with his/her orchestra, there is a tremendous difference of procedures. In the KTH system, the analysis-by-synthesis makes this particularly evident: The rules are not given a priori, but have to be derived via human criticisms and successive revision. This diachronic process is however not part of the grammar, it is a meta-theoretic construct. There is no trace of this successive improvement of rules in the system. History is annihilated by the uncontrollable criticism of a human expert (the violinist Lars Frydén in the KTH methodology). No trace of how the improved rules are produced from the old ones and their result is retained.

The KTH methodology is loaded by a meta-theoretical historical dimension in the analysis-by-synthesis loop. This dimension is not present in Todd's approach, although his hierarchical technique suggests a construction of the surface level (single beats in his prototypical example) from hidden levels. These hierarchies are more of a generative nature from global to more local structures, and not historically guided, though.

If one studies the way a musician rehearses a performance, it seems that a decisive component of this process is the successive improvement of performance which is built upon an added value with respect to the respective previous stages of perfection.

The formal theory of such a diachronic process is the following stemma theory. This theory is already prefixed in the definition of the LPS spaces, since the variables "mother" and "daughters" create the connection to inheritance structures. This theory cannot grasp the totality of diachronic strategies, though. While a stemma comprises genealogical LPS trees, the selection strategies for the construction of determined LPS are not formalized in the stemma concept, only the result is.

Also, the arsenal of performance operators is not under control in this setup, it is supposed that a number of operators is available, but their systematic construction is not dealt with in this framework. The classification of performance operators is far from settled, see chapter 44.7 for this subject.

We are also aware that the present stemma structure is far from general with respect to feedback options, although the circularity of the LPS definition includes quite a lot of feedback techniques. But there is a serious limit to generalization: If one is going to implement denotators for stemmata by way of LPS

G. Mazzola, *The Topos of Music II: Performance*, Computational Music Science,
https://doi.org/10.1007/978-3-319-64444-8_7

denotators, their finite character must be assured in order to create performance outputs in a finite calculation time.

38.1 Motivation from Practising and Rehearsing

Summary. This section analyzes the development of an artistic performance through practising and rehearsal. We exhibit structural ramification and hierarchy, together with shaping mechanisms.

– Σ –

When a performance is realized, this is never the result of a one-step process. Performance results from a development of successively improved intermediate performances. To understand performance, we need to understand its genealogy. A number of experiments have been undertaken in order to analyze the process of performance preparation via rehearsal and preparation. For example, Kacper Mikaleszewski has investigated this process via video recordings of the preparation of the *XII Prélude* from the *Second Book of Préludes* by Claude Debussy [738] by a professional pianist. Mikaleszewski introduces his investigations by the remark that it is not mandatory that pianists may clearly separate stages of their preparatory work of new compositions. In an investigation of A.A. Wicinski [1116], ten famous pianists[1] were interviewed on their strategies. Seven of them comprise the first group of pianists to distinguish separate stages in their work. The second group of three pianists were not able to separate any such stages. This does not demonstrate that the pianists of the second group do not really follow unconscious strategies. The incapability of verbal description of such strategies is in fact a problem known among musicians: Very often, they are not able to verbalize their activities.

The first observation about this experiment is that "the divisions of the musical material introduced by the subject (the pianist) agree with the basic formal units of the composition, here related tightly to its texture. This characteristic agrees with earlier notions about the role of the structure of music in performance, and the tendency to practise longer compositions divided into shorter units (...). At the same time, more complex textures led to the selection of shorter fragments for separate practice, and to making more divisions of the musical material. (...) we may say that what he (the pianist) was been doing is to prepare effective sub-routines of a more complex programme which in turn would make him able to perform the musical composition at a satisfactory level of proficiency." [1116]

Although the general picture is somewhat ambiguous, it can be deduced from these findings that the subdivision of the given score structure, together with a development of local strategies of performance, are crucial. This can be viewed as a strategy for the acquisition of a global "performance plan". The analysis is that "the majority of comments concerning the text of the composition, fingering, hesitations, error corrections, and memorization seem to be in agreement with the general objective of the first stage of work mentioned by Wicinski: to work out a general idea of the composition and to become able to perform it with sketchy interpretation." *In other words, the strategy is not to start working out details and local performance, but to go top-down from the overall picture (as sketchy as it might be) to more and more detailed aspects in the fragments of the subdivision of the given score. So one exhibits a hierarchy of performance development which starts with the global sketch and successively ramifies to sub-routines of local aspects.*

38.1.1 Does Reproducibility of Performances Help Understanding?

Summary. Psychologically, the structure and function of performance generation is far from inscripted within a conscious memory. We discuss the value of an explication and memorization of such a process: Why is reproducibility of performance processes of scientific interest? Memorization relates to the question of identity of a performance (process). This leads to the question whether human precision is different from "machine" or "mathematical" precision.

[1] Among them were: Sviatoslav Richter, Emil Gilels, and Harry Neuhaus.

– Σ –

As we have already mentioned above, a number of excellent musicians cannot (or do not want to) control their performance generation, they just rehearse by some instinctive activity and do not care about strategies and conscious plans. So why try to make such processes explicit, since a good number of artists just do not care. The question really is whether there can be a culture of performance without reflection of the conditions of good performance, good in the sense of Adorno: expression of contents that are discovered via analyses of the underlying text. Now, if this goal is accepted, we need to know about the unfolding of such a performance along the nerve of the inner logic of an artist's elaboration—supposing that such an inner logic subsists. We insist that the absence of such a logic would result in a random walk to performance, an agnosticism driven by blind admiration of an artist's instinct, genius, call it as you like. But then, understanding performance would reduce to plain admiration of a miraculous phenomenon which does not meet our concept of a performance culture.

The scientific treatment of performance culture seems to face still another objection, i.e., the problem of objectivization of performance structures: The uniqueness and magic of artistic performance is sometimes viewed as being in contradiction to objective description, of conceptualization in the framework of scientific experimentation where reproducibility of objectively given conditions is mandatory. This skepticism culminates in the claim that objective, "mathematical" precision misses the precision of a human artist, that the latter precision is of another nature.

This is to say that you may draw a faithful trace of a performance and miss the essence, because a human may reproduce the performance in another way, however maintaining the core of "human precision" as opposed to machine precision. This is however not an argument against "mathematical" precision since its claim is an invariance argument: Although mathematical changes in a performance may occur, the "human precision" is maintained. This means that we would have to search for invariants within the scientifically precise variety of performances. So this is not in contradiction to "mathematical" precision, it simply states that the latter may be too precise. But as long as we do not know what the "human precision" is about, we better stick to the "mathematical" precision which englobes a faithful trace of performance.

Another argument against scientific performance analysis is that its very concept of performance is illdefined. But this means that something has been forgotten in this conceptualization, i.e., one ought to include other components of a performance to grasp its identity. This could be: the audience, the space (concert hall, studio, etc.) of performance, the individual conditions under which a performance is perceived (listening in a bad mood, listening repeatedly to a recording, etc.), and the general historical and cultural background of such an experience. It is evident that the effect of a performance depends on these factors. And that therefore, a complete image of the phenomenon must include these factors. But this does not change the problem of giving a precise description and pursuing an in-depth analysis of the performance in the sense developed so far. *This is a fundamental objection we make against the methodology of the humanities: that they refuse to investigate parts of a phenomenon because they are related to other parts. We do not contend that understanding the whole can be reduced to the understanding of its parts, we contend that understanding the whole cannot refrain from understanding its parts, and that this latter task is the first step of any scientific procedure.*

38.2 Tempo Curves Are Inadequate

Summary. We discuss the conceptual and technical inadequacy of "flat" tempo structure for performance construction and derive the ramified hierarchical tempo trees as realized on the *presto*® software.

– Σ –

It is a commonplace in performance research that a single tempo curve cannot control non-trivial tempo configurations. We therefore implemented a module for hierarchical tempo configurations, called *AgoLogic*, within the composition software *presto*® which was developed for the Atari® computer from 1988 to 1994 [677, 680, 682]. Tempo hierarchies[2] are a preliminary version of stemmata in the onset domain.

In order to make clear the scope of tempo hierarchies, we first give three examples of tempo hierarchies which are of practical significance in the historical context. After that, we shall give a formal definition of a tempo hierarchy in terms of corresponding denotator spaces.

Example 52 The first example is an exercise in tempo curves taken from Carl Czerny's "Pianoforte Schule" [228], see also Figure 38.1 . Czerny's exercise proposes to play this short composition with different tempo

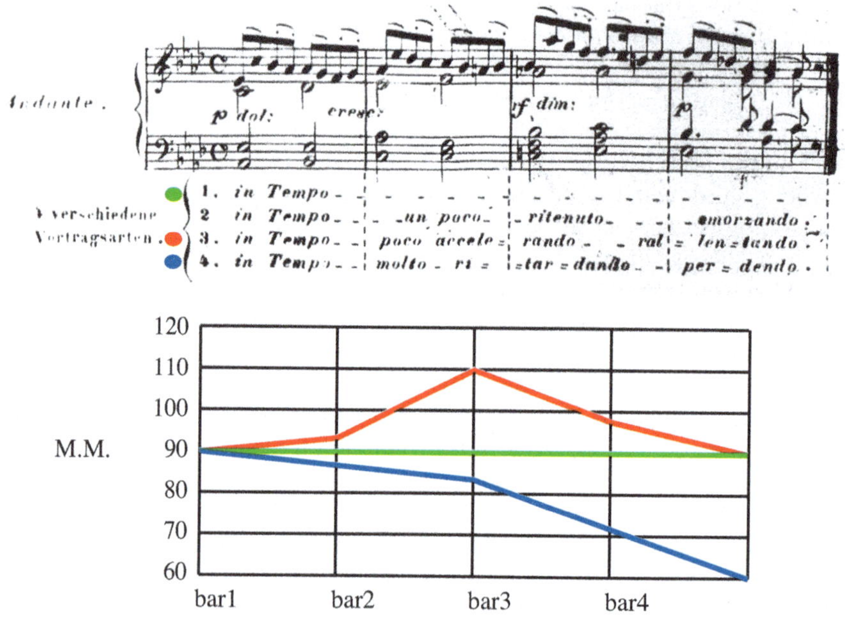

Fig. 38.1. Czerny's exercise for tempo curve testing.

curves: first without any tempo change, second with an accelerando in the middle, third with a strong rallentando at the end. We have simulated these proposals on *presto*®, and it turned out that the result is quite deceiving: No really relevant tempo experience results. In order to construct a less poor tempo structure, we have split the tempo levels in order to achieve the so-called *Chopin rubato*[3]. The three tempo setups as well as the Chopin rubato version can be heard in the first four samples on the audio-file `Czerny` on the book's CD-ROM, see page xlvii. This is a very classical technique of tempo shaping. In *presto*®, this works following a hierarchical construction. In our example, we want the left hand to play a constant master tempo T. The right hand is the slave in tempo; we ask that the right hand tempo may vary anyhow under the condition that both hands coincide on each bar-line. To do so in *presto*®'s AgoLogic module, we may split the onset domain $I = [a_0, a_n[$ of the right hand into a sequence of one-bar portions $I_1 = [a_0, a_1[, I_2 = [a_1, a_2[, \ldots I_n = [a_{n-1}, a_n[$. Then we have a hierarchy $I \to I_1, I \to I_2, \ldots I \to I_n$. On each bar portion I_i of the right hand, the user may reshape tempo via graphically interactive editing. The user can define any polygonal tempo curve T_i within the limits a_{i-1}, a_i of this interval, provided the integral is conserved, i.e.,

[2] This terminology is antiquated now, since we reserve hierarchies for cell hierarchies, and what we call a tempo hierarchy here is in fact a tempo stemma in the present terminology. We nonetheless conserve this antiquated terminology in this special discussion of tempo in the *presto*® software.

[3] Sometimes also called "bound rubato" since one hand is playing a trigger tempo, whereas the other is bound to cope with the master every bar-line onset, say. In contrast to this concept, "free rubato" means that both hands play rubato, but exactly the same, and this frees them from following a master tempo.

$\int_{I_i} \frac{1}{T} = \int_{I_i} \frac{1}{T_i}$, see Figure 38.2. The routine taking care of the boundary condition of invariance of the above

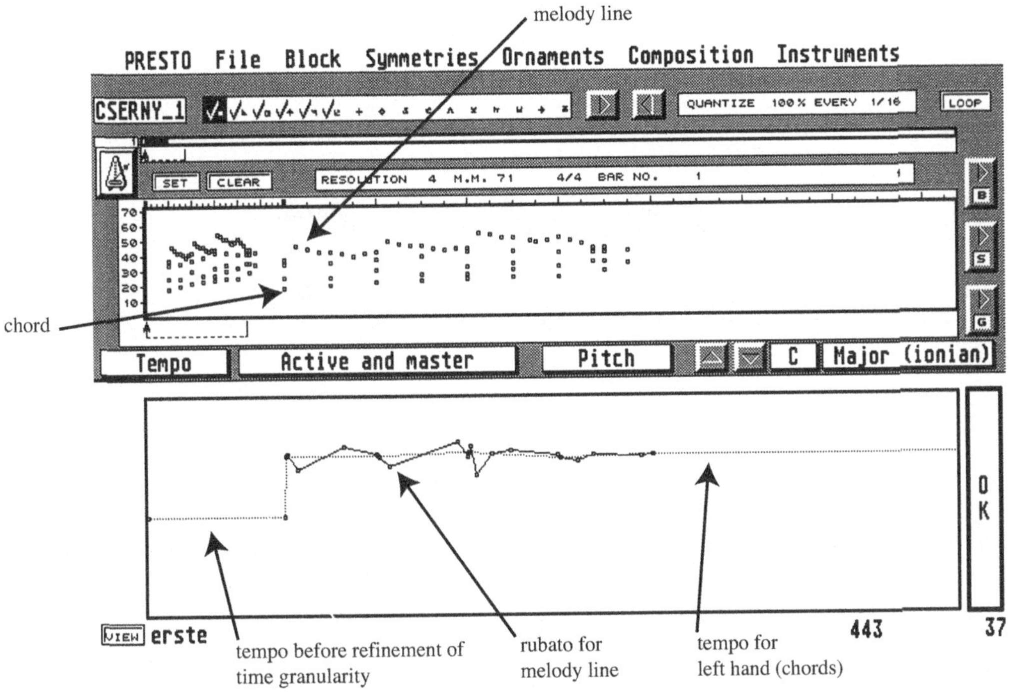

Fig. 38.2. Czerny's example with a depth 1 hierarchy of tempo curves where each right hand bar is a daughter tempo polygon of the right hand tempo which is constant here.

integrals was presented in Corollary 21 of section 34.2. Of course, the mother tempo of the left hand need not be constant, any polygonal tempo curve can be produced for the left hand, and the daughters intervals I_i can be given corresponding polygonal variations with the above boundary condition $\int_{I_i} \frac{1}{T} = \int_{I_i} \frac{1}{T_i}$.

Example 53 The second example is taken from the Chopin *Impromptu* op.29, refer to Figure 38.3, bars No.78-80. This situation shows a series of trills and arpeggi. The tempo of bars SS is confronted with the tempi of the trills and the arpeggi. This is not the situation of a bound[4] Chopin rubato, since the trill ornament is only one portion of a number of notes to be played by the right hand, it is more of a hierarchy in a scenic arrangement where the ornamental notes add a supplementary level of structure. The trill notes are not even explicitly denoted, the artist has to fill up the trill sign in the spirit of Chopin style tradition.

The second auxiliary structure in this example is the arpeggio. This is also an incomplete notation insofar as you have an anchor note and several "satellite" notes, i.e., the chord's notes attached to the anchor note. These must be played in a temporal succession. The temporal succession and the anchor note are not always clear: It could be that the arpeggio succession is read top-down in pitch, or vice versa. Also could the onset of the last or that of the first note be the anchorage onset. These things being selected, the speed and shape of temporal succession in the arpeggio are not well defined. As with the trill notes' temporal distribution, the arpeggio development is a "satellite" phenomenon in the note hierarchy.

---

[4] A *bound* Chopin rubato is one where one hand plays a trigger tempo whereas the other hand's tempo may vary locally, as a tempo slave, but coping with the right hand trigger tempo on a number of master events. A *free* Chopin rubato is one where the rubato is played synchronically for both hands.

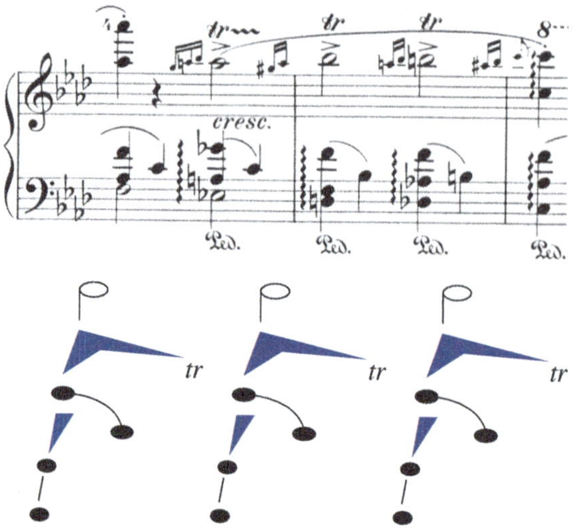

Fig. 38.3. Chopin's *Impromptu* op.29, bars No. 78-80 shows a series of trills and arpeggi.

We have chosen this hierarchy in *presto*®'s Agologic module (see again Figure 38.3): The mother tempo is present on the top level represented by the half notes in the graphic below the score. The top level has two daughter tempi: the trill daughter (to the right below the half notes) for the trill tempo, as well as the arpeggio daughter (to the left below the half notes). The latter controls the tempo of the descending interval notes. It has a daughter tempo for the expression of the arpeggio tempi.

This hierarchy does not determine the concrete tempo curves on the mother level, on the trill, and on the arpeggios; to this end, the graphically interactive input by the user is needed. This means in particular that the same tempo hierarchy can express very different ways of performing a piece: from the beginner to a virtuoso. On the book's CD-ROM (see page xlvii), four samples of performances of this tempo hierarchy are traced. They are heard on the last four samples (after the Czerny samples) of the audio-file `Czerny`.

Example 54 The third example illustrates the tempo hierarchies as they are needed in the performance of a large orchestra with special instrumental groups. Our example is a large orchestra which is controlled by a conductor. Within this orchestra, we suppose given a string group (violins, say) which has to obey the conductor's indications, but within two time windows (the curva1, and the curva2 windows) may follow an individual tempo curve (as indicated by the concert master, say). Within each such slice, on a small time window (curvetta3, curvetta4), there is a soloist part (played by a first violin, say) which may realize a cadence-like small expression, but has to cope with the string group (curva1 for curvetta4, curva2 for curvetta3, respectively). The tempo window of *presto*® shows the individual curves with their local deviations which, according to the implemented algorithm, yield the same total physical durations as their respective mothers.

38.3 The Stemma Concept

Summary. The formalism of performance stemmata is introduced. A stemma is a rooted directed graph which carries on its nodes local performance scores. The generative principles as well as the entire structure are modeled after a matrilineal scheme: There are only mothers and daughters on the stemma, starting with the primary mother. But the proliferation of these families is sexual: Fathers, formally represented by performance operators, do contribute to their daughters together with their mother. The detailed "sexual behavior" is described and turns out to be quite similar to the veritable biological/sociological behavior in life.

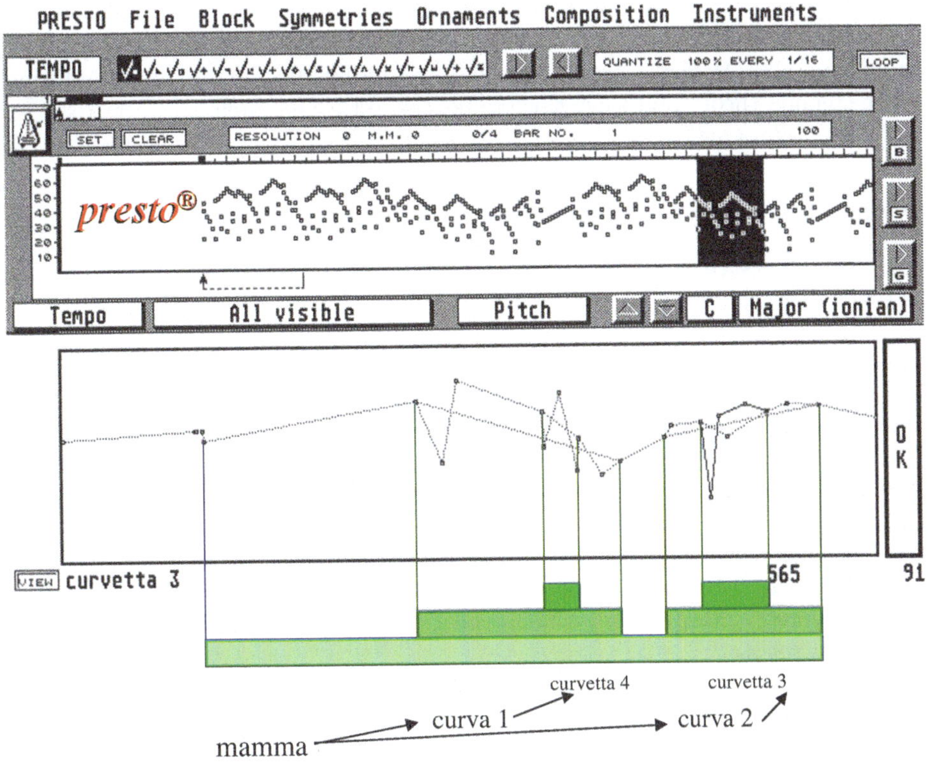

Fig. 38.4. The tempo hierarchy of an orchestra is shown. The mother tempo curve (mamma) is ramified into two daughters (curva1, curva2) which in turn have one daughter each (curvetta4, curvetta3, respectively). This interdependence of local and global tempi is used to differentiate roles in orchestral time control.

– Σ –

This section terminates the chapter with a formal construction of stemmata. This is a remarkable subject for three reasons:

First, the stemma concept is probably one of the first to grasp historical processes on the formal level in the mathematical sense, where rather than "formal" we should say: "precise", as contrasted to the notoriously poor conceptualization in the humanities. This has tremendous consequences for the experimental modeling of historical processes: One can now perform historical developments which never have taken place, more concretely: rebuild real and simulate fictitious performance history of the rehearsals of a pianist, compare these directions and draw conclusions on the quality of the real history against the way of fiction: Are we, and why are we living in the best of possible worlds.

Second, this formalization is not only formal in the mathematical sense, but also on the level of implementation in computer software. In fact, we have seen in the course of the intricate definition of an LPS in section 35.4 that this concept can only be defined by means of a circular form $LocPerfScore_{BP}$, since a standard mathematical definition would not allow circularity, circularity being only an accepted technique in implicit equations, not in conceptualization—although, as we have seen in chapter 9, set theory (probably unconsciously) conceptualizes sets in a circular form. But this nature perfectly fits the nomenclature of object-oriented programming languages: An instance variable can very well be an object of the same class as the object whose instance variable it declares. This observation once more evidences a turning point between mathematics and object-oriented programming, where the denotator and form concepts have been derived.

Third, the very nature of this formalization of historicity appears to fit with biological inheritance principle of sexual propagation. This is not only a happy coincidence, rather is it a mandatory direction if one wants to model learning: By sheer life experience, inheritance and evolution are the best proven models of successful learning.

Although it is possible from the preset concepts, we do not diverge on global performance score constructions for the stemma theory and leave this segment to future research.

All the LPS denotators will be situated at the zero address as long as we do not stress the contrary.

38.3.1 The General Setup of Matrilineal Sexual Propagation

Summary. This section describes the overall mechanism of stemma construction. In particular, we discuss the reason for the matrilineal approach.

– Σ –

Given a sequence of basis and pianola spaces B, P, recall the definition of the LPS space (35.2) with the six factors *Mother* μ, *Daughters* Δ, *CellHierarchy*$_{BP}$ $\mathfrak{h}$, *Instrument* ι, *Operator* Ω, *Weight* w.

Definition 106 *If* $\Lambda : 0 \rightsquigarrow LocPerfScore_{BP}(\mu, \Delta, \mathfrak{h}, \iota, \Omega, w)$ *is a local performance score, let* $\Lambda \downarrow$ *(resp.* $\Lambda \uparrow$*) the directed graph of all LPS that can be reached by finite descent from* Λ *to the daughters, to their daughters, etc. (resp. the set of all LPS that can be reached by finite ascent from* Λ *to the mother, to its mother, etc.), together with the mother-daughter arrows. Denote by* $\Lambda \updownarrow$ *the union of the graphs* $\Lambda \downarrow$ *and* $\Lambda \uparrow$*. Call* Λ^∞ *the directed graph of all LPS that can be reached from* Λ *by finite ascent and descent, together with the mother-daughter arrows.*

Definition 107 *A local performance score* $\Lambda : 0 \rightsquigarrow LocPerfScore_{BP}(\mu, \Delta, \mathfrak{h}, \iota, \Omega, w)$ *is called* a stemma[5] *if its graph* $\Lambda \updownarrow$ *is finite and defines an undirected tree (no undirected cycles), and if it is not the daughter of its mother, if that mother exists*[6]*. The leaves of* $\Lambda \updownarrow$ *(which are also the leaves of* $\Lambda \downarrow$ *in this case) are called* the leaves of Λ, *whereas* Λ *is called* the primary mother of the stemma. *If* Λ *has no mother, it is called* a prime stemma. *The set of leaves of* Λ *is denoted by* $\underline{\Lambda}$.

Intuitively speaking, a stemma is a primary mother, together with its daughters, its granddaughters, etc., until we reach the leaves which are the output LPS that will eventually yield the data to be performed. It is important to have the mother of a stemma being disconnected from its daughter, i.e., not pointing at the latter in its daughter list. The primary mother of a stemma is meant to define a new tree of unfolding LPS which cannot be accessed by any other stemma. Although it can access another stemma, it is invisible to this latter. We have seen a stemmatic structure in the previous discussion of "tempo hierarchies" which are, in the present terminology, a kind of tempo field stemma, although usually there is more in a stemma than just onset-related fields and onset kernels.

Exercise 79 *Restate the "tempo hierarchies" in terms of stemmata with*

$$B = \{Onset\}, P = \emptyset.$$

The matrilineal terminology is not really the whole truth here. In fact, each LPS contains its operator, the patrilineal component which is responsible for the generation of this LPS. But the result, more precisely: the hierarchy $\mathfrak{h}$, returns the output information for performance. And it is this result which will also be used to produce further daughters, and not the operator. This justifies the matrilineal terminology, whereas the operator is only a hidden generative instance.

Let us look at a historical example of a stemma: the stemma for the composition *Kuriose Geschichte*, the second *Kinderszene* in Robert Schumann's synonymous collection op.15 [954]. This stemma was constructed on the performance platform RUBATO® in 1996 at the *Staatliche Hochschule für Musik*, Karlsruhe by the author, Oliver Zahorka, and Joachim Stange-Elbe. It took us three days to realize the whole setup and

[5] "Stemma" is synonymous to "genealogical tree".

[6] This is a slightly irritating subtlety of our conceptualization.

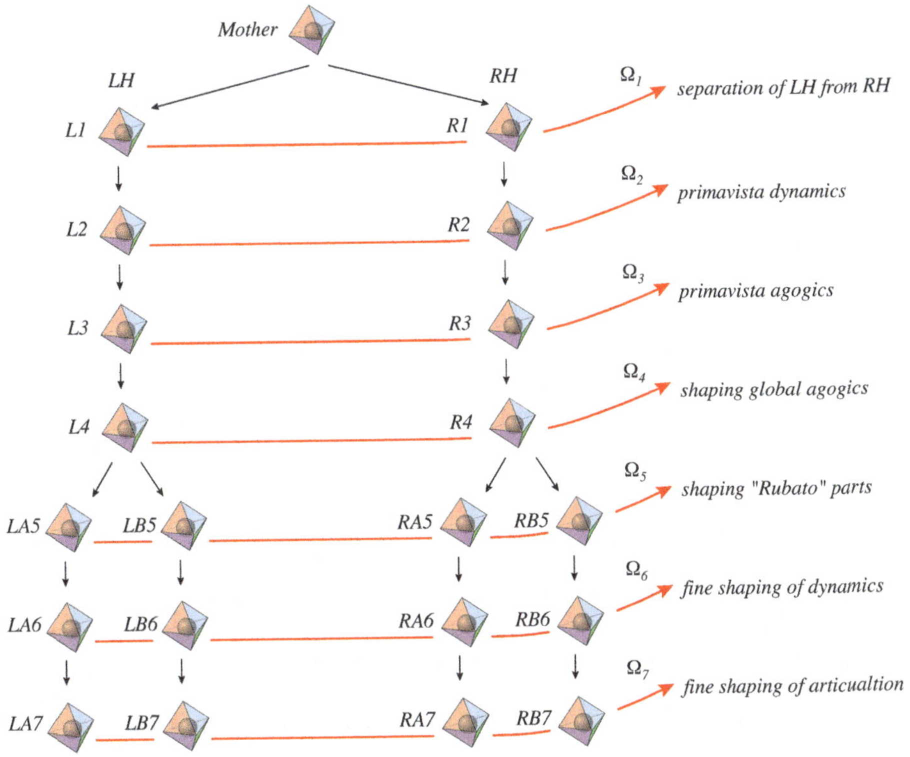

Fig. 38.5. The stemma of the first performance of Schumann's second *Kinderszene: Kuriose Geschichte* that was constructed and performed 1996 on the Boesendorfer grand piano at the *Staatliche Hochschule für Musik* in Karlsruhe.

performance on a Bösendorfer MIDI grand. The performance of the piece is documented on a CD, see [702], and in a broadcast of the Austrian TV [334]. Although the stemma is quite primitive, the shaping results were satisfactory and taught us a lot about the empirical aspects of computer-assisted performance research. The stemma is visualized in Figure 38.5. Although we see that each single refinement layer is controlled by one and the same operator (horizontal arrow), the layer did not form a grouping in the technical sense to be discussed in section 38.3.3: Each daughter had to be performed as an isolated instance, since no grouping methods were implemented at that stage. The construction of this stemma first follows the splitting of right (RH) and left hands (LH), then, after the shaping of primavista dynamics and agogics, global agogics is constructed on these two LH and RH symbolic kernels. The splitting for operators $\Omega_5, \Omega_6, \Omega_7$ regards a small number of bars which have to undergo a more differentiated rubato. The final shaping regards fine "tuning" of dynamics and articulation in all leaves.

38.3.2 The Primary Mother—Taking Off

Summary. The primary mother represents the performance score which is deduced from the score data as they are inscripted on the predicate level. We make the deduction process explicit, together with the set of prima vista parameters.

– Σ –

The primary mother Λ of the stemma is the starting point of a stemmatic evolution process, it is used to derive all the LPS in $\Lambda\downarrow$, and eventually leave set $\underline{\Lambda}$. There are two situations in such a primary mother: Either it has a mother LPS Λ_0 or else it is a prime mother with the score form mother denotator $S : 0 \rightsquigarrow ScoreForm(\sigma)$.

If it has a mother LPS, this means the following: The stemma which it defines is not an autonomous structure, but it is derived from another stemma Λ_0. This other mother stemma can however not access Λ, this one is invisible to Λ_0. The idea of this asymmetry is that we want to separate stemmatic processes in their information flow. A stemma is a closed unit of unfolding performance stages. It can be used to induce a new stemma, but it is not related by a daughter pointer to the new one. This prevents mixtures of stemmatic unfolding: Once a stemma is defined, it can help another stemma build refined performances, but it does not refer to this one, while vice versa, a new stemma may refer to the old one via the mother pointer of its primary mother. This technique creates a series of successively improved stemma $\Lambda_0, \Lambda, \Lambda', \Lambda'', \ldots$

If Λ has just the score denotator S, there is no reference to other previous stemmata, and *we are in the properly termed primavista situation*: There is a score, which is formally expressed in terms of the denotator S, and this score is the only reference to create the first performance. This is a complex operator which we shall now discuss.

The construction of all data of this prime mother is the scope of a special operator which we call the *PrimavistaOperator*[7]. The PrimavistaOperator is a macro-operator since it has to deal with all possible score signs incorporated in S, but this is more a question of software implementation than a mathematical problem.

The problem is this: recall that in the KTH theory, there was no such a thing as a prima-vista setup. This prime mother construction must be dealt with somewhere in the performance process. In the unfolding of a performance, it is the first action to be taken: to establish a first version of a performance which is uniquely based upon score data. It could be argued that the score data can be supposed to be introduced in advance via the symbolic kernel. This could even be admitted, but then, all the other data, such as rallentandi, fermatas, slurs, etc., where should they be piped in order to contribute to a primavista performance?

For example, a fermata sign must be taken into account while it defines a sensible tempo sink, and this must be done before any refined shaping activity of performance is set forth. Of course, it is not sufficient to just know that this fermata sign is situated at a determined place, we also have to provide the data for the exact shape of this specific tempo sink. So, first of all, the input must list all the possible relevant signs, i.e., the primavista predicates in the sense of section 18.3.3.

The input data are all based on the first information about the events of the score S in the different parameter spaces of BP that will participate in different special predicates. These may be ordinary notes, bar-lines, pauses, and the like. In order to produce a set of events that cope with the hierarchy induced by condition 1 in Definition 102, i.e., that in a cellular hierarchy, kernels project into kernels. This means that the symbolic kernel event set $Evt_S(U)$ associated with space $U \in BP$ must project into $Evt_S(V)$ if $V \subset U$. We shall here speak of "events x in space U" in the sense of zero-addressed denotators $x : 0 \rightsquigarrow U(\xi)$. This induces the Boolean predicate $Evt_S(U), U \in BP$, via

Definition 108 *For $U \in BP$ and an event x in U, we set $x/Evt_S(U) = \top$ iff $x : 0 \rightsquigarrow U(\xi)$ is such that there is an event $y : 0 \rightsquigarrow W(\eta)$ in the score denotator S, living in a (not necessarily strict) superspace W of U in BP, such that $p_U(y) = x$ (i.e., $p_U(\eta) = \xi$), else we set $x/Evt_S(U) = \bot$.*

This presupposes that the events of the score denotator S have been identified a priori. These are denotators in spaces of BP that can be found upon inspection of the S. This is a knowledge which ultimately exceeds the formalized knowledge base we are dealing with in our mathematical framework, it needs an instance which can create a score denotator S from the given score. This could be an optical character recognition (OCR) software for scores, or any machine that collects events from MIDI files, for example, or just a human expert in score reading. *In any case, we may suppose that the score is transformed into a score denotator S in a score form, and that the events are deduced from S. The latter is a standard task in logical and geometric motivation predicates.*

To define the prime mother LPS, we have to define all its constituents: Instrument, weights, operator, and hierarchy. As to mother and daughters, the first is S, and the latter will be added when the stemma is

[7] In accordance with code naming conventions in object-oriented programming, the nomenclature is this: every operator is named by a special name *Specialname*, directly followed by the postfix "Operator", yielding "*Special-name*Operator".

made explicit in a later stage of the historical process—presently, it is empty. The most important data is the hierarchy. We have to construct it by use of the PrimavistaOperator and the given weights. The hierarchy defines also the performance map on its top space, and we suppose that the instrumental specifications are sufficient to transform the physical parameter vectors of the image of the performance map $\wp$ into sound objects. So we may concentrate on the hierarchy construction here.

In the hierarchy $\mathfrak{h}$ to be constructed, we have to define the diagram of cells. This first of all means defining the projections of kernels. We instantiate all the event sets as the first bunch of predicates $Evt_S(U)$ in the weight list of our prime mother. For any space $U \in BP$, we select the kernel $K_U = supp(Evt_S(U))$, by definition of the predicates $Evt_S(U)$, the kernels map into each other. Also, the frames R_U may be pre-defined as the smallest cubes containing all the predicate supports K_U. Since we are starting with an instrumentally well-defined local situation (only one instrument), we may also suppose that there is a top space Top_S in BP where the kernel K_{Top_S} is not empty. To define the space hierarchy, start taking all spaces with non-empty kernel, these are just all subspaces of Top_S. This hierarchy is much too large, in general. We first have to restrict to the standard hierarchy requirement described in section 35.3, requiring that for any pianola space within a hierarchy space U, the corresponding basis space must also be in U. For example, in piano music, duration cannot be a reasonable hierarchy space. One further has to restrict this hierarchy to a subhierarchy which is reasonable for the given instrument. There is no general algorithm for such a procedure, one has to observe two things, however:

- the hierarchy must be standard with respect to basis and pianola spaces;
- if a K_U contains events that are not proper projections of other events, this space must be retained in $\mathfrak{h}$;
- there are default space hierarchies for specific instruments, such as the piano and violin hierarchies described in section 35.3.3.

With respect to these constraints, one will choose as space hierarchy the smallest standard subhierarchy $\mathfrak{h}$ of the total hierarchy of Top_S such that it contains the default space hierarchy and the spaces containing events that are not proper projections. For piano music this means taking notes in $EHLD$, pauses, slurs, accelerandi and similar events in ED, bar-lines, time and key signatures in E, but observe that these event types are all contained in one of the spaces of the default piano hierarchy. The selection of such boundary conditions for hierarchies is a typical system parameter in the PrimavistaOperator, a parameter which, together with the weights, creates concrete hierarchies.

A further set of system parameters: the selection of initial sets, performance fields, and initial performance, is a bit more delicate. The performance field to start with depends on the physical values attributed to the symbolic ones before any refined performance takes place. Typically, this can be solved by an affine isomorphism if we convene that on the physical level, we have the common space of pitch, corresponding to the logarithm of frequency, loudness, corresponding to the logarithm of intensity and time corresponding to onset units, etc., see Appendix A for such common spaces. Then the performance field is a constant one (the derivative of an affine isomorphism being constant, i.e., its linear part). Since in common situations, the linear part of $\wp$ is also diagonal (just some positive gauging constants), *the prime mother performance field is constant with positive coordinates.* This setup evidently guarantees that the projections commute with the constant performance fields $\mathfrak{Z}_U$ on the spaces U (condition 3 in Definition 102 of performance cell morphisms), which we have to define in their respective coordinates in each space U of the defined space hierarchy. This is what we now assume.

Consequently, the entire frame volume can be reached on these constant prime mother performance fields from the left "bottom walls" of the frames $R = [a_1, b_1] \times \ldots [a_n, b_n]$. There are n bottom walls for R, i.e., the $(2^{n-1} - 1)$-simplexes

$$W^R_{n,i} = \{(x_1, \ldots x_{i-1}, a_i, x_{i+1}, \ldots x_n), \text{with } x_k = a_k \text{ or } x_k = b_k, k \neq i\} \quad (38.1)$$

for $i = 1, \ldots n$; they are degenerate for $n > 2$. The initial set I on an n-dimensional space U with frame R is the family $(W^R_{n,i})_i$ of bottom-wall simplexes. Since we have $I \int^{\varepsilon} \mathfrak{Z}_U = R$ (with $\varepsilon = +$), condition 4 of the definition of a morphism of performance cells is also fulfilled. We are left with the initial performance condition 5 in the definition of a morphism of performance cells. This initial performance can be defined by the same data which we have used to define the performance fields. Therefore, condition 5 is automatically fulfilled. So we are left with the construction of the weight system and the PrimavistaOperator that acts on the defined hierarchy in order to integrate the information from the primavista score predicates.

For events with $x/Evts_S(U) = \top$, or for local compositions of such events, or for local compositions of such local compositions, etc., as required by the weight definition of an LPS, we now look for special predicates according to the musical notation conventions[8]. We have already listed the common prima vista predicates in section 18.3.3. Here, we give a selection to show how these predicates can be restated in terms of weights. If needed, such weights are added to the weight list of the LPS.

Slurs. There are two types of slurs: normal legato slurs and articulation slurs. Both are boolean predicates which are evaluated on sets of events in a space U. Call the legato predicate $LegatoSlur_U$ such that $x/LegatoSlur_U = \top$ iff x is a local composition in U which is embraced by a legato slur. As we are specifically interested in x being also a local composition stemming from S, we set $LegatoSlur_{U,S} = LegatoSlur_U \& Evts_S(U)$. We shall henceforth abbreviate the logical combination $?\&Evts_S(U)$, by an added index $?_S$; for example, the articulation slur is denoted by $ArtiSlur_{S,U}$.

Articulation. The following predicates are self-explanatory by their names and all relate to evaluation on single events of U in S:

$$Staccato_{U,S}, Staccatissimo_{U,S}, Marcato_{U,S}, Tenuto_{U,S}, Accent_{U,S}$$

Fermata. A fermata lives in the $U = E$ space of onset; no duration of the fermata is explicitly defined. The predicate weight is denoted by $Fermata_{E,S}$, and an onset x is a fermata iff $x/Fermata_S = \top$.

Value Change. A value change $p/q = r/s$ is a change in the time signature on a special bar-line, from $p/q = 2/4$ to $r/s = 3/8$, say. So we first need a time signature predicate for time signature p/q. This is an onset-located predicate, call it $TimeSig(p/q)_S$, and then a predicate for two-element onset sets $x = \{a, b\}$ of time signature predicates, i.e.,

$$\begin{aligned}
&x/ValCh(p/q = r/s)_S = \top \text{ iff}\\
&card(x) = 2\ \&\\
&\exists a \in x, a/TimeSig(p/q)_S = \top\ \&\\
&\exists b \in x, b/TimeSig(r/s)_S = \top,
\end{aligned}$$

a predicate that is motivated by a mathematical predicate and the already given time signature predicate. Observe that the structure of the mixed powerset spaces U enables us to use ordered pairs of any points of U as they are defined in classical set theory, see Appendix C, Definition 128. So it is also possible to redefine the above predicate via ordered pairs a, b of objects with $a/TimeSig(p/q)_S = \top, b/TimeSig(r/s)_S = \top$. The details are left as an exercise.

So the weights in our weight list of the prime mother LPS is the list of a) the symbolic kernels, and b) the appended list of all primavista predicates, as discussed above, with their restatement as weights with truth values in $\mathbb{R}$.

The *PrimavistaOperator* now has to add operations on the given hierarchy $\mathfrak{h}$ which stem from a paratextual meaning of the weight predicates, i.e., an interpretation beyond the abstract textual trace which the score denotator has induced. For example, a fermata predicate has to induce a tempo sink. This and all other actions of the *PrimavistaOperator* will be discussed in chapter 44.7 about operator theory.

38.3.3 Mono- and Polygamy—Local and Global Actions

Summary. The typology of actions that operators may take in order to unfold the stemma includes monogamic coupling with one mother and production of one or several daughters, or else polygamic cou-

[8] By that we mean the conventions of European tradition. For other traditions, special predicates and parameter spaces have to be introduced, but the procedure is the same, though not always an easy one, as the Japanese Noh nomenclature may illustrate (see the Noh example in 18.3.3.2).

pling with simultaneously several mothers. We give the formal description and its justification in terms of practising and rehearsal.

– Σ –

A performance operator Ω of an LPS $\Lambda : 0 \rightsquigarrow LocPerfScore_{BP}(\mu, \Delta, \mathfrak{h}, \iota, \Omega, w)$ has the function to define the cellular hierarchy and the initial performance ι of Λ. This data is calculated by use of the weight input w and—except for the primary mother—by the already calculated mother LPS μ. This is also represented by the equation

$$\Omega \propto \mu = \Lambda.$$

In general, a mother may have several daughters with the same father operator, i.e.,

$$\Omega \propto \mu = \Lambda_i,$$

which seems to be an inconsistent notation since the index is absent in the left part of the equation. In fact, we have to be more precise here. The operator for daughter Λ_i must have a system parameter i which specifies that daughter from one and the same mother μ. So the correct notation would be

$$\Omega(i) \propto \mu = \Lambda_i,$$

where the system parameter is specified. This is also what in software design is realized: For each parameter i, the corresponding daughter is instantiated as a function of that operator with the i-value in its instance variables.

This ordinary christian family life is however not the most economic and reasonable in many performance situations. For example, it may happen that one has already developed the stemma to a strongly ramified tree, and that one wants to apply the same refinement procedure to all or a large number of leaves $\Lambda_k, k = 1, \ldots K$. One could of course apply the same operator Ω to each Λ_k independently of the other leaves. But then, changing some parameters of this operator would require us to go through each leaf and alter the data step by step. In order to avoid this "copy and paste" process, it is better to let the leaves know that it is one and the same operator and not just a set of clones Ω^* of Ω which have to produce daughters $\Omega^* \propto \Lambda_k$.

The problem here lies in the definition of a denotator's identity. In fact, if we assign to all leaves Λ_k the same denotator Ω as their operator, then any change in this operator is simultaneously carried out on all these leaves. But this is not so trivial: If we had a situation as it is known from object-oriented programming, i.e., if the operator were an instance variable of a LPS class, then its change would be an automatic change in every instance Λ_k. In the context of denotators, however, we do not have object-oriented structures, and the change has to be declared explicitly according to some kind of concept surgery, as it was discussed in Section 6.9. The surgical intervention to be carried out here would be this: Search for all LPS Λ_k in the given stemma, or else in some more specific hierarchical position, such that their operator is named Ω. Then define new LPS Λ_k^* by replacement of the coordinate (named) Ω by a new coordinate (named) Ω^*. This only works if names identify denotators, otherwise, we have to search for other keys to retrieve the wanted denotators. One could denote this intervention by the symbol $\Lambda_k^* = \Lambda_k/\Omega \pitchfork \Omega^*$. A global notation of a grouping of LPS within a stemma is to write G/Ω or simply G for a set G of LPS which is operated by one and the same performance operator Ω, we call such sets *stemmatic groupings*, they are evident denotators in the powerset space of the present LPS space. Accordingly, we write $G/\Omega \pitchfork \Omega^*$ for the replacement of Ω by Ω^* in the grouping G. Clearly, any two groupings within a given stemma are disjoint. Observe also that the mothers of a grouping do not automatically define a grouping for *this* reason! Moreover, we ask for the following transitivity (collective responsibility) axiom:

Axiom 4 *If G is a stemmatic grouping, and if a daughter δ of a mother $\mu \in G$ is a member of a group H, then every daughter of a mother in G is also a member of the group H.*

For example, if G is grouped by a tempo operator which imposes a new tempo curve on each member of G, and if we create a group H of daughters with a refinement of their tempo curves, then, if a daughter $\delta \in H$ has its mother $\mu \in G$, it is reasonable to have the same refinement of the tempo curve for all daughters of all members of G, because the shared tempo curve from G should inherit its refinement.

Axiom 5 *It is always possible to resolve a grouping in the sense that any further changes to the stemma only affect the grouping's former members instead of the entire grouping. However, the descendants of a grouping's members are not affected in their grouping memberships.*

This is a typically historical process in the stemmatic construction: The final stemma is a sequence of intermediate stemmata, i.e., we really have the time parameter entering the world of denotators, or, rather (and more precisely) the world of predicates.

38.3.4 Family Life—Cross-correlations

Summary. Apart from genetic interaction as described by mono- and polygamic propagation, the shape of a single daughter can also be determined by cross-correlations with its sisters or other relatives. We again give the formal description and its justification in terms of practising and rehearsal.

– Σ –

Suppose that a mother μ in a stemma Λ has a number $\delta_1, \ldots \delta_m$ of daughters. What is the common situation for such a family? We have seen in the example of the composition *Kuriose Geschichte* (Figure 38.5) that such a set of daughters can occur if we have to split the whole composition or a part of it into mutually disjoint subcompositions, such as left hand and right hand, which must be treated in individual ways. Or if some bars require a different performance shaping than the other bars, as is the case for the splitting of left and right hand, respectively, in that stemma.

This procedure might also be applied in a more systematic way with respect to natural grouping structures of a composition: For example, in Schumann's *Kinderszene 7: Träumerei*, we have four periods A_1, A_2 (= the repetition of A_1), B A_3 (= recapitulation of A_1), eight bars each. This could be the basis of an operator which splits the entire composition into these four daughter. In each period, we have the grouping into the eight bars, and that can provide us with granddaughters $A_{1,1}, A_{1,2}, \ldots A_{1,8}, A_{1,1}$, $A_{2,2}, \ldots A_{2,8}$, $B_2, \ldots B_8$, $A_{3,2}, \ldots A_{3,8}$. This situation is however not what is really relevant in a realistic performance. The point is that we need operators that deal with relations between different sisters and not only between daughters and mothers. Musically speaking, this means that an artist must take into account what will be played in future bars and what has been played in previous bars of the present period, when shaping a particular bar performance. One must also take into account what is the performance of future and of previous periods *qua* periods. So we have to face operators which take into account more distant relatives in this large family of stemmatic nodes.

In fact, from the artistic point of view, a performance must testify a sophisticated coherence in the large and also within more restricted neighborhoods of musical events. This is a common place in musicological performance theory and in the feuilletonistic music criticism. But such a general insight is very difficult to make precise for several reasons.

First, it is not clear which performance structures should be involved when being set into a coherent whole. Must we think of coherence among local tempo curve segments, or, more generally: coherence among local parts of performance fields? Or should we separate coherence questions along the strata of cellular hierarchies? And if we are searching for such a coherence, how should the analytical background, more precisely: the weights which aliment operators, be screened for coherence? In other words: What is a coherent analysis? Evidently, without such a concept, coherent performance fields can be defined, but they risk failing in their rational task: to reflect the analytical background. Can they be coherent just with respect to themselves without expressing coherence of analyses?

Second, the mathematical variety (in the non-technical sense of the word) of coherence structures is virtually infinite: One could imagine linear, polynomial, differentiable, analytical, and—hélas—statistical, any kind of non-linear relations that would be involved to define coherence. The influence of other family members on a determined LPS could also come from its sisters, from cousins, from more distant relatives, anything is imaginable. We shall make explicit models of stemmatic coherence in chapters 44 and 47. These models are essentially linear models on every set of sisters, but in their combination among the whole stemmatic inheritance, they accumulate non-linear phenomena which lead to non-trivial algebro-geometric phenomena, and phenomena pertaining to second-order differential operators.

These perspectives should make clear that we scarcely understand the genuine concepts of performative coherence in their musical phenomenology and accordingly cannot construct mathematical models on the basis of such a blurred phenomenology. Adorno and Benjamin have given us the catchwords for a deeper investigation of performance, catchwords which we could very well transform into adequate mathematical concepts. But these theorists did not elaborate their conceptual germs to a degree of differentiation that could help describe and understand the concrete artistic shaping of performance. We can hardly understand and even less forgive the tremendous lack of musicological conceptualization and knowledge about performance in view of the overly fluffy and too often ridiculously blasé music criticism in the feuilletons of our newspapers.

39

Operator Theory

If I chance to talk a little wild, forgive me;
I had it from my father.
William Shakespeare, Henry VIII.

Summary. Operators are the substance of shaping performance. They refer to mothers, but they are the only instances capable of altering, refining, or ruining what has been achieved. They are also the pipes where exterior information, be it from score predicates, from analytical data, or from general system parameters, can be channeled and transformed into performance structures. We describe and motivate the concept of a weight (function). This is the turning point between "exterior" and "interior" strata to performance. We discuss different exigences for performance operators to cope with 'primavista' and 'analytical' data. A series of common primavista and analytical weights is discussed. We then expose a taxonomy of operators, followed by some special examples of the existing realizations, regarding tempo and articulation, as well as theoretically founded generalizations which are based upon Lie derivatives. The chapter concludes with a discussion of more "social" types of operators which correspond to "family life" correlations introduced in Section 38.3.4. The final subject is a prospective to 'continuous stemmata', i.e., generalized stemmata based upon infinitely small coupled space portions.

– Σ –

Whereas the previous theory culminated in the matrilinear stemma theory, we now have to face the masculine contribution to sexual propagation of performance shaping. This part is driven by the performance operators Ω which are a coordinate of each LPS δ and generate this LPS from its mother μ: $\delta = \Omega \propto \mu$.

The information which is used to feed an operator is a list of weights in the sense of the space form $WeightList_{BP}$ of Section 35.4. Recall that a weight is a real-valued predicate on an iterated powerset of local compositions derived from the kernel in the hierarchy spaces. We have to justify this approach and will do that in Section 39.1.

But from our discussion of expressive performance in Chapter 36 it is clear that operators have to use some rationale to shape their LPS' performance maps. In our following examples, we shall restrict such rationales to rational operators, i.e., such operators which use exclusively score-related primavista and analytical information.

Despite this restriction, we are far from understanding the general nature of rational operators. Clearly, the primavista operators are not the problem: they are straightforward (though not always easily formalized) translations of musical score competence into mathematically valid operations. The problem is rather on the side of general formalisms which could englobe the processing modes of analytical weights for producing reasonable (expressive) performance deformations of a given mother LPS data. A coarse qualitative classification of operators is nevertheless possible according to where the operator intervenes in the entire transformation process from the mental (symbolic) kernel, through the performance field (and the associated performance transformation), to the physical output data.

G. Mazzola, *The Topos of Music II: Performance*, Computational Music Science,
https://doi.org/10.1007/978-3-319-64444-8_8

It is also possible to show that a number of special formulas for rational operators, such as the tempo operator, or some types of articulation operators, are special cases of the so-called Lie-type operators (Section 39.7). "Lie-type" means that the Lie derivative of a weight function along a performance field is responsible for the field deformation of the daughter with respect to its mother. These Lie operators then specialize to basis operators and pianola operators according to their action being on basis or pianola parameters, respectively.

The chapter concludes with a subject that extends the discrete character of a stemma to continuous parameters for the description of families of daughters. It is in the same vein as the introduction of continuous performance maps to replace discrete maps on discrete sets of notes: Although the material is discrete, the mental construction behind the shaping of this material is more than that: Human cognitive activities can operate on a continuous (or differentiable, etc.) paradigm when generating discrete effects. In fact, we could state this as a very principle of musical activity:

Principle 26 *Music notation and also the very expression of musical material is essentially a trace of infinitesimal forces on a discrete reduction*[1].

This is, what Valéry was alluding to when he stated[2]: "La musique mathématique-ment discontinue peut donner les sensations les plus continues." In our understanding, this is also valid in the sense of a poietic principle, the words "donner les sensations" being replaced by "provenir des forces".

39.1 Why Weights?

Summary. Within the transformation process from abstract and conceptual analysis and representation of score data, weights are an *inevitable* instance charged with the production of numerical performance data.

– Σ –

Usually, in musicology, analysis does not end up with numerical data. A harmonic analysis, for example, yields a sequence of harmonic functions which take their values in abstract symbols, such as $D_F, Am7$. Or a motivic analysis ends up on a verbal description of certain important motives. However, when an artist has to perform a score, the abstract description level is of no direct use: Only the numerically quantized information can immediately help in defining shaping processes of instrumental parameters. *The final condensation of any analytical facts in performance has a numerical appearance.*

Our presentation of weights is in some sense an ideal compromise between abstract predicates, such as harmonic symbols, and the naive numerical evaluation of analytical properties. We understand that weights are predicates, i.e., truth-valued functions on local compositions and their powerset constructions, but the truth module is at the same time a numerical one, or at least a truth value can be associated with a numerical value if required.

39.1.1 Discrete and Continuous Weights

Summary. A priori, weights are functions on discrete sets of points within determined parameter spaces. In order to insert weights into general tasks of performance, in particular those referring to infinitesimal or continuous character, one has to consider extrapolation methods.

– Σ –

We recall that we have defined a cubic interpolation formula of class C^1 for a real-valued (discrete) weight w with zero derivative on local compositions K in an n-dimensional simple form space S over the reals, according to defintition 99 of Section 32.3.2. We also supposed that K is contained in the n-cube C^n and

[1] This reduction is in fact—among others—a reduction of originally gestural neumatic signs!

[2] See the catchword of Chapter 33.

that the interpolation is constantly equal to 1 outside C^n. In our present situation, we want to make a slight generalization to general weights with truth values in $\mathbb{T}_{\mathbb{R}}$. We suppose that the truth values $w(x), x \in K$ are all non-empty sets, i.e., do not have the value $\perp$, and that K is still in the cube C^n. Usually, this is the case since we are given weights which are open intervals $]-\infty, a[$ with non-negative upper bound a. Suppose that $w_x \in w(x)$ is a selection of values from all truth sets $w(x), x \in K$. Supposing that an interpolation function F has been chosen (e.g., a cubic interpolation with respect to a specific permutation of coordinates, or the weighted sum of all such functions, or still another adequate candidate), we have an interpolation function $F_{w.}$ for each such selection $w.$, which is constant with value 1 outside C^n and with zero derivatives in all points of K. Then we obtain a predicate on every element $s \in S$ by the definition

$$F(s) = \{F_{w.}(s)|w. \in \prod_{x \in K} w(x)\}. \tag{39.1}$$

In case the truth values are intervals $w(x) =]-\infty, a(x)[$, the interpolation yields the value intervals associated with the interpolation values stemming from the upper bounds $a(x)$.

The fact that such a generalized interpolation is not defined for higher powerset arguments is not really a problem. Some operators in fact take these weights and boil their information down to more common weights as discussed above. We may then apply interpolation formula (39.1) to these boiled-down weights.

Behind such an interpolation procedure there is the non-trivial question about the justification of the usage of continuous weights in performance operators. There are two reasons we can find for this: The first is a very practical one: calculation precision. Suppose that we are given a local composition K and that we want to calculate the weight $w(x)$ at a point $x \in K$. In the context of computer programs, it is often not clear whether we can really catch x in its numerical identification, for example for numerical calculation rounding effects. Therefore, it would be good to have the weight being a non-discrete function which varies little around a given argument. The second argument is that the continuous character of performance fields suggests continuous methods to shape such fields. For example, the commonly accepted continuous character of tempo would ask for a continuous weight function in order to determine a tempo change for all values where the tempo curve is defined. The inherent idea is that anaytical facts that are calculated on discrete events of a given score should have their presence in a neighborhood of such events, in the sense of a field action of such predicates.

39.1.2 Weight Recombination

Summary. We describe and motivate a set of procedures for building new weights from given ones.

– Σ –

A further advantage of the numerical truth value set $W \underset{Id}{\longrightarrow} \mathbf{Simple}(\mathbb{R})$ for weight predicates is that one may recombine such weights on standard operations from linear algebra. This type of building new predicates from given ones is a special item in the methodology of predicate calculus as exposed in Section 18.3.4, i.e., it is a special logical motivation since it operates on the codomain of predicates.

Given a scalar $\lambda \in \mathbb{R}$ and a weight w on a local composition K, or its differentiable extension F_w by a cubic interpolation formula as described above, we can define

$$\lambda.w(x) = \{\lambda.\nu|\ \nu \in w(x)\}, \tag{39.2}$$

$$\lambda.F_w(x) = \{\lambda.\nu|\ \nu \in F_w(x)\}, \tag{39.3}$$

and it is clear that $\lambda.F_w = F_{\lambda.w}$, and $(\kappa.\lambda).F_w = \kappa.(\lambda.F_w)$, $(\kappa.\lambda).w = \kappa.(\lambda.w)$, respectively. If λ is positive, and if w is a weight whose values are intervals $]-\infty, a[$, then $\lambda.w$ is still of this type, viz, its intervals are the shifted intervals $]-\infty, \lambda.a[$. If we are given two weights v, w, then we can define

$$(v + w)(x) = \{\mu + \nu|\ \mu \in v(x), \nu \in w(x)\}, \tag{39.4}$$

$$(F_v + F_w)(x) = \{\mu + \nu|\ \mu \in F_v(x), \nu \in F_w(x)\}; \tag{39.5}$$

$$(v.w)(x) = \{\mu.\nu|\ \mu \in v(x), \nu \in w(x)\}, \tag{39.6}$$

$$(F_v.F_w)(x) = \{\mu.\nu|\ \mu \in F_v(x), \nu \in F_w(x)\}, \tag{39.7}$$

and we also have $F_v + F_w = F_{v+w}, F_v.F_w = F_{v.w}$. Sum and product are associative, but $(\kappa + \lambda).F_w \neq \kappa.F_w + \lambda.F_w$, in general. If both, v and w (and therefore also the differentiable extensions) have intervals $]-\infty, a[$, then so is their sum.

The philosophy of these combinations is that the action of weights may be better if their influence is mixed and weighted by scalars such as with cooking, the delicate dosage can be controlled.

A more general type of recombination is a non-linear deformation of a weight according to a deformation function $\delta : \mathbb{R} \to \mathbb{R}$. With the previous notation, we define a scalar multiplication by

$$\delta.w(x) = \{\delta(\nu) |\ \nu \in w(x)\} \tag{39.8}$$

$$\delta.F_w(x) = \{\delta(\nu) |\ \nu \in F_w(x)\}, \tag{39.9}$$

and it is clear that for two such functions δ, ϵ we have $(\delta \circ \epsilon).F_w = \epsilon.(\delta.F_w)$.

In practical cases, we often have the weight w being defined by intervals $w(x) =]-\infty, a(x)[$ for $x \in K$, and therefore also $F_w(x) =]-\infty, \alpha(x)[$ for any $x \in S$. Since K is supposed to be finite, and since by construction $\alpha(x) = 1$ outside the defining cube C^n, the image set $\alpha(x), x \in S$ is a finite interval $[\alpha_{min}, \alpha_{max}]$. We then consider a continuous deformation function $\delta(\alpha_{min}, \alpha_{max}, \tau)$ such that $\delta(\alpha_{min}, \alpha_{max}, \tau)(t) = t$ for $t \notin [\alpha_{min}, \alpha_{max}]$. For $t \in [\alpha_{min}, \alpha_{max}]$, the deformation parameter τ describes a one-parameter family of continuous, monotonically increasing deformations of the interval $[\alpha_{min}, \alpha_{max}]$ with $\delta(\alpha_{min}, \alpha_{max}, 0)$ being the identity. Typically, the deformations for τ and $-\tau$ are related to each other by a reflection of their graph at the main diagonal in $\mathbb{R}^2$. For example, we may take affine images of hyperbolas $y = -1/x$ on intervals $[-u, -1/u]$, where $u = e^\tau$, yielding the typical formula $\delta(0, 1, \tau)(t) = \frac{t}{e^{2\tau} - t(e^{2\tau} - 1)}$ for non-negative τ, and the symmetric (diagonal reflection) formula for negative τ.

The philosophy of non-linear deformations is that often, the action of an analytical weight on performance is qualitatively correct, but its quantitative influence should be distorted in order to yield a good perception. This effect can then be achieved via non-linear deformation functions.

39.2 Primavista Weights

Summary. Many of the traditional score data are not codified in numerical values. Numerical quantification is, however, a conditio sine qua non for any performing artist. We give an overview of common transformation procedures for non-numerical score parameters, including their extrapolation to continuous weights.

– *Σ* –

39.2.1 Dynamics

Summary. We discuss the quantification and syntax of absolute and relative dynamics.

– *Σ* –

Here, we need some preliminary remarks concerning the parametric interpretation of dynamical score signs since these are very coarse and of different types. We distinguish three types:

- *Absolute* dynamical signs such as *ppp, mf, fff, sempre pp*, etc. They give information at a determined onset.
- *Relative punctual* signs such as *frz, sf*, etc. They indicate a momentous change of dynamics *as a function of* the momentous dynamical level.
- *Relative local* signs such as the *crescendo* signs, giving verbal or pictorial indications on their domain of validity or having the shape of short wedges.

To begin with, absolute dynamic signs are ambiguous verbal descriptions of intended loudness. The first step towards a performance-adequate representation of such symbols is to assign them numerical values in a symbolic loudness scale. The good thing to do here is to assign them values which are in an affine relation to physical (Cents) units[3]. The precise value relations are however not codified and must be left to the free decision of the formalizing instance (in instrumental practise: the performer, the conductor, in computer-systems: the user). So we start with the usually sufficient setup of an increasing sequence[4]

$$\begin{aligned}\gamma(ppppp) &< \gamma(mpppp) < \gamma(pppp) < \ldots \\ &< \gamma(mp) < \gamma(p) < \gamma(mf) < \gamma(f) < \ldots \\ &< \gamma(ffff) < \gamma(mffff) < \gamma(fffff)\end{aligned} \tag{39.10}$$

of 19 real numbers in the *Loudness* space. The naive setting would be an equidistant series, such as suggested by MIDI and other technological codes. If we could limit loudness quantification to such absolute signs, this would be the end of the story: We would just define a parser for verbal dynamics signs into the numerical format of the space *Loudness*.

However, the meaning of the absolute signs is more than the above gauging convention. In fact, when we are given a sequence of absolute dynamic signs, we are also facing relative dynamic signs which have extensions, i.e., onsets and "offsets" within the range of the given absolute signs. For example, we may see a crescendo wedge, followed by a second such wedge, and then by a decrescendo wedge. What is the loudness curve for this situation?

The basic data is a sequence of absolute loudness symbols which are specified by their onsets and the symbols, i.e., a sequence of denotators in the limit space

$$\begin{aligned}AbsDynamicEvents &\xrightarrow[Id]{} \mathbf{Limit}(AbsDyn, Onset, Duration) \\ AbsDyn &\xrightarrow[Id]{} \mathbf{Simple}(\mathbb{Z} < UNICODE >).\end{aligned} \tag{39.11}$$

The duration of such an event is the difference from its onset and the onset of the next event, except for the last event, whose duration defines the end of the given composition. So we have the sequence $AbsDynSequ =$

$$\begin{aligned}&Evt_1 : 0 \rightsquigarrow AbsDynamicEvents(Dyn_1, E_1, D_1), \\ &Evt_2 : 0 \rightsquigarrow AbsDynamicEvents(Dyn_2, E_2, D_2), \\ &\ldots \\ &Evt_m : 0 \rightsquigarrow AbsDynamicEvents(Dyn_m, E_m, D_m),\end{aligned} \tag{39.12}$$

with $D_1 = E_2 - E_1, \ldots D_{m-1} = E_m - E_{m-1}$. This data induces the weight $w_{AbsDynSequ}$ on onset events E with these values $w_{AbsDynSequ}(E) =]-\infty, a(E)[$:

$$\begin{cases} a(E) = -\infty \text{ if } E < E_1 \text{ or } E_m + D_m \leqslant E, \\ a(E) = \gamma(Dyn_i) \text{ if } E \in [E_i, E_i + D_i[, i = 1, 2, \ldots m-1. \end{cases} \tag{39.13}$$

Over this absolute dynamics data, we now tilt the relative dynamic events: On each interval defined by event Evt_i, we are given a sequence of events of this form:

$$\begin{aligned}RelDynamicEvents &\xrightarrow[Id]{} \mathbf{Limit}(RelDyn, Onset, Duration) \\ RelDyn &\xrightarrow[Id]{} \mathbf{Simple}(\mathbb{Z} < UNICODE >).\end{aligned} \tag{39.14}$$

The symbols in *RelDyn* are such as *crescendo,molto crescendo, decrescendo, molto decrescendo*, for example. Their onset and duration are visible from the position of the wedges or the dashed lines in the score notation. So we are given sequences $RelEvt_{i,.} =$

---

[3] See Appendix A.2.3 for this gauging question.

[4] We do NOT follow the unlogical standard notation for *mp*, which positions it above *p* and below *mf*. We use the prefix *m* always for being below the following loudness symbol.

$$RelEvt_{i,j} : 0 \rightsquigarrow RelDynamicEvents(RelDyn_{i,j}, rE_{i,j}, rD_{i,j}), \tag{39.15}$$

$j = 1, \dots m_i, 0 \leqslant m_i$, with $m_i = 0$ for empty sequences. This time, the conditions on the respective onsets and durations are

$$\begin{aligned} rE_{i,j} + rD_{i,j} &\leqslant rE_{i,j+1}, 1 \leqslant j < m_i, \\ E_i \leqslant rE_{i,1}, rE_{i,m_i} &+ rD_{i,m_i} \leqslant E_i + D_i. \end{aligned} \tag{39.16}$$

This is the framework for redefining the finer dynamics according to the relative signs sequences. The problem here is that we do not know how much a crescendo may increase absolute dynamics in order to remain within interval $E_i, E_i + D_i$ of the given absolute value $\gamma(Dyn_i)$ of the absolute event Evt_i. To this end, one needs other information which is implicit in the interpretation of a score: this is the *tolerance* of dynamical variation around $\gamma(Dyn_i)$ such that we still accept the label Dyn_i. So we have to provide each symbol Dyn of absolute dynamics with a tolerance number $0 < \tau_{Dyn}$, and this means that all values in the half-open interval $[\gamma(Dyn) - \tau_{Dyn}, \gamma(Dyn) + \tau_{Dyn}[$ will be accepted as variations of the label Dyn_i. This does not exclude that these intervals may overlap, so the labels can be ambiguous and contradictory: a high *pp* value can be higher than a low *mp* value.

Upon this tolerance system, we now have to define the meaning of relative dynamics in the sequences $RelEvt_{i,.}$. To this end, we have to fix a quotient $0 < \kappa_{RelDyn}$ of dynamical increase or decrease of each relative dynamical symbol. For crescendi, we suppose $1 < \kappa$, and for decrescendi, we want $\kappa < 1$. For each relative dynamical event $RelEvt_{i,j}$, we calculate the dynamical values (i.e., the upper bounds b of the predicate's intervals $]-\infty, b[)$ at $rE_{i,j}$ and $rE_{i,j} + rD_{i,j}$, and then, we interpolate linearly between these cornerstones, whereas the value remains constant inbetween two relative dynamics events.

The first relative event $RelEvt_{i,1}$ has the starting value at $rE_{i,1}$ equal to the absolute value $\gamma(Dyn_i)$ of Evt_i. We now suppose inductively, that the starting value v_j of the relative event $RelEvt_{i,j}$ has been defined within the *open* interval $]\gamma(Dyn_i) - \tau_{Dyn_i}, \gamma(Dyn_i) + \tau_{Dyn_i}[$. This defines the quotient $\frac{v_j - (\gamma(Dyn_i) - \tau_{Dyn_i})}{(\gamma(Dyn_i) + \tau_{Dyn_i}) - v_j}$ which we want to increase by the factor $\kappa_{RelDyn_{i,j}}$ defined by the relative dynamical event at index i, j, i.e.,

$$\kappa_{RelDyn_{i,j}} \frac{v_j - (\gamma(Dyn_i) - \tau_{Dyn_i})}{(\gamma(Dyn_i) + \tau_{Dyn_i}) - v_j} = \frac{v_{j+1} - (\gamma(Dyn_i) - \tau_{Dyn_i})}{(\gamma(Dyn_i) + \tau_{Dyn_i}) - v_{j+1}} \tag{39.17}$$

defines the new dynamical value v_{j+1} at the end of the (i, j)th relative sign. The new value is still in the open interval $]\gamma(Dyn_i) - \tau_{Dyn_i}, \gamma(Dyn_i) + \tau_{Dyn_i}[$, and we may go on inductively until all relative dynamical signs are parsed. Observe that a succession of an increase factor κ by its inverse neutralizes the dynamical value change. More generally, the succession of two factors κ_1 and then κ_2 results in a value change by factor $\kappa_1.\kappa_2$. So we can manage any succession of relative dynamical signs of this "crescendo/decrescendo" type without falling out of the prescribed tolerance interval. Denote the weight on all of the onset axis which is defined in this way by $w_{Evt.,RelEvt..}$.

So we are left with punctual relative dynamical signs *pct* at a note x, such as $pct = accent$ or $pct = marcato$. We may define the predicate

$$pct(x) = \begin{cases} \{1\} \text{ if } x \text{ is not marked by } pct, \\ \{\lambda(pct)\} \text{ else,} \end{cases} \tag{39.18}$$

where $\lambda(pct)$ is the increase factor for *pct*. Then, the punctual dynamical sign influences the already given absolute and relative dynamical weight *dyn* by the product *pct.dyn* of weights.

Of course this implementation need not live for ever, but it is one reasonable solution of a non-trivial problem of making blurred score signs precise. Observe that this kind of weight is not a performance map, it is just a weight on the symbolic (mental) events that must be used for operators to be defined later. Observe also that the above weights are not continuous functions of onset if several absolute dynamical signs are present.

39.2.2 Agogics

Summary. We discuss the quantification and syntax of absolute and relative agogical indications. A special attention is payed to general curve types for retards, fermatas, and general pauses.

– Σ –

These are the common agogical indications:

- *Absolute* tempo, such as Mälzel metronome, or anterior verbal indications of type *andante, adagio*, etc. Formally, they correspond to absolute dynamic signs.
- *Relative punctual* tempo signs such as fermatas and general pauses. Remarkably, there is no relative punctual acceleration sign corresponding to the fermata.
- *Relative local* tempo signs are the following:
  1. Coarse indications concerning agogics, e.g., *ritardando, rallentando, accelerando, stringendo* etc.
  2. notation of correspondence between two adjacent tempi, such as "2/4 = 3/8",
  3. rest signs such as *a tempo*.

The essential difference to dynamical signs is that symbolic onset and duration are precisely codified, so agogical signs relate to proper performance transformations, and not to making blurred signs precise.

We have already discussed some of the agogical predicates (fermata, value change) in the presentation of weights for the prime mother LPS in stemma theory, Section 38.3.2.

Let us first look at the Mälzel metronome sign $x\ quarter/Min.$. This is a predicate which takes its $\top$ values exactly on onsets E and tempi x where a Mälzel indication is $x\ quarter/Min.$, and $\perp$ else. The reader may easily make the underlying forms precise. Verbal absolute tempo indications need a parsing instance to render them in numerical (Mälzel) terms.

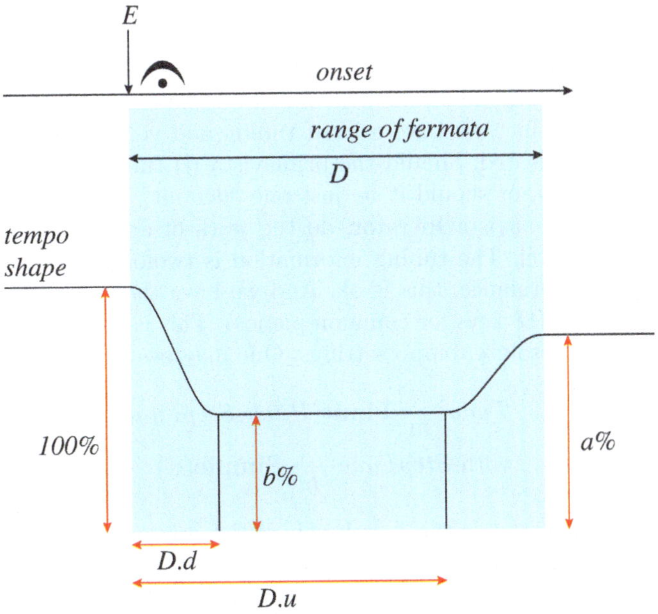

Fig. 39.1. The interpolation curve of a fermata weight. See the text for the explanation of the symbols.

A fermata has only its onset E made precise, the rest is blurred. We need several additional parameters to generate a viable weight. We first need a duration D, then a shape parametrization to describe the fermata's tempo sink. We may for example take this form:

$$Fermata \underset{Id}{\longrightarrow} \mathbf{Limit}(Onset, Duration, Bottom, Down, Up, After) \tag{39.19}$$
$$\text{with } Bottom, Down, Up, After \underset{Id}{\longrightarrow} \mathbf{Simple}(\mathbb{R}).$$

A fermata $Ferm : 0 \rightsquigarrow Fermata(E, D, b, d, u, a)$ is parametrized by (1) the percentage number B of maximal tempo lowering with respect to the given tempo, (2) the percentage d of the duration D from the beginning to reach the lowest tempo, (3) the percentage u from the beginning to restart getting back to the following tempo, (4) the percentage a of the original tempo which is resumed after getting back. This data is used to define a cubic spline interpolation as follows: We have four onsets: $E, E + b.D, E + u.D, E + D$ and corresponding relative tempo values $1, b, b, a$. This defines a discrete weight $w(Ferm)$ and a corresponding interpolation function $F_{w(Ferm)}$, see Figure 39.1. Of course, many different curves are possible, but with this shape, the main quality of a fermata can be imitated.

Weights for accelerando- or ritardando-typed predicates follow the building scheme of a fermata, except that the duration of these signs is defined. We leave this as an exercise to the reader.

Exercise 80 *Give an explicit description of accelerandi and ritardandi in terms of denotators and associated weights.*

The general pause (G.P.) predicate is also undetermined like the fermata. It does however not imply a smooth recapitulation of the original tempo: After the general pause, the previous tempo is reset. We will have to deal with this in corresponding operators. The same phenomenon of a tempo reset is the case for the *a tempo* predicate.

39.2.3 Tuning and Intonation

Summary. We discuss the weights for pitch values.

– Σ –

Tuning and intonation is a delicate subject for the PrimavistaOperator since it it not clear from the beginning how much the settings depend on the instrument and how they depend on individual instruments. For piano, the situation is easy since we have a fixed tuning and very often, it is even well-tempered. For violins, this is much more complicated: Should the primavista (!) tuning be a just tuning for each tonality which is encountered in the score, or should it be just one "default" just tuning? It is wise to let a special operator, not necessarily the PrimavistaOperator, do the work of delicate tuning and to just operate the minimum on the first process level. The tuning information is twofold: We have the chamber pitch which is the initial set and initial performance, this is ok. And we have the tuning data for all pitch events of a sufficiently large chromatic scale (88 keys for common pianos). This is completely analogous to a step tempo function given by a number of absolute tempo settings. One may use the form

$$Tune \underset{Id}{\longrightarrow} \mathbf{Limit}(Pitch, StepTune) \tag{39.20}$$
$$\text{with } StepTune \underset{Id}{\longrightarrow} \mathbf{Simple}(\mathbb{R})$$

being the tuning quantity that measures the "pitch velocity" between to neighboring pitches. Examples for default tuning for some classical cases are found in Appendix M. The common well-tempered case has the denotators

$$x : 0 \rightsquigarrow Tune(x, 1/100\ Semitone/Ct)$$

for each pitch x of the well-tempered chromatic scale.

39.2.4 Articulation

Summary. We present the parameters for articulation types, from molto staccato to molto legato.

– Σ –

On one hand, articulation is a predicate type which regards single notes, for example molto staccato, staccato, legato, or molto legato. All these predicate types can be encoded by weights on events x of a space U within the BP hierarchy which contain duration and take their values $]-\infty, a(x)[$ according to whether articulation stretches durations or compresses them. For staccato and other compressing signs, we take $0 < a(x) < 1$, whereas for legato and other stretching signs, we take $1 < a(x)$.

On the other hand, we have articulation in the sense of grouping of a set G of usually consecutive notes which are grouped by an articulation slur via form $ArtiSlur_{U,S}$ as introduced in Section 38.3.2. Such a predicate may be made more precise by a weight which takes into account the group G as well as the single notes x with their position within G. On a space U, this is covered by the form U_2 with the notation from formula (35.18). The weight $w_{GrpArti}(G, x) =]-\infty, a(G, x)[$ is parametrized by the number $a(G, x)$ which tells how much the note's x duration is altered relatively to its nominal duration.

39.2.5 Ornaments

Summary. Historically, ornaments form a complex set of constructions. We introduce a unified language for ornaments which is based on macro-events (see Section 6.7).

– Σ –

Recall from Section 6.7 that we had defined macro-events by the form $Makro_{Basic}$ which is based on an event form $Basic$. Here, we take $Basic = U$, one of the event spaces in the hierarchy of our stemma. An ornament (one of the many forms of a trill, for example [151]) of events in U can be described by a special denotator of the form $Makro_U$.

An ornament is first of all anchored at an event a of U, this is the note which is ornamented by a specific sign, with a reference onset a, see Figure 39.2. So we start our ornament denotator by a singleton:

$$Ornament : 0 \rightsquigarrow Makro_U(\{D\})$$

with

$$D : 0 \rightsquigarrow Knot_U(a, SmallAnchors).$$

Here, the next ramification denotator $SmallAnchors$ encodes the three anchor events of the three structural units O_1, O_2, O_3, each possibly with repetitions within:

$$SmallAnchors : 0 \rightsquigarrow Makro_U(O_1, O_2, O_3).$$

Each of these knots O_1, O_2, O_3 has a reference event a_1, a_2, a_3:

$$O_i : 0 \rightsquigarrow Knot_U(a_i, P_i), i = 1, 2, 3,$$

with macros P_1, P_2, P_3. Each of these macros is a set of knots:

$$P_i : 0 \rightsquigarrow Makro_U(q_{i,j}, j = 1, \ldots m_i)$$

with multiplicities m_i. Each such knot has the shape

$$q_{i,j} : 0 \rightsquigarrow Knot_U(d_{i,j}, o_i),$$

where the macros o_i are independent of the second index j: They signify a repetition of the same macro according to the shift quantities $d_{i,j}$. The terminal macros o_i have this meaning: The first, o_1, is the starting

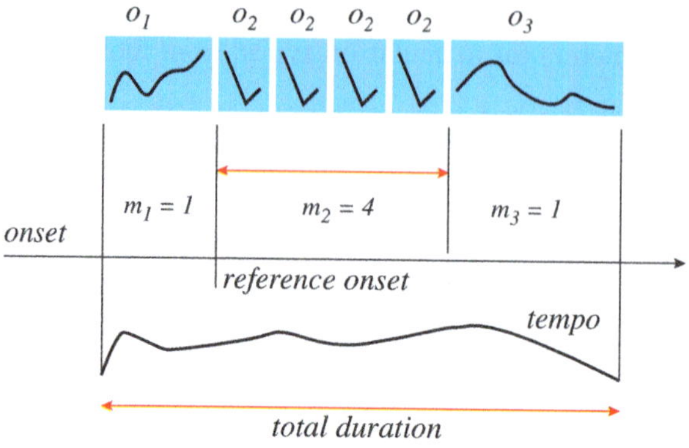

Fig. 39.2. The structure of an ornament, showing a start shape (sequence) o_1, a sequence of middle shapes o_2, and a final shape (sequence) o_3.

"micro-motif" of the ornament, it is played before the reference onset of a and ends with the beginning of the second micro-motif o_2 which usually starts from a. This inner motif is repeated several times, usually something like eight times, and very fast. The ornament terminates on the third micro-motif o_3 which is a tail to the ornament and may be played much slower than the middle sequence. If we apply successive flattening operations $Flatten^n, n = 0, 1, 2, 3$ to the *Ornament* denotator, we obtain a hierarchy of successively appearing U-events which gives rise to different refinement levels of tempo curves which an *OrnamentOperator* in the PrimavistaOperator has to manage.

This perspective suggests that the hierarchy spaces U in BP should really be extended to the spaces $Knot_U$ in order to couple stemmatic refinement with hierarchies of sounds, but this theory is not developed so far.

It transpires however that macro-events fit in the general weight scheme which is defined by successive powerset constructions from the spaces U in BP. So we can assign to each of the involved macro-events or knots a weight in $\mathbb{T}_{\mathbb{R}}$ in order to weight the performance role of the ornament's components.

39.3 Analytical Weights

Summary. We give a generic view on analytical weights. Their concrete shape depends on the analysis which is available. Several types are realized on analytical RUBETTE®modules, see Chapter 41.

– Σ –

Analytical weights are crucial to rational performance and intervene from analyses of the score S relating to metrical, rhythmical, motivic, thematic, harmonic, contrapuntal, grouping and other structural perspectives. We shall sketch the construction for metrical, motivic, and harmonic weights. In all these descriptions, we shall abbreviate the nomenclature of predicates and simply write down the numerical values a of such weights, the predicative statement being implicit, either in form of intervals $]-\infty, a[$ or in form of singleton predicates $\{a\}$, the information is the same, and the usage of the operators does not depend on which predicative encoding is chosen. Also we restrict to the discrete weights, the continuous associated weights being automatic from our previous discussions (Section 39.1.1)

1. *Metrical weights* w^S_{metro}: We have a sober weight on the onset space $U = Onset$ for which we have prepared a formula in example 41. In that formula, the weight $nW(\sigma)$ is calculated on a simplex σ of maximal local meters. For an onset E, we then take the simplex $\sigma_S(E)$ consisting of all maximal meters

within the onset kernel S_{Onset} of S which contain E and set $w^S_{metro}(E) = nW(\sigma_S(E))$ (with the evident singular value $nW(\emptyset) = \bot$ for onsets outside S_{Onset}).

2. *Motivic weights* w^S_{motif}: Following the preliminary discussion in Section 22.9, we take an interpretation $S^{\mu}_{Onset \oplus Pitch}$ of the onset-pitch space of BP by motives. More precisely, μ may be chosen as the set of all motives within a range of cardinality and extent of onsets between the first and last event. We define sober motif weight which is induced by the weight on motives from the atlas μ. So we are left with the definition of the weight on motives $M \in \mu$. To this end, we consider a more precise motif theory framework as defined in Chapter 22. We choose a shape type Γ_t and a distance d, on which an equivariant, isometric group action of a paradigmatic group P is defined, and such that we have the inheritance property fulfilled (for example, the elastic and diastematic types with Euclidean distance and counterpoint groups). We then have the epsilon topology $\mathfrak{T} = \mathfrak{T}_{t,P,d}$ and choose a neighborhood radius $0 < \epsilon$. For a motif $M \in \mu$, we consider the ϵ-disk neighborhood $D^{\mu}_{\epsilon}(M)$ of M in the relative topology $\mathfrak{T}_{\mu}$. We define the *ϵ-presence of M* by the weighted sum
$$p_{\epsilon}(M) = \sum_{N \in D^{\mu}_{\epsilon}(M)} 2^{card(M)-card(N)} \tag{39.21}$$
which visibly counts the elements of the ϵ-neighborhood of M with a weight according to the cardinality difference to M's cardinality. This is roughly speaking[5] the presence implemented in the MeloRUBETTE®.
We then also define a *ϵ-content of M* function which is the following weighted sum:
$$c_{\epsilon}(M) = \sum_{N, M \in D^{\mu}_{\epsilon}(N)} 2^{card(N)-card(M)}, \tag{39.22}$$
and we therefrom get the weight of M, i.e., the product
$$w^S_{motif}(M) = p_{\epsilon}(M).c_{\epsilon}(M) \tag{39.23}$$
which—again roughly speaking the presence function which is also implemented in the MeloRUBETTE®—measures the presence of M combined with the content of M. Intuitively, the weight of M is a measure of where M appears in other motives—up to ϵ distance—and how much motives are contained in M—up to ϵ distance. The coefficients are a weight for the cardinality distance form $card(M)$.

3. *Harmonic weights* w^S_{harmo}: This weight is also a predicate whose values are non-false only on the onset-pitch space $S_{Onset \oplus Pitch}$ of BP. To define it, we have to start from an interpretation $S^{\eta}_{Onset \oplus Pitch}$ by chord events. The construction of a chord atlas is however not uniquely determined since it is not clear what is a chord. We may just take those local compositions of all events having one and the same onset. But we may also consider all the onsets where some chord notes end and some others still last, and then take the still lasting notes as constituents for new chords at those ending onsets.
Given such an interpretation η, we take its abstraction in form of a temporally defined sequence $a. = a_0, a_1, \ldots a_k$ of length k of chords, i.e., local compositions in the *Pitch* space. For any such sequence $a.$, we have defined weight functions in the context of the Riemann algebra, as discussed in Section 27.2. There, formula (27.13), with its interpretation as a truth denotator-valued weight, as explained in the example of Section 27.2.2, gives us a criterion for finding a best path $p_{opt} = (v_0, f_0, a_0) \to (v_1, f_1, a_1) \to \ldots (v_k, f_k, a_k)$ in the Riemann quiver. We now use this best path to calculate the *tension* $t_{p_{opt}}(a_i)$ of each chord a_i with respect to this path. The immediate tension function would be $t_{p_{opt}}(a_i) = \Omega(p_{opt}|a_i)$, where we denote by $p_{opt}|a_i$ the optimal path from the beginning to (inclusively) chord a_i. However, the global tension is a problematic quantity as discussed in Section 27.2.1. We therefore renormalize the function $t_{p_{opt}}(a_i)$ by the unique shearing of its graph in $\mathbb{R}^2$ such that the first point $(0, t_{p_{opt}}(a_0))$ goes to a predefined initial tension $(0, t_0)$ whereas the last point $(0, t_{p_{opt}}(a_k))$ goes to $(0, t_k)$. This has the

[5] In our implementation of this function, we have also taken into account the multiplicity of submotives X of the N in the neighborhood with $card(X) = card(M)$, which are in the distance less than ϵ to M.

advantage that the global tension can be cast to initial and final values which the local methods cannot predict and control. Call $t^{t_o,t_k}_{p_{opt}}$ this new tension function.
In order to calculate the weight of single notes x within a chord a_i, we look at the Riemann matrix predicates $TF_{f,t}(a_i) = [0, \phi(a_i)[$ and $TF_{f,t}(a_i \backslash x) = [0, \phi(a_i \backslash x)[$ where $a_i \backslash x$ is the chord a_i after omission of x. We know that $\phi(a_i)$ is not zero. We may then map the interval $[0, \phi(a_i)]$ onto the interval $[IP, 1], 0 < IP < 1$ by an affine map Q for a normalization purpose. Then, we get the relative importance $rel(x, a_i)$ of x within a_i by the formula

$$rel(x, a_i) = \frac{1}{Q(\phi(a_i \backslash x))}. \tag{39.24}$$

This gives us the final expression for the weight of the note x: We take the tension of the underlying chord a_i and multiply it by the relative importance of x within this chord, i.e.,

$$w^S_{harmo}(x) = rel(x, a_i).t^{t_o,t_k}_{p_{opt}}(a_i). \tag{39.25}$$

This is the harmonic weight which is implemented in the HarmoRUBETTE®.

An important technique to produce new weights from analytical weights is the *boiling down* method: Often, a weight is given on a space where the operator at hand does not work. For example, the TempoOperator (to be discussed later in this chapter) needs a weight on the *Onset* space. If a weight lives in a space with more dimensions, we should be able to boil it down to a weight on *Onset*. Here is the procedure: We have two spaces U, V in the system BP, and a projection $p_V : U \to V$. A weight w is given on the kernel S_U, and we would like to get a boiled-down weight $BD_V(w)$. We make these two definitions:

$$BD_V(w)(x) = \sum_{y \in p_V^{-1}(x)} w(y), \tag{39.26}$$

$$BD_V^{mean}(w)(x) = \frac{1}{card(p_V^{-1}(x))} BD_V(w)(x), \tag{39.27}$$

where $x \in S_V$, and with the value $-\infty$ for empty fibers.

In general, we have this situation: we would like to have a weight w on a specific space V of the hierarchy, but the given weight lives either in a larger space U which projects onto V, or it lives in a smaller space U onto which V projects. The former case is solved by the boiled-down construction $BD_V(w), BD_V^{mean}(w)$, whereas the latter is straightforward by the formula $w_U(x) = w(p_U(x)), x \in V$. In the future, *whenever we use a weight which is possibly defined in the "wrong" space, these constructions are referred to, and we simply write w instead of the above correct symbols if no ambiguity is likely.*

39.4 Taxonomy of Operators

Summary. Though a general description of operator principles is risky (since there is no general theory of how a stemma can be altered), we want to give a preliminary classification of the ways an operator may choose to intervene in the existing configuration. This taxonomy is guided by the generic description of how performance works: as a transformation from mental to physical reality, and by means of its description via performance fields given in Section 33. This means that an operator can intervene on the level of mental or physical reality, on the frame of the local performance cell, see Chapter 35.1, and on the performance field together with its initial set.

– Σ –

The basic data of a performance is encoded in the performance map $\wp : K \to \wp(K)$ on the given performance cell. An operator has to change any of the involved structures: K, $\wp$, or $\wp(K)$. This is understood in the sense that K is a set of arguments upon which the prescription $\wp$ acts, and a set $\wp(K)$ which is the output of that map.

The first operator type acts on K. It does this in two ways: First, named splitting operator, it just divides the kernel K into sub-kernels, nothing else. Second, it alters the kernel's events and submits these new events to the given prescription $\wp$. This one is termed symbolic operator. It is as if the artist would play in the same "mood" with a changed score.

The third type takes the input K, the map $\wp$, and the output $\wp(K)$ as given and now changes the output without regard to the previous process. This is the physical operator type.

The fourth type takes the input kernel K for granted and just changes the map's "formula". For example it changes the tempo curve or the intonation, etc. This is the so-called field operator type. This is the most complex operator type and has not yet been understood in its different outfits.

Let us now have a closer look at these types.

39.4.1 Splitting Operators

Summary. A splitting operator is rather simple. It restricts to partitioning the performance score's kernel into sub-kernels without any further change on the performance score. Splitting may also be operated on a group of selected instruments in a global performance score.

– Σ –

The idea of a splitting operator is this: You are given a score with different groups of notes that you want to perform in an individual way. For example: right hand and left hand, or onset-driven grouping: split the score at a given parameter, such as onset or loudness or duration etc. This may happen at a grouping line such as the beginning and the end of an eight-bar period. Very often, this also happens if a special group needs a special performance procedure because of its inherent structure, such as a trill or another ornament. Whatever you will do to this group is irrelevant, you decide later. So the splitting operator is a propaedeutical operator intending to prepare more in-depth operator actions.

A prototypical realization of the splitting operator may be implemented by use of an essentially mathematical predicate $Split_{U,\nu}$, where U is one parameter (such as pitch, onset, etc.) in the given hierarchy $\mathfrak{h}$, i.e., of the top space BP, and where ν is a real number parametrizing a coordinate value in the U parameter. Then if X is an event in one space of the hierarchy, the predicate $X/Split_{U,\nu} = \top$ iff either X does not share this parameter U, or if it shares it and the U-value X_U of X verifies $\nu \leqslant X_U$. So $X/Split_{U,\nu} = \bot$ iff X_U exists and $X_U < \nu$. We then get the logical combination predicate $Split_{U,\nu,\mu} = Split_{U,\nu}\&\neg Split_{U,\mu}$ which selects those events with either no X_U coordinate or else with $\nu \leqslant X_U < \mu$. In the RUBATO® software, the splitting operator has been implemented such that it may perform any logical conjunction of splitting operators of the types $Split_{U,\nu}$ and their negations $\neg Split_{U,\nu}$. For example, this enables us to select all events within a half-open onset interval $[E_1 = \nu, E_2 = \mu[$ and a half-open loudness interval $[L_1 = \xi, L_2 = \chi[$, and having durations D with $D_1 = \delta \leqslant D$. So the splitting operator *Split* (we omit further specifications in this notation) is defined by a Boolean predicate which extracts a given set K' of events from a given kernel K. This means that *Split* produces two daughters, one with the K' kernel, the other (by logical negation of the former) with the remainder kernel $K - K'$. In the nomenclature of Section 38.3.3, we have a coupling

$$\Omega(i) \propto \mu = \Lambda_i, i = 1, 2$$

which is also written more intuitively as

$$Split \propto \mu \tag{39.28}$$

$$Remainder_{Split} \propto \mu. \tag{39.29}$$

This operator may also be applied to a GPS $(\mu_i)_i$ just by applying the operator simultaneously to all the member LPS μ_i of this GPS. We should add that all the other data of the mother(s) in μ (in $(\mu_i)_i$) are left as they are, and everything still works fine. The new LPS just have some points of the original kernel being removed.

In the formal setup of the weight system of an LPS, the predicate *Split* would be viewed as being a weight of both daughters and the operator would then act on the mother via the Boolean selection and its negation which is defined by this weight.

39.4.2 Symbolic Operators

Summary. A symbolic operator affects the score data *before* they are performed whatsoever. This means that the operator really changes the *composition* which is to be interpreted. Such an operator type is seemingly contradictory to the very objective of performance. Nonetheless, primavista weights suggest an intervention of operators before any real performance, i.e., performance is already initiated on the very level of the score's *interpretation*.

– Σ –

Symbolic performance operators are a delicate species since they intervene at a very early stage of the performance process: on the kernel level of the LPS hierarchy $\mathfrak{h}$. One could see this fact as a natural completion of the performance philosophy in that composition is the first stage of performance, so why not alter the kernels, i.e., make a new composition out of an old one.

We adopted this integrative point of view while developing RUBATO®. From the software engineering perspective, this is indeed tempting since it would in the limit yield an integrated software for performance *and* composition. In this generic setup the program routine must, however, recalculate virtually every ingredient of the given LPS. The change of the basic score elements in a recomposition entails that the frames, the initial sets, the initial performances, the fields, and the weight list must be updated, or, rather: rewritten from scratch. Potentially, nothing will be the same again. So the operator then would just be a reset of the complete LPS data for a new composition. And this is not what performance was meant to do. After all, having rehearsed on a given piece and then being told that you get a new score, but you may go on with the old tempo curves, is not the kind of thing you will enjoy, since it will not make sense to perform on this schizophrenic data. For example, if a tempo curve has been developed, and then all onsets and durations are reset to half of the former values, the tempo curve becomes useless unless you also redefine the curve by a so-called time stretching operation (multiply the time arguments by 0.5).

In order to avoid such risks, one should really recalculate all weights whenever a symbolic operator has changed the kernel data. But even then this would not necessarily be the right solution since some primavista weights of analytical nature really need to be calculated on an unrefined score, i.e., on symbolic data which do not yet share some sophisticated symbolic explications, for example in dynamics. This problem could be attacked by an implementation of weight calculation routines which average out the over-refined symbolic data, but this is not very elegant. It remains a fact that weights should also be given the parameter of the stemma LPS where the weight is instantiated. By this method, one would then be sure that a weight is related to a determined kernel in the stemmatic inheritance tree and not exclusively to the primary kernel data.

With these caveats in mind, we may nevertheless define some useful (though also risky!) symbolic operator. We name it *SymbolicBrueForceOperator* to remind you of this dangerous enterprise. It takes as arguments a "directional factor" $d \in W$, W a space of the hierarchy H, with components d_R for parameters R of W, and a weight w. Then, if $X \in U$ is an event of $Kernel_U$, we set

$$SymbolicBruteForceOperator_{d,w}(X) = w_d(X) \tag{39.30}$$

where $w_d(X)$ is the new event in U, whose R-coordinate $w_d(X)_R$ is the product $w(X).d_R.X_R$, if R is a coordinate of W, and X_R else. So we only change X-coordinates for the directional factor, and there, we scale the coordinates of X by the weight and by a fixed directional coordinate.

The risk here is that for d or $w(X)$ values far from 1, the image $w_d(X)$ is likely to fall out of the given frame. But one may then easily redefine the frame, the field, etc., since usually this symbolic operator is applied in a stemma stage where we are just given the default data. As we have seen in Section 38.3.2, this data can be adapted to special kernels without difficulty: The frame is extended to include new events, the performance field is the constant field, and the initial set and performance are not a function of special kernel data. Under general stemmatic conditions, this operator is risky, however, but it is also an elegant solution of some nasty parameter problems which arise from the fuzzy score notation. For example, if we are given a weight for primavista dynamics $w_{Evt.,RelEvt..}$ (see Section 39.2.1), then the directional factor

$d = (1) \in Loudness$ gives the complete dynamics values if the given loudness was set to a preliminary value 1, say. Same for other primavista weights relating to symbolic parameter values.

For dynamics, it seems to be clear that the symbolic operator is the right one to be applied. For other primavista weights this not so easy to decide. For example, the articulation weights for slurs, legati, or staccati could be seen as symbolic prescriptions to alter duration within the score event framework. We contend that this is not analogous to the dynamics situation because absolute dynamical signs (such as *ff, mf*) are not quantified and even less are relative dynamical signs (such as *crescendo*). We therefore have a clear mission to transform such verbal signs into quantitative data. In contrast, duration is a perfectly quantified parameter, so it needn't be generated on that level. We therefore decided to deal with these weights in the context of field operators which alter the mother's articulation field.

This minor dilemma demonstrates (once more) that one may have the same performance output with very different infrastructure in the transformation process from the score to the physical events. This subject has not been dealt with from a more theoretical point of view: It pertains to the inverse performance theory to be dealt with in part XII, i.e., the theory of the variety of performance scores leading to a given performance output. But there is no result on the variety of local or global performance scores with *variable* score data inducing a given performance output; everything to date supposes that we are given a fixed score. Especially in ethnomusicological contexts, where the very concept of a score is uncertain, this is not acceptable.

For completeness of this taxonomy, we should add that we also call symbolic operator an operator which changes the initial set of a performance cell. This may be necessary if we want to enrich certain boundary conditions on the performance of selected groupings of notes.

39.4.3 Physical Operators

Summary. Physical operators affect given results of performance which are inherited from the *mother's* data. They act in a very simplistic way since they do not relate to the deeper process of performance; they only take for granted what the mother had already produced as a performance output.

– Σ –

At the other end of the operator scale, we have the physical operators. While symbolic operators should be applied in the very beginning of a stemmatic unfolding, physical operators should be applied after all other operators have been deployed and no risk to have them applied once more is likely. An exception to this general rule is the case where after a physical operator another operator is applied, but not in any of the coordinates to which the physical operator has made changes. But the one who applies such exceptional stemmatic operations must know very well what is being done—as with pointers in C-style programming.

Formally, what we call the

$$PhysicalBruteForceOperator$$

is quite analogous to the operator

$$SymbolicBruteForceOperator$$

discussed in equation (39.30). We are again given a directional factor $d \in W$, W a space in the hierarchy H, with components d_R for parameters R of W, and a weight w. Then, if $X \in U$ is an event of $Kernel_U$, we set

$$PhysicalBruteForceOperator_{d,w}(X) = w_d^{\wp}(X) \tag{39.31}$$

where $w_d^{\wp}(X)$ is an event in the physical space u corresponding to U, whose r coordinate (r corresponding to R according to our general nomenclature) is the product $w(X).d_R.\wp(X)_r$, if R is a coordinate of W, and $\wp(X)_r$ else.

So this operator really changes the performance at the point X, but the symbolic operator only affects the symbolic point and does not intervene on the $\wp$-level or the output. There is another very pleasant difference: Whereas the symbolic operator is risky as to the frame, field, and initial conditions, the physical operator is completely indifferent against these conditions. It only needs the mother's performance and the rest will always work! This operator is also very fast on the programming level, it produces very little changes and therefore is quite comfortable to handle. This preconizes another application of *PhysicalBruteForceOperator* in the sense of a test operator. In some situations, we would just like to know how an alteration of the output looks. Instead of applying a complicated operator (acting on the performance field and consuming much calculation power, say), we first apply this physical operator and obtain a first impression of how the given weight might influence the mother's performance. If we like the effect, we may go on, first resetting the stemma to the mother's level, and then apply a more sophisticated operator with the weight in question.

For completeness of taxonomy, we should add that operators which alter the initial performance map are also called physical operators. They tell initial events where to be mapped in the physical space. This is however not completely satisfactory since it usually depends upon the way the mother told its initial events where to be mapped. And this could have been done on a more field-theoretic level. The actual change could also involve field-theoretic considerations.

39.4.4 Field Operators

Summary. Field operators are the very essence of shaping performance. They include a variety of more or less general operators according to the direction where the field has to be altered. The theoretical complexity is paralleled by a computational complexity (numerical integration of ordinary differential equations) which is very important for software implementation of such operators.

– Σ –

Whereas symbolic and physical operators are an easy chapter in performance classification, field operators constitute an entire world of various approaches to the definition and deformation of performance fields. There are several reasons for this. The first is the mathematical complexity of fields, the definition and deformation of a field requires differential-geometric tools of local and global nature. The second is that performance fields have to be inserted in the hierarchies of their LPS. Therefore, the change of a field is always coupled to the fellow fields in the hierarchy. The third reason is that the stemmatic inheritance enforces operators which are capable of acting upon given mother fields of quite general nature: the stemmatic framework has to face unexpected field inheritance. Therefore, the definition of field operators is not completed if a particular type of deformation is implemented, one must also allow for mother fields which do not fit in the naive setup of such a deformation type. The fourth reason is that there is no such a thing like a performance field theory in traditional performance theory, and not even in modern approaches, such as the KTH school or the Todd approach. Performance fields are a new paradigm to performance theory, although, as we have shown, they are a completely natural and mandatory setup to control performance on an artistic level and from the more esthetic point of view as put forward by Adorno and Benjamin.

We shall start our investigation of performance field operators by the classical tempo operator. We then generalize its setup to cope with more general mother fields. Then we generalize this approach to a more systematic framework, the so-called scalar operator.

Another approach will be discussed under the title of basis-pianola operators, which arose from a systematic generalization of a physical operator to quite general situations. It turns out that this type is an interesting general approach involving Lie derivatives of weights along the given mother fields. We then specialize this setup to operators for basis parameter field and pianola parameter field deformations.

Although the field operators look very attractive, they have a big practical drawback: If one has to compute integrals of ordinary differential equations in order to solve the integral curve problem for performance fields, one has to control the type of vector field. In fact, not every wild field can be integrated by one and the same numerical ODE routine. For example, we have implemented Runge-Kutta-Fehlberg routines in RUBATO®'s PerformanceRUBETTE®. We often experienced that the application of a field operator, such as the tempo operator or the scalar operator (see Sections 39.5 and 39.6, respectively, for these operators), causes a breakdown of these numerical routines. So the routines should be adapted to the field, but this is beyond this discussion and should be left to numerical ODE specialists.

39.5 Tempo Operator

Summary. The tempo operator is the classical field operator which alters the mother's tempo as a function of given weights. We expose the different variants of such an intervention.

$-\Sigma-$

This operator acts on the tempo field $\mathfrak{T}_E$ of a performance cell, so we suppose that E is a member of the space hierarchy H of $\mathfrak{h}$. We are given a weight w and set

$$TempoOperator_w(\mathfrak{T})_E(X_E) = w(X_E).\mathfrak{T}_E(X_E) \tag{39.32}$$

for a E-event X_E. This operator is only defined if the weight is a strictly positive continuous function on the given onset frame, which we shall from now on tacitly assume. If the tempo operator is clear, we shall write the shorthand

$$TempoOperator_w(\mathfrak{T}) = \mathfrak{T}_w.$$

This operator is as easy as problematic if we do not take into consideration other parameters which are canonically tied to onset. Duration is such a parameter. Let us look at the combined articulation field that is influenced by the weight in the tempo operator. Let us first start with the parallel articulation field (see Section 33.2.1), i.e., $\mathfrak{T}_D = 2.\mathfrak{T}_E \circ \alpha_+ - \mathfrak{T}_E$. So if we deform the tempo field $\mathfrak{T}_E$ to $\mathfrak{T}_{w,E}$, the D-component of the parallel field is deformed to a D-component of the parallel field with the weight contribution, i.e., $\mathfrak{T}_{w,D} = 2.\mathfrak{T}_{w,E} \circ \alpha_+ - \mathfrak{T}_{w,E} = w \circ \alpha_+ .2.\mathfrak{T}_E \circ \alpha_+ - w.\mathfrak{T}_E$. However, this formula is not satisfactory since it only works if we are given a parallel field situation. Else, we would destroy the given mother D-component and retain only the mother's tempo information.

In order to take the mother's duration field component into consideration, we need a more invariant formula. To this end, rewrite

$$2.\mathfrak{T}_E \circ \alpha_+ = \mathfrak{T}_D + \mathfrak{T}_E \tag{39.33}$$

and then get the formula

$$\begin{aligned}\mathfrak{T}_{w,D} &= w \circ \alpha_+ .(\mathfrak{T}_D + \mathfrak{T}_E) - w.\mathfrak{T}_E \\ &= w \circ \alpha_+ .\mathfrak{T}_D + (w \circ \alpha_+ - w).\mathfrak{T}_E\end{aligned} \tag{39.34}$$

or for the total articulation field $\mathfrak{T}_{w,ED} = (\mathfrak{T}_{w,E}, \mathfrak{T}_{w,D})$:

$$\mathfrak{T}_{w,ED} = \begin{pmatrix} w & 0 \\ w \circ \alpha_+ - w & w \circ \alpha_+ \end{pmatrix} \mathfrak{T}_{ED}. \tag{39.35}$$

The new articulation field therefore results from the old one by a linear action

$$Q_w = \begin{pmatrix} w & 0 \\ w \circ \alpha_+ - w & w \circ \alpha_+ \end{pmatrix} \tag{39.36}$$

on the tangent bundle of the ED-frame[6]: $\mathfrak{T}_{w,ED} = Q_w.\mathfrak{T}_{ED}$. The shape of the operator Q_w is precisely that which one obtains for the parallel field at tempo w. This means that instead of a real tempo, we take the 'weight tempo' w and the associated 'performance' $\wp_w$. Its inverse 'is' Q_w, and the new articulation field $\mathfrak{T}_{w,ED}$ is the inverse image of the old articulation field $\mathfrak{T}_{ED}$ under the 'weight performance' $\wp_w$.

Usually, the change from the given tempo curve $\mathfrak{T}_E$ to $\mathfrak{T}_{w,E} = w.\mathfrak{T}_E$ changes also the duration $\Delta = \int_{E_0}^{E_1} \frac{1}{\mathfrak{T}_E}$ to the duration $\Delta_w = \int_{E_0}^{E_1} \frac{1}{\mathfrak{T}_{w,E}}$. If the integration limits E_0, E_1 are initial onsets, this is bad news, and we have to adapt the tempo deformation to the condition that these integrals should coincide. Moreover, one usually requires also that the tempi at the initial points should not change. To this end, one introduces

[6] Recall that performance fields are defined on open neighborhoods of frames.

a continuous support function *supp* which vanishes outside $[E_0, E_1]$, is identically 1 in a slightly smaller interval $[E_0+\beta, E_1-\beta]$ for a relatively small positive real number β, and is the cubic spline with zero slopes on the boundary intervals $[E_0, E_0+\beta], [E_1-\beta, E_1]$. Then

$$s(\mathfrak{Z}_E, w, t) = \mathfrak{Z}_E(t)(1 + supp(t)(w(t) - 1))$$

coincides with w within $[E_0+\beta, E_1-\beta]$ and is 1 outside $[E_0, E_1]$. We then try the new duration

$$J(\mathfrak{Z}_E, w) = \int_{E_0}^{E_1} \frac{1}{s(\mathfrak{Z}_E, w, t)}$$

and compare the durations:

$$r(\mathfrak{Z}_E, w) = \frac{J(\mathfrak{Z}_E, w)}{J(\mathfrak{Z}_E, 1)}.$$

We then try the obvious correction $w_1 = r(\mathfrak{Z}_E, w).w$ and get a weight which gives us nearly the right duration since within the interval $[E_0+\beta, E_1-\beta]$, the error is corrected. We look for the new error $r(\mathfrak{Z}_E, w_1)$, set the new weight $w_2 = r(\mathfrak{Z}_E, w_1).w_1$, and repeat this procedure until the error becomes small enough. This procedure has been implemented in RUBATO®'s PerformanceRUBETTE®, but we do not know whether it converges[7].

39.6 Scalar Operator

Summary. The scalar operator is a first generalization of the tempo operator to articulation. The conceptual background is that tone parameters can be split into *basis* and *pianola* parameters. Onset, pitch, loudness are basis parameters. The corresponding pianola parameters are duration, glissando, crescendo. Pianola components of performance fields are coupled to the corresponding basis components of the performance fields. This coupling can be distorted by making operator weights act on specific parameter sets. The scalar operator does this job.

– Σ –

Suppose that we are given a space B of basis parameters, and that P is the corresponding space of pianola parameters. We suppose given a weight w on the B space. The scalar operator $ScalarOperator_w$ deforms a performance field $\mathfrak{Z}_{BP} = (\mathfrak{Z}_B, \mathfrak{Z}_P)$ to a new field $ScalarOperator_w(\mathfrak{Z}_{BP}) = (\mathfrak{Z}^*_B, \mathfrak{Z}^*_P)$ as follows. We have two parameters: parallel/not parallel, and a two-bit parameter B = yes/no, P = yes/no. The *notparallel* case is a deformation of the components of the given field according to the second parameter:

1. $B = P =$ no: We leave the field unchanged.
2. $B = P =$ yes: We set $ScalarOperator_w(\mathfrak{Z}_{BP}) = w.\mathfrak{Z}_{BP}$.
3. $B =$ yes, $P =$ no: We set $ScalarOperator_w(\mathfrak{Z}_{BP}) = (w.\mathfrak{Z}_B, \mathfrak{Z}_P)$.
4. $B =$ no, $P =$ yes: We set $ScalarOperator_w(\mathfrak{Z}_{BP}) = (\mathfrak{Z}_B, w.\mathfrak{Z}_P)$.

The *parallel* case runs as follows. It means that we refer to the parallel field structure in one way or another. The four cases run as follows:

1. $B = P =$ no: No weight influence, but the parallel structure is installed, i.e.,

$$ScalarOperator_w(\mathfrak{Z}_{BP}) = \partial\mathfrak{Z}_B \tag{39.37}$$

[7] Perhaps this could be settled by use of Banach's fixpoint theorem, we do however not know whether the map is a contraction.

2. $B = P =$ yes: This is the generalized tempo operator, i.e.,

$$ScalarOperator_w(\mathfrak{Z}_{BP}) = Q_w.\mathfrak{Z}_{BP} \tag{39.38}$$

with the matrix being understood as a block matrix of scalar endomorphisms on the basis and pianola spaces, respectively. The original tempo operator is the special case $B = E, P = D$.
3. $B =$ yes, $P =$ no: We set $ScalarOperator_w(\mathfrak{Z}_{BP}) = \partial(w.\mathfrak{Z}_B)$.
4. $B =$ no, $P =$ yes: We set $ScalarOperator_w(\mathfrak{Z}_{BP}) = w.\partial\mathfrak{Z}_B$.

We should remark that the *ScalarOperator* and a fortiori the *TempoOperator* do not change any standard hierarchy. But the initial data could be affected. First of all, the field changes in the articulation plane of onset and duration could throw some kernel event out of the reach of the given initial set, excluding them from being performed. Second, the initial performance could be turned into an incompatible state, a problem we have already dealt with at the end of the *TempoOperator* discussion. For the general situation of the *ScalarOperator*, one must develop specific routines to control initial data, not a trivial task.

39.7 The Theory of Basis—Pianola Operators

Summary. The tempo and scalar operators can be generalized to operators which act on the basis and pianola parameter grouping and their coupling on the level of performance fields. The formalism is deduced from the standard situation and stated in its general shape. The point of this approach is the introduction of Lie derivative as a general device for the production of deformations of performance fields by use of scalar potential funcions. The Lie formalism is applied when stressing on basis rather than on pianola parameters. Although it is a general formalism for operators, we do not know whether its application can generate all important deformation operators needed for a musically satisfactory performance construction.

– Σ –

The standard situation which motivates the basis-pianola operators is this: We are given a performance map $\wp(X,Y) = (x(X), y(X,Y))$ on variables X, Y of two disjoint spaces U, V, with a hierarchy projection $U \oplus V \to U$. For example, this could be an articulation ($U = E, V = D$) or a direct product ($U = EL, V = H$), etc. We now want to deform this performance on the second factor, and we are using a positive deformation weight function λ on U. The deformed performance map is defined by $\wp_\lambda(X,Y) = (x(X), \lambda(X).y(X,Y))$. So this is a kind of physical operator we are familiar with. Let us calculate the Jacobian of this transformation (supposing that everything is smooth). The Jacobian for $\lambda = 1$, the start situation is

$$J\wp_1 = J\wp = \begin{pmatrix} \partial_X x & 0 \\ \partial_X y & \partial_Y y \end{pmatrix} = \begin{pmatrix} A & 0 \\ B & C \end{pmatrix} \tag{39.39}$$

where $\partial_X x$ is the submatrix with the partial derivatives of all x coordinates with respect to the X arguments, etc. The inverse Jacobian is

$$J\wp^{-1} = \begin{pmatrix} A^{-1} & 0 \\ -C^{-1}BA^{-1} & C^{-1} \end{pmatrix}. \tag{39.40}$$

And the general inverse Jacobian is

$$J\wp_\lambda^{-1} = \begin{pmatrix} A^{-1} & 0 \\ -C^{-1}BA^{-1} - \lambda^{-1}C^{-1} \cdot d\lambda \otimes y \cdot A^{-1} & \lambda^{-1}C^{-1} \end{pmatrix} \tag{39.41}$$

with the dual gradient $d\lambda = grad\lambda^*$ and the identification[8] $Lin(U,V) \overset{\sim}{\to} U^* \otimes V$. The performance field is calculated by use of the diagonal unit vector $\Delta = (\Delta X, \Delta Y)$ and yields

[8] See Appendix E.3.2.

$$\mathfrak{Z}_\lambda = J\wp_\lambda^{-1}\Delta = \begin{pmatrix} A^{-1}\Delta X \\ -C^{-1}BA^{-1}\Delta X - \lambda^{-1}C^{-1}\cdot d\lambda \otimes y \cdot A^{-1}\Delta X + \lambda^{-1}C^{-1}\Delta Y \end{pmatrix}.$$

If we set $\mathfrak{Z}(X,Y) = (Z(X), W(X,Y))$, and $\mathfrak{Z}_\lambda(X,Y) = (Z(X), W_\lambda(X,Y))$, we get

$$\begin{aligned} W_\lambda(X,Y) &= W(X,Y) - \lambda^{-1}C^{-1}\cdot d\lambda \otimes y \cdot Z(X) + (\frac{1}{\lambda} - 1)C^{-1}\Delta Y \\ &= W(X,Y) - L_Z(\Lambda)C^{-1}y + (\frac{1}{e^\Lambda} - 1)C^{-1}\Delta Y \end{aligned} \tag{39.42}$$

with the Lie derivative $L_Z(\Lambda)$ of the function $\Lambda = \ln(\lambda)$, knowing that λ is a strictly positive function. Everything is taken at the argument X, Y, or (X, Y), respectively.

This gives us formula

$$\mathfrak{Z}_\Lambda = \mathfrak{Z} - L_Z(\Lambda)i_V C^{-1}y - (1 - e^{-\Lambda})i_V C^{-1}\Delta Y \tag{39.43}$$

where i_V is the canonical injection $V \to U \oplus V$. So we obtain a deformation of the original field $\mathfrak{Z}$ by two additive terms $L_Z(\Lambda)i_V C^{-1}y$ and $(1-e^{-\Lambda})i_V C^{-1}\Delta Y$, which we want to interpret in the following sections.

Both terms involve the weight function. The first term $L_Z(\Lambda)i_V C^{-1}y$ involves the basis field Z, while the second term only involves contributions from the Y component of the field. Therefore, we call the first term the *basis deformation*, whereas the second is called the *pianola deformation.*

39.7.1 Basis Specialization

Summary. The general basis-pianola theory is specialized to basis parameters and explicated through concrete formulas.

– Σ –

If we have a weight $\Lambda \in U_\varepsilon(0)$ in a very small neighborhood of 0 (or else λ in a small neighborhood of 1), we may neglect the pianola term in formula (39.43). If further the performance map in the second variable is affine, i.e., $y = A.Y + B$, and does not depend on the first variable X as is the case with pianola coordinates Y, then $C^{-1} = const. = A^{-1}$, i.e., $C^{-1}y = Y + const.$ So we have a deformation of type $L_Z(\Lambda)i_V C^{-1}y = L_Z(\Lambda)i_V(Y + const.)$.

For example, consider the two-dimensional situation $U = E, V = L$ of two basis spaces. The performance map for a weight $\Lambda \in U_\varepsilon(0)$ on E is $\wp_\Lambda(X,Y) = (x(X), e^\Lambda.y(Y))$, where we start from a primavista performance map $\wp = \wp_0$. We may therefore suppose that $y = A.Y$. Then, one has the primavista product field $\mathfrak{Z} = T \times D$ of tempo T and dynamics D. Then the deformed field is

$$(T \times D)_\Lambda = T \times D - L_T(\Lambda)i_L Y = T \times D - T.\frac{d\Lambda}{dX}i_L Y.$$

In other words, we deform the given tempo by the gradient of Λ and project this component to the loudness field component via a field $i_L Y$. So musically speaking, the loudness field deformation is essentially controlled by the change of Λ and the mother tempo.—Let us now propose a generalization of the basis deformation: We are given

- two not necessarily disjoint spaces U, V of the space hierarchy,
- a weight Λ on the first (the basis) space U,
- an affine *directional endomorphism* $Dir(Y) \in V@V$ of the V space.

With these data, we define a field deformation of the performance field $\mathfrak{Z}$ on a superspace of $U \cup V$ in the hierarchy by the formula

$$\mathfrak{Z}_{\Lambda,Dir} = \mathfrak{Z} - L_{\mathfrak{Z}_U}\Lambda.i_V Dir \tag{39.44}$$

where the argument of Dir is the V-component, and that of the Lie-derivative is the U-component of the total argument, whereas i_V is the embedding of V in the space of $\mathfrak{Z}$. The following is immediate from the linearity of the Lie derivative in both arguments.

Lemma 53 *With the above notation:*

(*i*) *The basis operator is linear in the performance fields, i.e.,*

$$(\mu\mathfrak{Z}_1 + \nu\mathfrak{Z}_2)_{\Lambda,Dir} = \mu\mathfrak{Z}_{1\Lambda,Dir} + \nu\mathfrak{Z}_{2\Lambda,Dir}.$$

(*ii*) *If U and V are disjoint, and if Λ_1, Λ_2 are two weights, then*

$$(\mathfrak{Z}_{\Lambda_1,Dir})_{\Lambda_2,Dir} = \mathfrak{Z}_{\Lambda_1+\Lambda_2,Dir}.$$

Example 55 An example of a non-disjoint union of spaces U, V is the elementary deformation of tempo by a positive C^1 weight γ. On onset arguments X in a frame of positive onsets, we set $\Lambda = \int \frac{1-\gamma}{X}$. This implies $\gamma T = T - L_T(\Lambda).1_E$, and we have another example of a basis operator.

Example 56 For this example, we suppose given a two-dimensional C^2 performance map $\wp_{EH} : EH \to eh$ on the plane of onset and pitch with a hierarchy $EH \to E$ over the onset performance $\wp_E : E \to e$ with tempo curve T. So the performance field $\mathfrak{Z}_{EH}$ is C^1. Setting $I_\wp = (\partial_H h)^{-1}, T = (\partial_E e)^{-1}$, the field reads

$$\mathfrak{Z}_{EH} = (T, I_\wp(1 - \partial_E h.T)). \tag{39.45}$$

We want to view such a field as a deformation of the primavista product field $T \times I$ which is composed of the same tempo factor T and an intonation factor I, which is also C^1 and strictly positive, as usual for primavista fields. We are looking for a weight Λ on the onset-pitch frame R and a pitch shift $e^{-Y_0} \in H@H$ such that the deformation $L_{T\times I}(\Lambda)i_H e^{-Y_0}$ of the primavista field yields $\mathfrak{Z}_{EH}$. This defines a linear partial differential equation (PDE) in the onset variable X and the pitch variable Y:

$$T\partial_E\Lambda + I\partial_H\Lambda = (Y - Y_0)^{-1}(I_\wp(\partial_E h.T - 1) + I) = Q. \tag{39.46}$$

This equation has C^1 coefficients T, I, Q, if we let Y_0 be smaller than the lower boundary value of the pitch frame interval. Using the method of characteristic curves (see Appendix I.6) in pseudo-linear PDEs, we can see that there is a solution of equation (39.46). In fact, the characteristic curve projection ODE reduces to the pair $d_tX = T(X), d_tY = I(Y)$, whereas the third curve component $Z(t)$ is defined by the ODE $d_tZ = Q(X, Y)$. This means that the characteristic curve projection onto the XY-space is an integral curve of the primavista field $T \times I$. Clearly, a transversal curve Γ to the characteristic curves exists since the Jacobian criterium $T\Gamma \nparallel T \times I$ can be met for the non-vanishing field $T \times I$. We therefore have this result:

Proposition 59 *Let $\wp_{EH} : EH \to eh$ be a two-dimensional C^2 performance map on the plane of onset and pitch with a hierarchy $EH \to E$ over the onset performance $\wp_E : E \to e$ with tempo curve T. For the primavista product field $T \times I$ which is composed of the same tempo factor T and an intonation factor I, which is also C^1 and strictly positive, there is a C^1 weight Λ on the EH-frame of this performance such that the performance field $\mathfrak{Z}_{EH}$ is a deformation of the primavista field $T \times I$ by a basis operator:*

$$\mathfrak{Z}_{EH} = T \times I - L_{T\times I}(\Lambda)i_H e^{-Y_0}. \tag{39.47}$$

The basis operator is designed for distributing weight information from the "basis" subspace U over any other space V, independently whether this one is also in the hierarchy or not. This creates a considerable freedom of shaping performance. The way this shaping is related to the given weight is the Lie derivative with respect to the performance field $\mathfrak{Z}_U$ on U. This one also measures the angle between the weight's gradient and $\mathfrak{Z}_U$. If their mutual position is perpendicular, the derivative vanishes and the operator has no effect.

39.7.1.1 Deforming Hierarchies

Summary. We discuss the change in an existing hierarchy after the application of a basis operator.

– Σ –

The most dramatic effect of the basis operator is the deformation of the given hierarchy. For example, in the above example of Proposition 59, the projection $EH \to H$ is destroyed since the weight Λ is not only a function of H or of E, but of both, in general. So only the projection $EH \to E$ survives this deformation. Let us describe more systematically which hierarchy spaces disappear a priori by a basis operator deformation.

Suppose that we have a hierarchy space U, a weight Λ on U any other space V within the top hierarchy space, a directional endomorphism $Dir = e^B.A$, and any hierarchy space W which we want to test for survival after the subtraction of the basis deformation $L_{\mathfrak{Y}_U}(\Lambda)i_V Dir$. Consider the projection $p : V \to V \cap W$ onto the intersection space, including the empty intersection which then defines the projection onto the zero space. Let $[p, A] = p.A - A.p$ denote the commutator endomorphism of p and A on V.

Proposition 60 *With the above notation, the mother hierarchy space W remains a priori alive (i.e., member of the daughter hierarchy deduced from the mother hierarchy) after the basis deformation $L_{\mathfrak{Y}_U}(\Lambda)i_V Dir$ iff either $p = 0$, i.e., $V \cap W = \emptyset$, or $p[p, A] = 0$ and $U \subset W$. In particular, for $[p, A] = 0$ and $U \subset W$, W remains alive.*

Proof. Suppose that W remains alive. If $p \neq 0$, we have at least one coordinate that is common to V and W. On that coordinate, the functional dependence of U is inherited via $L_{\mathfrak{Y}_U}(\Lambda)$. Therefore we must have $U \subset W$. Since no functional dependence from $W - V$ arguments can be the case on the coordinates in $V \cap W$, and since the constant part of Dir is irrelevant here, we must have $p.A.(1 - p) = 0$. On the other hand, if $p = 0$ everything is clear, and under the conditions $p[p, A] = 0$ and $U \subset W$, the deformation arguments all stem from W, and we have saved the life of W, QED.

Example 57 Let the top coordinate space be all six usual basis and pianola parameters $EHLDGC$. Let a weight Λ act on $U = EH$, take $V = ED$, whereas the directional endomorphism is

$$e^B.\begin{pmatrix} 0 & 1 \\ -1 & 0 \end{pmatrix}$$

with a rotation A as its linear part. Suppose that the hierarchy is the parallel hierarchy, i.e., the hierarchy generated by the basis hierarchy TID, TI, TD, T and the parallel fields $\partial TID, \partial TI, \partial TD, \partial T$ with the corresponding projections. Since the rotation A has no proper invariant subspaces, we must have either $ED \cap W = \emptyset$ or else (because of $p[p, A] = 0$) $ED \subset W$. But every field in our hierarchy contains E, hence $U \cup V = EHD \subset W$. This is only true for the sub-hierarchy $EHD, EHDG, EHLD, EHLDG, EHLDC, EHLDGC$.

39.7.1.2 Lie Derivatives

Summary. Basis-pianola operator theory leads to Lie derivatives as a device for operator definition. We discuss the Lie formalism in its realization as a component of a performance grammar.

– Σ –

The appearance of the Lie derivative in this context is quite surprising. Its usage is well known from classical mechanics, for example, where the Lie derivative of a function with respect to a Hamiltonian vector field is related to the Hamiltonian function H via the Poisson bracket [2]. Presently, we do not know of any analogous structures of Hamiltonian or Lagrangian type in performance field theory. But it is good to have this perspective in mind for a future 'dynamics of performance'. We should however observe that the Lie derivative $L_{\mathfrak{Y}}$ associated with a vector field $\mathfrak{Y}$ induces an isomorphism $L_? : \mathfrak{X}_R \xrightarrow{\sim} Der(\mathcal{F}(R))$ between the real vector space $\mathfrak{X}_R$ of smooth vector fields over the frame R and the vector space $Der(\mathcal{F}(R))$ of derivations on the real algebra $\mathcal{F}(R)$ of smooth functions (see Appendix I.2.4). So $L_{\mathfrak{Y}}$ can be identified with $\mathfrak{Y}$, and the basis deformation means taking into account the 'weight' $L_{\mathfrak{Y}}(\Lambda)$ deduced from the weight Λ. In this sense, weights become natural mathematical objects associated with performance fields: They are just the natural objects, these fields act upon qua derivations, they are not only justified by the quantification argument given when we introduced weights in Section 39.1. In other words:

Thesis 6 *A performance field is not only a construction principle for the performance map, but equivalently an 'interpretation of weights'—this is effectively the mathematical transfiguration of the rational approach to performance.*

This thesis suggests that one should study the natural properties of the Lie algebra $Der(\mathcal{F}(R))$ with respect to performance theory, in particular the question of what is the musical interpretation of the Lie bracket $[\mathfrak{Z}_1, \mathfrak{Z}_2]$ of two performance fields. We have to pass this subject to future research.

Exercise 81 *Consider the performance* $\wp : ED \to ed : (X, Y) \mapsto (x(X), f(d)), d = x(X+Y) - x(X)$, *which is a functional change in duration, built upon the parallel performance, with a* C^1 *invertible deformation function* f *of physical duration* d. *This is an example of a physical operator. Show that its field is*

$$\mathfrak{Z} = \begin{pmatrix} T \\ \frac{(f'+1)}{f'} T \circ \alpha_+ - T \end{pmatrix} = \begin{pmatrix} 1 & 0 \\ \frac{(1-f')}{f'} \frac{T \circ \alpha_+}{T} & 1 \end{pmatrix} \partial T, \tag{39.48}$$

i.e., a linear automorphism of the tangent bundle of which we have already seen an example for the tempo operator. Observe however, that the automorphism is also a function of the tempo curve T*! Show by use of the characteristics method for quasilinear PDEs that this deformation of the parallel field* ∂T *can also be obtained by a basis operator with* $U = ED, V = D$.

39.7.2 Pianola Specialization

Summary. The general basis-pianola theory is restricted to pianola parameters and explicated through concrete formulas.

– Σ –

Compared to the basis deformation, the second contribution $(1 - e^{-\Lambda}) i_V C^{-1} \Delta Y$ in the general deformation formula (39.43) plays a different role. Whereas the basis contribution is sensitive to the gradient of Λ, i.e., its local changes, the second contribution is sensitive to the absolute values of the weight, so this contribution is relevant if the weight changes little, but has values different from zero. We give a more precise interpretation of this contribution in the case of a basis-pianola-space situation, i.e., U is a space of basis parameters, and V is the corresponding space P_U of pianola parameters. We then have the alterator $\alpha_+ : U \oplus P_U \to U$, and we have a canonical isomorphism $\tau : U \xrightarrow{\sim} P_U$. If the original field is defined on $U \oplus P_U$ and is a parallel field $\partial \mathfrak{Z}_U$, then we have $C^{-1} \Delta Y = \tau \circ \mathfrak{Z}_U \circ \alpha_+$. Therefore we obtain the *pianola operator* for this special space configuration:

$$\mathfrak{Z}_{\Lambda,U} = \mathfrak{Z} - (1 - e^{-\Lambda}) i_{P_U} \circ \tau \circ \mathfrak{Z}_U \circ \alpha_+. \tag{39.49}$$

This formula only involves a weight on *any subspace* of the top space and a hierarchy space U consisting of basis parameters, together with the corresponding pianola space P_U which must also be a subspace (but not necessarily hierarchic!) of the top space.

39.8 Locally Linear Grammars

Summary. According to Section 38.3.4, interaction between "inherited" performance score structures of sisters—or farther relatives—can be envisaged. We describe this formalism which is a basic approach in inverse performance theory (see Section 47.2).

– Σ –

Until now, we have only considered operators which are directly related to the mother data, and not to farther relatives, such as sisters, or daughters of sister, etc. In the following discussion, we shall present an essentially linear model for such a more global interconnection of a stemma's LPS.

We start with a stemma, i.e., a local performance score Λ whose graph $\Lambda \updownarrow$ is an undirected tree. We want to forget about the mother of Λ and concentrate on the tree $\Lambda \downarrow = T_0$. For the following construction, we need the *stemma quiver*[9] $T = T_\Lambda$ associated with Λ:

Definition 109 *The* stemma quiver *is a finite directed graph $T = (V, A)$ with vertex set V and arrow set A, including multiple arrows and loops. It is constructed as follows. We start with a directed tree $T_0 = (V, A_0)$ with root, i.e., each vertex can be reached by a unique path starting from the root. If $x \to y$ is an arrow of T_0, we say that x is the* mother *of y and y is a* daughter *of x (in combinatorics they are known as 'father' and 'son', respectively, here we try to be politically correct). If we have a path $x \to y \to z$, then z is a* granddaughter *of x, while x is the* grandmother *of z, and so on. For a vertex x of T_0, the set of vertexes which are daughters of x is denoted by $D_x(T_0)$. The vertexes x which are not mothers, i.e., $D_x(T_0) = \varnothing$, are called* final *(the 'leaves' in graph theory). Similarly for each vertex $x \in V(T)$ we define $M_x(T_0) \subseteq V(T_0)$ as the set of vertexes lying on the unique path from x to the root (x included).*

In order to define the stemma quiver T the directed tree T_0 is enriched by the following set of arrows (no vertexes added): First, each vertex x is given a loop $x \circlearrowleft$, and for any couple of sisters x_1, x_2, i.e., of daughters of a common mother y, we add an arrow $x_1 \to x_2$. The resulting quiver $T = (V, A)$ is called a stemma quiver of the tree T_0, which is uniquely determined by the stemma quiver and is called the stemma tree. *We therefore may define $D_x(T) = D_x(T_0)$ and $M_x(T) = M_x(T_0)$.*

Definition 110 *With these graph-theoretical data, a* locally $\mathbb{R}$-linear grammar *is a family of $\mathbb{R}$-linear representations*[10] *of a stemma quiver T which is defined by the following data:*

1. *For each vertex $x \in V(T)$ let A_x, B_x be two real vector spaces, B_x of finite dimension s_x.*
2. *Each mother-daughter arrow $x \to y$ is represented by a surjective linear map $r_{x,y} : A_x \to A_y$.*
3. *For all $x \in V(T)$ let $\varphi_x : B_x \to End(A_x)$ be an affine map, i.e., a linear map φ_x^0, followed by a displacement by $\varphi_x^t \in End(A_x)$. So each loop $x \circlearrowleft$ is represented by a family of endomorphisms $(\varphi_x(b) : A_x \to A_x)_{b \in B_x}$ parametrized by the* parameter space *B_x.*
4. *For all pairs of sisters $x, y \in D_m(T)$, the sister arrow $x \to y$ is represented by an isomorphism $i_{x,y} : A_x \to A_y$ with $i_{y,x} = i_{x,y}^{-1}$ and $i_{x,x} = Id_{A_x}$.*

For concrete performance field configurations associated with Λ this axiomatic setup is realized as follows. For the LPS of the stemma tree, we suppose a number n of real parameters for the top space S of the LPS hierarchies. We concentrate on this top space S of all the LPS, and do not discuss the other cell data of these space hierarchies. For each vertex x of T_0, we have the frame $R_x = [l_{x_1}, u_{x_1}] \times \ldots [l_{x_n}, u_{x_n}]$ of the top space of its hierarchy. For each mother m and daughter x, we suppose that $R_x \subset R_m$, and that for each couple of sisters x_1, x_2, $R_{x_1} \cap R_{x_2} = \varnothing$. This corresponds to a restriction of a larger portion of a musical score to a disjoint grouping of smaller portions. Here is the realization of our above system (properties 1–4) of quiver representations:

Consider the vector space $\mathcal{F}_x$ of C^∞ functions on R_x. We then set $A_x = Der(\mathcal{F}_x)$, the space of derivations, i.e., the C^∞ vector fields on R_x. The surjective maps $r_{x,y}$ are defined as the restrictions of vector fields on the mother's rectangle R_x to the daughter's rectangle R_y. To define the representations for a sister arrow $x_1 \to x_2$, consider the unique affine morphism $a_{x_2,x_1} : R_{x_2} \to R_{x_1}$ on the sisters' rectangles such that the respective vertexes are mapped onto each other. Then the sister arrow representations are isomorphisms $i_{x_1,x_2} : A_{x_1} \to A_{x_2}$ defined by the transport of a vector field Z on R_{x_1} to $Z \cdot a_{x_2,x_1}$ on R_{x_2}.

To define the operation of a parameter family, first take a vector field $Z \in Der(\mathcal{F}_x)$, and a weight function $\Lambda \in \mathcal{F}_x$. For $Dir \in S@S$, consider the corresponding vector field $\Gamma_{Dir}(t) = (t, Dir(t))$ on R_x. Then

[9] See Definition 137 in Appendix C.2.2 for the quiver concept.

[10] These are representations of the quiver algebra over $\mathbb{R}$, see Appendix E.2.1 for the concept of a linear representation.

we have a new vector field $Z - L_Z\Lambda.\Gamma_{Dir}$, corresponding to a basis deformation. This is an $\mathbb{R}$-linear operator on Z, and the 'deformation' part $L_Z\Lambda.\Gamma_{Dir}$ is $\mathbb{R}$-bilinear in the weight function and the affine endomorphism.

We now take a finite dimensional subspace W_x of $\mathcal{F}_x$ which in performance theory represents the weight functions issued from analyses of metrical, motivic, and harmonic structures of the given score. We now set $B_x = W_x \otimes_{\mathbb{R}} S@S$, and we obtain an $\mathbb{R}$-linear map $\varphi_x^0 : B_x \to End(A_x)$ defined by $\varphi_x^0(\Lambda \otimes Dir)(Z) = -L_Z\Lambda.\Gamma_{Dir}$. Setting $\varphi_x^t = Id_{A_x}$, we have defined the required affine map $\varphi_x = \varphi_x^t + \varphi_x^0 : B_x \to End(A_x)$ with

$$\varphi_x(\Lambda \otimes Dir)(Z) = Z - L_Z\Lambda.\Gamma_{Dir}.$$

Let $m \in V(T)$ be a mother. Then for each daughter $x \in D_m(T)$, we define a triaffine (affine in each argument) map

$$f_x : A_m \times \mathbb{C}^{\# D_m(T)} \times \prod_{y \in D_m(T)} B_y \to A_x, \tag{39.50}$$

by

$$\begin{gathered}(a_m, (c_{y,x}^m)_{y \in D_m(T)}, (b_y)_{y \in D_m(T)}) \mapsto \\ \textstyle\sum_{y \in D_m(T)} c_{y,x}^m i_{y,x}(\varphi_y(b_y)(r_{m,y}(a_m))) = \\ \textstyle\sum_{y \in D_m(T)} c_{y,x}^m i_{y,x}(\varphi_y^0(b_y)(r_{m,y}(a_m))) + \sum_{y \in D_m(T)} c_{y,x}^m i_{y,x}(\varphi_y^t(r_{m,y}(a_m))).\end{gathered}$$

Referring to the above example, this formula describes the following: In order to determine the performance field on the frame F_x, we use the field of its mother m, we first restrict that field to any of its daughters y and get the fields $r_{m,y}(a_m)$. These sister fields to daughter x are then deformed under the endomorphisms $\varphi_y(b_y) = \varphi_y^t + \varphi_y^0(b_y)$ induced by the system parameters b_y. These deformed fields are then transported to x and weighted by the factors $c_{y,x}^m$. These sister fields influence the final value of the field at daughter x. Musically, this means that the performance at x is influenced by surrounding sister fields, which are typically the fields of past and future times (past or future periods, bars, etc.).

We shall pursue this model in the course of the inverse performance theory of Chapter 47. A final word on a perspective of generalized stemmata which seem to be suggested by the locally linear grammars as discussed above: One could envisage *continuous stemmata*. They are based on a generalization of the stemma's ramification structure to one-parameter families and narrowing of the daughters' extents of 'infinitesimal' quantities. This construction would take care of the fact that psychologically, interaction between neighboring performance moments is continuously 'updated'. A theory of continuous stemmata is still pending.

Part X

RUBATO®

40

Architecture

The most rigorous test of the efficiency of theories in modern cognitive science is the production of a working computer program whose external behaviour mimics that to be explained.
John Sloboda [983]

Summary. RUBATO® is a metamachine designed for representation, analysis, and performance of music. It was developed on the NEXTSTEP environment during two SNSF grants from 1992 to 1996 by the author and Oliver Zahorka [690, 689, 692, 699, 1146, 1148]. From 1998 to 2001, the software was ported to Mac OS X by Jörg Garbers in a grant of the Volkswagen Foundation. RUBATO®'s architecture is that of a frame application which admits loading of an arbitrary number of modules at run-time. Such a module is called RUBETTE®. There are very different types of Rubettes. On the one hand, they may be designed for primavista, compositional, analytical, performance stemma or logical and geometric predication tasks. On the other, they are designed for subsidiary tasks, such as filtering from and to databases, information representation and navigation tasks, or else for more specific subtasks for larger "macro" Rubettes. A RUBETTE® of the subtask type is coined OPERATOR and implements, for example, what we have called performance operators in Section 44.7. The RUBATO® concept also includes distributed operability among different peers. This software is conceived as a musicological research platform and not a hard-coded device, we describe this approach. Concluding this chapter, we discuss the relation between frame and modules.

– Σ –

In the original concept of RUBATO®[687], we had defined RUBATO® as being a software for analysis and performance, divided into two submodules: one for "structuring" a score, and the other for "shaping" this score. This meant that structuring would yield analytical structures, whereas the other would yield a shaped performance transformation, alimented by analytical data from the structuring process.

In the course of the software development, we learned that no data model for music objects known to the developers at that time would be sufficient for all requirements of a comprising music analysis and performance. This led to the concept of denotators and forms, as realized under the title of the "PrediBase" database management system (DBMS) of the first RUBATO® implementation in 1994, as described in [1147]. At that time, it became clear that under such a universal data model, RUBATO® would split into an application framework comprising the "PrediBase" DBMS and a series of dynamically loadable software modules as implemented in the Objective C language of those NEXTSTEP OS driven NeXT computers. According to the diminutive convention for modules, such a module was coined "RUBETTE®".

In 1996, at the end of a grant of the Swiss National Science Foundation where RUBATO® was realized, three analytical Rubettes and one for performance have been developed[1], which will be described in

[1] For NEXTSTEP, RUBATO® as well as these Rubettes are available on the internet, see [699]. The source code is GPL and is contained in the book's CD-ROM, see page xlvii.

G. Mazzola, *The Topos of Music II: Performance*, Computational Music Science,
https://doi.org/10.1007/978-3-319-64444-8_9

Chapter 41. The PerformanceRUBETTE® is connected to five OperatorRubettes (at those times still named "OPERATORS"). The PrimavistaRUBETTE® takes care of paratextual score predicates.

In 2001, this software has been ported (and improved in many data management aspects) to Mac OS X by Jörg Garbers and is now available as an open source project on the internet [699], or on the CD appended to this book, see page xlvii. The RUBETTE® screenshots in Chapter 41 are all taken from this version of RUBATO®. Figure 40.1 shows the info panel of the Mac OS X version. However, the version included in the book's CD-ROM is the latest version before the book went into production, whereas the screenshots are somewhat older. We hope that the reader will excuse us for this slight asynchronicity.

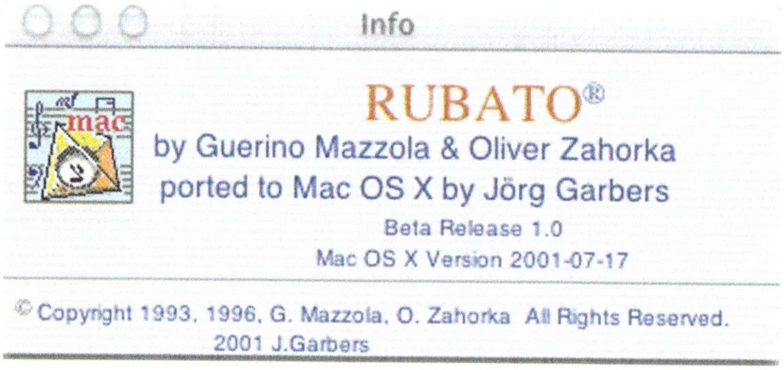

Fig. 40.1. The info panel of the Mac OS X version of RUBATO®.

In the following sections of this chapter we shall however not describe the Mac OS X implementation, we will rather expose the more advanced and flexible architectural principles of the ongoing Java-based implementation of the distributed RUBATO® version.

40.1 The Overall Modularity

Summary. RUBATO® is a modular engine for metamachine rationales and because research is itself increasingly modular. Built upon the denotator language, the RUBATO® concept is fully modular, all parts that can be split into modules *have* been split in this way.

– Σ –

The modularity of RUBATO® has two aspects: First, it shows a composition of the software from an arbitrary, a priori undetermined, number of functional units—the RUBETTE® modules. Second, the available Rubettes are a dynamic factor: According to the research progress, new Rubettes of any flavor may be added to the existing arsenal.

Modularity is an old principle, in fact, the traditional disciplinarity of science preconizes modules of scientific activities which have or pretend to have a relative autonomy in knowledge production. What is new with respect to the traditional disciplinarity is that this modularity is a dynamical one, at any time new modules of knowledge processing may be added or old ones removed. Discipline becomes a task-driven decision instead of being a rigid preset splitting.

Such a modularization can only work on the common ground of unrestricted cross-commu-nication among any subgroup of knowledge modules. Without a common language ground, which in our case is the denotator and form data model, dynamic disciplinarity would inevitably collapse since a new module would require language modifications, adaptations, and extensions.

To be clear, we view dynamic disciplinarity as the idealized version of inter- and transdisciplinarity. The unity of knowledge cannot be achieved without a temporary and task-driven compartmentation of research fields, grouping and regrouping is inevitable; there is no direct path to the unity of knowledge. This credo can however not be realized without a common language basis. Otherwise, language engineering frictions would paralyze any major effort of dynamical grouping of disciplines. Evidently, the present RUBATO® environment is limited to musical and musicological scopes. Here dynamic disciplinarity is not that utopic. After all, the denotators and forms are universal language approaches issued from music(ologic)al requirements. But it has been shown in [918] that denotators are not only a priori applicable to non-musical concept modeling, but also in concrete cases such as geographic information systems. This suggests that dynamic disciplinarity could be realized on RUBATO® for modules of completely general scopes as long as the denotator data model is joined. This is one of the most intriguing vectors of future developments concerning the RUBATO® environment.

The principle of dynamic disciplinarity has its social form: a so-called collaboratory. According to Bill Wulf, this is "a 'center without walls' in which the nations researchers can perform their research without regard to geographical location, interacting with colleagues, accessing instrumentation, sharing data and computational resources, and accessing information in digital libraries"[554]. To collaborate in this way requires adequate software platforms, and RUBATO® is precisely this type of software in the field of musicology.

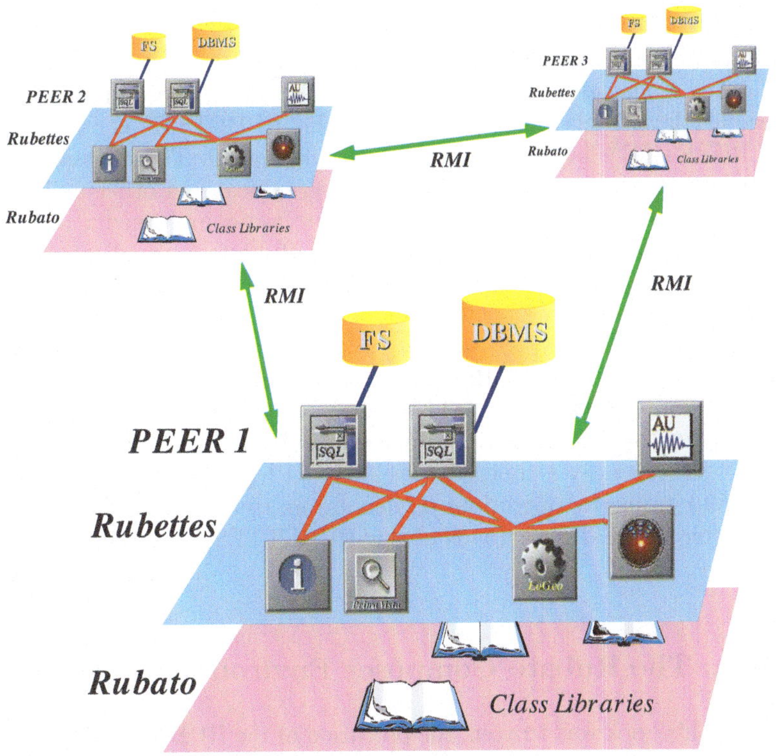

Fig. 40.2. The RUBATO® layers of Rubettes with denotator communication, and the RUBATO®Framework layer with class libraries that are related to different Rubettes. Each peer has one such configuration. Different peers are interconnected via remote method invocation (RMI).

40.2 Frame and Modules

Summary. Modularity has been realized on the basis of a frame application which offers interfaces to an arbitrary number of modules. This is one of the technical core features in the realization of a metamachine. We describe its splitting interfaces and their functional positions.

– Σ –

The RUBATO® platform consists of a number of installations of the software on different peers which may communicate via Java's remote method invocation protocol (RMI, see SUN's Java documentation on the internet). For each peer, RUBATO® contains two layers: the RUBETTE® layer and the RUBATO®framework layer. The first contains a number of Rubettes which are autonomous Java applications that communicate with each other and with Rubettes of another peer exchanging denotators via RMI. These are instances of the denotator class. The class library on the RUBATO® framework layer contains corresponding basic Java classes for denotators, forms, diagrams of presheaves, and modules. It also contains other classes which provide Rubettes with the necessary routines. The concept of these libraries is that they should contain all classes and methods that are of *general interest*, while classes and methods with specific interest for a RUBETTE®'s functionality should be installed in that RUBETTE®.

There are a number of mandatory Rubettes: The InfoRUBETTE® (with an "i" in Figure 40.2) is the initialization RUBETTE®. It is automatically started when RUBATO® starts and informs the user about available peers and Rubettes on the distributed environment. The visualization of all Rubettes' content and manipulation structures is managed by the PVBrowserRUBETTE®, whose functionality is to visualize any denotator in 3D space via Java3D classes, in Figure 40.2, this RUBETTE® is represented by a lens symbol. The concept of this RUBETTE® has been described in Chapter 20. The advantage of this centralization is that no other RUBETTE® designer has to take care of graphical and other multimedia representations, every denotator is piped to the PVBrowserRUBETTE® in case a multimedia representation is required. And every such representation is uniform according to this Rubette's visualization routines, which makes orientation much easier than individual design for every RUBETTE®. Nonetheless, as explained in Chapter 20, the flexibility of the *Satellite* form for multimedia objects allows an unlimited multiplicity of shape and behavior. Two further Rubettes are devoted to the storage of denotators. The first, to the upper left of the RUBETTE® layer, we have the DenotexRUBETTE®. It takes care of the storage and editing of denotators which are given in the Denotex ASCII format. The second storage RUBETTE® is a filter to a SQL DBMS such that SQL databases can be transformed into denotators for RUBATO®. A last central RUBETTE® is the LoGeoRUBETTE®, as shown by a toothed wheel symbol to the right front of the RUBETTE® layer. It manages the logical and geometric operations on denotators (see Section 18.3.4) and can be used by any RUBETTE® for its specific needs. The methods of this RUBETTE® are encoded in the class library of the RUBATO® layer.

40.3 Postscriptum: The Rubato Composer Environment

When the first edition of this book was written, the portation of the RUBATO®software to Java had started, but it was only with Gérard Milmeister's excellent PhD thesis *The Rubato Composer Music Software* [739] that the transfer to Java could be fully realized in the sense of a persistent completion of the original ideas of modular collboration of components for musical analysis, composition, and performance.

We refer to Chapter 69 for a detailed account, and to the subsequent Chapters 71 to 74 for a thorough discussion of the powerful new BigBang rubette that was written by Florian Thalmann in his gesturally conceived groundbreaking PhD thesis [1042].

41

The RUBETTE® Family

Vögel, Vieh und alles, was auf Erden kriecht,
die laß heraus mit dir, daß sie sich tummeln auf der Erde
und fruchtbar seien und sich mehren auf Erden.
The Holy Bible, Genesis 8

Summary. We give an overview of the analytical MetroRUBETTE®, MeloRUBETTE®, HarmoRUBETTE®, the PerformanceRUBETTE®, and the PrimavistaRUBETTE®, which have been realized on the NEXTSTEP and then on the Mac OS X environment.

– Σ –

Originally, the Rubettes as such were not the central concern of the RUBATO® project, this was rather to establish their collaboration and the realization of the whole transmission process from analytical data to the performance shaping operators of the PerformanceRUBETTE®. Each of these Rubettes was more an experimental prototype without the claim of a high-end tool in the specific domain.

The interest in such experiments lies in the fact that when one starts the design of a RUBETTE®, it turns out that musicology and music theory do not offer any reasonable support, be it in conceptual, be it in operational aspects. The path from the given score to a specific analysis reveals an incredible complexity of what in musicology and music theory looks like an easy enterprise. For example, in the design of the HarmoRUBETTE®, the mere definition of what is a chord cannot be traced from traditional literature. Should we only look for local compositions of pitches that stem from notes with a common onset, or should one also consider durational aspects? The standard answer—or rather: excuse—states that it depends on the particular context, and the context of the context, but this is no way out if one has to implement clear concepts and methods rather than rhetorics. So the design of a RUBETTE® is always a very good test of the validity of a model and of its adequacy with traditional fuzzy understanding. But it is also a test for the tradition: After all, who decides what is a good model for harmony? Here, the alternative between general speculative nonsense theories and concrete, but possibly non-sufficiently general implementation and operationalization becomes dramatic. At least, one can hope that this confrontation will force everybody to rethink ill-defined approaches.

41.1 MetroRUBETTE®

Die Zählzeiten (Schlagzeiten, rhythmische Grundzeiten)
gewinnen unter allen Umständen erst
reale Existenz durch ihre Inhalte.
Hugo Riemann [904]

G. Mazzola, *The Topos of Music II: Performance*, Computational Music Science,
https://doi.org/10.1007/978-3-319-64444-8_10

Summary. The MetroRUBETTE® is an elementary analysis module which shows that seemingly simple approaches yield complex but informative results. We also make evident that operationalization of abstract concepts reveals unexpected insights into generically not foreseeable structures.

– Σ –

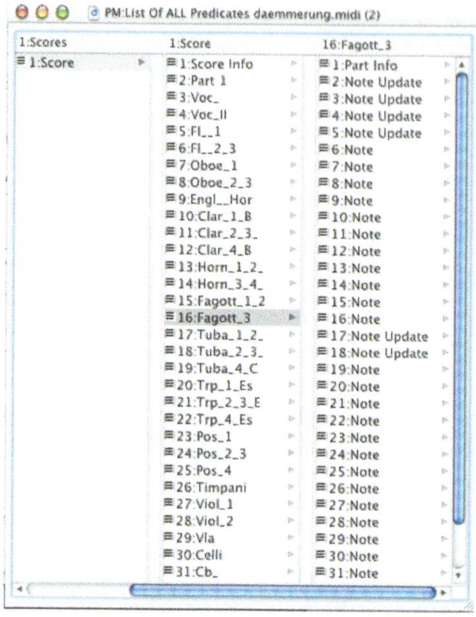

Fig. 41.1. The 31-part score denotator deduced from Richard Wagner's composition "Götterdämmerung".

The MetroRUBETTE® is built on the sober weight calculation which we developed in Chapter 21 in the frame of the maximal meter nerve topologies. The formula in Example 41 of Section 21.2 is realized, except that there is no upper length limitation (i.e., it is put to ∞). The mixed weight formula $W(x) = \sum_{X_i \in Sp_I(x)} W_i(x)$ is also realized in this RUBETTE®. The input is a score denotator (although still in the early shape of the PrediBase data model, which is the very special denotator form of a list of lists, which start from the simple form of strings). Figure 41.1 shows the 31-part score deduced from Richard Wagner's composition "Götterdämmerung".

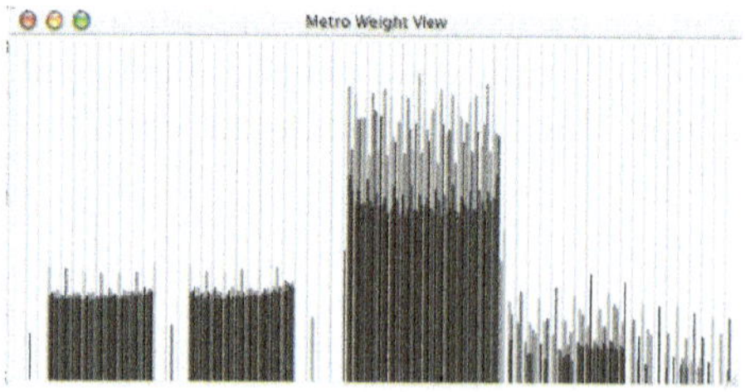

Fig. 41.2. he weight graphics of the above composition with profile = 2, minimal length = 2. Horizontal axis: onset, vertical axis: (relative) weight. The unit of the onset grid in the graphical representation of the weight is 1/8.

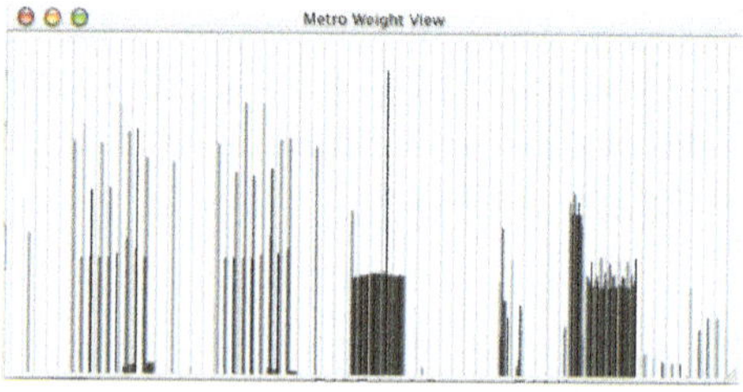

Fig. 41.3. The mixed weight graphics of four parts clarinet (part 11), bassoon (part 16), horn (part 14), and violin (part 27) with scaling factors 3, 2.5, 1, and 0.1 for the respective parts. The unit of the onset grid is again 1/8, as in Figure 41.2.

This data is imported to the RUBETTE® and then evaluated according to the said formulas. For example, the weight of the union of all the onsets of the 31 parts is shown in Figure 41.2. Its parameters are: profile = 2, minimal length = 2, and the unit of the onset grid in the graphical representation of the weight is 1/8.

The mixed weight is also an option of this RUBETTE®. Figure 41.3 shows the mixture of the parts: clarinet (part 11), bassoon (part 16), horn (part 14), and violin (part 27). The parameters are same as above, the weights are scaled by the factors 3, 2.5, 1, and 0.1 for the respective parts. For a more in-depth application of this RUBETTE®in musicology, we refer to [325]. This paper is an excellent example of an unexpected musicological application of a RUBETTE® which was not intended to be interesting on its own. The metrical analysis turns out to be quite sophisticated, although the concept of a local meter is a very elementary one. The surprising effect of this setup is that the combination of the simplistic concept reveals unprecented insights into the time structure of classical works. This is a good hint to the musicologists, teaching them that interesting insight can result from a complex aggregation of simple ingredients. In this case, the simple elements are provided by the maximal local meters, whereas their complex aggregation is conceived by the nerve of the covering they define. A second very interesting application of this RUBETTE® has been presented in [691]. It was recognized that the longest possible minimal lengths of local meters for the left hand part of Schumann's "Träumerei" yields a 3+5 quarters periodicity over two bars, whereas the same analysis of the right hand yields the expected periodicity of 4 quarters with stress on the bar-lines. The sonification of this fact can be heard on the audio example in the book's CD-ROM, see page xlvii.

In this RUBATO® version, however, the output is a weight which is by no means a denotator. This output is a final data and must be used in its special format. This will be the case for operators of the PerformanceRUBETTE®. For the Java-based distributed RUBATO®(Figure 40.2, such a restrictive usage would be forbidden.

41.2 MeloRUBETTE®

In general, the author does not believe in the possibility or even desirability of enforcing strict musical definitions.
Rudolph Reti on the concept of a motif [892]

Summary. The MeloRUBETTE® is an excellent example of the tension between abstract concepts and operational implementation. We expose the routines for motivic analysis, the interface concept, and we discuss the performance problem, including proposals for performance improvement and their theoretical limits.

– Σ –

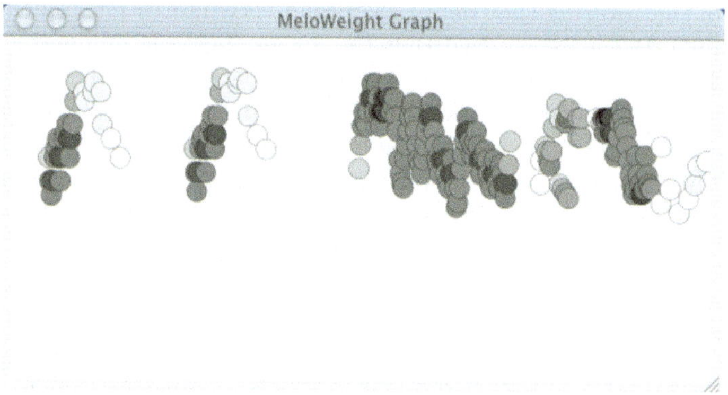

Fig. 41.4. The weight graphics for the celli part 30 in the above score denotator deduced from Richard Wagner's "Götterdämmerung".

The MeloRUBETTE® refers to the theory of motivic topologies in Chapter 22 and, in particular, to Section 22.9 about motivic weights. The score is loaded as for the MetroRUBETTE®, its projection to the onset-pitch space is then analyzed and yields a numeric weight for each onset-pitch event. One such weight is visualized in Figure 41.4. It corresponds to part 30 (celli) of the above composition "Götterdämmerung". The weight values are encoded in gray-levels of the discs which represent the events in onset and pitch.

The calculation relies on these parameters which relate to melodic topology:

- Symmetry Group. This is the paradigmatic group of the shape type. In each group, we include the translation group in pitch and onset. The choice is then between the translation group (encoded by "trivial"), the one generated by the translations plus the retrograde, or the one generated by the translations plus inversion, and the full counterpoint group, i.e., generated by the inversion and retrograde over the translations.
- Gestalt Paradigm. This is one of three possible shape types: diastematic, elastic, and rigid. The first candidate is in fact the type which we called "diastematic index shape type" in Chapter 22. Observe that we have no topology for the diastematic type, but may nevertheless define neighborhoods!
- Neighborhoods. This is the neighborhood radius ϵ which was used in the expression $D^{\mu}_{\epsilon}(M)$ in Section 22.9.
- Span. This is the maximal admitted onset difference between motive events.
- Cardinality. This is the maximal admitted cardinality of motives. Together with the Span, this condition, defines the selection μ of motives which is addressed in the approach from Section 22.9.

The presence and content functions are defined as follows:

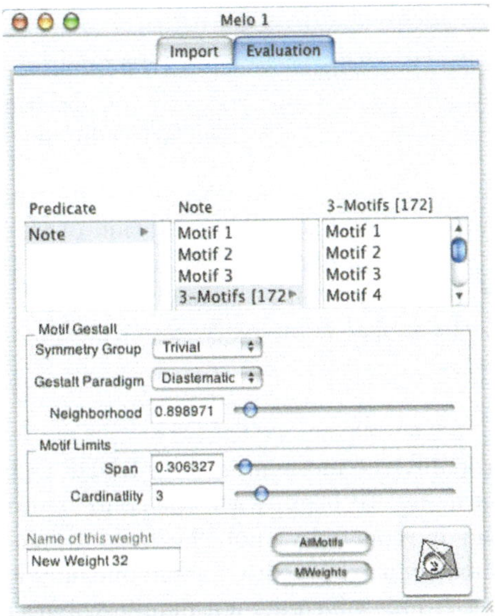

Fig. 41.5. The main window of the MeloRUBETTE®.

1. Presence. For a motif M, the presence value $pr_{\mu,\epsilon}(M)$ is the sum of all these numbers: For each $N \in D^\mu_\epsilon(M)$, we count the number m of times where M has a submotif M' of N that at a distance less than ϵ from M. We also look for the difference of cardinalities $d = card(N) - card(M)$. This gives a contribution $p(N) = m.2^{-d}$, and we add all these numbers.
2. Content. Similarly, for each motif $N \in \mu$ such that $M \in D^\mu_\epsilon(N)$, we take $p(N) = m.2^{-d}$ with $d = card(M) - card(N)$ and m the number of times, where N has a submotif N' of M that at a distance less than ϵ from N. We add all these numbers $p(N)$ and obtain the content $ct_{\mu,\epsilon}(M)$.
3. Weight. Given a motif $M \in \mu$, this is the product

$$nW_\epsilon(M) = pr_{\mu,\epsilon}(M).ct_{\mu,\epsilon}(M),$$

i.e., taking the function $\omega(x, y) = x.y$ from Section 22.9.

We have already given musicological comments on this construction in Chapter 22. It is however remarkable to see the overwhelming amount of calculations which arise in this routine. For example, we have calculated the number C of comparisons of motives (for distance measurements) Schumann's "Träumerei" ("Kinderszene" number 7) which comprises 463 notes. If we take Span $= 1/2$ bar and Cardinality $= 4$, we obtain $25'745$ motives and $C = 1'023'490'904 \sim 1.023.10^9$. This is beyond any explicit human calculation power. It demonstrates that the task of finding a dominant motif is a very hard one, and that this one, if it is recognized by a human listener, can at most be present in a very hidden layer of consciousness. This becomes even more dramatic for larger pieces, such as an entire sonata, say! Here, the combinatorial extent of motivic units exceeds any calculation power of humans and machines, as is easily verified. This means that a huge composition bears a motivic complexity that will escape to (human or machine-made) classification forever. Nonetheless, the usage of statistical methods, of simplified approaches to motivic topology, or of topological invariants that are more easily perceived could help find a rough orientation in the virtually infinite motivic variety of music.

This implementation also makes evident the tension between fuzzy concepts in musicology and implementation of a precise model. Although the concept of a motif is rather elementary, it entails a very sophisticated motivic analysis which could eventually converge to the intensions hidden[1] behind those fuzzy motive theories.

[1] We are not sure whether they are really hidden and not only faked...

41.3 HarmoRUBETTE®

Eine Theorie aber, die gerade dort versagt, wo auch das Phänomen, das sie erklären soll, ins Vage und Unbestimmte gerät, darf als adäquat gelten.
Carl Dahlhaus on Hugo Riemann's harmony [231]

Summary. The HarmoRUBETTE® makes clear that a vague theoretical approach does not reflect a vague phenomenon but an extremely complex one. The implementation of this RUBETTE® reveals several deep deficiencies of traditional "messy" analysis in harmony. We account for this on the level of preferences that have to be defined in order to get off ground with the analysis. The chapter concludes with a discussion of combinatorial problems due to the global complexity of harmony, and to the local character of tonal paradigms.

– Σ –

The HarmoRUBETTE® is probably the most interesting RUBETTE®, since it is situated on a turning point of several critical issues in harmony. To begin with, the context problem in harmony is a multilayered and ramified one which is (we said it repeatedly) not clarified by music theory. This is manifest in the preliminary question of what is a chord in a given score. Should one only look at groups of notes as a common onset, should one also consider onset groups which are not manifest, but can be deduced from plausible rules, or is the selection of the relevant set of chords within a score also a function of the harmonic statements which could result thereof? We have implemented two variants. The first one takes as the sequence $(a_i)_i$ all maximal zero-addressed local compositions $a_i = \{a_{i,j} | j = 1, \dots t_i\}$ of pitches of note events with identical onsets. The second one is less naive. Within the given score, we take all local compositions $a_i = \{a_{i,j} | j = 1, \dots t_i\}$ of pitches with this property: There is at least one onset which is the offset time of an event, and the chord a_i is the non-empty set of pitches of all note events which either start or still last at this offset time. This option is chosen by the button "use Duration" on the RUBETTE®'s main window, see Figure 41.6. This

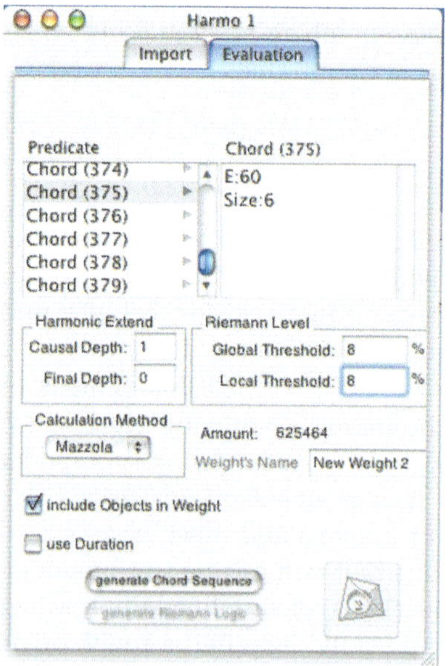

Fig. 41.6. The main window of the HarmoRUBETTE®.

second variant encodes all changes of chord configurations, not only the onset-commonalities. Using either of these methods, the generated chord sequence is the basis for the following analysis which at the end will yield a harmonic weight for each note event, and for which we refer to the harmonic tension theory presented in Section 27.2.2.

According to that approach, each chord a_i of the chord sequence $(a_i)_i$ must be given a Riemann matrix $(TF_{f,t}(a_i) = \phi_{f,t}(a_i)^\frown)_{f,t}$, from which we deduce the weights $\omega(f,t,a_i) = \ln(\phi_{f,t}(a_i))$ and also call this data the Riemann matrix of a_i. According to that discussion, we may also downsize weights below a threshold ϕ_{min} to $-\infty$. This is what the user sets when defining the "Global Threshold" in the main window. The local threshold is just the same for relative weights within a given Riemann matrix. The percentages in the main window mean that we downsize values below a defined percentage relative to the global or local (only within a fixed Riemann matrix) value range. Following the rules for the value $-\infty$, we may neglect any path through a chord which has this value. So this singular value means that a chord is "inharmonic" insofar as it cannot contribute to a positive harmonic evaluation. This is a mathematical rephrasing of the classical, but fuzzy concept of *inharmonic chords*. Here it just means that the harmonic weight of a chord is too small to be considered as a contributon to the global harmonic path, and that the minimal size of allowed weights is set without further theoretical justification. It is a regulatory limit for the sensitivity of the path maximization with respect to the involved weights. We should stress that this matrix $(\omega(f,t,a_i))_{f,t}$ defines Riemann weights without any contextual considerations. This is information that comes along from the isolated calculation on the chord a_i. There are different calculation methods for this matrix, one by the author, and using chains of thirds, as discussed in Section 25.3.3, and one by Noll, using self-addressed chords—the user may choose his preferred method by a pull-down button on the RUBETTE®'s main window. The Riemann matrix $(\omega(f,t,a_i))_{f,t}$ is visualized on the "ChordInspector" window for each chord (see Figure 41.7). In this window, the non-logarithmic values $\phi(a_i)$ are shown. The window also shows the chord's pitch classes as well as all its minimal third chains.

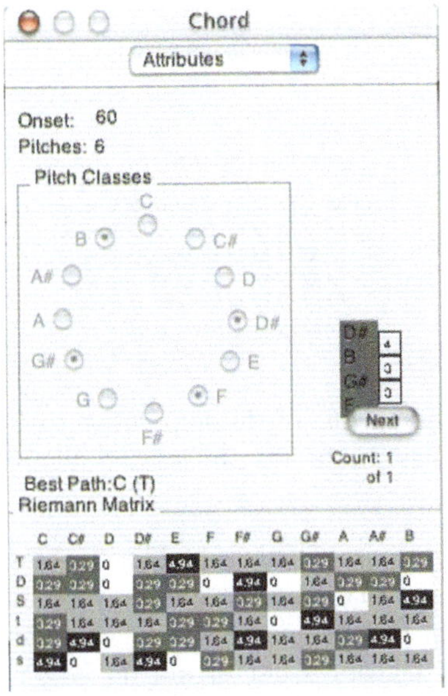

Fig. 41.7. The ChordInspector of the HarmoRUBETTE® shows each chord of the chosen chord sequence with its pitch classes, the third chains, and the Riemann matrix according to the chosen calculation method. The grey level of the values is proportional to their relative size.

Once the Riemann matrices $(\omega(f,t,a_i))_{f,t}$ are calculated, the optimal path in the Riemann quiver is calculated. This one follows the method discussed in 27.2.2. To this end, we need preferences for the matrices $T_{\mathrm{VAL}_{type}}, T_{\mathrm{VAL}_{mode}}, T_{\mathrm{TON}}$. The matrices $T_{\mathrm{VAL}_{mode}}, T_{\mathrm{TON}}$ are defined in the windows shown in Figure 41.8. To the left, we have T_{TON}, however as a distance value according to the amount of fourth between pairs of tonalities, to the right, we have $T_{\mathrm{VAL}_{mode}}$. The matrix $T_{\mathrm{VAL}_{type}}$ is shown in the left lower corner of the

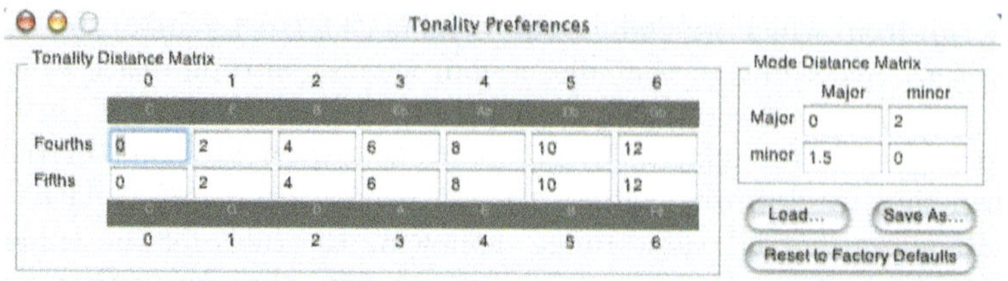

Fig. 41.8. This preference window (for the author's third chain method) shows the tonality distance matrix $T_{\mathrm{VAL}_{mode}}$ (left) and the mode matrix T_{TON} (right).

window in Figure 41.9. In the middle lower part of that window, we have check buttons for every Riemann locus (i.e., the position in the Riemann matrix), meaning that if the check is disabled, no path is possible through such a locus. The large upper matrix encodes the third weights needed for chord weight calculations according to the formulas (25.8) and (25.9).

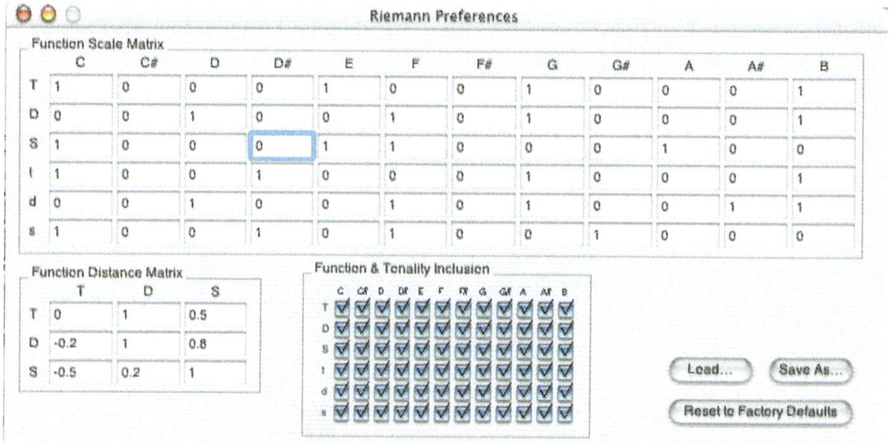

Fig. 41.9. The upper matrix encodes the weights of thirds (relative to a fixed reference tonic, the lower left shows the matrix $T_{\mathrm{VAL}_{mode}}$. The lower middle matrix encodes the a priori allowed Riemann matrix locus position.

With these settings, the best path is calculated. This is however a tedious task, to say the least. In fact, if we are given 200 chords (a very small example), we may choose from a number of $12^{200} \sim 6.8588169039290515.10^{215}$ paths. This number exceeds any calculation power of present computers. This exuberant number is due to different factors. First of all, no larger paths are taken into account, i.e., we have not implemented cadences as preferred paths, nor have we implemented modulatory constraints. More precisely, we do not give preference to maximal subpaths within a fixed tonality. We only take into account tonality changes a posteriori, i.e., via the weights of paths of length 1, when they show a tonality change. So we have to calculate the entire path and then hope that the negative points for tonality changes rule such paths out.

It is also not clear whether human harmonic logic really can take into account such global path comparisons. In other words, it is more likely that humans only consider local optimization of paths. This is what we have in fact implemented in the following sense. In each index i of a chord a_i, we consider only a local part of the entire chord sequence. Such a part is defined by two non-negative entire variables $CD =$ *Causal Depth*, $FD =$ *Final Depth*. This means that we look at the subsequence of chords from index $i - CD$

to $i + FD$ (inclusive) and therein select an optimal path $p_{i,CD,FD}$. Within this path, chord a_i is positioned at a determined Riemann locus $(f(i), t(i))$. The path which we finally select is the path p through all triples $(f(i), t(i), a_i)$. The causal part is a tribute to the influence of preceding chords down to index $i - CD$ on the harmonic position of chord a_i. The final part influences the harmonic position of a_i relating to the future chords up to index $i + FD$. The result is visualized in the *Riemann Graph* which is shown in Figure 41.10.

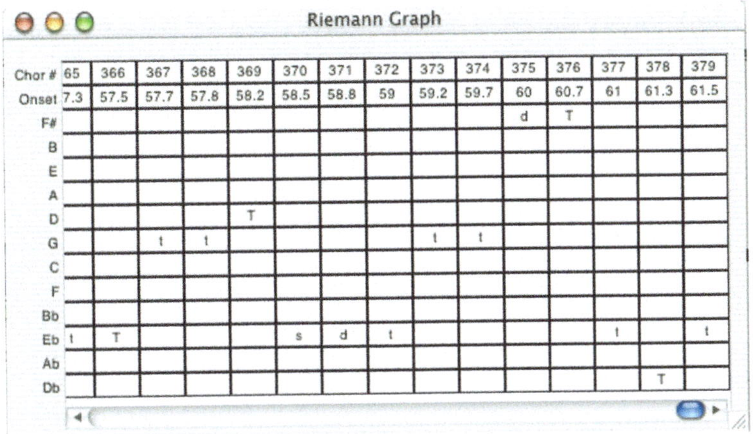
Riemann Graph

Chor #	65	366	367	368	369	370	371	372	373	374	375	376	377	378	379
Onset	7.3	57.5	57.7	57.8	58.2	58.5	58.8	59	59.2	59.7	60	60.7	61	61.3	61.5
F#											d	T			
B															
E															
A															
D					T										
G			t	t					t	t					
C															
F															
Bb															
Eb	t	T				s	d	t					t		t
Ab															
Db														T	

Fig. 41.10. The Riemann graph is the sequence of chords, together with their functional values as they result from the optimal path.

This is however not the end of the job. We do not have the weights of the single notes. To this end, we first need the globally calculated weight of each chord a_i. In a provisional form, the weight of chord a_i is defined by the weight $\omega(p_i)$ of the chosen path p from the first chord to a_i. This value is finally modified by a global slope σ preference such that the final and the first weight can be set to build a defined slope. Thereby, we meet the requirement of a control over the global tension which cannot be deduced from the locally (only 0 and 1 lengths of sub-paths) defined weights. Let us denote by ω_i the weight $\omega_\sigma(p_i)$ corrected by the slope preference.

Using the weight of a chord a_i, we finally may define the weight[2] $\omega(x)$ of a determined event x in a_i. To this end, the weight $\omega(f(i), t(i), a_i)$ is compared to the weight $\omega(f(i), t(i), a_i - \{x\})$ of the chord a_i which doesn't contain x. With a positive preference quantity $0 < d \leqslant 1$, we consider the factor

$$\lambda(x) = \frac{1}{d + (1-d)e^{\omega(f(i),t(i),a_i-\{x\})-\omega(f(i),t(i),a_i)}} \tag{41.1}$$

which measures the weight differences. It evaluates to 1 for the difference zero and yields $1/d$ for the weight $\omega(f(i), t(i), a_i - \{x\}) = -\infty$. This means that the influence of x in the building of the chord's weight is accounted for. If the weight decreases after omission of x, its influence is important and the factor increases $\lambda(x)$. So we finally get the weight

$$\omega(x) = \lambda(x).\omega_i.$$

The graphical representation of this weight is the same as for the MeloRUBETTE®, and we may omit this window of the HarmoRUBETTE®.

Whereas the Riemann graph is conformal to the usual function-theoretic analysis (although it need not provide the common data in general), the weights of chords and events are far beyond the usual harmonic analysis and therefore cannot be compared without caution to established knowledge in harmony. It is however a common approach to harmony in its performance aspects to weight chords or notes in a more or

[2] In this notation, we omit all the preferences.

less metaphoric way. Our present approach in the HarmoRUBETTE® is a concretization of these metaphors and also a point to be discussed with traditional performance theorists.

41.3.1 A Set of New Harmonic Analysis Rubettes on RUBATO® Composer

Most of the present HarmoRUBETTE®'s functionality has been reimplemented in a series of rubettes on the Java-based RUBATO® Composer environment by Ruhan Alpaydin, see Section 69.2.2.

41.4 PerformanceRUBETTE®

All words,
And no performance!
Philip Massinger (1583–1640)

Summary. The PerformanceRUBETTE® is a 'macro' RUBETTE®: It manages the stemma generation, the weight input and recombination, the operator instantiation, and the production of output of performance data on the level of music technology.

– Σ –

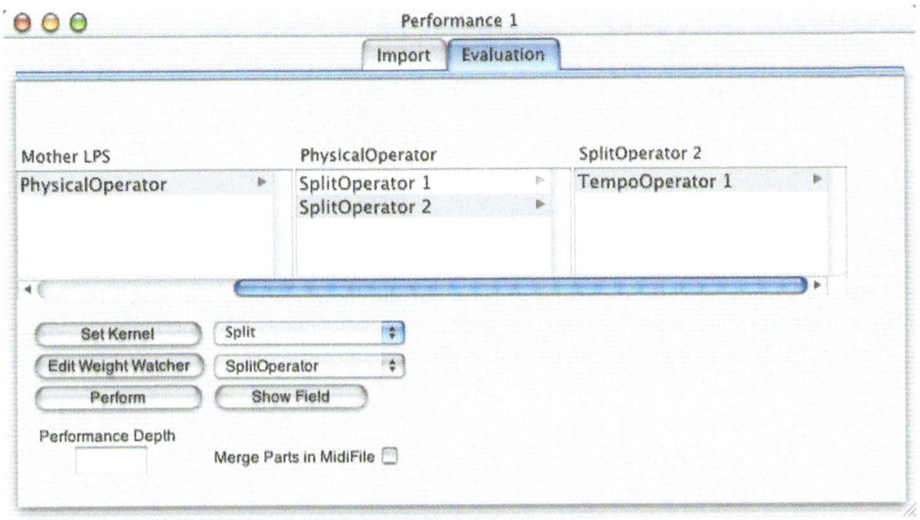

Fig. 41.11. The main window of the PerformanceRUBETTE®shows the stemmatic inheritance, descending from left to right. In the figure, the "Mother LPS" has a daughter named "PhysicalOperator", and this one (all mothers are indicated on top of the daughters' column) has two daughter "SplitOperator 1", "SplitOperator 2", generated by the SplitOperator, etc.

Originally, the PerformanceRUBETTE® was the very focus of RUBATO®. Its purpose was the implementation of a type of performance logic with arguments from an analytical output. Although the analytical Rubettes have earned a growing importance, one of the cornerstones of analysis is its success in the construction of a valid performance. In fact, *playing a good performance is a way of demonstrating one's understanding of music.* Therefore the performance theory implementation is important beyond its autonomous interest.

The PerformanceRUBETTE® implements the stemma theory of Chapter 38. The starting point is a selection of a score denotator. This will play the primary mother's role, i.e., we are constructing a primary mother LPS in the sense of definition 35.3. The score form is provided in the same format that we have known as input for the other Rubettes. For this RUBETTE®, the kernel is always given as a (zero-addressed) local composition in the space form $EHLDGC$. This local composition is the kernel in the top space of a cellular hierarchy pertaining to the primary mother's LPS. The primary mother's LPS is instantiated according to "hard coded" default parameters in the Objective C source code[3]. However, for each specific performance operator, the LPS data are adopted and define specific daughters. After setting the kernel (see Figure 41.11) of the primary mother, the main window shows the stemmatic ramification with individual names for each

[3] Objective C is a programming language for the NEXTSTEP-, OPENSTEP-, and Mac OS X-based RUBATO® projects.

LPS and arranged in a browser, stemmatic inheritance running from the left to the right. In the figure, the "Mother LPS" has a daughter named "PhysicalOperator", and this one (all mothers are indicated on top of the daughters' column) has two daughters, "SplitOperator 1", "SplitOperator 2", generated by the SplitOperator, etc.

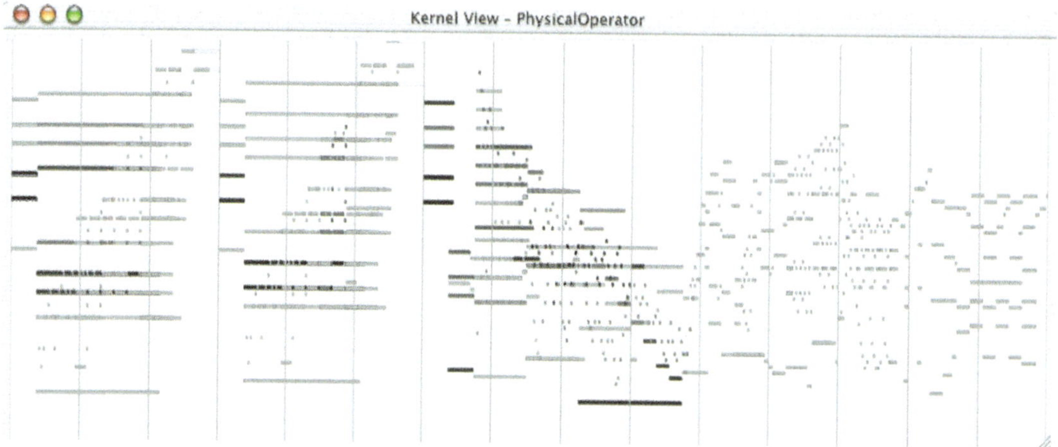

Fig. 41.12. The Kernel View window shows the top kernel of the hierarchy of a selected LPS (here the LPS named "PhysicalOperator" in the above stemma browser) in common pianola (piano roll) rectangles, loudness being codified by grey levels. The vertical bars are set to four bar intervals in the given score.

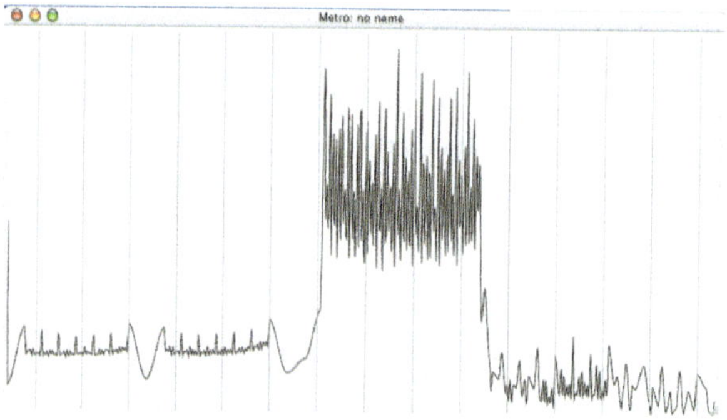

Fig. 41.13. In the PerformanceRUBETTE®, weights are used in their splined interpolation shape. Here, we see a metrical weight issued from the MetroRUBETTE®'s analysis.

For each ramification, one or two daughters are generated according to a chosen operator. For example, in Figure 41.11, the SplitOperator generates two daughters, whereas the TempoOperator always produces one single daughter. The operators need only be loaded at run-time as they are needed. So this RUBETTE® is non-terminal in the sense that it allows further ramifications via an arbitrary number of dynamically loadable performance operators. For every highlighted LPS in the stemma browser of the main window, we can visualize the top kernel of its hierarchy on the Kernel View window, as shown in Figure 41.12, by means of the usual pianola graphics. Here, the grey level indicates the loudness. In order to apply an operator to a given LPS, one next needs a list of weights, this is conformal with the operator theory exposed in Chapter 44.7.

The management of available weights as well as their concrete application are the business of the Weight Watcher system. Figure 41.13 shows a metrical weight in its splined interpolation shape. The weights to be used for a given operator can be loaded into the Weight Watcher, see top of Figure 41.14. The loaded weights are then added or multiplied (according to the Boolean flag button "Combine as Product" to the right, below the weight list), and the resulting weight combination is applied to the given operator. For each weight, one can set the upper and lower limit of range (High Norm, Low Norm), the non-linear deformation quantity (Deformation), the inversion/non-inversion flag (Inverted Weight button to the lower left corner), the influence in a combination of several weights (Influence), the slope of decrease to weight value 1 as the arguments tend to infinity (Tolerance). The moral of this Weight Watcher system is a gastronomic one: Weights may be mixed and dosed at will in order to experience their influence on a given operator. This is not merely a lack of theory, it is above all an experimental environment for effective performance research. In fact, since virtually nothing is known about the influence of weights on performance, we have provided the user with a great number of possibilities in order to realize an optimal testbed for future theory.

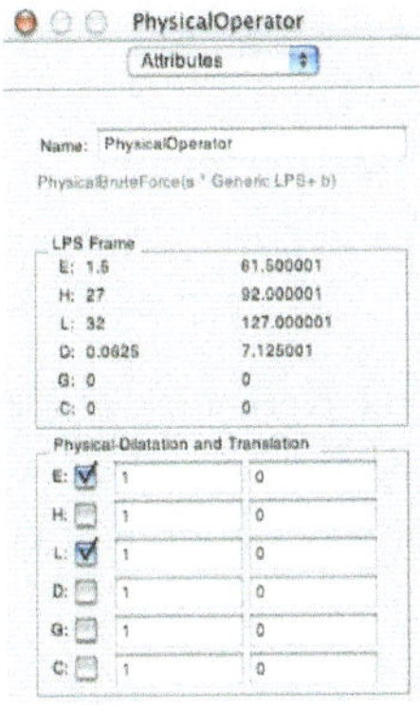

Fig. 41.15. The PhysicalOperator Inspector allows us to select a number of physical output parameters where the weight changes the values.

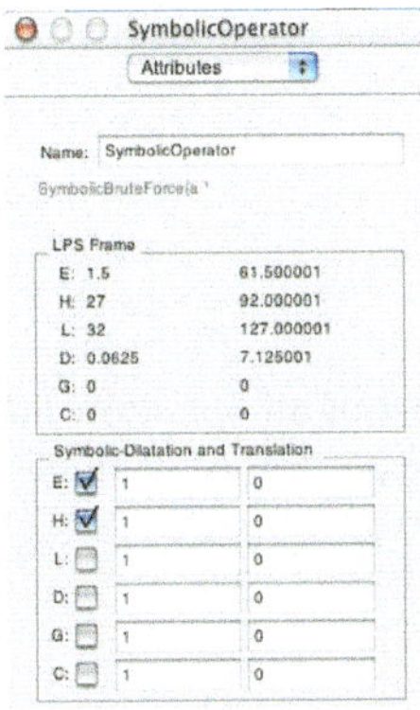

Fig. 41.16. The SymbolicOperator Inspector allows us to select a number of symbolic input parameters where the weight changes the values.

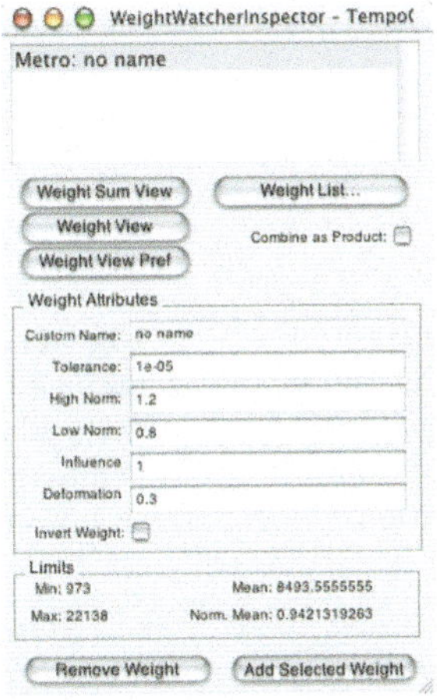

Fig. 41.14. The Weight Watcher window shows the loaded weights (top), the upper and lower limit of their range, the non-linear deformation, and the Boolean flag of inverting/non-inverting the weight (button in the lower left corner).

Now, given a weight watcher combination of weights, an operator is fed by this combined weight and acts on the mother LPS to yield a new daughter LPS (or two in the case of the SplitOperator, where however no weight is needed). The detailed operation of a specific operator has already been described in Chapter 44.7, we need not repeat these details here.

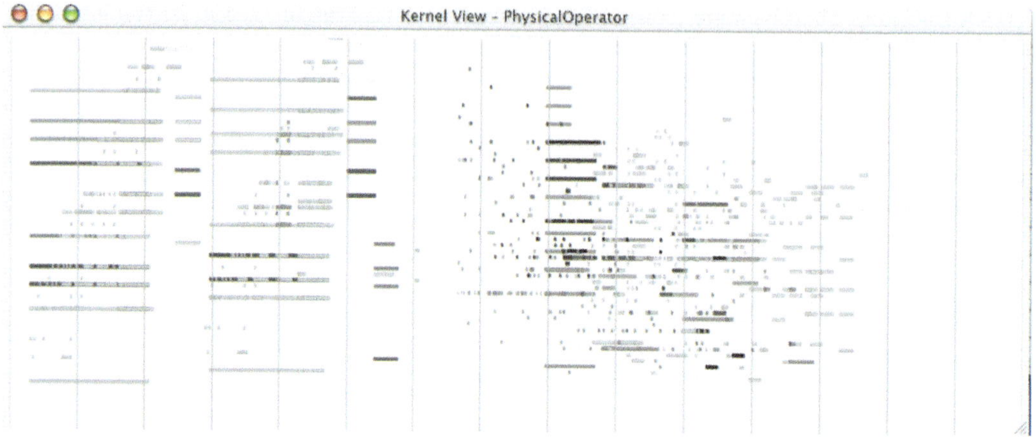

Fig. 41.17. The effect of an operator, here a PysicalOperator, is shown in the Kernel View of the performed kernel. This figure shows the performed symbolic kernel as shown above in Figure 41.12.

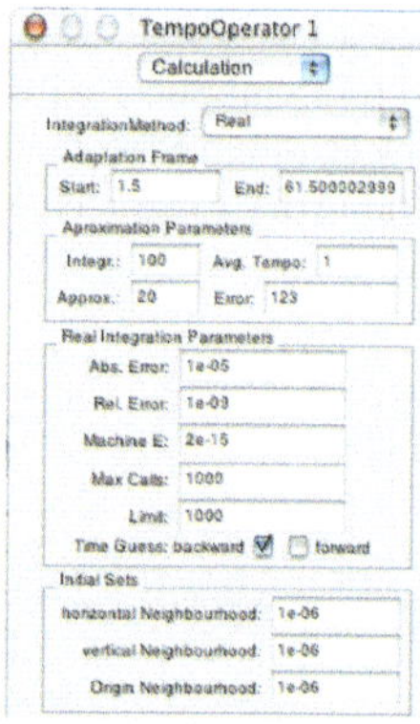

Fig. 41.18. The TempoOperator inspector allows us to select different integration methods. The "Real" method uses Runge-Kutta-Fehlberg routines, whereas the "Approximate" method uses simple numerical integration.

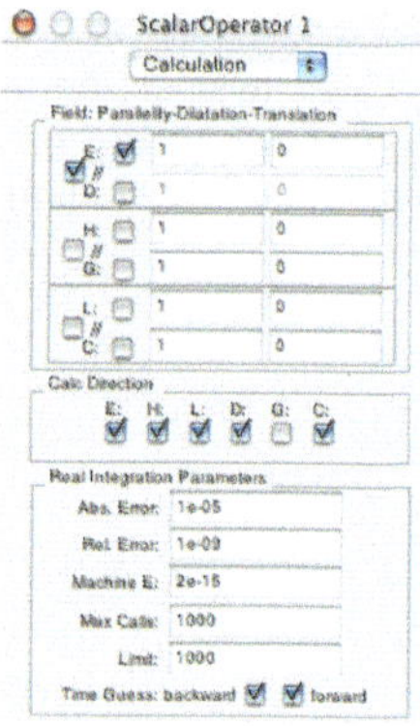

Fig. 41.19. The ScalarOperator inspector allows us to select different options as defined in the scalar operator theory, but Runge-Kutta-Fehlberg ODE integration is mandatory in this situation.

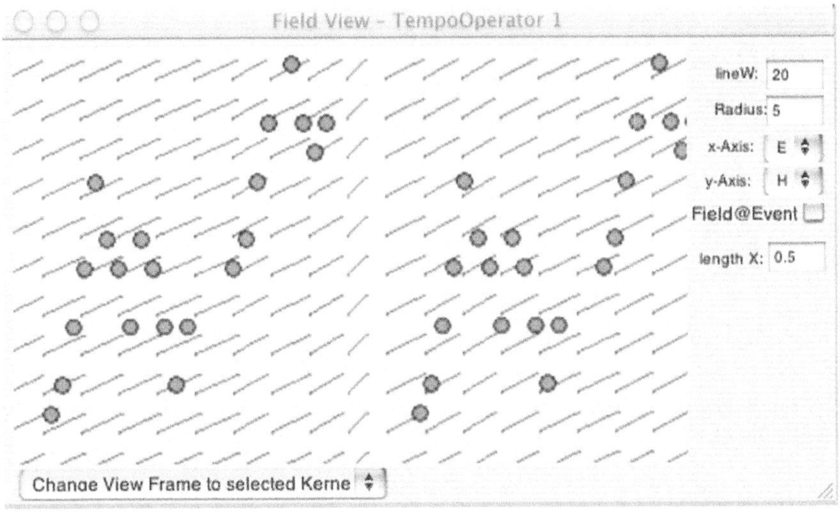

Fig. 41.20. The performance field of a selected LPS can be visualized. The user may select two parameters whereon the six-dimensional field is projected.

Figure 41.16 shows the inspector of the SymbolicOperator. The weight acts on selected parameters which are defined by Boolean buttons. The same procedure is performed for the PhysicalOperator, whose inspector is shown in Figure 41.15. The action of an operator on the symbolic score is shown in the performed kernel in the same pianola representation as for the symbolic kernel. The action of a physical operator is shown in Figure 41.17. A funny application of this operator to Schumann's "Träumerei" can be heard on the book's CD-ROM under the title `Alptraeumerei`, see page xlvii. This piece is the dead-pan version of the score with the melodic weight being applied to pitch via the physical operator, and everything being played with Schumann's original tempo indication. The inspector windows of the tempo-sensitive operators, the TempoOperator and the ScalarOperator, are shown in Figures 41.18 and 41.19, respectively. The TempoOperator implements basically two methods: "Approximate", and "Real". The former is a direct integration method, whereas the latter uses Runge-Kutta-Fehlberg numerical ODE routines, including different parameters for numerical precision. The ScalarOperator uses exclusively Runge-Kutta-Fehlberg numerical ODE routines since it is an operator that acts on two or more parameters, where the naive approximation method cannot work. For the visualization of performance fields, the window shown in Figure 41.20 is available.

Finally, the parameters for the SplitOperator are determined on the window shown in Figure 41.21. Here, one may define those lower and upper parameter limits of the total six-dimensional frame, where the subframe of the split daughter is cast.

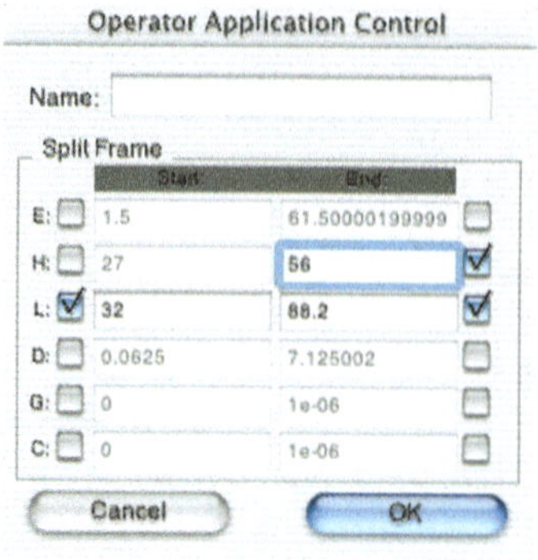

Fig. 41.21. On this window, the user may define those lower and upper parameter limits of the total six-dimensional frame, where the subframe of the split daughter is cast for the SplitOperator.

41.5 PrimavistaRUBETTE®

It is hard if I cannot start some game on these lone heaths.
William Hazlitt (1778–1830)

Summary. Several musical predicates from score notation are paratextual. The PrimavistaRUBETTE® takes care of the paratextual signification for the most important predicates regarding dynamics, agogics, and articulation.

– Σ –

The PrimavistaRUBETTE® serves a different task insofar as it is neither analytic nor performance oriented. It deals with paratextual information as it is provided by verbal indications for dynamics, tempo, and articulation. It basically does this: it transforms verbal information into weights which may then be used to shape the symbolic data and the tempo before performance in the proper sense is shaped.

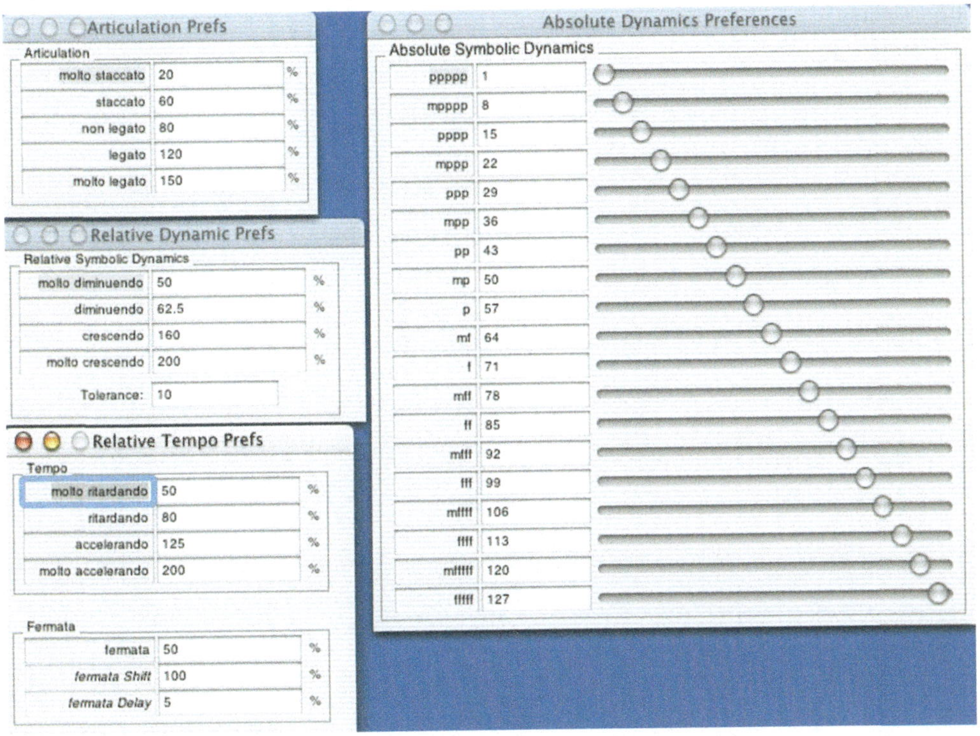

Fig. 41.22. A number for preference windows for dynamics, articulation, and tempo, allow us to define numerical values of paratextual predicates.

The input of this RUBETTE® is a local composition whose elements are events with verbal specification such as absolute dynamics (Figure 41.22 right preference window), relative dynamics (Figure 41.22 left middle preference window), articulation (Figure 41.22 left upper preference window), and relative tempo (Figure 41.22 left lower preference window). The functionality of this RUBETTE® is to transform these data into weights, this is performed on the window for primavista operations as shown in Figure 41.23, and according to the numerical data that are defined the above preference windows. These methods have been discussed in detail in Section 39.2.

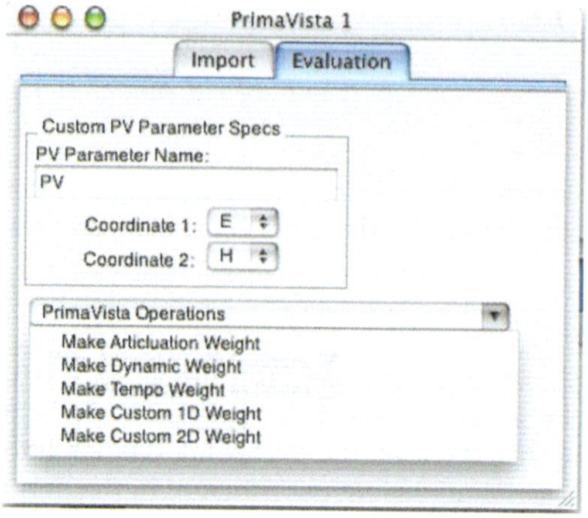

Fig. 41.23. The main window of the PrimavistaRUBETTE® manages the transformation of verbal (paratextual) predicates into weights.

42

Performance Experiments

Learning by Doing.

Summary. This chapter traces the analyses and syntheses processes which led to the historically first full-fledged RUBATO®-driven performance in July 1996 on the MIDI-Boesendorfer at the Staatliche Hochschule für Musik in Karlsruhe, as well as to the qualitatively high performance of contrapunctus III in Bach's *Kunst der Fuge*. We report the technical prerequisites, the analytical background generated by RUBATO®, and the step-by-step realization of the stemma and the overall parametrization.

– Σ –

42.1 A Preliminary Experiment: Robert Schumann's "Kuriose Geschichte"

Nicht eine Geschichte wird hier erzählt,
sondern der Eindruck, den eine solche
bei einem Zuhörer weckt,
wird charakterisiert.
Thomas Koenig [543]

The first realistic experiment with RUBATO® took place in July 1996 on the MIDI-Boesendorfer grand piano at the Staatliche Hochschule für Musik in Karlsruhe. The experiment lasted three days and was led by the author, together with his assistant Oliver Zahorka and the musicologist Joachim Stange-Elbe. The experiment was executed on a NeXTStation. The results were digitally recorded on DAT and everything was protocolled. The technical support for the MIDI-Boesendorfer was offered by Sabine Schäfer. In what follows, we want to give a very sketchy account of that experiment, however recalling the essentials of initial experiments in performance theory and the consequences thereof. The more elaborate discussion of such experiments is left to the following Section 42.2.

The preparatory work for this experiment was above all the analytical part, i.e., the metrical, melodic, and harmonic analyses on the respective Rubettes. We also experimented on the stemmatic and operator strategy, including the WeightWatcher mixtures as discussed in Section 41.4. The selection of the MIDI-Boesendorfer was also a consequence of preliminary experiments on the MIDI-Yamaha grand insofar as this instrument turned out to respond in a much too coarse way to the input. In contrast, the Boesendorfer has a refined calibrating interface allowing a realistic MIDI input (with roughly a thousand dynamical values instead of 128 from MIDI), in particular for soft dynamics, where the Yamaha grand is completely inappropriate.

The first day was used to adapt the different technical conditions, such as the calibration of the Boesendorfer, the weight ranges, and the dynamical limits. The second day was devoted to the evaluation of our preparatory material and strategy. The third day was devoted to the production of all stemmatic levels, as well as their recording up to the final output level.

G. Mazzola, *The Topos of Music II: Performance*, Computational Music Science,
https://doi.org/10.1007/978-3-319-64444-8_11

The most significant experiences were twofold: On the one hand, the judgment of the three experts concerning which weight mixture to use at which stemmatic knot, and the way to use it, with all the variables from range to inversion/non-inversion and deformation, was very precious! We learned that the effects of such strategies were almost constantly judged in three different ways—except for trivial failure or success situations. This means that performative coherence is a very personal affair, even when using very explicit techniques, analyses, and shaping tools.

On the other hand, we had to face a very disagreeable side effect of such an extensive performance shaping: Let me call it *the zoom of supersensitivity*. The effect is this: After having listened to performances of a selected part of the piece, the ear begins to recognize a steadily refined differentiation in the different parameters, such as agogics, dynamics, and articulation. For example, if in the beginning phase, one only recognized a huge change of a weight influence to agogics, with progressing trials, one would believe that the slightest further change in weight influence could change the performance of tempo in an unsupportable way. Maybe one was eventually incapable of really identifying the details of a performance, i.e., it is either simply too complex to be grasped by humans because the attention to different parameters is always selective, or it is not possible to memorize the samples and to compare them, or—and this is the worst case—our perception really changes so much form case to case that we cannot rely on the individual samples. The latter would mean that performance is greatly co-determined by the performance of the perceptive system.

The result was nonetheless acceptable and strongly differs from the so-called dead-pan version, see [702, CD attachment], or this book's CD-ROM (cf. page xlvii), where the performance and the dead-pan version can be inspected.

42.2 Full Experiment: J.S. Bach's "Kunst der Fuge"

Die Kunst der Fuge erschien wohl etliche Monate nach Bachs Tode und kostete vier Taler. Sie fand keinen Absatz. (...) Enttäuscht verkaufte der Sohn (Emanuel) die Platten, auf denen das letzte Werk seines Vaters geätzt war, um den Metallwert. Das war das Schicksal der Kunst der Fuge.

Albert Schweizer [956]

This experiment is fully accounted in [1003]. Here, we give a concise presentation, the version of RUBATO® used in this experiment is the one compiled for OPENSTEP/Intel. The "contrapunctus III" in Bach's *Kunst der Fuge* has these characteristics: It is a four-voice composition, comprises 72 measures, has time signature 4/4, and tonality *d*-minor. The main theme of *Kunst der Fuge* is only used in its inversion and appears the first time in a rhythmically dotted and syncopated variant; the fugue starts with the theme in its comes shape and contains three complete developments (measures 1-19, 23-47, and 51-67).

42.3 Analysis

Summary. We give an account of the metrical and melodic analyses, whereas the harmonic analysis has not been done.

– Σ –

42.3.1 Metric Analysis

For the metric analysis of the "contrapunctus III", the calculations were made for each single voice, including the sum of the voice weights, and for the union of all voices. Please, refer to the discussion of the MetroRUBETTE® in Section 41.1 for the following discussion.

The settings of the weight parameters are these: *Metrical Profile* is 2; *Quantization* is 1/16; *Distributor Value* is 1. Since the metrical profile of all voices should be viewed under the same valuation, the *Distributor Value* was set to a common neutral value; the value 2 for the *Metrical Profile* resulted from several trials of analyses and yields a well-ordered distribution of the weight profile.

The value for *Minimal Length of Local Meters* was successively decremented starting from the length of the largest local meter, descending until value 2 where the smallest cells are caught in their signification for the metrical overall image.

When considering single local meters, it is above all the onset time and the step size of the single time grids[1], which matters. Based upon the summation formula, this can however not be fully understood since from a certain superposition of periodical onset sequences, the observation of a single local meter becomes very difficult if not impossible (see also the consideration of the union of all onset times).

42.3.1.1 Single Voices

Like all other contrapuncti, "contrapunctus III" shows no regularities in the compositional structure, in the repetition of whole parts or single measures. At least for the four measures of the theme, the onsets of themes in the developments of fugues do however create structural incisions, which only become relevant for the motivic analysis—the metric analysis remains unaffected. As to the analysis of single voices, it is interesting to observe that the longest found local meters nearly exclusively relate to relatively short parts, including only a few measures. These parts usually consist of uninterruptedly pulsating sequences of quavers or semiquavers, which in their regular sequence of notes differ significantly from the otherwise pronounced principle of tying stressed measure onsets to unstressed precursors. This principle, having its origin in the shaping of the main theme with its sequences of quavers that lead to the interluding sections, can be recognized as a rhythmical pattern in the interludes and the countersubjects of the theme. By the tying of this sequence of quavers with the preceding half note, the theme breaks the hitherto confirmed measure accent scheme, a peculiarity which will be further confirmed in the sequel of this contrapunctus. It even has effects on the theme onsets which—unusual in a 4/4-measure—are partly shifted by a quarter note (see also the theme variant which is extended by dots and quaver transitions in the measures 23, 29, and 35).

When considering the single voices, one realizes that most of the longest grids (i.e., those with highest weights, according to our choice of system parameters) catch exclusively onsets on unstressed measure times in the middle voices. If the initial value lies on a stressed measure time, a dotted duration value as a step length for local meters also weights more onsets on unstressed times. Only the metrical weight of the bass voice acts in a contrary way, in fact an increasing confirmation of the measure meter is observed, see Figure 42.1.

42.3.1.2 Weight Sums of All Voices

When considering the weights of single voices, and in particular the sum of their weights, observe that the longest weights are the most important contributions. This means that the most important weights stem from the bass, followed by the alto, the tenor, and then the soprano voice. The latter has a weakened contribution since in the middle of the contrapunctus, there is a pause of thirteen measures, which breaks down the coherence of the soprano to the metrically weakest voice, see Figure 42.2 for the lengths of local meters for the four voices. Further, we observe that the bass voice prevails against the other voices in the confirmation of the measure meter since it creates high metrical weights by the longest metrical grids. Nearly all peaks of the metrical weight sum—except some cases in the middle part—coincide with stressed measure times.

Generally speaking, the ambitus of the metrical weight profile increases towards the middle of the contrapunctus—an observation which is valid for all pieces of the "Kunst der Fuge" which we have analyzed, see also Figure 42.3. The reasons for weak initial metrical profile lies in the formal construction of a fugue, where the voices do not appear simultaneously, but one after another, such that the full voicing only appears after a number of measures. Another reason can be seen in the structure of the theme of the fugue, which

[1] In this discussion, "grid" is synonymous with "local meter".

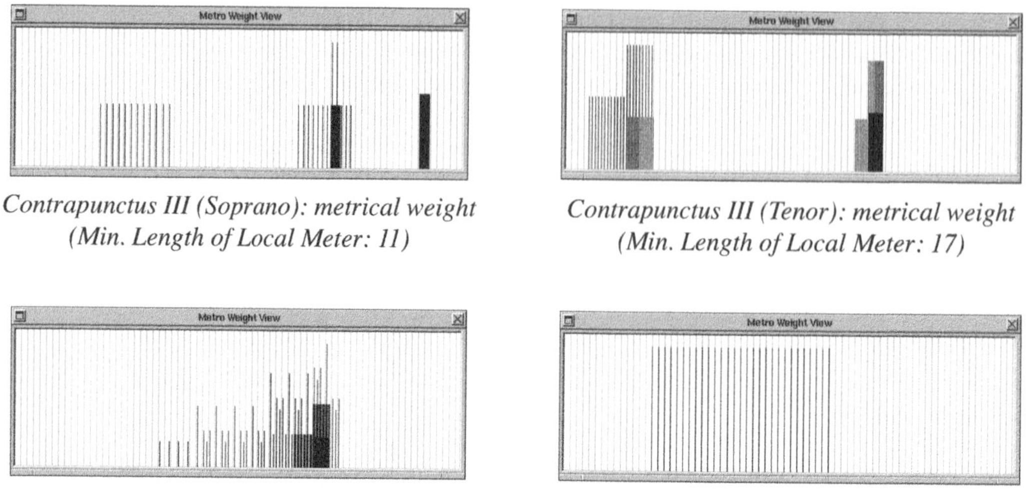

Contrapunctus III (Soprano): metrical weight (Min. Length of Local Meter: 11)

Contrapunctus III (Tenor): metrical weight (Min. Length of Local Meter: 17)

Contrapunctus III (Alto): metrical weight (Min. Length of Local Meter: 19)

Contrapunctus III (Bass): metrical weight (Min. Length of Local Meter: 28)

Fig. 42.1. *Contrapunctus III*: metrical weights for the four voices.

S														12	11	...	2
A			23	22	21		19	18	17		15	14	13	12	11	...	2
T						20		18	17				13	12	11	...	2
B	28	26	23	22		20			17	16	15	14	13	12	11	...	2

Fig. 42.2. The lengths of maximal local meters for the four voices.

does not develop its full motional impulse before the last five notes. These are used to shape—until the next thematic onset—counter-subjects and interludes. The equally very frequent decrease of the metrical profile

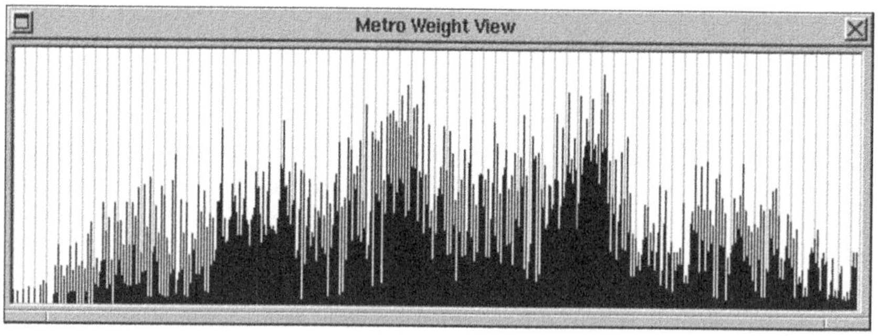

Fig. 42.3. The ambitus of the metrical weight profile increases towards the middle of the contrapunctus; minimal local meter length is 2.

towards the end of a piece is explained by different formal facts. In the case of the "contrapunctus III", this is due to long pauses of the single voices—in particular bass, soprano, and tenor—as well as the pedal point in the bass voice (measures 68-71).

42.3.1.3 Union of All Voices

Relating to the union of all onsets, the "contrapunctus III" shows at first sight dotted step lengths in metrical grids. However, by the two longest grids (*Minimal length of Local Meters* being 502 and 271), starting at the final notes of the exposition of the theme in the tenor and at the end of the exposition of the theme in the alto, an uninterrupted pulse of quavers and quarter notes is established, and the only remaining grid (*Minimal length of Local Meters* is 128) with step length a half note also drops at the non-stressed measure times of 4th and 8th quaver. All other grids have step lengths of different dotted durations, however dominated by the doubly dotted half note, the said dotted quarter note, and the half note with tied quaver, as well as once the dotted half note. What is remarkable in this contrapunctus is the three 'double grids' of *Minimal*

Length of metrical grid	Start of metrical grid (bar, note)		Step length	End of metrical grid (bar, note)	
502	9	4th Quarter	1/8	72	3rd Quarter
271	4	4th Quarter	1/4	72	3rd Quarter
172	8	1st Quarter	3/8	72	3rd Quarter
171	8	2nd Quarter	3/8	72	4th Quarter
128	8	4th Quarter	1/2	72	4th Quarter
108	4	8th Quarter	5/8	72	4th Quarter
104	7	3rd Quarter	5/8	72	3rd Quarter
102	8	2nd Quarter	3/8	72	1st Quarter
	8	3rd Quarter	3/8	72	2nd Quarter
91	4	1st Quarter	3/4	72	2nd Quarter
78	3	4th Quarter	7/8	72	1st Quarter
75	6	3rd Quarter	7/8	72	2nd Quarter
74	7	3rd Quarter	7/8	72	2nd Quarter
	7	4th Quarter	7/8	72	3rd Quarter
73	8	1st Quarter	7/8	71	8th Quarter
	8	3rd Quarter	7/8	72	4th Quarter
71	1	1st Quarter	4/4	72	1st Quarter

Fig. 42.4. Maximal local meters for the union of all four voices.

length of Local Meters equal to 102, 74, and 73, which within the same measure are shifted to each other by a quarter note, and two quarter notes in the last case.

Therefore, in the *contrapunctus III*, the tendency of 'metrical instability', which was already encountered in the single voices, persists, since by summing up the weight contributions by the grids with dotted step width, a metrical profile is established which breaks the measure meter as well as the pulse of uninterrupted quavers and quarters.

42.3.2 Motif Analysis

For the calculation of motivic weights each single voice of the "contrapunctus III" was analyzed separately. We refrained from a motivic analysis of the union of all voices since by the contrapuntal structure of the single and autonomous voices within the polyphonic setting, a motivic setup across the voices seems rather unlikely and therefore was omitted.

The settings for the motivic analysis were chosen as follows: *Symmetry Group*: counterpoint; *Gestalt Paradigm*: elastic; *Neighborhood*: 0.2. By the choice of the counterpoint symmetry group, the theme forms *recta* and *inversa*, as well as their (possibly appearing) retrogrades, were considered as being of equal weight. The neighborhood value has been chosen as based upon analytical experiments during the development period of RUBATO®. The elastic gestalt paradigm was preferred against the rigid and diastematic ones in order to obtain a more elastic point of view.

As to the values for *Motif Limits*, compromises with the calculation power had to be made. By the choice of *Span* equal to 0.625 and *Cardinality* from 2 to 7, motives within a span of a half note plus a quaver were captured; this corresponds exactly to the duration of the theme where the transition of the virtual theme to the interludes must be recognized.

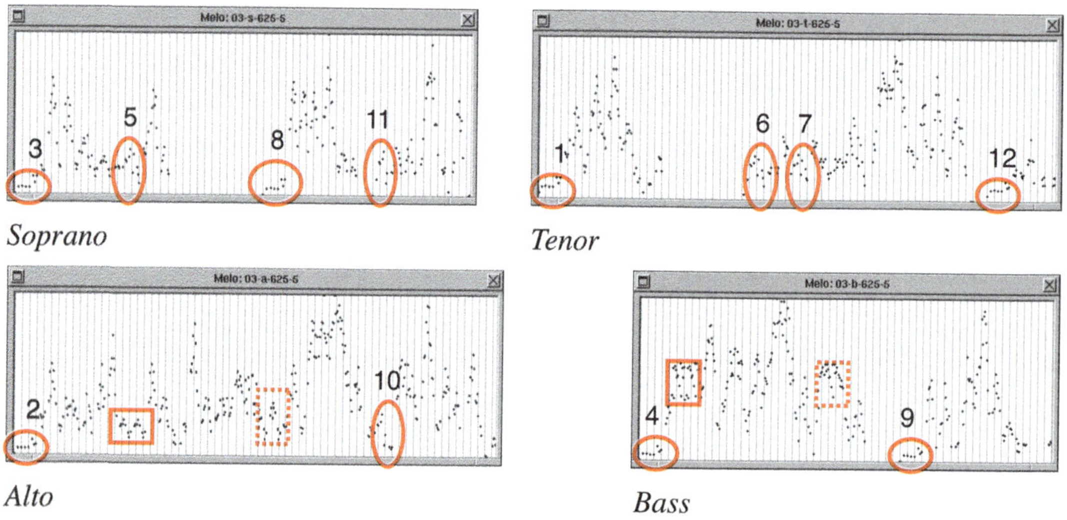

Fig. 42.5. Motivic weights for "contrapunctus III".

With the results of the metrical analysis, some regularities in the microstructures can be read at first sight; herein we find in particular the onsets of the theme within a particular development. The representations of the single motivic weights in each voice with a mean value of 5 for the *Cardinality* may be elucidated in detail, see Figure 42.5. Within these graphics, the theme onsets are numbered according to their temporal order; equal weights can be recognized on the onsets 1,2,3,4,8,9,12, and 5,6,7, as well as on 10 and 11.

While further considering these weights, the overly long pauses in the soprano, tenor, and bass voices attract attention. Further, in the length proportion of the single weight representations, the succession of onsets of the single voices (tenor-alto-soprano-bass) is reflected. Moreover, a significantly lower motivic profile at the beginning and after the longer pauses of the respective weights can be observed—due to preceding pauses, this is the case of exposed thematic onsets.

For the weight values, a neat exposition of the inverted gestalt of the original theme is observed, bearing nearly identical weights at the beginning of every motivic weight, here even the differences of comes and dux forms are visible, since the weights of the tenor (first appearance) and soprano (third appearance) differ slightly by the different initial interval of the theme (descending fourth in the comes, and descending fifth in the dux form) from the weights of the alto (second appearance) and bass (fourth appearance), see Figure 42.6.

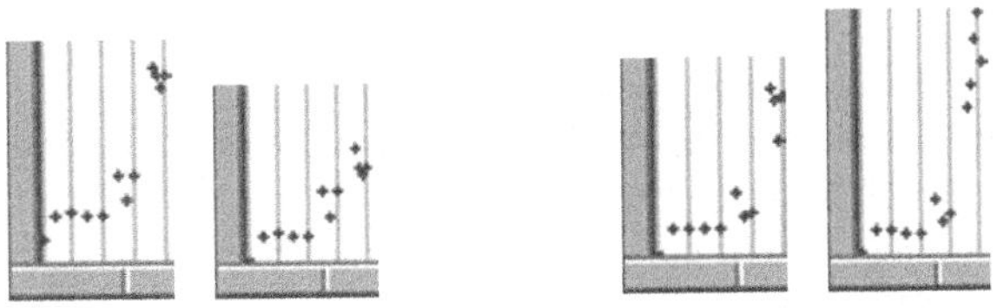

Fig. 42.6. Motivic weights for "contrapunctus III", measures 1-4.

Other clearly visible onsets of the theme in inverted shape are recognized after the long pauses in the soprano (eighth appearance), bass (ninth appearance), and tenor (twelfth appearance). Characteristically, the inverted shape always appears after pauses.

At first sight these observations may seem to be tautological. However, if these weights are viewed with respect to their sense and purpose, their force to shape performance, then the transition from a quantitative to a qualitative information content becomes evident: Thus the different onsets of themes can be shaped by these weights in one and the same way; if these weights are used—in inverted form—for the dynamic shaping, then the thematic onsets can be stressed with plasticity.

When trying to evaluate motivic data more in detail, a problem appears that is inherently founded in RUBATO®'s intended purpose: since the platform was originally conceived for performance research, it is sufficient to present the analytical material for its performance shaping and to ascertain its reasonable construction. But these analyses also yield information about the musical structure, which in the case of metrical analysis—by a certain extra effort apart from RUBATO®—can be traced without difficulties. However, for the MeloRUBETTE®, any practical support beyond the brute calculation of weights—which is also sufficient for performance—is absent; the "introspection" of the details of weight calculation is hidden to the user. Thus for the calculation a motif's weight it cannot be known which other motives are responsible for its *presence* and *content*. Although all motives are arranged in the motif browser in their temporal order, but another structured introspection of motives, for example with regard to their weights, is not possible for this version of RUBATO®. This makes the introspection and evaluation of the 'heaviest' motives a difficult task. All analytical information is present, but it is hidden; a particular difficulty may be seen in the exuberant effort in complexity which a detailed motivic analysis, which extracts knowledge from the score in an immanent way, is loaded.

42.3.3 Omission of Harmonic Analysis

A harmonic analysis was omitted in this situation since—to our mind[2]—the approach of Hugo Riemann which is implemented in the HarmoRUBETTE® is not really suited for Bach's harmonies. By use of the Riemann theory which was developed from the Viennnes classics, the specific harmonic structures of a contrapuntal maze, where harmony does not result from progression of fundamental chords but from the linearly composed voices, can only be captured in an incomplete way.

[2] Decision by Joachim Stange-Elbe.

42.4 Stemma Constructions

Summary. This section discusses the performance construction via the stemmatic paradigm, and using the analytical results investigated in Section 42.3.

– Σ –

Before the stemmatic construction and thereby in particular the problem of performing "contrapunctus III" are discussed, some general remarks regarding the various performance strategies are necessary. In the course of the single performance parcours, two different approaches resulted which would turn the given analytical weights into expressivity: the *target-driven* and the *experimental* strategy.

The *target-driven strategy* has its roots in the knowledge about existing performances, it is stamped by a preliminary experience of how the piece should sound and has been performed. With this procedure, the weights are used in a way which targets a predefined performance. One—just to name a pithy example—was oriented towards Glenn Gould's Bach interpretation; the corresponding weights were selected according to these targets to obtain particular effects. In this procedure, however, the intrinsic structural meaning of analytical weights was ignored! Stamped by the knowledge and the expectation of the existing performances, this strategy did not allow one to judge and categorize those performance constructions which did not suffice for the music-esthetic exigencies.

The other approach, the *experimental strategy*, moves the analytical weight to the center in order to investigate how this weight could 'sound', and which analytical insight it could convey in the listening. With this procedure, which views the main performing agent entirely within the weight, one has to free oneself completely from horizons of expectation for any particular performance target. The working process on such performances, the acquaintance of experience with the most different weights, and the playing with their effects taught us in the course of many experiments that this strategy would give rise to much more interesting performance aspects. Here we also have the freedom to admit extremal positions which disclose more about the inherent musical structure and as 'daring ingredients' may evoke lively musical expression.

Moreover, the experimental approach to single performance aspects, which starts from curiosity about the sonic realization of analytical weights, conveys a deeper insight into to score's musical structure. This path has its take-off in a "sonic analysis", or else in "the sonic analytical structure" and aims at a "musically reasonable performance". It is centered around the researcher's curiosity for a sounding and interpretational realization of analytical weights and for "the never heard", and it is paralleled by a liberation from expectational presets. Moreover, this strategy tries to apply as few weights as possible in order to couple the clearest possible analytical statements with the resulting performance.

42.4.1 Performance Setup

Summary. We discuss the detailed performance construction.

– Σ –

The performance of "contrapunctus III" took place in three parcours, out of which we only report the last two in more detail. The first one was entirely devoted to a target-driven strategy whereas the subsequent ones switched to an experimental strategy which yielded much more successful and conclusive results. Nonetheless, all these approaches contributed results that influenced the final result in a significant way.

Generally speaking, the procedure in all these parcours first focused on isolated single aspects of performance (articulation, dynamics, agogics) and then were put together for the final parcours. For the complete description of all these steps, see [1003].

42.4.1.1 Results From First Performance Parcours

As a result of the first parcours which was executed under the paradigm of a target-driven strategy, the usage of motivic weights for a determined shaping of dynamics has been recognized. Using the inverted

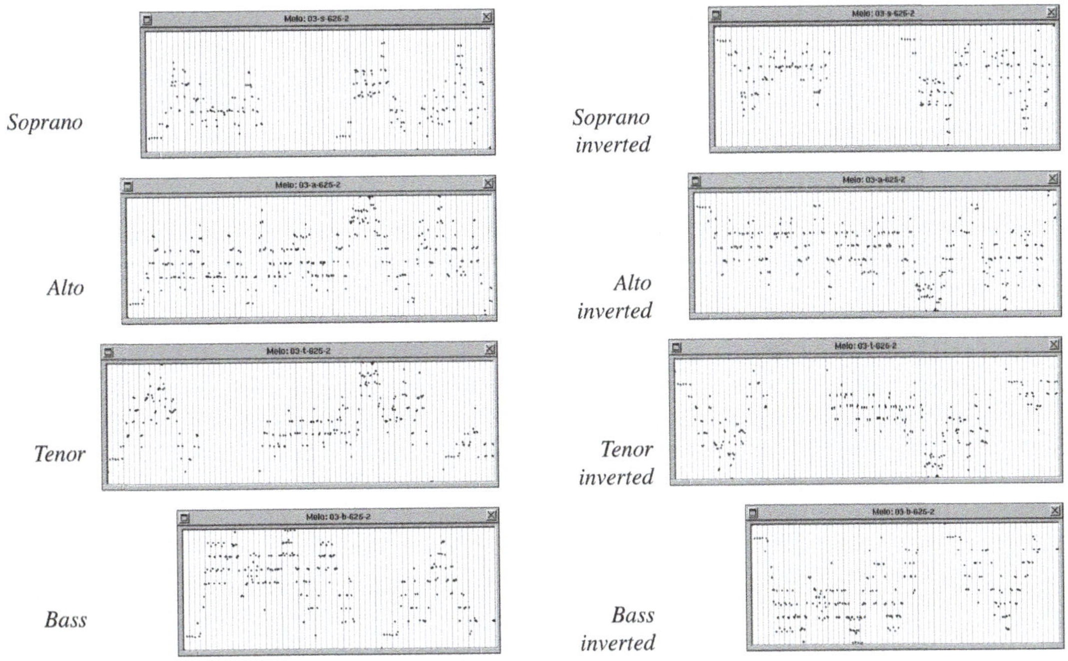

Fig. 42.7. Metrical weights for "contrapunctus III".

form of weights in the WeightWatcher for each voice, the thematic onsets for each development of the fugues could be modeled in an excellent way. Simultaneously however, the global usage of these weights lead to a completely disequilibrated passage so that this type of application was eliminated in the subsequent performance construction.

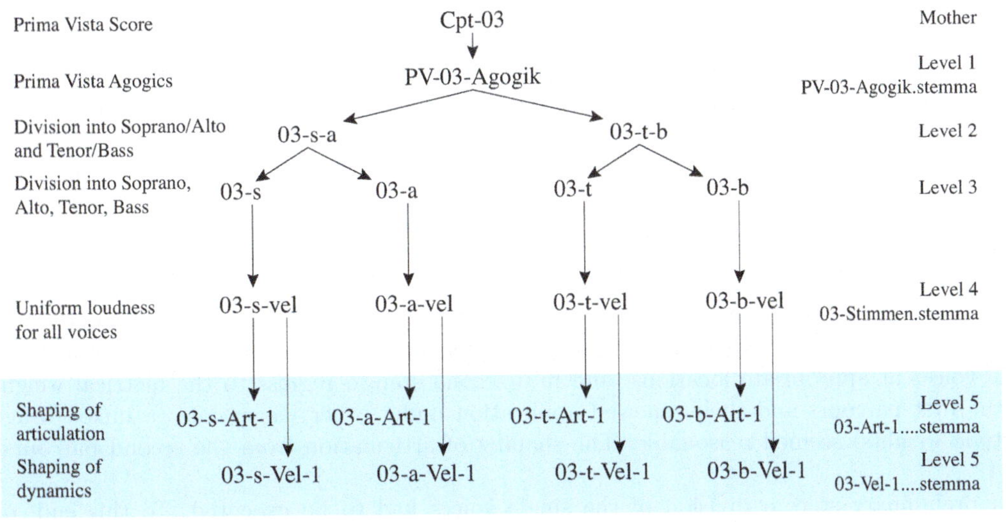

Fig. 42.8. Stemma construction for the second parcours.

42.4.1.2 Construction of Second Performance Parcours

In the second parcours, whose rather simple stemma (see Figure 42.8) we are going to discuss hereafter, we changed to the experimental strategy, with the aim to investigate to which degree motivic weights can contribute to the elaboration of a determined performance aspect. Those motivic weights of a voice were used which cover a minimal number of motives (Cardinality: 2) within a dotted half note (Span: 0.625); all these weights were applied in inverted form.

As with the subsequent third performance parcours, the core shaping work by means of weights (grey shaded levels in Figure 42.8) must be preceded by further steps. To begin with, this concerns the shaping of the score's primavista components, such as dynamic, articulatory, and agogical prescriptions. In the case of *contrapunctus III*, which contains no performance prescription by the author, this task consisted only in the primavista shaping of a short ritardando before the junction to the final tonical chord (Level 1).

The next steps within the stemma (levels 2 to 4) relate to the horizontal separation into the four voices (by the SplitOperator), which were identified by specific (artificial!) different loudness values (Levels 2 and 3), and which in the fourth step are reset to a uniform loudness. Upon this basis, an attempt to shape articulation and dynamics was undertaken.

In the course of six subsequent performance experiments it turned out that the global application of a single weight was capable of producing three slightly different variants of articulation shaping, however one had to pay attention to change the parameter for the weights' influence (in the WeightWatcher) in a minimal and systematic way. Figure 42.9 shows the final choice. Built upon these insights we tried to use

	High Norm	Low Norm	Influence	Deformation	Invert
Soprano	1.6	0.2	1	-0.75	YES
Alto	1.5	0.3	1	-0.75	YES
Tenor	1.5	0.4	1	-0.5	YES
Bass	1.4	0.6	1	-1	YES

Fig. 42.9. Choice of WeightWatcher parameters for motivic weights.

the same motivic weights, the same systematic and a similar handling of the change of intensities in order to produce comparable results for the shaping of dynamics. Without going into detail, it remains to be stated as a result that these attempts all failed. Typically, the dynamical relations between the voices turned out to be disequilibrated, and although the dynamics was perfectly modeled within single voices, the dynamical profile of thematic parts was sensibly worse. To obtain conclusive dynamics, one had to find another shaping procedure.

42.4.1.3 Construction of Third Performance Parcours

Because of these different dynamical profiles, the principle of former performance experiments—the exclusive usage of a weight and its global extension—had to be given up. In a first step it was recommended to split the single voices at appropriate locations, and in a second step, a regress to the metrical weights already used in the first parcours and their renewed application under other viewpoints (a mixed usage together with motivic weights) seemed reasonable. The shaping of articulation from the second parcours would be conserved.

In a preliminary step, a division of the single voices had to be executed. To this end, one had to find structurally legitimate points from the musical context, such as articulation by harmonic incisions or thematic groupings for developments and interludes.

The first division of all four voices took place in measure 39, legitimated by a harmonic close to the major parallel of the minor dominant (C-major); at the same time this is viewed as a possible ending of the second (however incomplete) development and a beginning of a four-measure interlude.

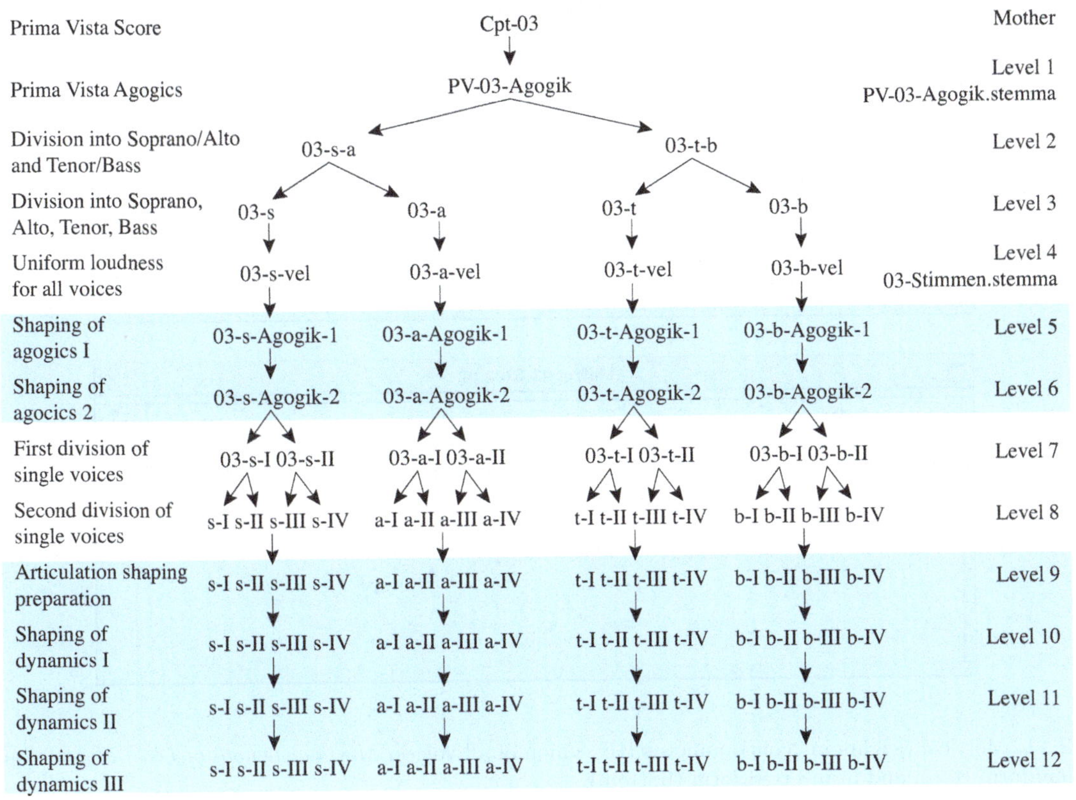

Fig. 42.10. The stemma of the third parcours.

In order to equilibrate the dynamical unbalances relating to the interludes from measures 19 and 46, a further division of the two halves of the fugue were necessary. A division of the first half was recommended in measure 19, having a close of the first development (exposition of fugue) and its half close on the dominant (A major).

Because of the too strong dynamic sink of the three-voice interlude from measure 46/47, the division of the second half had to take place not later than at this point. This division was legitimized by the half close on the minor dominant (A-minor) beginning in measure 46 on the one hand, and the simultaneous ending of the second (then complete) development according to the three-part construction of the fugue.

For the subsequent performance shaping, consider Figure 42.10. Besides the already known preparatory steps—horizontal division into single voices (Level 3) and equalizing of loudness (Level 4)—two performance steps for the later shaping of global agogics were inserted (Levels 5 and 6). This trick is applied, because agogics needs long calculation time on the global level of single voices and should be calculated after the stemmatically subsequent shaping articulation and dynamics. The vertical division of the single voices is applied in the previously described steps (Level 7 and 8). For the subsequent shaping of articulation and dynamics, each voice had to receive its separate and individual performance shaping for the four sections. This enabled us to apply different parameter values for the intensity effects, one per used weight.

For the shaping of articulation, the three already elaborated performance steps were inherited.

As is seen in the stemma (Figure 42.10), the shaping of dynamics was realized in three consecutive steps. Here, besides the known motivic weights, two additional metrical weights were applied. For the first step (Level 10), we applied the metrical weight from the union of all voices with *Minimal Length of Local Meters* equal to 2, in inverted form, and without deformation, see Figure 42.11.

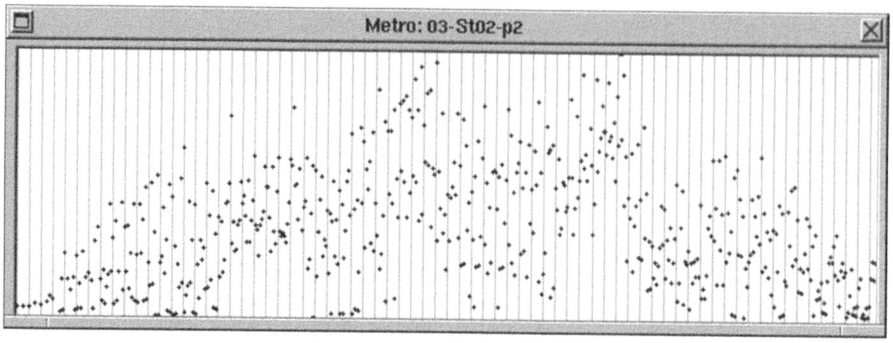

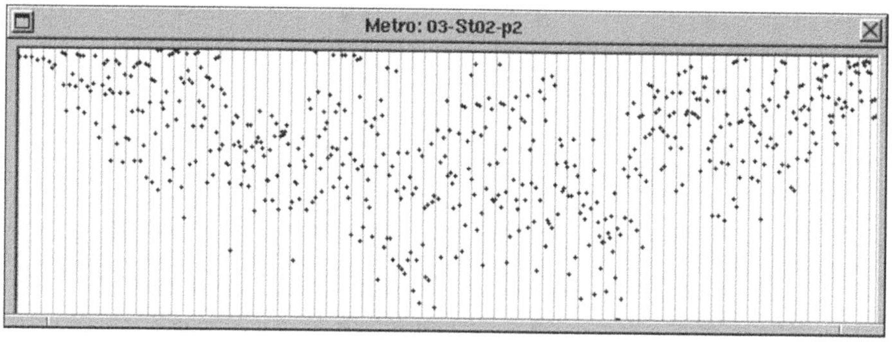

Fig. 42.11. Metrical weights in "contrapunctus III", union of all voices, *Minimal Length of Local Meters* equal to 2, in original form (top), and in inverted form (bottom).

Upon this stemma, the second step (Level 11) applied the metrical weights with value 5 for *Minimal Length of Local Meters* for each individual voice in inverted form and also without deformation (the weight graphics were comparable to those from the second performance parcours described above).

For the concluding shaping of dynamics, the already known motivic weights were applied in order to give the thematic onsets a plastic relief. From the interplay of the various intensity values, we got the constellation documented in Figure 42.12. The result of this performance communicates a relatively balanced dynamics, spread over the whole contrapunctus, the thematic onsets gain a profile, which can also be confirmed in the slight crescendo that leads to the beginning of the third development after the three-voiced interlude (from measure 46/47).

Bringing together the dynamic and the already elaborated articulatory aspects, the result can be stated as a complementary shaping of both performance aspects, which on top of that reveals a musical sense in the elaboration of thematic onsets and the three-voiced passages of the interludes.

For the shaping of agogics, the said levels 5 and 6 of our stemma were reserved. We did two different subsequent performance parcours with two different metrical weights: the sum of all voice weights (*Minimal Length of Local Meters*: 2) and the weight of the voice union (*Minimal Length of Local Meters*: 91 (!)), see Figure 42.13. The intensity values are shown in Figure 42.14.

As a result we may state that according to the weight structure and a minimal intensity effect a slight increase of the basic tempo towards the middle of the piece happens, but it is balanced by a slightly slower tempo for the initial and final measures. For the resulting performance as traced on the audio track on the CD-ROM attached to this book, see page xlvii. The corresponding stemmata are also found on this location of the CD-ROM.

	Soprano				Alto				Tenor				Bass			
	Part 1	Part 2	Part 3	Part 4	Part 1	Part 2	Part 3	Part 4	Part 1	Part 2	Part 3	Part 4	Part 1	Part 2	Part 3	Part 4
Level 10	Metro: St02-p2				Metro: St02-p2				Metro: St02-p2				Metro: St02-p2			
Weight																
High Norm	1.1	1.1	1.1	1.1	1.1	1.1	1.1	1.1	1.1	1.1	1.1	1.1	1.1	1.1	1.1	1.1
Low Norm	0.6	0.6	0.6	0.8	0.6	0.6	0.6	0.8	0.6	0.6	0.6	0.8	0.6	0.6	0.6	0.8
Influence	1	1	1	1	1	1	1	1	1	1	1	1	1	1	1	1
Deformation	0	0	0	0	0	0	0	0	0	0	0	0	0	0	0	0
Invert	Y	Y	Y	Y	Y	Y	Y	Y	Y	Y	Y	Y	Y	Y	Y	Y
Level 11	Metro: S05-p2				Metro: A05-p2				Metro: T05-p2				Metro: B05-p2			
Weight																
High Norm	1.0	1.0	1.0	1.0	1.0	1.0	1.0	1.0	1.0	1.0	1.0	1.0	1.0	1.0	1.0	1.0
Low Norm	0.7	0.7	0.7	0.8	0.7	0.7	0.7	0.8	0.7	0.7	0.7	0.8	0.7	0.7	0.7	0.8
Influence	1	1	1	1	1	1	1	1	1	1	1	1	1	1	1	1
Deformation	0	0	0	0	0	0	0	0	0	0	0	0	0	0	0	0
Invert	Y	Y	Y	Y	Y	Y	Y	Y	Y	Y	Y	Y	Y	Y	Y	Y
Level 12	Melo: s-625-2				Melo: a-625-2				Melo: t-625-2				Melo: b-625-2			
Weight																
High Norm	1.0	1.2	1.4	1.2	1.0	1.2	1.05	1.2	1.0	1.4	1.2	1.2	1.3	1.7	1.3	1.3
Low Norm	0.5	0.4	0.4	0.5	0.7	0.7	0.7	0.7	0.75	0.7	0.7	0.75	0.7	0.8	0.8	0.7
Influence	1	1	1	1	1	1	1	1	1	1	1	1	1	1	1	1
Deformation	0.5	0.5	0.5	0.5	0.25	0.25	0.25	0.25	0.25	0.25	0.25	0.25	-0.25	-0.25	-0.25	-0.25
Invert	Y	Y	Y	Y	Y	Y	Y	Y	Y	Y	Y	Y	Y	Y	Y	Y

Fig. 42.12. Intensity values for the concluding shaping of dynamics.

42.4.1.4 Local Discussion

Starting from the experimental strategy, the performance of "contrapunctus III" was first centered around the shaping with a single weight in order to sound the potential of a single weight. The shaping of articulation showed a reasonable local profile for every voice, i.e., a single analytical structure was capable of giving one performance aspect a reasonable expression.

The global application of weights and the usage of a single weight showed its limits, as we have learned from the dynamical shaping of "contrapunctus III". For example, the global application of weights failed in the different grades between the contributions of the four voices. Especially with the motivic weights of the tenor and bass voices, different weight profiles become visible which cannot be eliminated even by suitable deformations. These differing profiles of weights do result from the compositional structure. As this one splits into a number of parts—developments and interludes, groupings by harmonic closes and semi-closes—the division of the voices according to such compositional criteria is legitimized. Within these parts, the selected weights can be applied with different intensities and thusly equalize the disparate shapings.

Dynamics received a special significance in the shaping process: the elaboration of thematic onsets by an inverted motivic weight—this was justified in the structure of the theme. More precisely, the long durations of the initial notes resulted in a weak weight profile, which could be used to stress these incipits by the weight's inversion. This principle which remains valid for almost all contrapunti, can however only be proved for the "Kunst der Fuge".

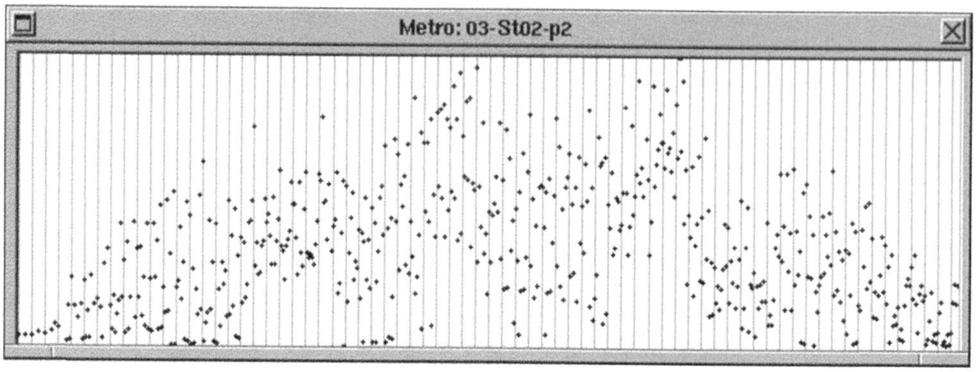

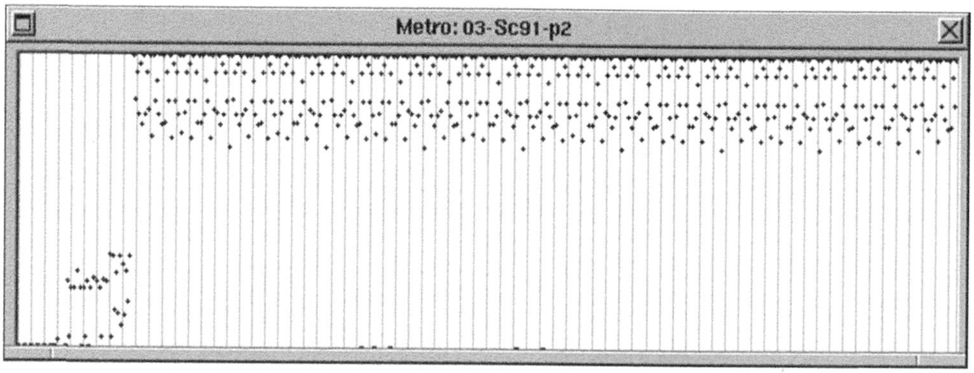

Fig. 42.13. "Contrapunctus III"; top: metrical weights, sum of all voice weights (*Minimal Length of Local Meters*: 2); bottom: weight of voice union (*Minimal Length of Local Meters*: 91).

	High Norm	Low Norm	Influence	Deformation	Invert
Total Contrapunctus	1.025	0.975	1	0	NO
Total Contrapunctus	1.05	0.95	1	0	NO

Fig. 42.14. Intensity values for the metrical weights.

42.4.2 Instrumental Setup

Summary. The conditions and influence of the instrumental setup are discussed.

– Σ –

A discussion of adequate instrumentation must be in the forefront of virtual performance work, since the behavior of the respective sound generator is an essential basis for shaping of the single musical parameters. Since the performance results from RUBATO® are encoded in a MIDI file, we have the possibility to access a MIDI-driven acoustical piano or else to use a corresponding digital device. Here, we should observe some relevant differences which—besides the basic difference in sound—the repetition, the dynamic response, the resonance behavior, and the spatial environment which we cannot discuss here. These differences are not only present between the acoustical and the digital instruments, they also act within each category.

For pianos and grand pianos, the MIDI-driven access is offered by Boesendorfer and Yamaha models. At the time of our experiments, the comparison between these two brands could a priori not be made in a serious way. Even if the shaping of dynamics admits a limited bandwidth of variation within the 128 MIDI velocity values, the *piano* and *pianissimo* ranges (velocities between roughly 30 and 1) for the Boesendorfer were much finer to tune than for the Yamaha piano and grand piano, where these dynamical values make the keys move silently. This defect as well as the extremal differences in the repetition mechanics show the futility of a comparison between such Boesendorfer and Yamaha instruments and imply that the Boesendorfer is the only reference for a reasonable performance. In this sense, the Boesendorfer was already chosen for the first experiment with Schumann's "Kuriose Geschichte" as briefly reported in Section 42.1.

In principle, an acoustical piano should be chosen as a reference for computer-assisted performance, but because of restricted availability of such instruments[3], a digital piano had to be selected in our case.

Our experiences with digital pianos and their sounds showed significantly different behavior for identical dynamics and articulation when applying identical weights. And it must be remarked that at the present state of performance work, it is difficult to obtain exact information about the dynamics and articulation shaping in the sense of a possibly reliable performance grammar since the judgment of the dynamical and articulatory aspects *as a function of the available instruments*, see [1004] for a detailed presentation of sound experiments. Any conclusion regarding the sources and rationales for the performance shaping of a score are vitally influenced by these instrumental conditions.

In other words, within a strictly scientific framework, the MIDI-encoded performance data enforce an instrumental selection. If research can be realized by means of one and the same MIDI piano, comparative statements can be made exactly for this instrument.

In a somewhat broader sense, this is also true for digital maps of the acoustical piano, i.e., one has to restrict the research to one and the same digital piano, such as Kurzweil's *Micro Piano* as it was used in [1003]. Any conclusion from the digital to the analog piano or vice versa is impossible, even among different digital or analog pianos no comparison is possible, see also [1004].

One solution out of this dilemma could be the application of the "physical modeling" or "virtual acoustics" technology (see Appendix A.1.2.4), where a direct access of the physical technology of an instrument by the virtualization (i.e., the software modeling) of the physical system of an instrument is enabled. This methodology offers flexible instruments which can be deformed seamlessly and can respond without delay. Although only first experiments have been initiated, this perspective opens an encompassing approach of computer-assisted performance research. The sound and resonance environment with its consequences for the shaping of performance aspects would not only yield new insights in the functioning of musical instruments, but also insights in the practice of instrumental playing. In this context, an interdisciplinary research team consisting of instrumentalists/interpreters, musicologists, computer scientists, mathematicians, and physicists would be required.

42.4.3 Global Discussion

Summary. We summarize the insight drawn from this second experiment.

$- \Sigma -$

In the course of the performance experiments, two different approaches and performance strategies crystallized. We tried to give the score's text an immanent shaping by means of two approaches:

- what is the sound of the analytical structure?
- can the sounding analytical structure yield a musically reasonable performance?

and two contrary performance strategies:

- the target-driven strategy,
- the experimental strategy.

[3] In Germany, the Boesendorfer MIDI grand exists only in two locations: at the conservatories in Lübeck and Karlsruhe. In both cases, access to this instrument is virtually impossible since it is located in rooms that are used for ordinary school activities.

In contrast to objective analytical approaches, for performance, subjective ingredients cannot be completely eliminated. They are present in their feedback with the performance result, while weights and intensity parameters in the WeightWatcher are determined, but they play a fairly reduced role.

From the first performance experiments, which have not been discussed in detail here, until the complete performance as described above, we have known milestones which demonstrated several problematic issues: It was not easy to eliminate the impression of an existing performance—in our case by Glenn Gould, say—and to stick strictly to what is written in the score; the performed version of the piece automatically resonates as a comparison while doing the performance work. This was the situation where we started these experiments with the ambitious task of approaching an artistical and esthetical performance as far as possible.

Therefore, the *target-driven* strategy was to a certain degree determined by the comparison with traditional human performances. Under these conditions, weights were applied and results were judged. This turned the tradition into an obstruction, it positioned the expected performance in the foreground and the shaping weight in the background.

Only the consequent questioning of the analytical structure and the systematic liberation from traditional performance expectations led to a performance strategy which positioned the analytical weights in the center of the investigation. This *experimental* strategy was coined by an as unbiased as possible sounding realization of analytical structures, centered around the question of how a weight, when applied to a particular performance aspect, would sound. Within this procedure it was possible to insert 'unheard' results, to admit purposed over-subscriptions in the sense of the 'still more clear', whereas the question whether an interpreter would play in this way turned out to be completely irrelevant.

From this point of departure, how a determined analytical structure would sound, the experimental approach to shaping a musically reasonable performance was sought. This qualitative determination of what is a "musically reasonable" performance is inevitably a subjective one which as such decided upon the subsequent steps towards the final performance. Similarly to the interpreter who puts up for discussion his provisionally final version while performing in concert—where in the last analysis it is more his personality than the musical performance which is judged—in the case of computer-assisted performance, the subject who works with the performance workstation RUBATO® presents his results as a provisionally final contribution to the ongoing discussion.

When judging all these performances, one has to take into account that only metrical and motivic weights were applied and the effects of harmonic passages were not included in the shaping of performance (except of the motivations for the not machine-made subdivisions from global to more local applications of weights in the third parcours). Furthermore a certain economy in the choice of weights and their application was applied. In this sense, we first had to check out which weights would entail what type of shaping consequences, and how the change of intensity parameters would influence the musical expressivity. It was only after this preliminary work that a systematic application of the weights and a partially purposed work with their intensity parameters became possible.

The portability of the presently described performance technique must be deduced from the compositional structure (a fugue in general and the thematic structure of the "Kunst der Fuge" in particular) as well as from the instrumental context. In nuce it can be said that such systematic statements are still premature. Many more analyses and performances would be necessary, but these can only be realized as soon as RUBATO® has become a common tool of musicology. Then the question of whether general recipes which are valid beyond the limits of single compositions can be stated, or whether performance is rather bound to each individual composition, could be restated.

Whatever is true for the transformation of the analytical structure in a scientific work targeting an artistically valid esthetic performance, one should not forget about the elimination of (and nonetheless omnipresent) emotional and gestural aspects. The realization of a sonification of analytical structures during the interaction with the computer always bears a degree of emotionality, a phenomenon that should be taken into account as a kind of "uncertainty relation".

The judgment of the performance results took place in the same line as the judgment of a human performance, and the work with RUBATO® was also proposed as a provisionally final contribution to the work's discussion.

While describing the performance results, the stress of a scientific analytical performance was central. The feedback to the analysis has a particular significance in that possibly, the conclusive character of a performance could yield an analytical criterium. This implies an absolutely serious attitude towards analysis, and no disclosure from emergent new aspects and innovative analytical ways of hearing.

Therefore we refrain from a discussion of subjects such as "prejudices against results which are produced by a machine", or "performance and the soul of music versus soulless performance machines". Instead, we favor representations of procedures and performance strategies, the exemplary demonstration of connections between analyzed structures, performed results, and the attempt at a generalization of these insights in the form of a performance grammar in its dependency on the instrumental conditions.

Part XI

Statistics of Analysis and Performance

43

Analysis of Analysis

O sancta simplicitas!
Jan Hus (1370–1415)

Summary. Not unexpectedly, weight analysis turns out to be complex information that cannot always be handled intuitively. This suggests techniques that help analyzing weight analysis. The problem may be tackled by use of statistical methods. We expose the subject and Jan Beran's approach based upon hierarchical decompositions of weights, together with an application to comparison of analyses of Bach, Schumann, and Webern.

– Σ –

Although the MetroRUBETTE® seems to implement a very simple analysis, the metrical weights turn out to encode quite complex information. For performance applications such as RUBATO®, this may be acceptable, but for an analytical understanding per se, the weights are too complex to be used directly, except for direct visual inspection of evident surface properties. The prejudice that musical analysis should be a simple, intuitive affair, is therefore banned to the fairy tales of auto-incompetent humanities.

In this chapter, we give an account of a statistical approach to understanding weights. More specifically, Jan Beran's method of hierarchical smoothing (see also [105]) is presented.

43.1 Hierarchical Decomposition

Summary. This section describes and motivates the hierarchical decomposition of weights by use of decreasing sequences of time bandwidths, the so-called *hierarchical smoothing.*

– Σ –

43.1.1 General Motivation

Can additional structural insight into weights be gained by suitable analysis of the analytic weight curves? The idea of the following method is to find a "natural" decomposition of the weight functions in order to find hidden regularities. In time series terminology, the general problem can be stated as follows: Let $\{x_s(t_i),\ t_i \in R,\ s = 1, ..., k, i = 1, ..., n\}$ be a collection of k time series, measured at the time points t_i. The aim is to find a decomposition

$$x_s(t_i) = \sum_{j=1}^{M} x_{j,s}(t_i)$$

such that the components $\{x_{j,s}, s = 1, ..., k\}$ reveal a maximal amount of 'regular structure'. One of the difficulties is to define what is meant by 'regular structures' and to define corresponding meaningful measures

G. Mazzola, *The Topos of Music II: Performance*, Computational Music Science,
https://doi.org/10.1007/978-3-319-64444-8_12

of the amount of 'regular structure'. Here, a pragmatic approach is taken, in that the amount of 'regular structure' is judged visually. Clearly, more formal definitions could be used.

Before introducing the idea of hierarchical decomposition, a few general remarks should be made:

Remark 18 Traditionally, one of the main structures of interest for time series is periodicity. In particular, spectral decomposition based on sines and cosines may be used for this purpose (see e.g., *[863], [154]*). In our context, this is not applicable, because many compositions are likely to have much more interesting structures than just periodicities. In fact, some scores may not contain any nontrivial periodicities at all. More generally, the problem is that using the same basis of functions, irrespective of the structure of the score, results in focusing on a very limited number of predetermined features that may in fact not be present.

Remark 19 As a consequence, a nonparametric approach based on kernel smoothing will be proposed here. In a traditional setting, the bandwidth b is chosen by minimizing a criterion such as the mean squared error as n tends to infinity. In particular, b tends to zero with increasing sample size. This concept is not directly applicable in our context. The main reasons can be summarized as follows:

1. Based on the definitions given above, the metric, melodic and harmonic aspects of a score are characterized respectively by one weight function only. In contrast, a composer is likely to have a hierarchical view. For instance, a piece has on one hand a global harmonic shape that makes the piece coherent as a whole, and on the other hand more local structures. Some composers in fact consciously write a score using a hierarchical approach, first defining a global shape and then refining more and more local structures. Similarly, while rehearsing, a performer is likely to focus first on global features of the score and then successively refine more and more local features. This fact was also used in RUBATO to design the process from a primavista performance to the refined artistic result *[699]*. Here, a genealogical tree, the *stemma* of the performance process, is responsible for successive refinement and localization of the performance. In order to obtain a better picture of the structure of a score it is therefore necessary to "extract" the hierarchy that is hidden in the weight functions. For smoothing, this means that there is not just one optimal bandwidth that is of interest. Instead, there is a hierarchy of relevant bandwidths $b_1 > b_2 > ... > b_M$. Moreover, the structure of the score, rather than an omnibus statistical criterion (such as the mean squared error), is likely to yield the key information about which sets of bandwidths could be interesting.
2. The weight functions obtained from the analysis above are generally rather complex. In particular, the weights often jump abruptly up and down between very small and very large values (see Figure 43.4). This can certainly not be carried over linearly to musical performance. For instance the tempo of a "musically acceptable" performance is unlikely to change up and down *drastically and repeatedly* within a few seconds. It is therefore reasonable to assume that a performance is not a linear function of the weights but rather a weighted sum of non-linearly deformed smoothed versions of these functions. Again, there may be a hierarchy of several bandwidths that need to be considered.

These general considerations motivate the idea of *hierarchical smoothing* and *hierarchical decomposition* described below.

43.1.2 Hierarchical Smoothing

Let $\{x_s(t_i), t_i \in R, i = 1, ..., n, s = 1, ..., k\}$ be a k-dimensional time series observed at time points $t_1, ..., t_n$ and K_b a smoothing kernel with bandwidth b and support $[-b, b]$. Applying the smoothing operator

$$K_b x_s(t) = \sum_{i=1}^{n} K_b(t, t_i) x_s(t_i)$$

($t \in R$) for a hierarchy of bandwidths $b_1 > ...b_M$, we obtain a hierarchy of $k-$dimensional curves

$$\{x_{j,s}(t) = K_{b_j} x_s(t), s = 1, ..., k\},\ j = 1, ..., M.$$

Here, the Naradaya-Watson kernel

$$K_b(t, t_i) = \frac{K(\frac{t-t_i}{b})}{\sum_{j=1}^{n} K(\frac{t-t_j}{b})}$$

with a triangular function $K(s) = 1\{|s| \leqslant 1\} \cdot (1 - |s|)$ was used. For $b = 0$, we have $K_b x_s(t) = x_s(t)$. Figures 43.1 and 43.2 display hierarchies of smoothed curves for Schumann's "Träumerei" and for Bach's

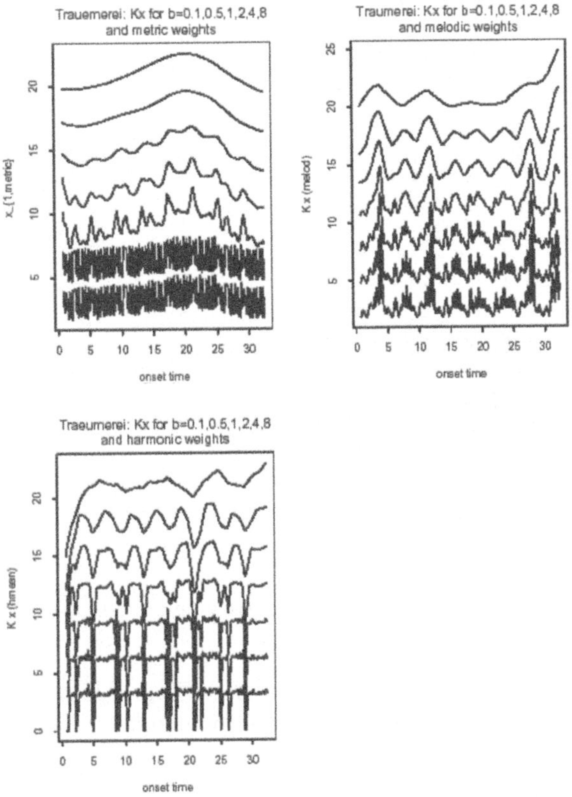

Fig. 43.1. Smoothed version of metric, melodic, and harmonic weights for Schumann's "Träumerei".

"canon cancricans" (see also Figure 8.7), resulting from the metric, melodic and harmonic weights. The figures illustrate that different bandwidths make different features more visible. In particular, for the metric weights, smoothing highlights places where high values occur more frequently. Also, some remarkable similarities between the metric, melodic and harmonic weights become apparent after smoothing.

Remark 20 The statistical technique of using smoothing kernels deserves a comment from the point of view of stemma and operator theory (Section 39.8), a comment which strongly relates to the inverse performance theory to be exposed later in part XII. To begin with, taking into account neighboring values of the analyses by kernel smoothing has a musical meaning: The interpreter is rightly supposed to be conscious of what happened and will happen within the time bandwidth b. In inverse performance theory, the idea of kernel smoothing is introduced in the locally linear performance grammars, see Section 39.8. There, the mutual influence of different local parts C_i and C_j of the composition C for the performance shaping process is formalized by use of *interaction matrices* $(c_{i,j})$. The coefficient $c_{i,j}$ quantifies the influence of C_i on C_j. The general theory of locally linear performance grammars deals with the description of the de facto algebraic

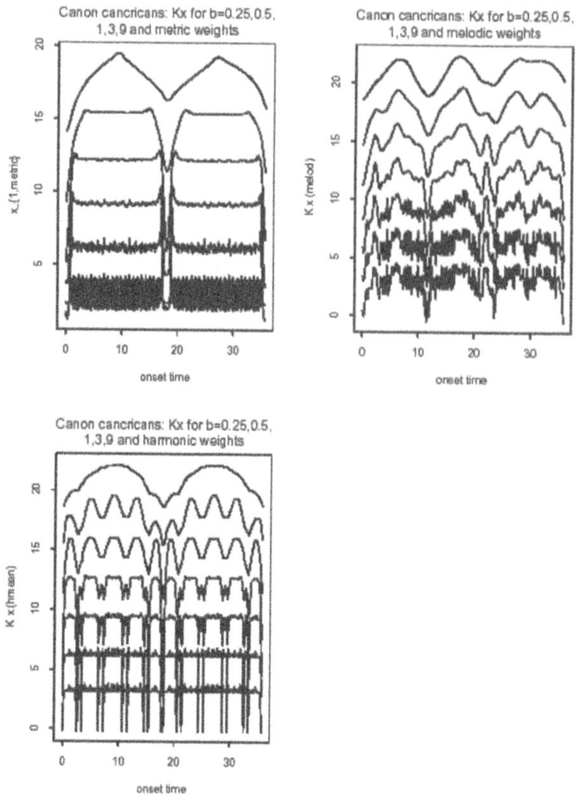

Fig. 43.2. Smoothed version of metric, melodic, and harmonic weights for Bach's "canon cancricans".

variety of interaction matrices inducing a fixed performance. From this point of view the triangular kernel smoothing means a selection of the a priori shape of interaction matrices, showing a peak around the diagonal.

43.1.3 Hierarchical Decomposition

The approach of hierarchical smoothing suggests a decomposition of the weight function into components of varying smoothness. Thus, let $\{x_s(t_i),\ t_i \in R,\ s = 1, ..., k, i = 1, ..., n\}$ be a collection of k time series. As discussed above, the aim is to find a decomposition $x_s(t_i) = \sum_{j=1}^{M} x_{j,s}(t_i)$ such that the components $\{x_{j,s}, s = 1, ..., k\}$ reveal a maximal amount of "regular structure". Structure can be, for instance: symmetry, repeated shapes/periodicities, relationship between different components etc. Note that with respect to cross-correlations, a number of methods are known in the literature for testing dependence between stationary time series (see e.g., [364], [448], [479], also see [863] and references therein). A direct adaptation of these methods is not possible for the following reasons: 1) The series considered here are not stationary in a nontrivial way and can, in particular, not be reduced to white noise by applying a linear filter. 2) The time points are not equidistant. 3) The aim is not only to obtain high cross-correlations but also to highlight regular features of the individual series. 4) Not only cross-correlations between "residual" but between all components are interesting. 5) The musical context suggests that the decomposition should be hierarchical in the sense that, with increasing index j, $x_{j,s}$ should contain increasingly local features.

We thus define the following decomposition:

1. Define a hierarchy of bandwidths $b_1 > b_2 > ...b_M = 0$, based on structural information from the score.
2. Define the smoothed function $x_{1,s} = K_{b_1} x_s$ and for $1 < j \leqslant M$, $x_{j,s} = K_{b_j}(x_s - \sum_{l=1}^{j-1} x_{l,s})$.

It should be noted that this decomposition is only one of many possible decompositions of x_s. The problem of choosing a meaningful decomposition of a time series is not new. In particular, in the context of regression analysis (see Chapter 44), it is a special case of the general problem of defining meaningful explanatory variables in regression models. Here, subject-specific considerations provide important guidelines. From a pragmatic point of view, a chosen decomposition can be considered reasonable if the subsequent regression analysis leads to meaningful interpretable results. In our context, the above decomposition appears meaningful, since it decomposes x_s in a simple additive way into components of decreasing smoothness. This translates, in a straightforward way, the generally accepted fact that a musical composition as well as a performance may be considered as a superposition of a hierarchy of local and global "shaping features", obtained by different degrees of "zooming in or out". For a given sequence of bandwidths $b_1 > b_2 > \ldots b_M = 0$, the first component $x_{1,s}$ represents the most global view of the score (or more specifically of the metric, harmonic or melodic structure, respectively), $x_{2,s}$ represents the next step of refinement by considering, in a more detailed fashion with a smaller bandwidth $b_2 < b_1$, the remaining information (obtained by subtracting the "global information" $x_{1,2}$, and so on.

43.2 Comparing Analyses of Bach, Schumann, and Webern

Summary. The statistical method developed in the preceding section is applied to a comparative study of RUBATO-analyses of works by Johann Sebastian Bach, Robert Schumann, and Anton Webern.

– Σ –

Each of Figures 43.3.a through 43.3.d displays the melodic (dotted, middle), metric (full, lower) and harmonic (dashed, upper) weights for Schumann's "Träumerei" op. 15/7 (Kinderszene No.7), Webern's "Variationen für Klavier" op. 27/II, the "canon cancricans" from Bach's "Musikalisches Opfer" BWV 1079, and Schumann's "Kuriose Geschichte" op. 15/2 (Kinderszene No.2). For onset times with more than one value of the melodic and harmonic weight respectively, the average of the values was taken.

It is also interesting to look at scatterplots of the three types of weights against each other. The example for Bach's "canon cancricans" is displayed in Figure 43.4. For each of the compositions, some simple regular features of the weights are visible:

- *Träumerei*: From the score it is clear that this composition may be divided into four parts P_j, $j = 1, 2, 3, 4$, corresponding to the onset intervals $I_1 = [0, 8]$ and $I_j = ((j-1) \cdot 8, j \cdot 8]$, $j = 2, 3, 4$, respectively. Also is it obvious that these four parts are similar to each other, and that P_3 differs most from the other parts. In fact, P_2 is, by definition, an exact replicate of P_1 (except for the slightly different up-beat). In Figures 43.5.a through 43.5.c, the weights for the four parts are plotted on top of each other, i.e., onset time is taken modulo 8. The weights are indeed almost identical to each other. Interestingly, the fact that P_3 differs most from the other parts shows only for the melodic weights. Also, the scatter plots do not indicate any strong relationship between the three weight functions. The sample correlations are all in the range $[-0.01, 0.09]$.
- *Variation op. 27/II*: With respect to the melodic and harmonic weights and from the score, it is clear that the composition can again be divided into four parts $P_j, j = 1, 2, 3, 4$, corresponding to a division of the onset time into four intervals of equal length. Clearly, the first two parts are almost identical with respect to the melodic and the harmonic weights. The same is true for the last two parts. For the metric weights, however, P_2 is not a simple replicate of P_1. The same is true for the last two parts. Also, for P_1 and P_2, the maximal values of the metric weights are much higher than for P_3 and P_4. Again, no apparent relationship seems to exist between the three weights (not shown). However, the largest correlation (in absolute value) is much higher than in the previous example, namely -0.31 between metric and harmonic weights.
- *Canon cancricans*: As expected for a retrograde canon, there is an almost exact time symmetry with respect to the middle of the onset axis. The symmetry is not exact, because the retrograde is not just a reflection of onsets but rather a transvection in the onset duration space parallel to the onset axis (see

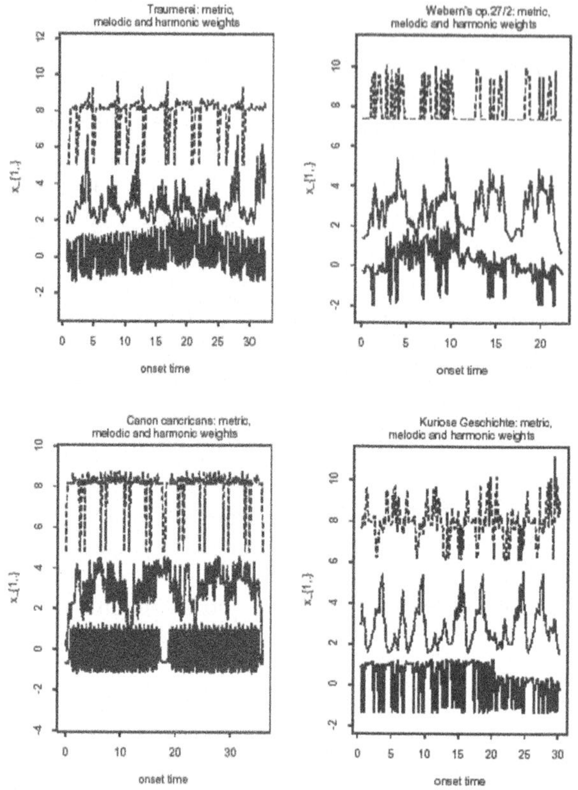

Fig. 43.3. Metric, melodic, and harmonic weights for Schumann's "Träumerei", Webern's "Variationen für Klavier", Bach's "canon cancricans", and Schumann's "Kuriose Geschichte".

Section 8.1.1, Example 7). Also striking is the clustered nature of the weights and the apparently very regular high frequency oscillation of the metric curve. A high metric weight is almost always succeeded by a low weight and vice versa. Because of the clustered nature of the weights, scatter plots are not very useful in this case. The correlations between the weights are again very small, ranging between 0.03 and 0.04.

- *Kuriose Geschichte*: Here, the score is again divided into four parts corresponding to the onset intervals [0,6], (6,12], (12,21], (21,30], with P_1 equal to P_2 and P_3 equal to P_4. Again, it is difficult to tell in how far the three different curves may be related to each other. Note however that the metric weights are much lower for onset times above 21. Thus, for the metric weights, the correspondence between P_3 and P_4 is much weaker. The reason is the breakdown of local meters at bar 21. Similarly to Webern, the strongest correlation between the weights is quite remarkable, namely -0.33 between melodic and harmonic weights.

It is also interesting to compare the four compositions with each other. The weights of Bach's "canon cancricans" exhibit an extreme high frequency oscillation that is not observed for the other scores. Ignoring that onset times are not exactly equidistant, this can be seen for instance very clearly by comparing the sample autocorrelations of the metric weights (Figure 43.6).

Another property of interest is the marginal distribution of the weight functions. Eliminating global 'trends' by taking first differences $x(t_j) - x(t_{j-1})$ the histograms are given in Figure 43.7 for the metric weights. For the compositions by Schumann and Bach, the first difference of the metric weights can essentially be classified into three clusters (low, medium, high). For Webern's score, the distribution is completely different and in fact rather close to a normal distribution. In contrast, the distributions of the differenced

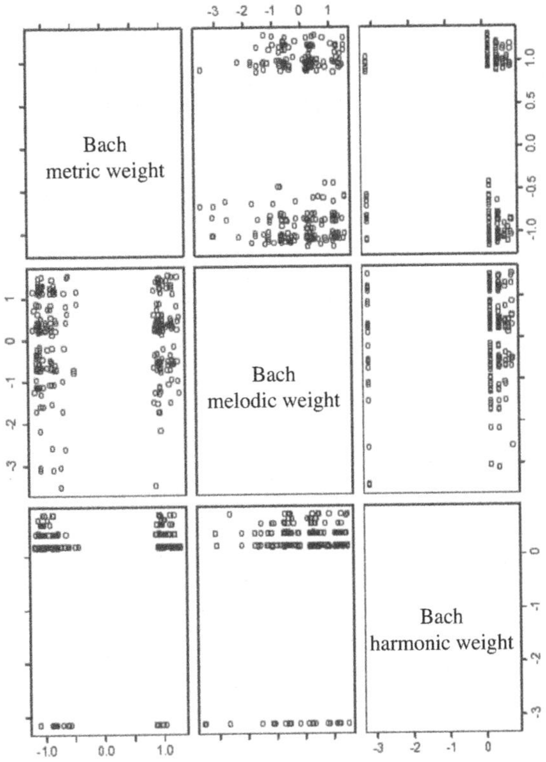

Fig. 43.4. Scatterplots of analytical weights for Bach's "canon cancricans".

melodic weights are qualitatively similar for all four scores. For the harmonic weights, all distributions appear to be essentially symmetric. However, while for Schumann's "Träumerei" and the score by Webern there appear to be three clusters, the histograms for Bach and the "Kuriose Geschichte" are essentially unimodal.

In summary, a first look at the weight functions reveals certain elementary features of the score. In the following it will be demonstrated that a more thorough analysis leads to further new insights about the structure of the scores. In particular, note that the three weight functions were defined in a completely different way. It may therefore be expected that there is no strong relationship between the curves. The scatter plots of the weights seem to support this conjecture. But the following analysis will show that certain components of the weight functions are indeed closely related.

Specifically, application to the four examples was carried out using $M = 4$. This choice was based on musicological considerations (time signature and bar grouping) as explained in the following. In this sense, the analysis here is exploratory, since no statistical selection criterion was used for choosing M. The following notation will be used here: $x_1 = x_{metric}$ =metric weight, $x_2 = x_{melod}$ =melodic weight, $x_3 = x_{hmean}$ =harmonic (mean) weight, $x_{j,metric} = x_{j,1}$, $x_{j,melod} = x_{j,2}$, $x_{j,hmean} = x_{j,3}$. The choice of the bandwidths was based on the time signature and bar grouping information. Example Schumann/Träumerei is written in 4/4 signature, the grouping is $8+8+8+8$. The chosen bandwidths are therefore 4 (4 bars), 2 (2 bars) and 1 (1 bar). Example Webern is written in 2/4 signature, its formal grouping is $1+11+11+11+11$; however, Webern insists on a grouping in 2-bar portions [1108], suggesting the bandwidths of 5.5 (11 bars), 1 (2 bars) and 0.5 (1 bar). Example Bach is written in 4/4 signature, the grouping is $9+9+9+9$. The chosen bandwidths are 9 (9 bars), 3 (3 bars) and 1 (1 bar). For example Schumann/Kuriose Geschichte, the

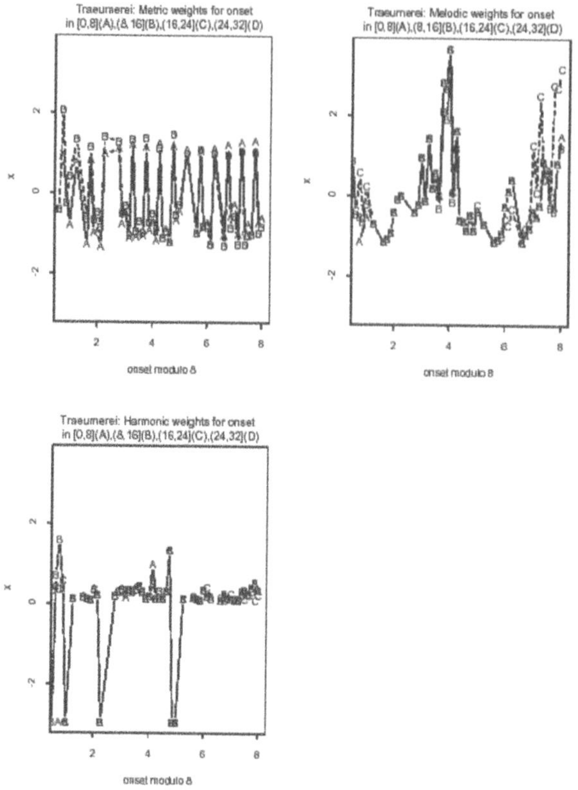

Fig. 43.5. Analytical weights for Schumann's "Träumerei" against onset time modulo 8.

time signature is 3/4, the grouping is 8 + 8 + 12 + 12. The chosen bandwidths are 3 (4 bars), 1.5 (2 bars) and 0.75 (1 bar).

Figures 43.9 ("Träumerei") and 43.8 ("canon cancricans") show remarkable regularities that have not been observed for the original weights (same for Webern and Schumann/Kuriose Geschichte, which we omit here). In particular, for all four compositions, much stronger similarities between the metric, melodic, and harmonic components can be observed than for the original weights, especially for $j = 2, 3$. Moreover, for the first two scores, the same kind of relationship can be observed for $j = 2, 3$, namely: positive correlation between $x_{j,melod}$ and $x_{j,hmean}$, negative correlation between $x_{j,melod}$ and $x_{j,metric}$, and negative correlation between $x_{j,hmean}$ and $x_{j,metric}$. Particularly surprising is the fact that Webern's score shows the same type of association as Schumann's "Träumerei". This leads to new insights into different approaches to composition. The weight functions are in fact very complex data and deserve a refined "analysis of analysis". Hierarchical smoothing is a possible approach to this problem.

Webern's piece is written in a completely dodecaphonic way, and thus breaks with harmonic and homophonic tradition. *This deserves a special methodological comment.* The fact that we have nevertheless applied harmonic analysis could be viewed as being in contradiction to Webern's rupture with harmony. Now, we do not claim that this analysis corresponds to Webern's poietic position when composing his "Variationen". Nonetheless, an objective analysis according to the Riemann approach is reasonable for two reasons: (1) Riemann intended to *attribute tonality to any possible chord.* The fact that he did not succeed in his goal is no reason for refraining from completion of his sketch. This is what the HarmoRUBETTE® is about: It is a proposal to discuss possible completions of Riemann's theory. (2) Therefore it is also very interesting to discuss its application to apparently atonal compositions. Such an experiment is likely to yield a testbed for the universality of Riemann's approach. These considerations suggest that *the following fact*

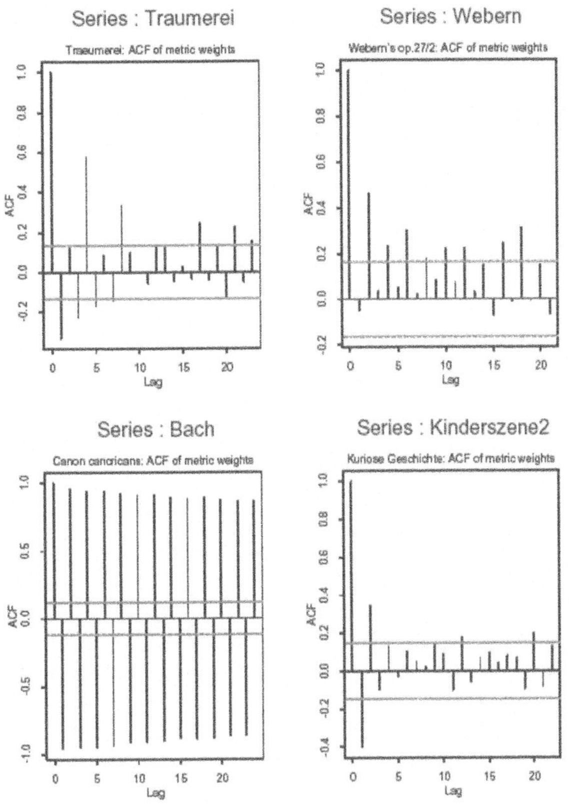

Fig. 43.6. Autocorrelograms of metric weights for Schumann's "Träumerei", Webern's "Variationen für Klavier" op.27/2, Bach's "canon cancricans", and Schumann's "Kuriose Geschichte".

is not completely surprising, although it has not been established explicitly elsewhere in the literature: The correspondence between metric, melodic, and harmonic structure in Webern's "Variationen" is very similar to Schumann's "Träumerei". It should be emphasized that this conclusion and in particular its quantitative demonstration is new in the musicological literature.

Schumann's "Kuriose Geschichte" also shows a strong correspondence between the three curves for $j = 1, 2$ and 3. But this time, the relations are different:

For onset times below 12, we have the following:

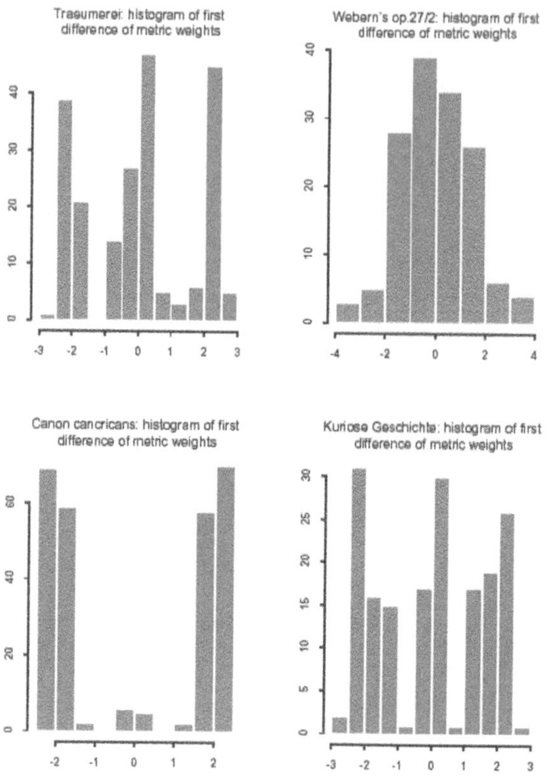

Fig. 43.7. Histogram of first difference of metric weights for Schumann's "Träumerei", Webern's "Variationen für Klavier", Bach's "canon cancricans" and Schumann's "Kuriose Geschichte".

1. For $j = 1$,
   $cor(x_{1,metric}, x_{1,melod}) = 0.83$,
   $cor(x_{1,metric}, x_{1,hmean}) = -0.71$,
   $cor(x_{1,hmean}, x_{1,melod}) = -0.63$.
2. For $j = 2$, we have the following (rounded) correlation values:
   $cor(x_{2,metric}, x_{2,melod}) = 0.00$,
   $cor(x_{2,metric}, x_{2,hmean}) = -0.31$,
   $cor(x_{2,hmean}, x_{2,melod}) = -0.82$.
3. For $j = 3$, we have
   $cor(x_{3,metric}, x_{3,melod}) = -0.67$,
   $cor(x_{3,metric}, x_{3,hmean}) = -0.20$,
   $cor(x_{3,hmean}, x_{3,melod}) = -0.61$.

Observe in particular that, in contrast to the other scores, melodic and harmonic components are negatively correlated.

After onset time 12, the correlations are:

1. For $j = 1$:
   $cor(x_{1,metric}, x_{1,melod}) = 0.10$,
   $cor(x_{1,metric}, x_{1,hmean}) = -0.38$,
   $cor(x_{1,hmean}, x_{1,melod}) = -0.29$.

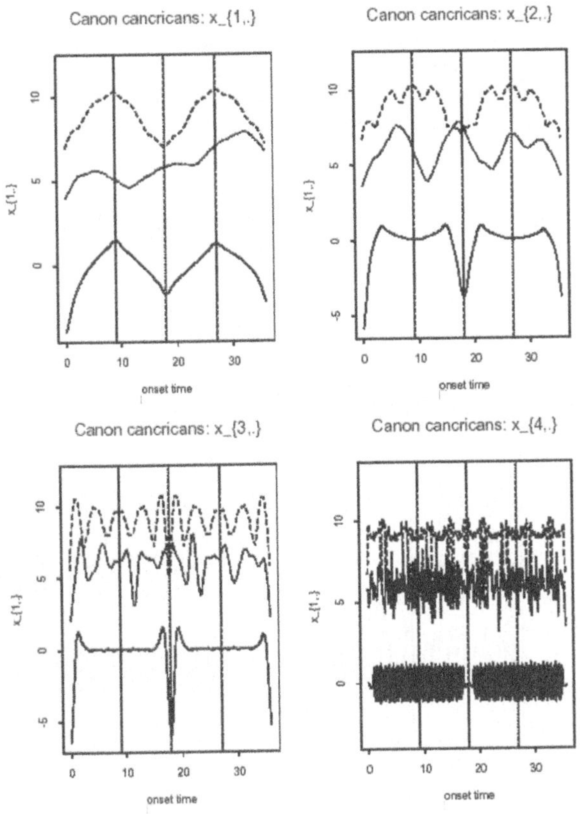

Fig. 43.8. Hierarchical components of metric (solid lines), melodic (dotted lines), and harmonic (dashed lines) weights for Schumann's "Träumerei", as defined in Section 43.1.2: (a) $b = 4$; (b) $b = 2$; (c) $b = 1$; (d) remaining (residual) series.

2. For $j = 2$:
   $cor(x_{2,metric}, x_{2,melod}) = -0.47$,
   $cor(x_{2,metric}, x_{2,hmean}) = -0.14$,
   $cor(x_{2,hmean}, x_{2,melod}) = -0.11$.
3. For $j = 3$:
   $cor(x_{3,metric}, x_{3,melod}) = -0.75$,
   $cor(x_{3,metric}, x_{3,hmean}) = 0.58$,
   $cor(x_{3,hmean}, x_{3,melod}) = -0.69$.

Finally, for Bach's composition, the only noticeable correlations occur between metric and harmonic weights, namely:

1. For $j = 1 : cor(x_{1,metric}, x_{1,hmean}) = 0.94$,
2. For $j = 2 : cor(x_{2,metric}, x_{2,hmean}) = 0.63$,
3. For $j = 3 : cor(x_{3,metric}, x_{3,hmean}) = 0.61$.

With respect to the shapes of $x_{j,.}$, for $j = 2$ and 3, the two scores by Schumann and the one by Webern are clearly more similar to each other as compared to Bach's shapes. From the point of view of music history, this is quite plausible, since Webern's organic composition principle is more related to Schumann's rankly growing romanticism than to Bach's self-disciplined architectural setup (see also the following remarks).

Finally, note that the scatterplots in Figure 43.4 show that Bach's harmonic weights are highly clustered and the smoothed curves in Figures 43.8.a through 43.8.d are more 'edgy' than for the other compositions.

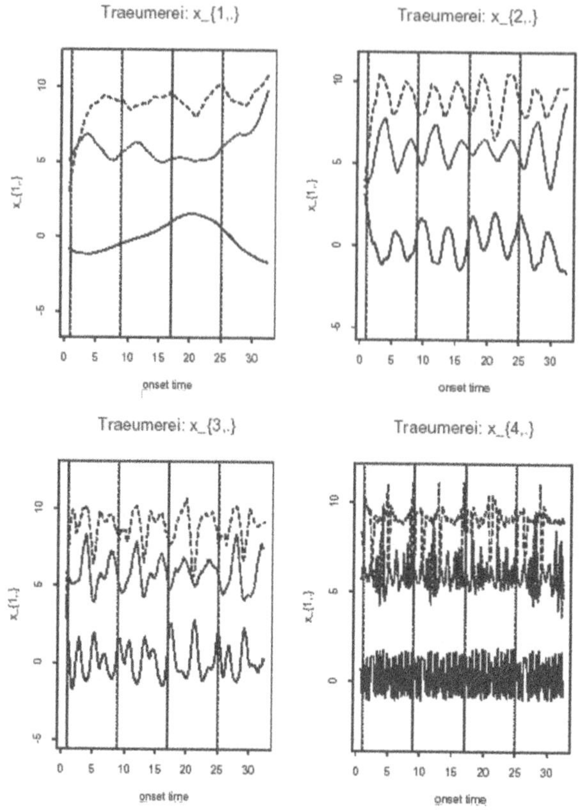

Fig. 43.9. Hierarchical components of metric (solid lines), melodic (dotted lines), and harmonic (dashed lines) weights for Schumann's "Träumerei", as defined in Section 43.1.2: (a) $b = 4$; (b) $b = 2$; (c) $b = 1$; (d) remaining (residual) series.

In this sense, *Bach's composition exhibits a high degree of organization.* This confirms the general belief that the principle of architectural rather than processual construction plays a dominating role in Bach's music.

Overall, we may conclude that hierarchical decomposition reveals interesting properties, in particular strong similarities between the metric, melodic and harmonic weights, that were not visible in the original series. The results are musically plausible in that the analysis of Bach's score turns out to be the most regular one and the analyses of Webern and Schumann appear to be closer to each other than to Bach's. The results are surprising in that (the analysis of) Webern turns out to be closer to (the analysis of) Schumann than expected. Also, the strong relationship between the three analytic curves could not be expected a priori, since the three weights were calculated using completely different aspects of the score and the scatter plots of the original curves did not show almost any association.

Based on the results, one may conjecture that appropriate matching of metric, melodic and harmonic structure plays an important role in music, independently of musical style. The tools introduced here provide the possibility of investigating which types of relationships may exist in which musical and historical contexts. An important task for future research will be to investigate such aspects for a larger variety of compositions.

Check for updates

44

Differential Operators and Regression

Rejection by common sense, for whatever reason, proves nothing.
Other fields of science are built on propositions that
seem absurd but in fact are true.
Donald O Hebb [451]

Summary. We give statistical evidence from 28 performances of Schumann's "Träumerei", as measured by Bruno Repp [886] that the rhythmic, motivic, and harmonic analyses provided by RUBATO® are shaping structures for the agogical streams. The statistical model is based on regression analysis and realizes shaping of agogics by a second degree linear differential operator as a function of analytical weights which are averaged over a natural grouping hierarchy (as described in Chapter 43) of the score.

– Σ –

At present, the best investigated aspect of performance theory—including appropriate software—is timing microstructure, i.e., agogics on the level of tempo curves and their hierarchies, see [546, 688] for further reading. This chapter deals with this topic: agogics as an expression of harmonic, melodic and rhythmic structures. So observe that we do not consider emotional or gestural rationales for agogics. This does not mean that these factors are negated. We merely restrict our investigation to the question whether and how strongly agogics could be explained by exclusive causal reference to structural analysis.

Even in this neat reduction is the question neither trivial nor even well defined, since musicology does not offer precise tools for rhythmical or melodic analysis, and even harmonic analysis is far from effective. Therefore, the question is only a scientific one if one specifies the analyses and their output data format. For the general setup for such an explicit and operationalized analysis framework, namely the RUBATO® analysis and performance workstation, we refer to part X. The main concern is an *empirical study* regarding the basic question *whether agogics (the tempo curve of the "Träumerei", also called "timing microstructure") may be expressed in mathematical terms by use of structural data obtained from a specific set of musicological analyses offered by* RUBATO® *'s modules.* The experimental data are taken from Repp's timing measurements of 28 famous performances of Schumann's "Träumerei" [886]. Summarizing our results, we can state that

Result 2 *There is strong statistical evidence for the timing microstructure in 28 famous performances of Schumann's "Träumerei" as being an expression solely of harmonic, melodic and rhythmic structures furnished by the* RUBATO® *analysis. In other words: These structural rationales are sufficient for explaining agogics, and emotional or gestural rationales can be disregarded for the present experimental material.*

In the framework of RUBATO®, the analytical output of any type of analysis is always a (smooth) weight, i.e., differentiable function of one or several note event parameters, such as onset E, pitch H, loudness L, duration D, etc. With this specification, the above result can be restated in more mathematical but less intuitive terms:

G. Mazzola, *The Topos of Music II: Performance*, Computational Music Science,
https://doi.org/10.1007/978-3-319-64444-8_13

Result 3 *The tempo curve (timing microstructure) can be generated by an* agogical operator Ω *which is essentially a linear differential operator of second order as a function of harmonic, melodic and rhythmic smooth weights.*

Statistically speaking, this means that in our empirical context, the fiber (in the sense of Todd's approach) of the chosen performance transformation is not empty. Observe that the operator Ω which plays the role of the transformation Π in Todd's theory [1058] (see also Section 36.3) does not rely on general encoding tempo curve functions. Intuitively, this means that our approach generates agogics from smooth weights as *score-specific* functions, and not general, score-independent, curve types, as proposed in [1022, 1060]. This does not contradict usage of general encoding tempo functions, it simply suggests that agogics is a superposition of more "primavista"-like tempo functions and of a strong and differentiated timing microstructure stemming from analytical data.—Thirdly, the statistical results suggest

Result 4 *Essential commonalities and diversities among tempo curves may be characterized by a relatively small number of analytical weight curves. There is in general no unique way of attributing features of the tempo to exactly one cause (harmonic, metric or melodic analysis). Results depend on which of the three analyses is given priority. However, there appear to be a certain number of* canonical curves *that are essentially independent of the priority.*

Overall, a large variety of musically meaningful results is obtained. This is in particular due to the fact that a score-specific basis of curves is used on which the tempo curves are projected. We thus may conclude

Result 5 *The analytical curves obtained from (1) score-specific harmonic, melodic and rhythmic smooth weights and (2) a score-specific hierarchical decomposition of these weights, yield a natural* score-specific linear basis *in the space of tempo curves, for performances of the considered score.*

We should stress that all our results are intimately related to the concrete analyses which RUBATO® produces—together with the underlying theories. There is no unique analysis, and therefore, specification and numerical representation of musical analysis is not secondary and will in any case (!) influence the results. We should also remind the reader trained in natural sciences that musical analysis is not a neutral tool but pertains to the unavoidable "artifacts" of analysis in the humanities. *There are no objective laws in human creations which subsist beyond interpretative interaction.* This is a caveat to those who believe that performance can be more than a relation between what we understand (rationally, emotionally, gesturally) and how we express this understanding.

44.0.1 Analytical Data

Summary. This section describes the analytical weights used in this analysis.

– Σ –

We shall omit the notification of the used parameter lists *Param* which are related to the specific RUBETTE®, and of the predicate $P_{Träumerei}$ of Schumann's score and abbreviate

$$\begin{aligned}
x_{metric} &= x^{P_{Träumerei}}_{MetroRubette,Param_{metric}}\\
x_{melodic} &= x^{P_{Träumerei}}_{MeloRubette,Param_{melodic}}\\
x_{hmax} &= x^{P_{Träumerei}}_{HarmoRubette,Param_{harmonic/max}}\\
x_{hmean} &= x^{P_{Träumerei}}_{HarmoRubette,Param_{harmonic/mean}}
\end{aligned}$$

to denote the four following weights used in our context. We are going to give their description when evaluated on onsets E of note events occurring in $P_{Träumerei}$:

- $x^{P_{Träumerei}}_{MetroRubette, Param_{metric}}$
This is a metrical weight which measures the rhythmic relevance or "weight" of every onset of a note event in the composition in the lines of Riemann [904], Jackendoff-Lerdahl [496], and Mazzola [682]. A detailed description of the MetroRUBETTE® was given in Section 41.1.
- $x^{P_{Träumerei}}_{MeloRubette, Param_{melodic}}$
This is a "boiled-down" melodic weight[1] which measures the sum of the *melodic weights* $w(Evt)$ of all note events Evt at a given onset E. The calculation is extremely complex and time-consuming and goes back to theories of Reti [892], see Chapter 22. See Section 41.2 for a detailed description of the MeloRUBETTE®.
- $x^{P_{Träumerei}}_{HarmoRubette, Param_{harmonic/max}}$
This harmonic weight measures the harmonic relevance of a chord ch which occurs at an onset E in $P_{Träumerei}$. It is calculated by the same method as the fourth weight. The only difference is that this weight captures the harmonic relevance of the most important note in ch whereas the fourth weight represents the average harmonic relevance among all notes of ch. See Section 41.3 for a detailed description of the HarmoRUBETTE®.
- $x^{P_{Träumerei}}_{HarmoRubette, Param_{harmonic/mean}}$
As already mentioned, this weight is a variant of the third one, the only difference being an averaging instead of maximizing procedure. We refer to the previous discussion for the basics.

44.1 The Beran Operator

Summary. In this section, we define the conceptual setup for the following statistical analysis.

$- \Sigma -$

The general idea is that agogics is to be shaped by use of smoothed versions of the boiled-down weights, their first and second derivatives, and corresponding kernel-smoothed versions with respect to hierarchical (triangular) kernel functions.

44.1.1 The Concept

As we know, the general RUBATO® concept of shaping performance is built on smooth (actually C^2 in this context) weights x where $d_E x$, $d_E^2 x$ denote the first and second derivatives with respect to symbolic time E. The kernel smoothing process relates to *kernel* functions $\hat{b}(s) = 1/b \cdot \chi\{|s| \leqslant b\} \cdot (1 - |s|/b)$ with triangular, zero-symmetric support of extent $\pm b$, and characteristic function $\chi\{P\}$ for a predicate P. The linear smoothing operator $b \diamond f$ on a function f is defined by the convolution

$$b \diamond f(E) = \int \hat{b}(t - E) \cdot f(t). \tag{44.1}$$

It averages f around E with weighted center E and *bandwidth* b. If this function is a weight, this means that the weight's analysis within the entire bandwidth neighborhood of a given onset is included instead of spiking the analysis to the singular onset. In the following process, this kernel smoothing process has been applied to a hierarchy of bandwidths, starting with $b = 4$ (= eight bars), then $b = 2$, then $b = 1$. The averaging process is taken to define successive remainder functions as follows:

$$f_1 = 4 \diamond f,\ f_2 = 2 \diamond (f - f_1),\ f_3 = 1 \diamond (f - f_1 - f_2),\ f_4 = f - f_1 - f_2 - f_3 \tag{44.2}$$

This means that the decomposition

$$x = x_1 + x_2 + x_3 + x_4 \tag{44.3}$$

for a smooth weight x defines a "spectrum" of that weight with respect to successively refined neighborhoods of its ambit.

[1] See formula (39.26) in Section 39.3.

Remark 21 *Musically speaking, as already observed before, this kernel smoothing process is completely natural.* In fact, the kernel function alters the original time function $f(E)$ by a weighted integration of f-values in the kernel neighborhood of a given time E. This means that we now include the information about f from the neighboring times to make an analytical judgment. This latter is a well known and common consideration in musical performance: The interpreter looks up a full neighborhood of a time point to derive what has to be played in that point. Moreover, the repeated application of the kernel smoothing process with increasingly narrowed neighborhoods is understood as a succession of a refinement in local analysis: First, the interpreter makes a coarse analysis over eight bars ($b = 4$), then he/she looks for the remainder $f - f_1$ and goes on with refined actions, if necessary.

This procedure is applied to the metric, melodic and harmonic weights and to their first and second derivatives. This gives the following list of a total of 48 *spectral analytical* functions:

$$\begin{array}{llll}
x_{metric,1} & x_{metric,2} & x_{metric,3} & x_{metric,4} \\
d_E x_{metric,1} & d_E x_{metric,2} & d_E x_{metric,3} & d_E x_{metric,4} \\
d_E^2 x_{metric,1} & d_E^2 x_{metric,2} & d_E^2 x_{metric,3} & d_E^2 x_{metric,4}
\end{array}$$

$$\begin{array}{llll}
x_{melodic,1} & x_{melodic,2} & x_{melodic,3} & x_{melodic,4} \\
d_E x_{melodic,1} & d_E x_{melodic,2} & d_E x_{melodic,3} & d_E x_{melodic,4} \\
d_E^2 x_{melodic,1} & d_E^2 x_{melodic,2} & d_E^2 x_{melodic,3} & d_E^2 x_{melodic,4}
\end{array}$$

$$\begin{array}{llll}
x_{hmax,1} & x_{hmax,2} & x_{hmax,3} & x_{hmax,4} \\
d_E x_{hmax,1} & d_E x_{hmax,2} & d_E x_{hmax,3} & d_E x_{hmax,4} \\
d_E^2 x_{hmax,1} & d_E^2 x_{hmax,2} & d_E^2 x_{hmax,3} & d_E^2 x_{hmax,4}
\end{array}$$

$$\begin{array}{llll}
x_{hmean,1} & x_{hmean,2} & x_{hmean,3} & x_{hmean,4} \\
d_E x_{hmean,1} & d_E x_{hmean,2} & d_E x_{hmean,3} & d_E x_{hmean,4} \\
d_E^2 x_{hmean,1} & d_E^2 x_{hmean,2} & d_E^2 x_{hmean,3} & d_E^2 x_{hmean,4}
\end{array}$$

For which *musical reasons* are these derivatives added to the analytical input data? The first derivatives measure the local change rate of analytical weights. Musically speaking, this is an expression of transitions from important to less important analytical weights (or vice versa), i.e., a transition from analytically meaningful points to less meaningful ones (or vice versa). This is crucial information to the interpreter: It means that he/she should change expressive shaping to communicate the ongoing structural drama. In the same vein, information about second derivatives is musically relevant since it lets the interpreter know that the ongoing structural drama is being inflected. Evidently, one could add higher derivatives but we argue that an interpreter is already highly skilled if he/she can take care of all these functions, also because different analytical aspects from metrics to harmonics must be observed simultaneously.

Besides these analytical input functions, we add three types of 'sight-reading' functions. They regard the following three instances: ritardandi, suspensions[2], and fermatas. It is clear that any text-sensitive performance should be aware of such information.

1. **Ritardandi**
   The score shows four onset intervals R_1, R_2, R_4, R_4 for ritardandi, starting at onset times $E_o(R_j)$ $(j = 1, 2, 3, 4)$ respectively. We define the four linear functions

$$x_{rit_j}(E) = \chi(R_j) \cdot (E - E_o(R_j)), j = 1, 2, 3, 4. \tag{44.4}$$

[2] Suspensions are notes which are tied by a slur while the harmony changes; we may attach to such events the time interval where the suspension does not start until the harmony change is terminated.

2. **Suspensions**
   The score shows four onset intervals S_1, S_2, S_4, S_4 for suspensions, starting at onset times $E_o(S_j)$ $(j = 1, 2, 3, 4)$ respectively. We define the four linear functions

$$x_{sus_j}(E) = \chi(S_j) \cdot (E - E_o(S_j)), j = 1, 2, 3, 4. \tag{44.5}$$

3. **Fermatas**
   The score shows two onset intervals F_1, F_2 for fermatas. We define the two support functions

$$x_{ferm_j}(E) = \chi(F_j), j = 1, 2. \tag{44.6}$$

Summarizing, we have a total of 58=48+4+4+2 onset functions of analytical and primavista types. Call X the *analytical vector* of these 58 functions listed in a fixed order. The present approach is to define the tempo function at onset E as being a linear function of these 58 variables.

For a 'shaping' vector $\omega \in \mathbb{R}^{58}$, the shaping operator Ω_ω^X for the tempo curve is defined by the canonical scalar product of X with the shaping vector ω,

$$\Omega_\omega^X = (X, \omega). \tag{44.7}$$

This means that for every onset E, we have $\Omega_\omega^X(E) = (X(E), \omega)$. Recapitulating the meaning of the analytical vector X, we are dealing with a second order differential operator which we call "Beran operator" since it was introduced by Jan Beran in [107].

On this basis, the central question of the following is whether tempo curves T of the "Träumerei" as they appear in the context measured by Repp in [886] may be approximated via Ω_ω^X by appropriate choice of the shaping vector ω. The main result of this approach states that there is strong statistical evidence for the equation

$$\ln(T) = \Omega_\omega^X + C \tag{44.8}$$

for the given analytical vector X, a suitable shaping vector ω, and a constant C.

This means that the 58 coefficients of the shaping vector ω are random variables and that we prove a significant statistical correlation—in the mathematical form described by the Beran operator—between a certain subset of the analytical vector X and tempo as it is measured for the 28 performances by Repp.

Thesis 7 *One may therefore try to use the above formula* (44.8) *to define tempo as a function of analytical score data in the sense of a general performance grammar as described in Chapter 37.*

44.1.2 The Formalism

44.1.2.1 Tempo Information

In the following, a more detailed description of the tempo data used for the analysis is given.

- **Onset times:** The onset times are on a grid of $1/8^{th}$ beats. Thus, for instance, grace notes are excluded. From this set of onset times, we consider only onset times where at least one note is actually played. This results in a set T of $n = 212$ not equidistant onset times t_i $(i = 1, ..., n)$ which are multiples of $1/8$.
- **Log-transformation:** Instead of the original tempo y we consider its natural logarithm $\ln y$. Intuitively this can be justified by the expectation that a performer may control the tempo in a relative rather than an absolute way. Also, the statistical results were more satisfactory on the logarithmic scale. In the following we refer to the logarithmic tempo as 'the tempo curve'.
- **Standardization of individual curves:** The data consist of tempo measurements (or tempo curves) for $m = 28$ performances. In the current analysis, the interest lies in investigating the shape of the tempo curves rather than the absolute tempo values. Therefore, each of 28 tempo curves is standardized. More

specifically, let $y^*(t_i, j)$ be the (natural) logarithm of the tempo of the jth performance at onset time t_i $(i = 1, ..., n; j = 1, ..., m)$. Then the standardized tempo data are defined by

$$y(t_i, j) = \frac{y^*(t_i, j) - \bar{y}^*(j)}{s^*(j)}$$

where $\bar{y}^*(j) = n^{-1} \sum_{i=1}^{n} y^*(t_i, j)$ and

$$s^*(j) = [(n-1)^{-1} \sum_{i=1}^{n} (y^*(t_i, j) - \bar{y}^*(j))^2]^{\frac{1}{2}}.$$

44.1.2.2 The Explanatory Variables

The following notation is used: Let A be a $p \times q_1$ matrix and B a $p \times q_2$ matrix, then $C = (A, B)$ denotes the $p \times (q_1 + q_2)$ matrix obtained by 'attaching' B on the right-hand side of A. The following steps describe the definition of the matrix of explanatory variables in more detail.

According to the concept in Section 44.1.1, the score data (metric, harmonic and melodic weights, additional score information) are given in the form of a design matrix that is used subsequently in a regression analysis. The following definitions are used:

1. **Derivatives.** "Derivatives" are defined as finite differences divided by the difference of the onset times. Thus, for instance,

$$d_E x_{metric,j}(t_i) = \frac{x_{metric,j}(t_i) - x_{metric,j}(t_{i-1})}{(t_i - t_{i-1})}$$

and

$$d_E^2 x_{metric,j}(t_i) = \frac{d_E x_{metric,j}(t_i) - d_E x_{metric,j}(t_{i-1})}{(t_i - t_{i-1})}.$$

2. **Hierarchical smoothing.** Each of the weights and their first and second (discrete) derivatives are decomposed into four components of different smoothness as defined by equations (44.2) and (44.3).
3. **Additional variables.** Additional variables modeling ritardandi, suspensions and fermatas were defined by (44.4), (44.5), and (44.6). The aim is to model these musical events in a "minimal" way. For instance, the resulting linear model for a ritardando is only a crude approximation to a "true" ritardando. The reason for using only the simplest parametrization is that the main purpose here is to examine to what extent the metric, melodic, and harmonic weights alone, together with only *absolutely necessary* additional information from the score, contain enough information to "explain" the tempo of a performance.
4. **Initial design matrix.** Using the definitions above, we define for $j = 1, 2, 3, 4$ the $n \times 4$ matrices

$$X_j(harmo) = (x_{hmean,j}, x_{hmax,j}, x_{metric,j}, x_{melod,j})$$
$$X_j(metric) = (x_{metric,j}, x_{hmean,j}, x_{hmax,j}, x_{melod,j})$$
$$X_j(melod) = (x_{melod,j}, x_{hmean,j}, x_{hmax,j}, x_{metric,j})$$

Clearly, there are more possibilities of permuting columns. Here, we consider only the representative permutations above. The first column of $X_j(harmo)$ is $x_{hmean,j}$ so that, due to the orthonormalization to be described in the following section, the main emphases is put on the harmonic mean weights. Similarly, the metric and melodic emphasis are chosen with $X_j(metric)$ and $X_j(melod)$. Furthermore, we define

$$d_E X_j(harmo) = (d_E x_{hmean,j}, d_E x_{hmax,j}, d_E x_{metric,j}, d_E x_{melod,j})$$
$$d_E X_j(metric) = (d_E x_{metric,j}, d_E x_{hmean,j}, d_E x_{hmax,j}, d_E x_{melod,j})$$
$$d_E X_j(melod) = (d_E x_{melod,j}, d_E x_{hmean,j}, d_E x_{hmax,j}, d_E x_{metric,j})$$

$$d_E^2 X_j(harmo) = (d_E^2 x_{hmean,j}, d_E^2 x_{hmax,j}, d_E^2 x_{metric,j}, d_E^2 x_{melod,j})$$
$$d_E^2 X_j(metric) = (d_E^2 x_{metric,j}, d_E^2 x_{hmean,j}, d_E^2 x_{hmax,j}, d_E^2 x_{melod,j})$$
$$d_E^2 X_j(melod) = (d_E^2 x_{melod,j}, d_E^2 x_{hmean,j}, d_E^2 x_{hmax,j}, d_E^2 x_{metric,j})$$

and the $n \times 10$ matrix

$$X_{add} = (X_{rit}, X_{sus}, X_{ferm})$$

where $X_{rit} = (x_{rit_1}, x_{rit_2}, x_{rit_3}, x_{rit_4})$, $X_{sus} = (x_{sus_1}, x_{sus_2}, x_{sus_3}, x_{sus_4})$ and $X_{ferm} = (x_{ferm_1}, x_{ferm_2})$. Finally define the $n \times p$ matrices (with $p = 58$)

$$X(harmo) = (X_1(harmo), X_2(harmo), X_3(harmo), X_4(harmo), d_E X_1(harmo), d_E X_2(harmo), d_E X_3(harmo), d_E X_4(harmo), d_E^2 X_1(harmo), d_E^2 X_2(harmo), d_E^2 X_3(harmo), d_E^2 X_4(harmo), X_{add}),$$

$$X(metric) = (X_1(metric), X_2(metric), X_3(metric), X_4(metric), d_E X_1(metric), d_E X_2(metric), d_E X_3(metric), d_E X_4(metric), d_E^2 X_1(metric), d_E^2 X_2(metric), d_E^2 X_3(metric), d_E^2 X_4(metric), X_{add}),$$

$$X(melod) = (X_1(melod), X_2(melod), X_3(melod), X_4(melod), d_E X_1(melod), d_E X_2(melod), d_E X_3(melod), d_E X_4(melod), d_E^2 X_1(melod), d_E^2 X_2(melod), d_E^2 X_3(melod), d_E^2 X_4(melod), X_{add}).$$

5. **Orthonormalization.** Each of the design matrices $X(metric), X(harmo)$, and $X(melod)$ turned out to be singular, since the last column can be expressed as a linear combination of the previous ones. Hence, we omit the last column. For simplicity of notation, the new $n \times 57$ matrices will also be denoted by $X(metric), X(harmo), X(melod)$.
Figure 44.1 shows that the corresponding columns of the three matrices are closely related, at least for $j = 1, 2, 3$. Intuitively this means that *it is not possible to distinguish exactly whether certain characteristics of the tempo curve stem from the metric, the harmonic or the melodic analysis.* Thus, results may depend on the sequence of orthogonalization. This sequence reflects whether, in our view, the harmonic, the metric or the melodic has priority. Moreover, instead of focusing on names such as "metric weight", "first derivative of the metric weight", etc., we will also try to extract typical weight curves (canonical curves) that appear to be important for the tempo, independently of which of the three analytic approaches (metric, harmonic, melodic) have priority.
More specifically, the three design matrices are defined in the following way: The columns of $X(harmo)$, $X(metric)$, and $X(melod)$ respectively are orthogonalized and standardized successively. We thus obtain three $n \times 57$ matrices which will be denoted by

$$Z(harmo), Z(metric), \text{ and } Z(harmo).$$

Each of these matrices has orthonormal columns. The reason for computing three different matrices is that orthonormalization depends on the initial sequence of the columns. An artificial preference of the variables that are accidentally in the first (or first few) column(s) is avoided by carrying out three separate regression analyses with the respective matrices $Z(harmo), Z(metric)$, and $Z(melod)$, and by comparing the common features of the three results.

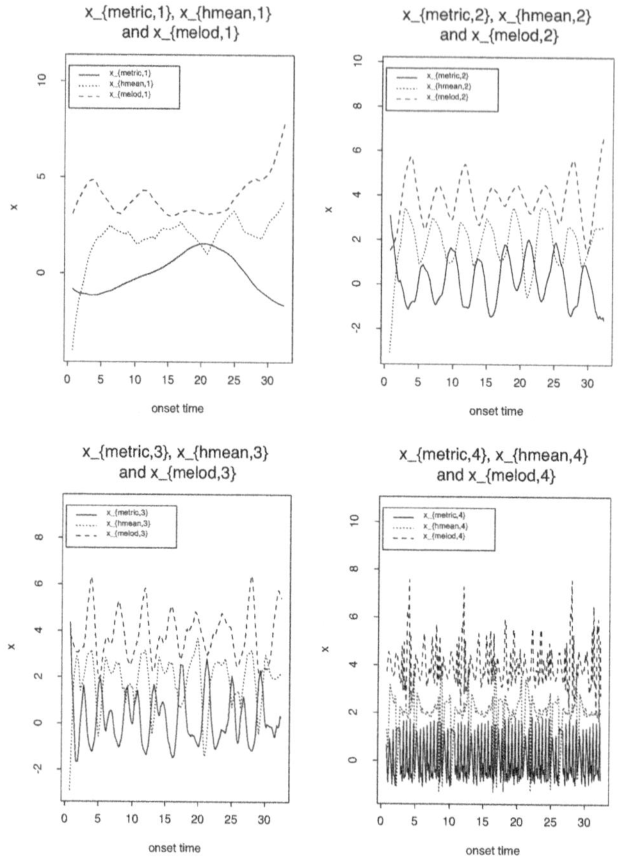

Fig. 44.1. The four hierarchical levels of metrical, harmonic, and melodic analyses.

44.2 The Method of Regression Analysis

Summary. Inspired by visual comparison of logarithmic tempo cures, where strong similarities between the different performances become visible, we applied the following regression model.

$-\Sigma-$

44.2.1 The Full Model

Let Z be one of the three matrices $Z(harmo)$, $Z(metric)$, or $Z(melod)$, respectively. The full (i.e., biggest possible) model for the jth *individual tempo curve* is

$$y(t_i, j) = Z_i \beta(j) + \epsilon(t_i, j), \ (t_i \in T),$$

where Z_i is the ith row vector of Z, $\beta(j) = (\beta_1(j), \ldots \beta_{57}(j))^t$ and $\epsilon(t_i, j)$ $(t_i \in T)$ are (for each fixed j) identically distributed zero mean random variables. This means that we assume each performance to be essentially characterized by a 57-dimensional parameter vector $\beta(j)$. Under the present orthonormalized conditions, this vector corresponds to the shaping vector ω introduced in Section 44.1.1. Note that we do not assume the residuals ϵ_i to be independent, since corrected p-values are used that take into account serial dependence. Also, due to standardization of y and of the columns of Z, there is no intercept in the model.

The vector $\beta(j)$ is the parameter vector corresponding to the performance number j. Therefore, $\beta(j)$ is assumed to be a random vector, sampled from the space of all "possible" interpretations, with expected value $E[\beta(j)] = \beta$. We then may write

$$\beta(j) = \beta + \eta(j)$$

where $\eta(j)$ is a random vector with $E[\eta(j)] = 0$ and

$$y(t_i, j) = Z_i\beta + Z_i\eta(j) + \epsilon(t_i, j).$$

Intuitively this means that, up to a small unexplained deviation $\epsilon(t_i, j)$, the (logarithmic) tempo of the jth performance at onset time t_i can be expressed as a "mean performance" $Z_i\beta$ plus an individual deviation from the mean that is equal to $Z_i\eta(j)$. For the mean tempo curve

$$\bar{y}(t_i) = m^{-1}\sum_{j=1}^{m} y(t_i, j),$$

we then have

$$\bar{y}(t_i) = Z_i\beta + \tilde{\epsilon}_i(t_i)$$

where $\tilde{\epsilon}(t_i)$ $(t_i \in T)$ are identically distributed zero mean random variables.

44.2.2 Step Forward Selection

In the following, the main focus is on the individual curves. Some comments on the mean curve are also given.

In order to decide which components of β or $\beta(j)$ respectively are not zero (i.e., which explanatory variables contribute "significantly" to the tempo curve), stepwise forward selection [191] is carried out with F-to-enter level of significance 0.01. For the individual curves, a separate stepwise regression is carried out for each individual. The statistics software S-Plus [923] was used for the calculations.

44.3 The Results of Regression Analysis

Summary. In the following discussion the main questions are: 1) Is there a relevant association between the analytical weights computed from the score and the observed tempo curves? 2) How complex is the relationship? 3) Are there commonalities and diversities; how can they be characterized?

– Σ –

44.3.1 Relations Between Tempo and Analysis

Summary. The statistical analysis of the relation of tempo data and analytical weights via the Beran operator is carried out.

– Σ –

To begin with, it is interesting to learn how much can be 'explained' at most by the analytical weights. Recall that in regression, R^2 denotes the proportion of the variability of y that is explained by the estimated regression function. Ideally, R^2 would be equal to 1.00 which would mean that the (log-)tempo curve can be expressed exactly as a linear function of the analytic information encoded by the design matrix Z. Such an exact correspondence between the analytic curves and each tempo curve can hardly be expected. The maximal achievable values of R^2, which are obtained by using the full matrix Z (i.e., without eliminating nonsignificant variables), are quite high however for the mean curve as well as the individual curves. For the mean curve, R^2 is equal to 0.84. For the individual curves, we have $0.65 \leqslant R^2 \leqslant 0.85$.

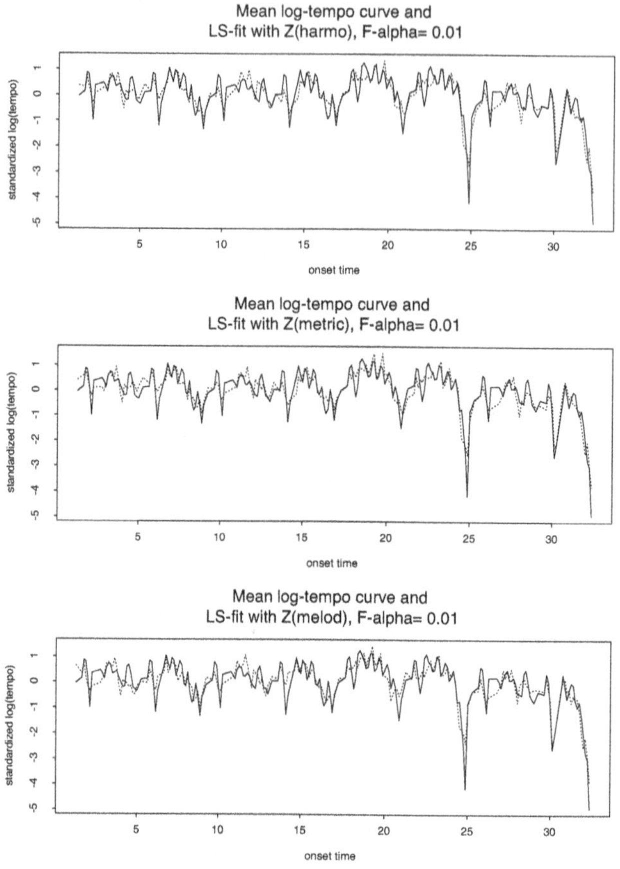

Fig. 44.2. Mean tempo curve and LS-fit with orthonormalized Z matrices with harmonic, metric, and melodic emphasis.

It is well known that R^2 can be increased by simply including a sufficiently large number of explanatory variables, even if these variables have nothing to do with y. It is therefore necessary to investigate which explanatory variables contribute significantly to the response y. To do this, we first applied stepwise forward selection with F-to-enter $\alpha = 0.01$. In all cases, all coefficients in the resulting model turned out to be significantly different from zero at the 5% level of significance, even after taking into account the possibility of serial correlations in the residuals.

The values of R^2 for the mean curve and the individual curves respectively are still remarkably high: For the mean curve, R^2 is equal to 0.79 for $Z(harmo)$, 0.79 for $Z(metric)$ and 0.77 for $Z(melod)$. Figure 44.2 shows that the fit (dotted line) to the mean tempo curve (full line) is very good, even if only significant coefficients are used. For the individual curves, we have $0.46 \leqslant R^2 \leqslant 0.79$ for $Z(harmo)$, $0.48 \leqslant R^2 \leqslant 0.78$ for $Z(metric)$ and $0.36 \leqslant R^2 \leqslant 0.77$ for $Z(melod)$. The low value of 0.36 is obtained for Kubalek. Excluding this performance, we have 0.51 as the lower bound $Z(melod)$. Thus, the quality of the fit is also good in general, but varies individually. This is illustrated further by figure 44.3 to 44.3 for $Z(melod)$. The results on commonalities and diversities given below in Section 44.3.3 yield further evidence for the existence of a meaningful association between y and Z.

44.3.2 Complex Relationships

Summary. We discuss the complexity of the relationships between the individual results.

– Σ –

Even when using only significant coefficients, the estimated models are very complex. As an example, consider the performance by Brendel. With $Z(melod)$, the R^2 is in this case equal to 0.76. No. 4.4 in Figure 44.3-44.6 confirms the good fit.

The following table A summarizes the result:

Table A: Coefficients of explanatory variables chosen by stepwise forward selection with F-to-enter=0.01 and $Z(melod)$, for the logarithmic tempo curve of Brendel. (The P-values given here do not take into account serial correlations.)

$z-$variable	est. coefficient	std. error	t-statistic	P-value
$z_{melod,1}$	-0.3136	0.0353	-8.8760	0.0000
$z_{hmean,1}$	-0.2737	0.0353	-7.7443	0.0000
$z_{hmean,2}$	0.2781	0.0353	7.8713	0.0000
$z_{hmax,2}$	0.2659	0.0353	7.5248	0.0000
$z_{melod,3}$	-0.1258	0.0353	-3.5597	0.0005
$z_{hmean,3}$	0.1303	0.0353	3.6882	0.0003
$z_{hmax,3}$	-0.2800	0.0353	-7.9244	0.0000
$z_{melod,4}$	-0.1562	0.0353	-4.4193	0.0000
$d_E z_{melod,1}$	-0.2663	0.0353	-7.5371	0.0000
$d_E z_{melod,2}$	0.1567	0.0353	4.4337	0.0000
$d_E z_{hmean,2}$	0.1143	0.0353	3.2356	0.0014
$d_E z_{melod,3}$	0.2938	0.0353	8.3143	0.0000
$d_E z_{hmean,3}$	0.1308	0.0353	3.7024	0.0003
$d_E^2 z_{metric,1}$	-0.1361	0.0353	-3.8508	0.0002
z_{rit_3}	-0.1952	0.0353	-5.5249	0.0000
z_{rit_4}	0.1032	0.0353	2.9208	0.0039
z_{ferm_1}	-0.1425	0.0353	-4.0331	0.0001

The number of significant coefficients is very large. The model contains all four weight functions, first and second derivatives of various degrees of smoothness and also two ritardando-variables and one suspension. Also note that all degrees of smoothness are used. Formally, even after adjustment for serial correlations, all p-values are below 0.05. (As a cautionary remark, it should be noted however that p-values obtained after model selection can be used as guidelines only.)

44.3.3 Commonalities and Diversities

Summary. In spite of the high complexity of the selected models for 28 individual tempo curves, there are interesting commonalities and diversities. They are characterized in this section.

– Σ –

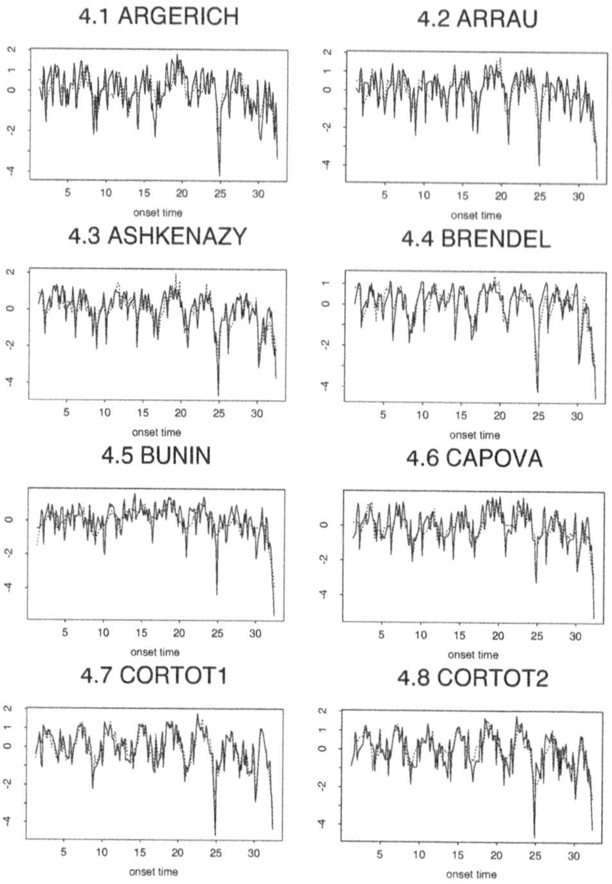

Fig. 44.3. Fit for $Z(melod)$.

Before going into the details of commonalities and diversities among the 28 given performances, we should make a remark on the performance selection and the performers as made available by Repp [886]. Above all, we should emphasize that Repp succeeds in a choice of first quality pianists, among others the celebrated "romantic virtuoso" Vladimir Horowitz, the "analytical mannerist" Alfred Brendel, or the "perfect but utterly cool" omnipresent Vladimir Ashkenazy, to name just three of them[3]. So from the point of view of performance culture, the election is unprecedented and representative. But in this scientific context, we must refrain from further judgments. This discussion is not about journalistic criticism. However, we should encourage critics to review their understanding of performance by focusing on the question whether and to what degree the analytical structure of a score may be responsible for agogical expressivity. Our present answer to this question—incomplete as it must remain—may seem to position some of the artists in unexpected relative position to each other. But this is not surprising since we do not claim that the overall judgment from common criticism really does represent the strictly analytical perspective of our approach. As mentioned in Chapter 36, one should absolutely add emotional and gestural components to reach a complete description, an objective which was out of reach of this discussion.

In spite of the high complexity of the selected models for individual tempo curves, there turn out to be interesting commonalities and diversities that can be characterized by either of the matrices $Z(harmo)$, $Z(metric)$ and $Z(melod)$ respectively. Recall that using $Z(harmo)$ corresponds to an understanding of the

[3] For a complete list of all performers, see [886]. We shall only name selected performers who are relevant to this analysis.

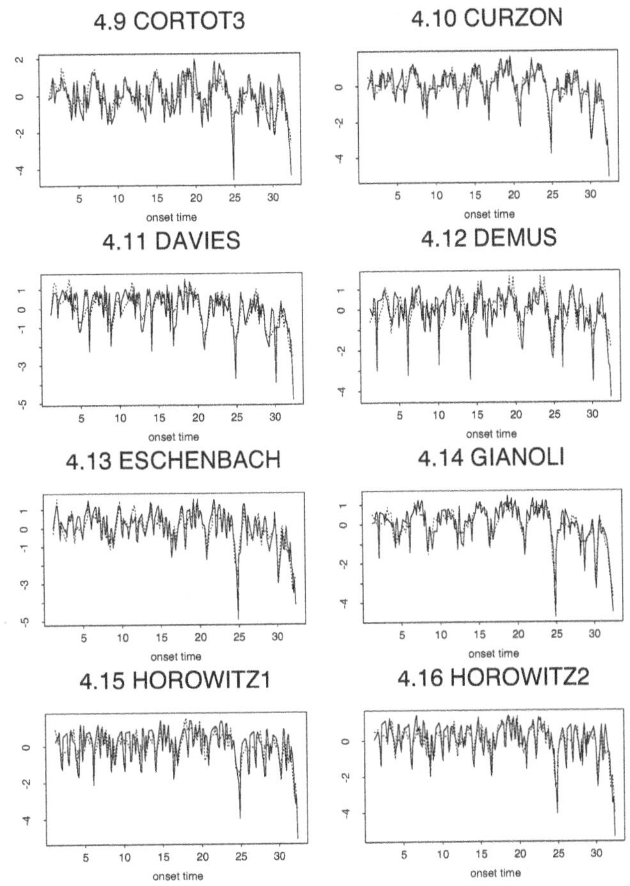

Fig. 44.4. (Cont.)

score that put a first priority on the harmonic structure. Using $Z(metric)$ corresponds to putting first priority on the metric structure. Using $Z(melod)$ corresponds to putting first priority on the melodic structure. Therefore, depending on which of the three matrices is used, somewhat different results should be expected. The fundamental problem is the ambiguity of a performance. In general, based on one performance, it cannot be decided with certainty whether certain features of the tempo are 'due to' the harmonic, the metric or the melodic content. Nevertheless, the results below show a strong similarity between the three regressive analyses. Thus, there appears to exist at least a core of tempo features that are unambiguously attributable to specific weight functions.

A number of different aspects of commonality and diversity can be considered. Here, three possible aspects are described.

44.3.3.1 Signs of Coefficients

We ask the following question: For which $k = 1, \ldots p$ do we have either $\hat{\beta}_k(j) \geqslant 0$ for all $1 \leqslant j \leqslant m$, or $\hat{\beta}_k(j) \leqslant 0$ for all $1 \leqslant j \leqslant m$? In other words, which coefficients have the same sign for all performances? *The result is quite amazing: For $Z(metric)$, $Z(harmo)$, and $Z(melod)$, all except 3, 2, and 1 coefficients (out of 57) respectively have the same sign for all performances.* Thus, the sign of the coefficients is a very strong commonality. The analytic curves 'act' in the same direction. In particular, the following general tendency can be observed:

- The tempo decreases as the original (not orthogonalized) harmonic weight increases.

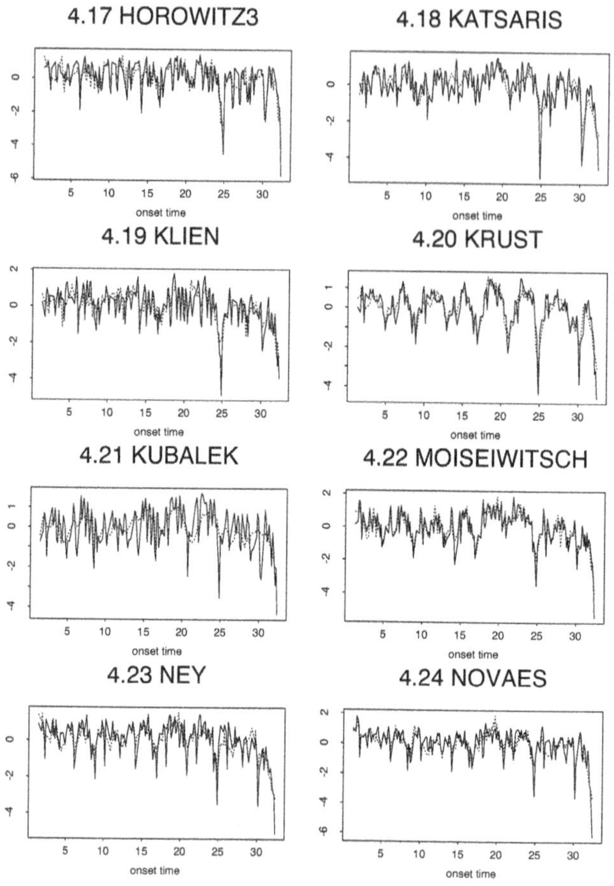

Fig. 44.5. (Cont.)

- The tempo increases as the original (not orthogonalized) metric weight increases.
- The tempo decreases as the original (not orthogonalized) melodic weight increases.

It should be noted however that these conclusions are valid under the assumption that all other variables are kept fixed. As we saw above, for the score considered here, the original weights are strongly correlated. This makes the actual relationship between weights and tempo much more complicated.

44.3.3.2 Frequency of Variable Inclusion

For $k = 1, ..., p$, let $n_k = \sum_{j=1}^{m} 1\{\hat{\beta}_k(j) \neq 0\}$ be the number of performances for which the explanatory variable number k was included in the model.

Figures 44.7-44.9 show, for $Z(harmo)$, $Z(metric)$, and $Z(melod)$ respectively, the curves of variables that were chosen at least 24 times (out of 28). The curves are multiplied by the sign of the coefficient. At least two types of curves are common to practically all performances, independently of the matrix that is used: 1) very smooth 'global' curves, such as $z_{melod,1}$, that shape the overall tendency of the tempo; 2) almost periodic curves, with a period of about four measures, corresponding to the approximate periodicity of the harmonic curve $z_{hmean,2}$.

Comment on $Z(harmo)$. Note that $z_{hmean,1}$ is identical with $x_{hmean,1}$, see Figure 44.1a. Moreover, $z_{hmean,2}$ is almost the same as $x_{hmean,2}$ in Figure 44.1b. Also, $d_E z_{hmean,3}$ exhibits features that are very similar to $x_{hmean,3}$, see Figure 44.1c. Thus, analytical weights obtained by local averaging without orthogo-

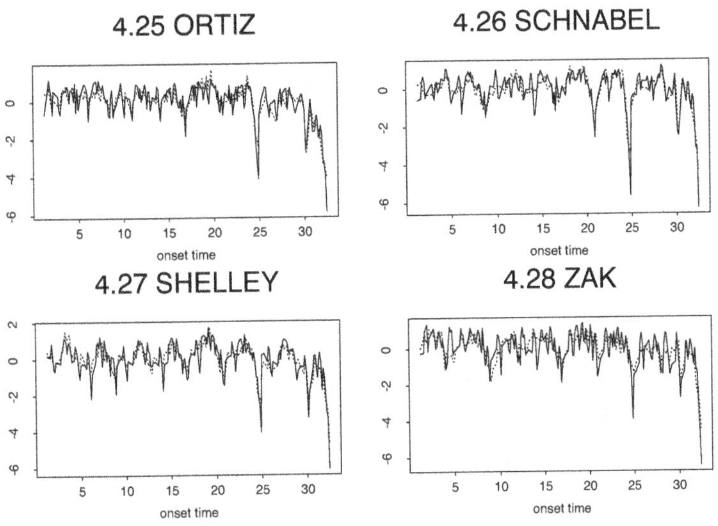

Fig. 44.6. (Cont.)

nalization have a direct impact on the performance. In fact, by the above, the orthonormal curves selected as most relevant by the regression turn out to be closely related to the original curves.

Comment on $Z(metric)$. Here, resemblance of curves is similar to $Z(harmo)$. Namely: $z_{hmean,1}$ is almost the same as $x_{hmean,1}$, $z_{metric,1}$ and $x_{metric,1}$ are identical, and $z_{hmean,2}$ is very similar to $x_{hmean,2}$. Also $z_{melod,1}$ is almost the same as $z_{melod,1}$ in $Z(harmo)$, compare Figures 44.7 and 44.8. Again, we conclude that the original averaged weight curves influence the performance directly.

Comment on $Z(melod)$. Similar comments as for $Z(harmo)$ and $Z(metric)$ apply for $Z(melod)$. Here, $z_{melod,1}$ is the mirror image of $x_{melod,1}$. Further, $z_{hmean,1}$ and $z_{hmean,2}$ are very similar to $x_{hmean,1}$ and $x_{hmean,2}$, respectively.

Comment on non-linear deformations. The melodic weights seem to play the most prominent role. Independently of the emphasis, $z_{melod,1}$ is chosen for all 28 performances. For the melodic emphasis, $z_{melod,1}$ is obviously identical with $x_{melod,1}$. For the harmonic and metrical emphasis, the corresponding $z_{melod,1}$-curves turn out to be non-linear deformations of $x_{melod,1}$. The method of non-linear deformations of analytical weights as arguments of refined shaping of performance is also implemented in the PerformanceRUBETTE® of RUBATO®.

In summary, we conclude that for the score of "Träumerei", *there is a small number of "canonical" analytical weight curves that are relevant for most performances and essentially do not depend on the analytical emphasis.*

44.3.3.3 Largest Coefficients

Since the design matrix Z is orthonormal, the importance of the kth explanatory variable may be assessed by the absolute value of the corresponding k^{th} estimated coefficient (ranked in comparison to the other coefficients). (Also note that, due to orthonormality, all estimated slope components are uncorrelated and their standard deviations are the same.) For fixed j, let $r_k(j)$ be the rank of $|\hat{\beta}_k(j)|$ among all coefficients $\hat{\beta}_s(j)$ $(s = 1, ..., p)$. Furthermore, for $1 \leqslant l \leqslant p$, let $f_k(l) = \sum_{j=1}^{m} 1\{r_k(j) > p - l\}$. Thus, $f_k(l)$ is the number of performances for which $|\hat{\beta}_k(j)|$ is at least the lth largest.

Consider first $l = 1$. Thus, $f_k(1)$ is the number of performances for which the kth variable is most important. It turns out that $f_k(1)$ is not zero for a very small set of variables. The results for the three matrices are:

$Z(harmo)$: Figure 44.10 displays the four curves for which $f_k(1) \neq 0$, i.e., variables that are most important for at least one performance. Table B shows the clusters of columns H1 to H4 of performances for which the corresponding variables $z_{hmean,2}$ ($\sim$ H1), $z_{hmean,1}$, $z_{metric,1}$, and $z_{melod,4}$ are the most important ones.

Table B: Overview of clusters as derived by the above criterion with $l = 1$.

Artist	H1	H2	H3	H4	M1	M2	M3	M4	M5	M6	ML1	ML2	ML3
ARG	●							●			●		
ARR	●							●			●		
ASH	●				●								●
BRE		●			●						●		
BUN			●				●				●		
CAP	●						●				●		
CO1	●					●						●	
CO2	●					●						●	
CO3	●					●						●	
CUR	●					●					●		
DAV	●							●			●		
DEM	●					●					●		
ESC		●			●						●		
GIA			●				●				●		
HO1				●						●	●		
HO2		●			●						●		
HO3		●			●						●		
KAT	●					●					●		
KLI		●			●						●		
KRU	●							●				●	
KUB	●						●				●		
MOI		●			●						●		
NEY		●			●						●		
NOV		●			●						●		
ORT		●			●						●		
SCH	●								●		●		
SHE	●					●					●		
ZAK		●			●						●		

Thus, using $Z(harmo)$ and therefore an approach that gives priority to the harmonic structure, the first cluster of performances has the 4-measures periodicity of the harmonic structure as the dominating feature. In particular, all Cortot performances are included. For the second cluster that includes in particular

Horowitz1 and Horowitz2, the more global shaping curve $z_{hmean,1}$ is most prominent. For Bunin and Gianoli, a global curve with a peak around the 15th measure is most important. Finally, for the first performance by Horowitz1, the very detailed local structure of the melodic curve $z_{melod,4}$ seems to dominate.

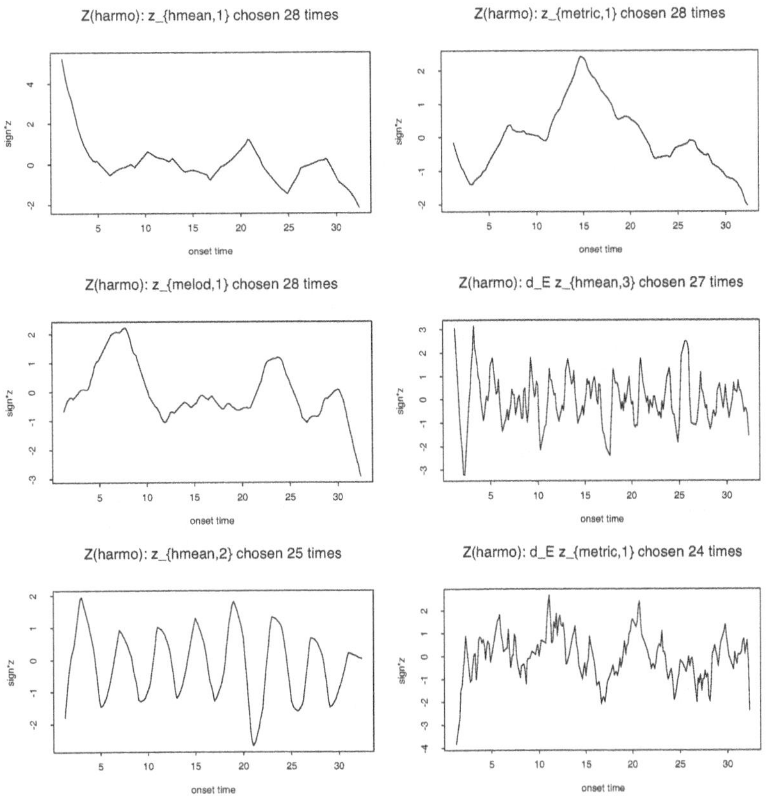

Fig. 44.7. Frequently selected curves for $Z(harmo)$.

$Z(metric)$: The clusters of columns M1 to M6 in Table B show the results for $Z(metric)$ (see Figure 44.11). These six columns correspond to the variables $z_{hmean,1}$, $z_{metric,2}$, $z_{metric,1}$, $z_{hmean,2}$, $z_{hmax,3}$, and $d_E z_{metric,3}$.

Hence, by use of an approach that gives priority to the metric structure, the cluster with Horowitz2 and 3 is exactly the same as for $Z(harmo)$, with the exception of Ashkenazy who was not included before. Also, it corresponds essentially to the same curve. Similarly, the cluster with the three Cortot performances corresponds to a very similar curve. It is however smaller since it is a proper subset of the previous 'Cortot' cluster above. Bunin and Gianoli are again in the same separate cluster, this time together with Capova and Kubalek. The peak of the curve is now around measure 20. The cluster with Argerich contains several of those performances that were previously in the Cortot cluster. The dominating curve is still almost periodic with a period of about four measures. Finally note that Horowitz1 builds again a separate cluster with a locally very refined metric curve corresponding to the derivative of $z_{metric,3}$.

$Z(melod)$: For $Z(melod)$ (Figure 44.12), we obtain the clusters shown in columns ML1 to ML3 in Table B. The corresponding variables are: $z_{melod,1}$, $z_{hmean,2}$, and $z_{melod,2}$.

The melodic approach yields very simple clusters. For almost all performances, including all Horowitz performances, the global shape of $z_{melod,1}$ is the most important feature. For Cortot1 through 3, and Krust, the 4-measures periodicity of $z_{hmean,2}$ is most important. For Ashkenazy, the 4-measures (almost) periodic curve $z_{melod,2}$ is dominating. Note in particular that the minima and maxima of this curve do not occur at

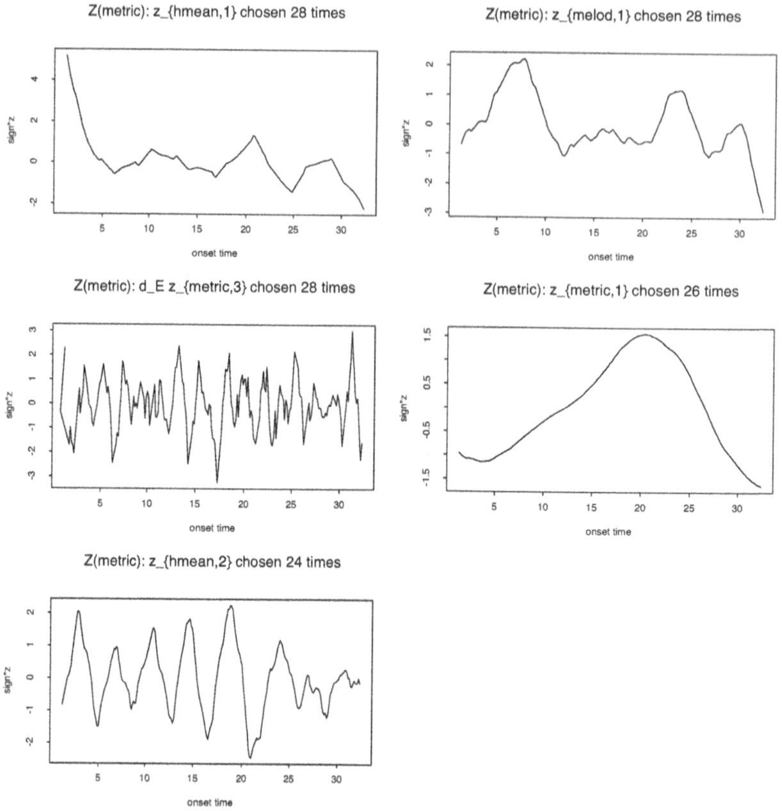

Fig. 44.8. Frequently selected curves for $Z(metric)$.

the same place as for $z_{hmean,2}$. In comparison, $z_{melod,2}$ appears to be shifted to the right. Also, $z_{melod,2}$ has an extreme local minimum around the beginning of measure 30.

44.3.3.4 Argerich "Versus" Horowitz

The remarkable first performance Horowitz1 from 1947 evidences a preference of very detailed local information, be it from the melodic or metrical analysis—in typical contrast to the highly coherent Argerich performance. This observation is confirmed by an investigation of the correlation coefficients in the algebro-geometric analysis of the performance genealogy in the sense of RUBATO®'s stemma theory, see Chapter 38. When translated into common language these quantitative results are in perfect coincidence with the judgments of experts on Argerich's and Horowitz' specific differences in performance [88]. Let us therefore make these findings more meaningful to the common understanding. How would an interpreter such as Horowitz experience his performance? He would look up a few notes ahead and remember just a few of the past note events when hitting a couple of keys in a given moment. He would then realize a couple of neighboring analytical facts, such as a harmonic step or a melodic contour, in his imagination, and then shape the present note event in its tempo, dynamics, and articulation within this minor context, and thusly express his analytical consciousness. Metaphorically speaking this resembles a near-sighted man who can only see and recognize nearby objects. It is as if he had no significant memory of what was happening several bars ago, or of what will happen in the larger time span ahead. In contrast, a performer of the Argerich type would be aware of lots of long-range facts in the overall analytical stream of the piece which is being played. She would then remember and plan everything and therefore hit the present notes in full consciousness of what was and what will be. This is what the semantics of the data tells.

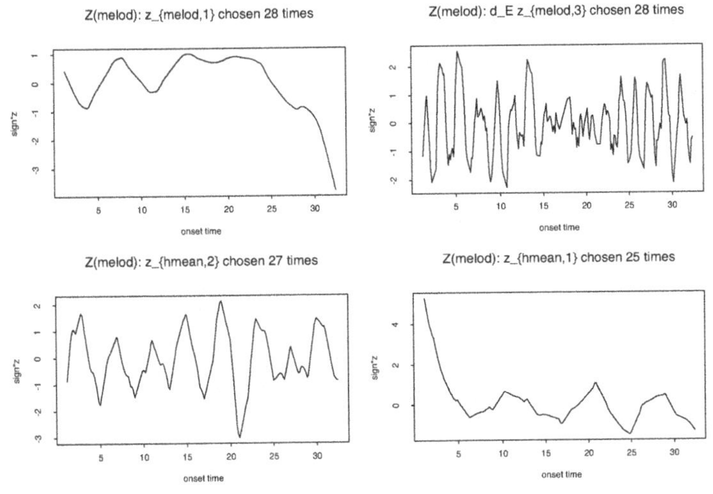

Fig. 44.9. Frequently selected curves for $Z(melod)$.

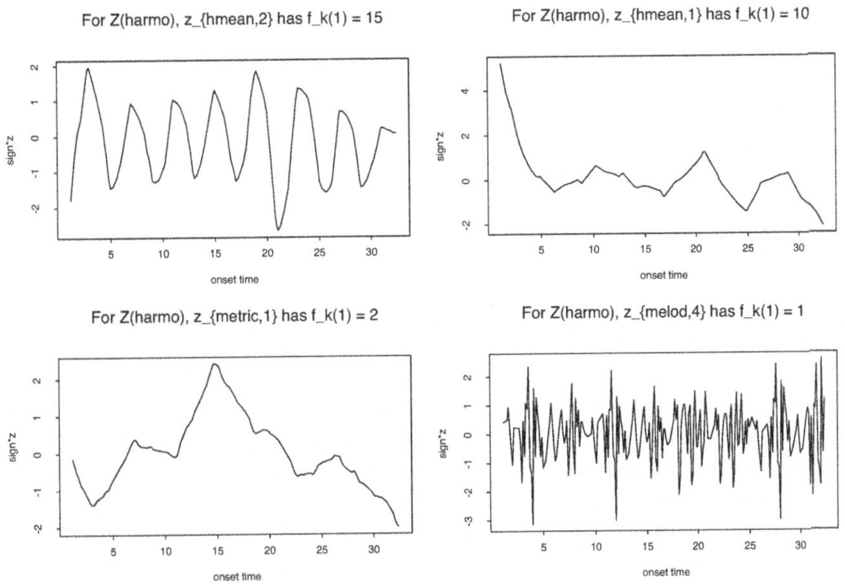

Fig. 44.10. Important variables for $Z(harmo)$.

Similar results can be obtained for $l > 1$. Here, $f_k(l)$ is the number of performances for which the kth variable is among the k most important ones. Consider, for instance, $l = 3$. The following partially overlapping clusters corresponding to variables with $f_k(3) \neq 0$ are obtained (see Table C). The columns correspond to the following variables: M1 $\sim z_{melod,1}$, H2 $\sim z_{hmean,2}$, H1 $\sim z_{hmean,1}$, DM3 $\sim d_E z_{melod,3}$, M2 $\sim z_{melod,2}$, DM2 $\sim d_E z_{melod,2}$, HM2 $\sim z_{hmax,2}$, HM3 $\sim z_{hmax,3}$, M4 $\sim z_{melod,4}$, DM1 $\sim d_E z_{melod,1}$, DDM1 $\sim d_E^2 z_{melod,1}$, and SS2 $\sim z_{sus2}$.

The first cluster consists of all performances except Ashkenazy. Thus, using the 'melodic approach', apart from Ashkenazy, the global melodic curve $z_{melod,1}$ is one of the three most important factors for all tempo curves. Again, the 4-measure periodicity determines clusters with Cortot performances. It is remarkable that, in spite of the large number of overlapping clusters, there is no cluster—except the first

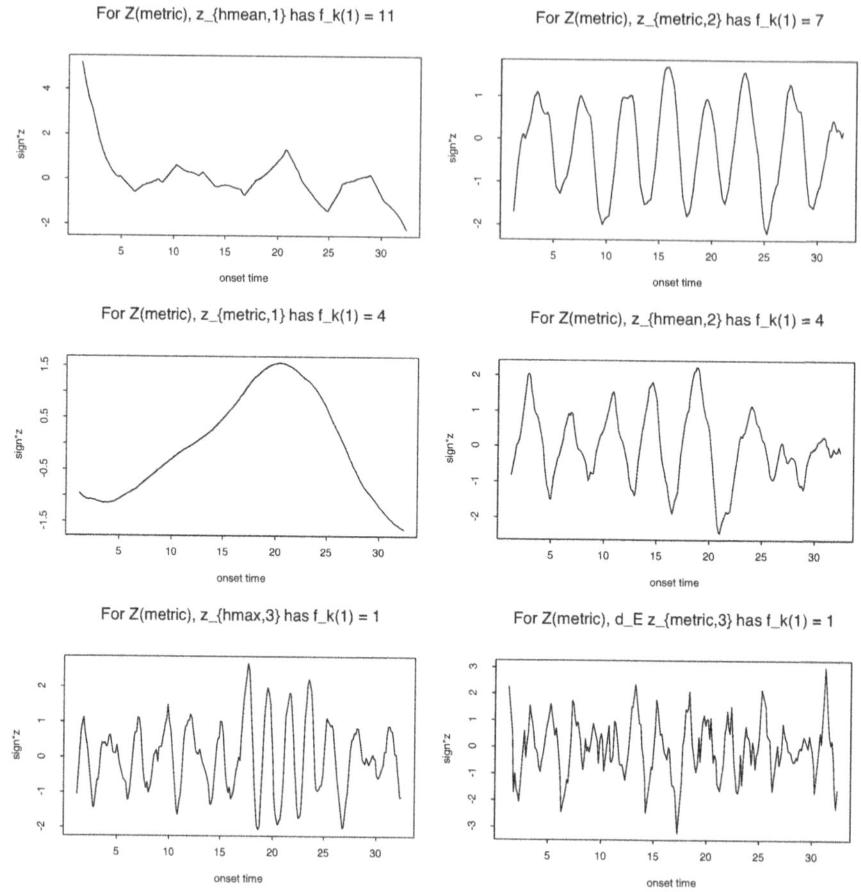

Fig. 44.11. Important variables for $Z(metric)$.

one—that contains at the same time Cortot and Horowitz. Moreover, there is one cluster consisting solely of Horowitz1 through 3 corresponding to the complex local melodic structure of $z_{melod,4}$.

Evidently, many more detailed comments about Figures 44.10-44.12 could be added. We conclude the analysis by noting that the relative size of the coefficients suggests a natural way of obtaining simplified tempo curves that contain the most important features. For given j and $1 \leqslant q \leqslant p$, let the $p \times 1$ vector $\gamma_q(j) = [\gamma_{q,1}(j), \gamma_{q,2}(j), , ..., \gamma_{q,p}(j)]^t$ be defined by $\gamma_{q,k}(j) = \hat{\beta}_k(j)1\{r_k(j) > p-q\}$. Then $y_q(t_i, j) = Z\gamma_q(j)$ is a simplified tempo curve that corresponds to using the variables (analytic curves) that are among the q most important ones for tempo curve j, importance being measured by $r_k(j)$. Thus, the resulting tempo curve is a simplified curve obtained superposition of the q most important features only. Note that, for $q = p$, this yields the complete curve fitted by stepwise regression. Nos. 44.13.1-44.16.28 in Figures 44.13-44.16 display y_q $(q = 1, ..., p)$ for $Z(melod)$ for all performances.

44.3.4 Overview of Statistical Results

The main statistical conclusions from the analysis above can be summarized as follows:

- There is a clear association between metric, melodic and harmonic weights and the tempo.
- The exact relationship between the analytic weights Z and an individual tempo curve is very complex. However, a large part of the complexity can be covered by our model.

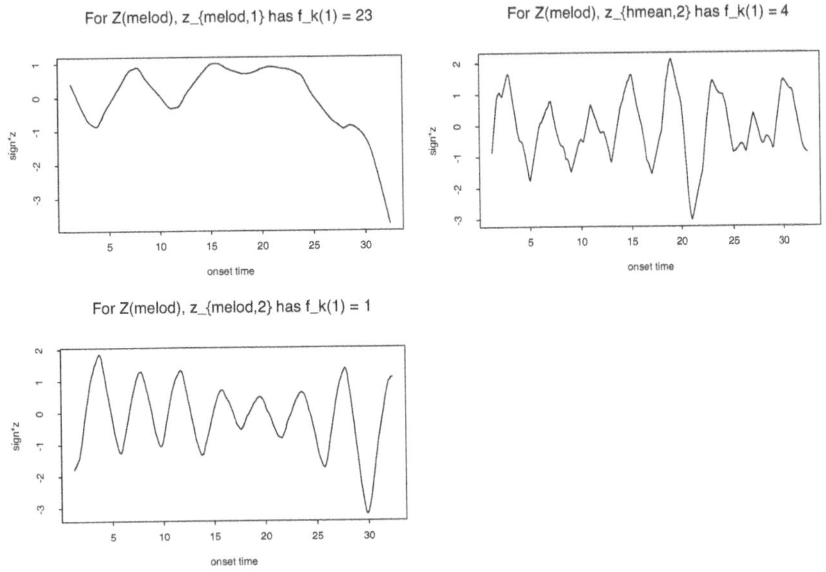

Fig. 44.12. Important variables for $Z(melod)$.

- Commonalities and diversities among tempo curves may be characterized by a relatively small number of curves. There is in principle no unique way of attributing features of the tempo to exactly one cause (harmonic, metric or melodic analysis). Which curves need to be used depends partially on which of the three analyses (harmonic, metric, melodic) has 'priority'. However, there seems to be a small number of canonical curves that are essentially independent of the priorities and which determine a large part of the commonality and diversity among tempo curves. Natural clusters can be defined.
- There is a natural way of reducing an individual tempo curve to a series of simplified tempo curves containing an increasing number of features.

Overall, the proposed method yields a variety of results that are interpretable from the point of view of music and performance theory. In particular, the hierarchic approach of decomposing each of the weight functions into components of different degrees of smoothness seems to be appropriate. The different choices of the bandwidth h correspond to a hierarchic approach to musical performance, starting with the most global features of the score and refining the performance successively in greater detail.

The results here are closely related to Repp's work [886]. Repp applied principal component analysis to the 28 tempo curves. One of his main results is that Cortot and Horowitz appear to represent two extreme types of performances. Thus, in a heuristic way, Repp suggested classifying the performances according to their factor loadings into a Cortot and a Horowitz cluster respectively. Our regression analysis confirms the basic findings. Due to the use of weights obtained from a musical analysis of the score, we obtain further information about the nature of the commonalities, diversities and clusters. For instance, as discussed above, for the Horowitz cluster, the most important feature appears to be the overall descending line of $z_{1,melod}$ and very local variations of the tempo that correspond to the local variations of the analytic weights. On the other hand, for the Cortot cluster, the up and down movement of $z_{3,hmean}$ with a period of about four measures is the most important characteristic.

Result 6 *More generally, this approach reveals a set of canonical curves whose combination yields the most important features of a tempo curve. It should be emphasized that* these curves are score-specific. *Thus, for each score (in our case "Träumerei"), a new set of essential curves is obtained. The original weights as well as the decomposition into parts of different smoothness are based on the specific score that is performed. This is a crucial feature that is in sharp contrast to traditional mathematical 'omnibus-decompositions' such*

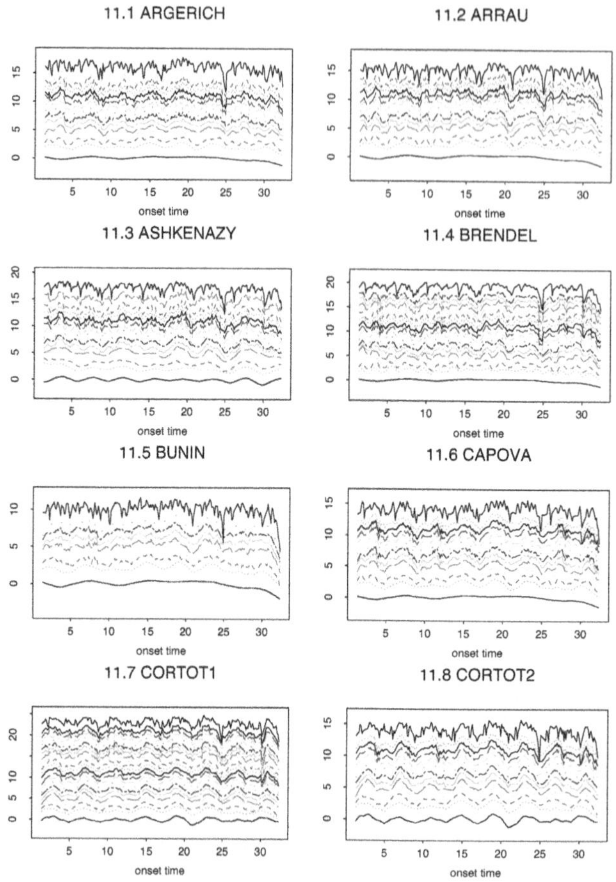

Fig. 44.13. (44.13-44.16) Superposition of the q most important features.

as provided by Fourier or wavelet analysis. The score-specific choice of the $Z-$matrix enables us to relate statistical results directly to the musical/analytic content of the score.

The main point here is this: We argue that to understand the character of tempo, it is above all important to refer it to a "basis" of score specific analytical curves and not to curves—such as sinoidal curves in Fourier representation—which have a generic type that tells nothing about the particular context of the genealogy for the given tempo curve.

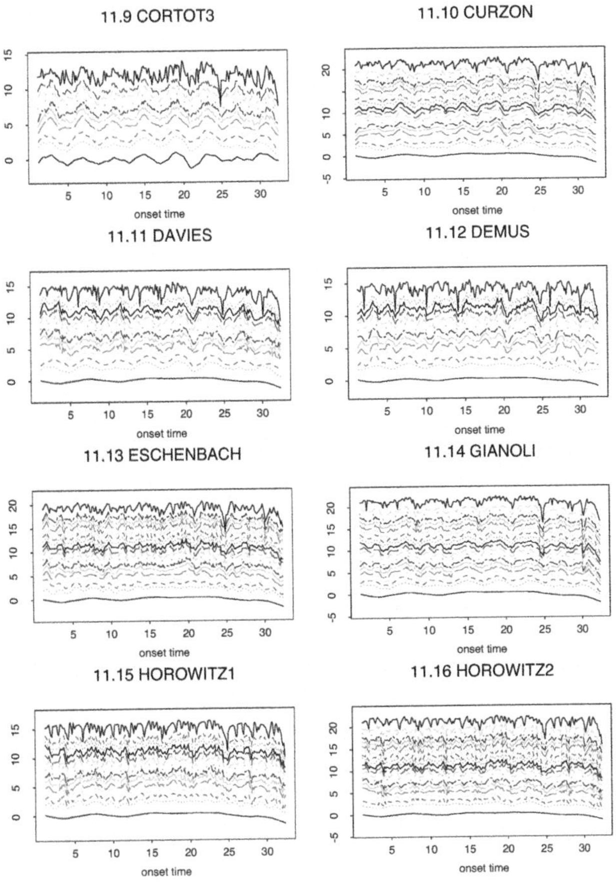

Fig. 44.14. (Cont.)

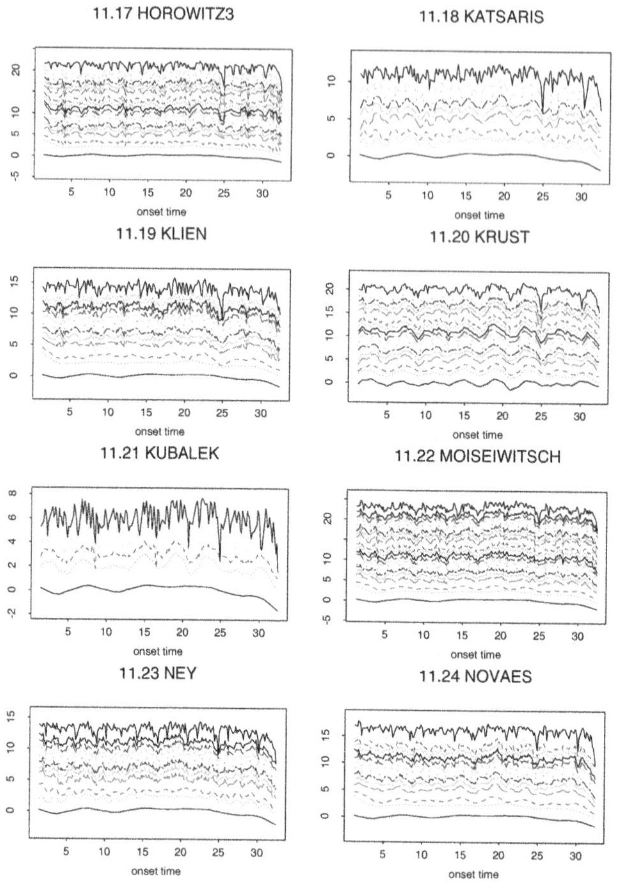

Fig. 44.15. (Cont.)

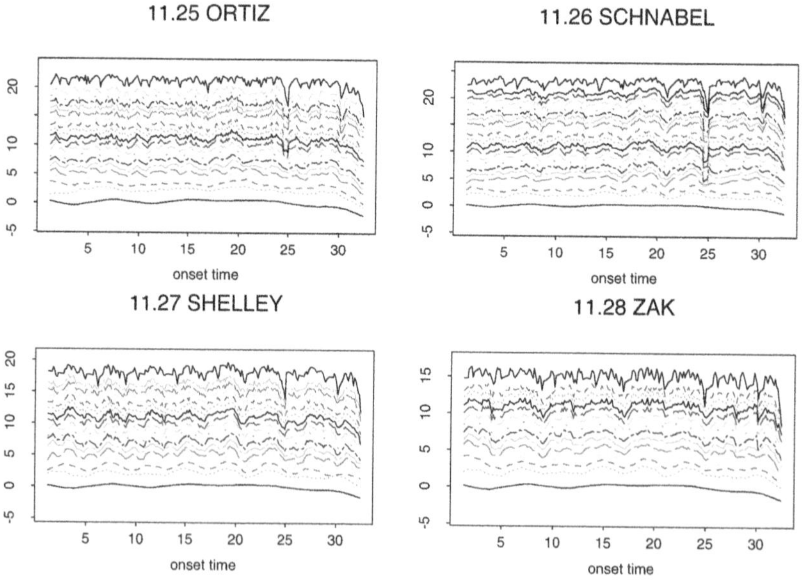

Fig. 44.16. (Cont.)

45

Relating Tempo to Metric, Melodic and Harmonic Analyses in Chopin's Prélude op. 28, No. 4

Musik ist ohne Begriffe.
Hans Heinrich Eggebrecht [232, p.192]

Summary. We investigate the relationship between analytical information extracted from the score of Chopin's *Prélude* op. 28, No. 4 and tempo curves obtained from 30 recordings by well known pianists. The score related information consists of metric, melodic and harmonic weight functions. A hierarchic regression approach is used. The statistical analysis yields strong evidence for the approximate validity of an operator defined in terms of hierarchically decomposed weight functions, as originally introduced in Beran and Mazzola (1997) in the context of Schumann's *Träumerei*, see also Chapter 44. Further analysis based on these findings reveals some typical features of the performances. This chapter only summarizes the methods and results and refers to the paper [110]. The reader is asked to consult that paper for a thorough discussion.

– Σ –

45.1 Introduction

If there is any truth in Leibnitz's famous claim *musica est exercitium arithmeticae occultum nescientis se numerare animi*, performance must play a crucial role as it mediates between rational structures that are hidden in a score's text, and emotional responses provoked by the performative communication to the listener. Performance is a critical pivot when investigating the relation between rational and emotional layers of musical reality and communication. The scientific truth of this thesis is however far from confirmed. One straightforward strategy towards such a proof is to break it down to two partial statements: First, that performance is strongly related to emotional responses, and second, that performance is strongly related to hidden rational score structures.

In a recent paper [984], music psychologists John A. Sloboda and Andreas C. Lehmann have investigated the first thesis for Chopin's *Prélude* op. 28, No. 4. They have tracked listeners' emotional responses to ten pianists' performances, comparing performance parameters of interonset intervals (for tempo), dynamics of the melody, and asynchronies between melody and bass notes to emotional responses while listening to the music and also after the performances were terminated. In their conclusion, the authors state: "This study showed that musical structure had a very important impact on the emotionality ratings made by listeners during performance." We should emphasize that "musical structure" means structure of the performed music.

The present investigation relates to the second thesis. And it happily does so for the same Chopin composition. Understanding musical performance as an expression of analytical rationales (apart from emotional or gestural rationales) is a problem that has gained attention from performance research since the KTH software-implemented model of performance was introduced by Johann Sundberg and his collaborators [1021]. Understanding can be attempted by computer-driven models of performance (such as Sundberg's software or our own approach using the RUBATO® performance rubette [689]) or by empirical research based

G. Mazzola, *The Topos of Music II: Performance*, Computational Music Science,
https://doi.org/10.1007/978-3-319-64444-8_14

on recorded performances. The latter approach was highlighted by Bruno Repp's influential paper [886] on agogic performance in Schumann's *Träumerei*, and we also used Repp's tempo data to initiate our research on musical performance [104, 105, 106, 107, 108] (also see [714, 102, 725] and references therein).

Our approach extended Sundberg's idea by introducing three analytical tools from RUBATO® the Metrorubette for metric analysis, the Melorubette for melodic analysis, and the Harmorubette for harmonic analysis. We thus focused on performance as an expression of metric, melodic and harmonic curves provided by *Rubato*. This lead to collaborative interdisciplinary research between a statistician (Jan Beran), a musicologist (Robert Goswitz), a mathematical music theorist (Guerino Mazzola), and a pianist (Patrizio Mazzola).

Let us shortly summarize the main results obtained from Repp's tempo data of 28 performances and our analytical data extracted from Schumann's *Träumerei*. Apart from the general result that *there is strong statistical evidence for a correlation between performed agogics and the Rubato-generated analytical data*, a more quantitative description of the relationship could be given as follows (in [714] the formula (1) below is called *Beran operator*). Using the idea of hierarchical smoothing, analytical functions derived from metric, melodic and harmonic curves were defined, together with their first two derivatives, and some prima vista data (performance commands written in the score), yielding a 58-dimensional function $X(t)$ (of onset time t). Denoting the logarithm of tempo by $y(t)$, statistical evidence was found for the approximate validity of the relationship

$$y(t) = \beta_0 + X^T(t) \cdot \beta \tag{45.1}$$

where β_0 is a constant and $\beta = (\beta_1, \ldots, \beta_{58})^T \in \mathbb{R}^{58}$ (thereafter called the *shaping vector*).

It should be noted however that equation (45.1) was obtained for one single composition only. Though one may conjecture that similar operators may apply to performances of classical music in general, additional data need to be analyzed before any general conclusions can be drawn. In the present investigation, a new data set consisting of 30 performances of Chopin's *Prélude* op. 28, No. 4 is considered. The following statistical analysis relating metric, melodic and harmonic weights to the logarithmic tempo curves provides further support for the conjecture. The main qualitative result of this experiment is that *there is strong evidence for a relationship of tempo curves with weight functions of all three types, metric, melodic, and harmonic.*

The statistical analysis consists of several steps. In a first step, univariate tests provide some statistical evidence for an association between weight functions and tempo curves. The results show however also that the relationship is likely to be nonlinear, and may depend on scale and location. This, together with general theoretical considerations, motivates the approach of hierarchical regression introduced in [104, 105, 106, 107, 108] (also see [714, 102, 725]). In comparison with the original definition in the context of Schumann's *Träumerei* [104] a slightly modified method is used. A minor change is that Naradaya-Watson kernels are replaced by Priestley-Chao kernels. Since for Chopin's *Prélude* onset times are equidistant, this does not make any difference. The second, more significant change is an improved variable selection procedure. Instead of relying on standard stepwise regression based on the assumption of uncorrelated errors, a backward selection procedure is defined where temporal dependence of residuals is taken into account. This is of particular importance for Chopin's *Prélude*, because errors are likely to contain (almost) periodic components. This dependence needs to be taken into account in order to obtain correct error bounds for estimated regression parameters. Finally, compared to the operator defined in [714, 102, 725], a simplified operator is used in that the first two derivatives of the weight functions are used directly, without decomposing them further. In spite of this simplification, the statistical analysis yields strong evidence for the approximate validity of a suitable version of equation (45.1), including derivatives.

Beyond the statistically oriented results some comments on philosophical and aesthetical aspects of the composition which are supported by our analysis are added (see Section 45.4), thereby presenting a remarkable congruence of the quantitative approach and a qualitative musical analysis in the realm of humanities.

45.2 Data

Two types of data are derived from Chopin's *Prélude*: (1) Analytical data regarding metric, melodic, and harmonic structures; (2) Agogical data of tempo curves measured from 30 pianists as shown in the following table.

Martha Argerich, 2008	Rudolf Kerer, 1998	Sviatoslav Richter, 2008
Shelia Arnold, 2009	Walter Klien, 2013	Artur Rubenstein, 1991
Claudio Arrau, 1974	Nikolai Lugansky, 2002	Russell Sherman, 1986
Tzimon Barto, 2003	Juan Carlos Martins, 1995	Grigory Sokolov, 1990
Idil Biret, 1992	Ivan Moravec, 1993	Jean-Yves Thibaudet, 1999
Alfred Cortot, 2008	Natalia Nikolai, 2010	Paris Tsenikoglou, 2012
Jeanne-Marie Darre, 1960	Garrick Ohlsson, 2008	Helene Tysman, 2008
Nikolai Demidenko, 2008	Alain Planes, 2001	Llyr Willians, 2006
Jakob Gimpel, 1976	Maurizio Pollini, 2012	Irina Zaritzkaya, 1988
Leticia Gomez-Tagle, 2010	James Rhodes, 2010	Houchen Zhang, 2009

45.2.1 Analytical Data

To obtain analytical data, Chopin's *Prélude*, coded in MIDI format, was fed into the analytical modules (rubettes) Metrorubette (local metric analysis), Melorubette (topological melodic analysis), and Harmorubette (Riemannian harmonic analysis) of *Rubato*. For the theoretical background and the implementation in *Rubato* we refer to [714, Parts V, VI, X]. The common format of the resulting analytical data consists of three *weight curves* $x_{metric}, x_{melodic}, x_{harmonic} : O \to \mathbb{R}_{\geqslant 0}$, where O denotes the set of onset times. In other words, we obtain functions that measure the metric, melodic and harmonic "weight" of notes played at (onset) time t.

The metric weight curve x_{metric} was calculated using the following tuning parameters: quantization 1/8, metrical profile 2, minimal local meter length 2. For the melodic weight curve $x_{melodic}$ the parameter settings were as follows: trivial symmetry group, rigid gestalt paradigm, motive neighborhood 2, span 0.374, maximal motive cardinality 5. This yielded a total of 20,022 motives. The harmonic weights $x_{harmonic}$ were based on duration-included calculation and final/causal depth 1/1.

45.2.2 Tempo Data

The tempo data were obtained from 30 CD recordings. Table 2 shows the list of all performances. Each recording was imported from CD to the computer hard drive and then processed by the digital audio workstation Audiosculpt. Each recording was normalized to -1 dB FS. The process of audio normalization has usually been performed in advance by the music producer/engineer in every recording. The goal of a second normalization was to increase the average amplitude of each transient[1]. This is necessary because during the mastering process, each recording is viewed in the context of the larger group of recordings that will comprise the album. A good producer or engineer will therefore normalize each recording relative to the others to create a cohesive sound. By performing a second normalization, the average amplitude of each transient was increased beyond the original recordings. The second step was to detect transients. Within Audiosculpt there is a function called "generate markers." This process involves FFT analysis that detects the frequency of the waveform. The function was set to detect a possible transient every 20 ms, allowing the computer to identify almost every note of the recording and then place a marker at the beginning of waveforms. In a third step we verified every marker and placed markers where a softer transient was ignored

[1] A transient refers to the very first compression of air from a given sound, the very beginning of a waveform.

by the computer. In the visualization provided by Audioscuplt, waveforms are broken up into different colors to accentuate different pitches and volumes. This increases the accuracy of hand-placed markers because each pitch is separated visually and sonically. If the computer did not detect a soft transient, that section could be highlighted and repeated until the transient was detected. To further ensure accuracy, each highlighted section was systematically shortened until the desired pitch was not audible. It was then far more certain at which millisecond each transient occurred. The final step within Audiosculpt was to export the markers as a text file. The text file contains a list in numerical order of each marker and its corresponding millisecond value.

Throughout this chapter the logarithm of the original tempo measurements will be considered. Thus, "tempo" will always refer to the logarithmic values. The right half of Figure 45.1 shows all 30 tempo curves. The mean of all 30 curves is shown in the left half Figure 45.2. The right half Figure 45.2 displays the standard deviation curve, obtained by calculating the standard deviation of the tempo for each onset time separately. This curve indicates where the main differences between the curves occur.

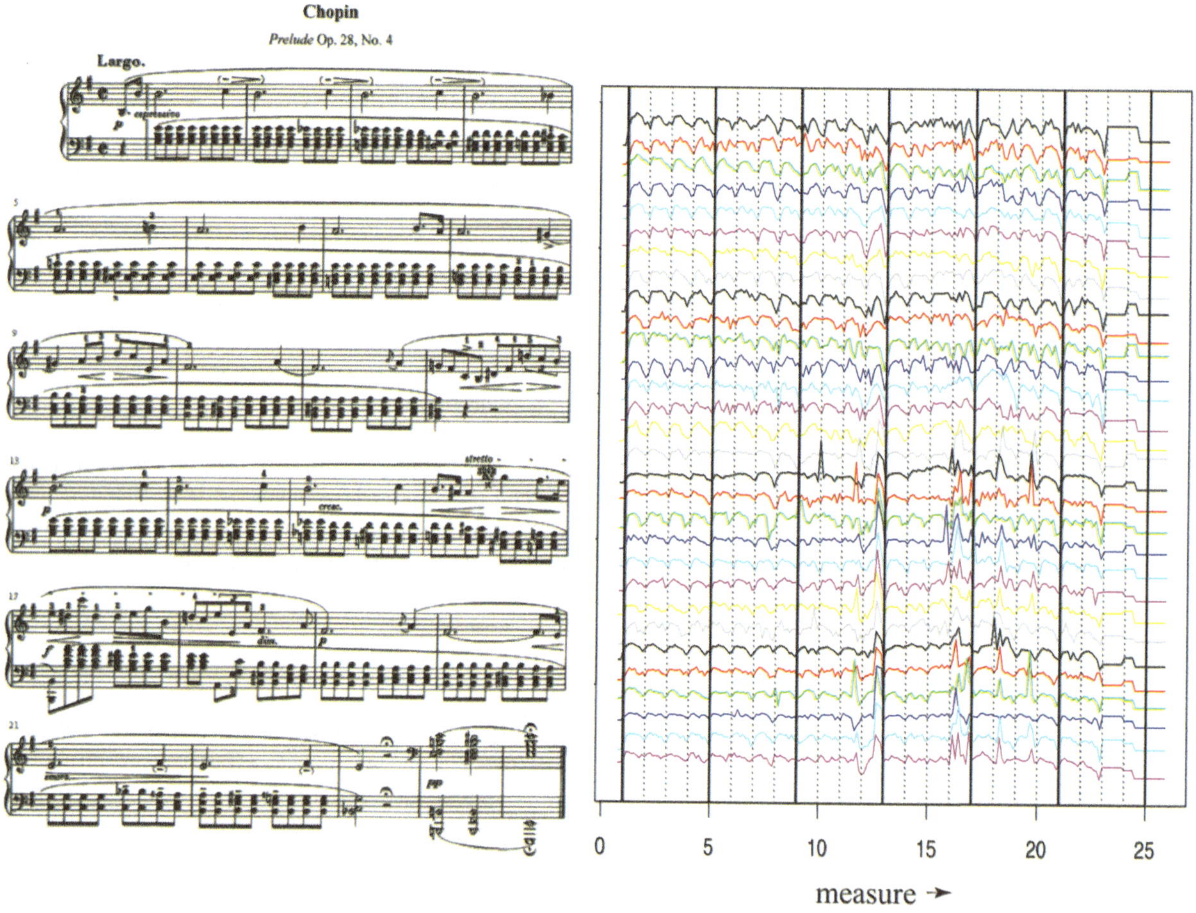

Fig. 45.1. Left: Chopin's *Prélude* op. 28, No. 4 (score by Finale notation software); right: log-tempo curves.

The onset times for the tempo measurements correspond to the regular grid G given in the score, i.e. $G = \{t_j = \frac{3}{4} + (j-1)\,\frac{1}{8},\ j = 1, ..., n\}$ where $n = 202$. The original metric, melodic and harmonic weights were calculated for irregularly spaced grids that differed partially from G. In a first step, the original weights were replaced by weights on the grid G, with values obtained by linear interpolation. The resulting curves $x_{metric}(t)$, $x_{melodic}(t)$, $x_{harmonic}(t)$ $(t \in G)$ are shown in 45.3.

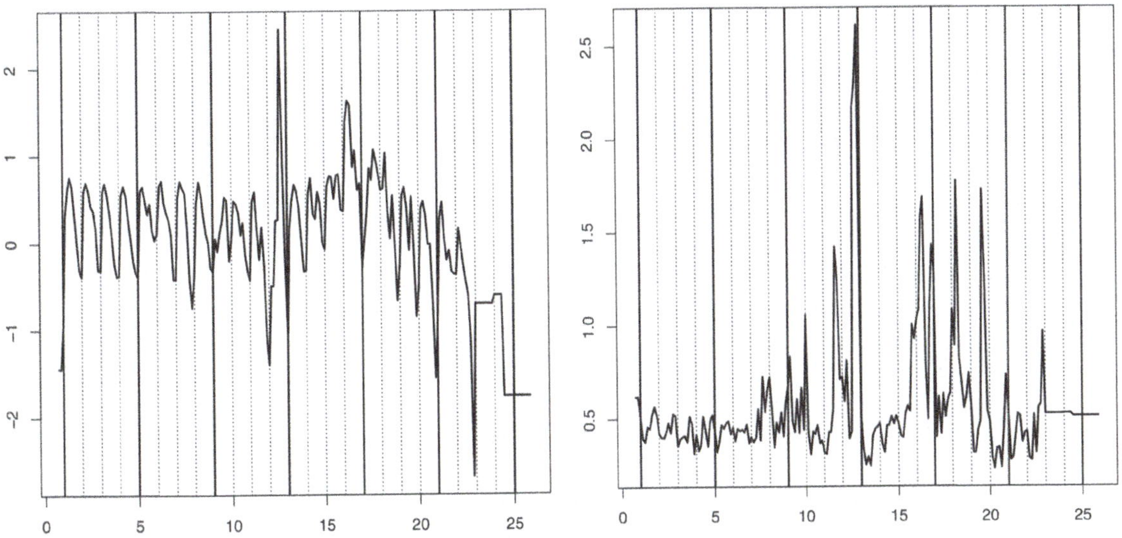

Fig. 45.2. Left: mean tempo of all 30 tempo curves; right: standard deviation of the tempo.

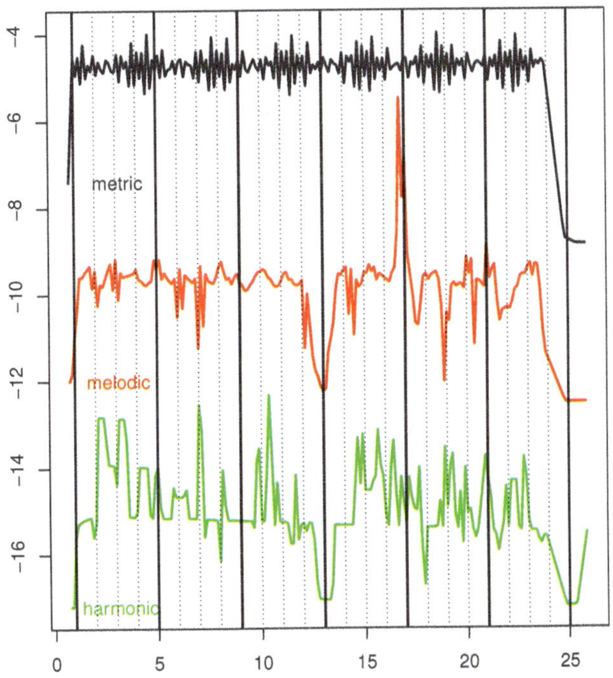

Fig. 45.3. Grid-adapted weight curves $x_{metric}(t)$, $x_{melodic}(t)$, $x_{harmonic}(t)$.

45.3 Short Summary of the Results

Chopin's prélude was analyzed using the same hierarchical smoothing method that was applied to the analysis of Schumann's *Trämerei* as described in Chapter 44. And here is a short summary of the results of this statistical investigation:

The statistical analysis yields strong evidence for the approximate validity of an operator defined in terms of hierarchically decomposed weight functions, as originally introduced in Beran and Mazzola (1997) in the context of Schumann's Träumerei.

45.4 Some Philosophical Comments

In this short composition by Chopin, we can experience in realtime how a initially floating and vaguely searching voice unfolds its consciousness in a growing melodic autonomy. This illustrates to our ears Arthur Schopenhauer's idea[2] of a human consciousness as reified in the melody. To Schopenhauer, the harmonic series with its integer multiples of interval relations represents the building block of the evolutionary universe[3] which in its limitation to harmony does not allow for self-awareness yet. The latter only appears through melody (in soprano), which represents the peak of the thinking mind. But it is more than mere representation, it is the very substance, "thinking music" so to speak. This exemplifies Schopenhauer's idea that in music, the essence of will is immediately present, more than in other arts, where only the will's shadow appears. Johann Wolfgang von Goethe supported these considerations with his idea that the arts do not only map nature but also act similar to nature. This general philosophy immediately points to the Sanskrit *nada brahma* (the world is sound), where the world's beginning and ending are rooted in sound. Schopenhauer even points out the extra-temporal nature of musical sound whose existencee is given beyond the world's substance.

In Chopin's *Prélude* we witness the deployment of self-awareness *in nuce*: In the first eight measures, the soul floats without real melody "over the waters" in search of its (re)incarnation. What initiates a movement towards a comprehensive self-awareness in measures 9-12 densifies towards a self-awareness of the *ego* as a subjective peak (*principio individuationis*) with a short and dramatic climax after a non-melodic recapitulation in measures 13-15. It now represents a fusion of the self (*ego*) with its object (nature). After that (measures 16-20) the soul has evolved into a thinking subject before it dissolves into eternity in measure 21.

Concluding this short philosophical excursus, we find ourselves confronted with the big mystery, similar to Goethe's Gretchen who remained untouchable for Faust and Mephistopheles. Chopin's *Prélude* is at the same time a musical and divine revelation in its reconciliation of *ratio* and *emotio* in their mediation by the performative action.

[2] Exposed in his chief opus *Die Welt als Wille und Vorstellung*, Das Objekt der Kunst, § 52.

[3] This coincides with Gottfried Wilhelm Hegel's thought that God is not an accomplished being but the worldspirit's ("Weltgeist") awakening that defines itself through the evolution of intelligent human subjects.

Part XII

Inverse Performance Theory

46

Principles of Music Critique

Never trust the artist. Trust the tale.
The proper function of a critic is to
save the tale from the artist who created it.
David Herbert Lawrence (1885–1930)

Summary. Inverse performance theory deals with the critic's problem of how to extract in his critique what is hidden behind a performance output. To initiate this theory, we therefore should question the perspectives which a critic has to envisage. To begin with, we inquire the intriguing task of feuilletonistic critique: why is it a never ending story? Is it substantially necessary—beyond music business? We then position the critic within the sociological context: How do norms intervene in critique? This is exemplified by Glenn Gould's performative redefinition of classics. The chapter terminates with an ethnomusicological view on historicistic performance as it is typically undertaken by Nicolas Harnoncourt.

– Σ –

46.1 Boiling Down Infinity—Is Feuilletonism Inevitable?

Summary. It is not accidental that music critique has stuck to feuilletonism. Its scope is an infinite one in several dimensions: It has to cope with the infinite interpretative work with respect to a given text and with the infinity of expressive nuances of each given interpretation. We analyze the necessity of boiling down this infinite challenge in view of the poor tools of traditional musicology.

– Σ –

We have learned in previous chapters, in particular in Section 13.4.1 and Section 32.2, fact 18, that performance conveys an infinite message. And that this infinity is a double one[1]: that of interpretative perspectives as they are realized in music analysis, and that of performative shaping variants as they may be expressed on the infinitesimal vocabulary of performance fields. It is not clear whether critics are aware of such a variety of background that may produce concrete performances. In particular with respect to (analytical) interpretation, they preferredly stick to the traditional canon of how the structure of a composition should be viewed and interpreted or analyzed, respectively. Of course, it is not clear whether music critics should be cognizant of possibly new interpretations, but once they have gone into their business, a creative dealing with analytical problems should be mandatory. One may understand that this is not necessary ante rem, but after the event, a re-reading of the text should be considered, be it only for comparative handling of the present performance: Could it be that the artist discovered a new interpretation of the given text? In practice, the selection of an analytical interpretation (in the best case, autoincompetent critics excluded...) is just a matter of limitations of time, energy, and interest, besides ignorance of the infinite variety of interpretations. As to

[1] Leaving aside the gestural and emotional rationales for the time being.

G. Mazzola, *The Topos of Music II: Performance*, Computational Music Science,
https://doi.org/10.1007/978-3-319-64444-8_15

the infinity of performance nuances, this is beyond the vocabulary of music critics and it is also beyond the present measurement technology for such data: In a common concert, no performance field reconstruction is feasible. So critics are nolens volens limited to describing performance by use of common language expressions ("elegant diminuendo blended by a mysterious pedaling cloud...") which beyond its imprecision cannot relate expression to interpretation.

So is feuilletonism inevitable? Or rather: Is such a bad feuilletonism inevitable? Is it necessary to play the game of a unique "best" interpretation whose expression has to move along unreflected paths of prejudices? The alternative would be to embed one's judgment in the potential infinity of analytical interpretation and expressive performance. And to keep this embedding omnipresent in the critical discourse. We argue that the most precious role of a music critic would be that of putting the infinity of perspectives on a musical work into evidence in every concert or CD review. These would be the crucial points:

- Infinity of analytical interpretations,
- infinity of expressive performances,
- infinity of correlations between interpretative rationales and expressive performance shaping.

And it is not the question about bad or good quality in these specifications, *in the limit, the only quality is to teach us something about the work in question, and about the relativity of each perspective. Suggesting a boiled-down finitistic or even unicorned view of art is a destructive way of reduction and hinders every deepening or progress in the arts.*

46.2 "Political Correctness" in Performance—Reviewing Gould

Summary. Given the infinity of critical understandings of artistic performance, norms are easily infiltrated against unlimited variation of expressivity. We make the point concrete with the example of Glenn Gould's eccentric (ab-normal...) performance of classics from Bach to Beethoven.

– Σ –

A testing ground for a valid music critique is Glenn Gould's performance of classical works from Bach to Webern. His technically unprecedented performances have evoked strong reactions which unveil a number of limitations in common critique styles.

Whereas Gould's Bach performances may be non-conformist, but still acceptable and adequate for Bach's compositions, his performances of Beethoven's sonatas is beyond the supportable deviation from common taste. The famous critic Joachim Kaiser has described in [517] the most famous "mis-performance" of a Beethoven sonata on the example of Gould's presentation of op. 57 "Appassionata":

> *Bei Goulds Wiedergabe des allegro assai dürfte es sich um die verrückteste, eigensinnigste Darstellung handeln, die jemals ein Pianist einem Bethoven-Satz hat angedeihen lassen; und das will etwas heißen. Gould hält es für richtig, demonstrativ langweilig und gelangweit den Kopfsatz so zu bieten, als ob ein Beethoven-Verächter seinen Plattenspieler nur mit halber Geschwindigkeit ablaufen ließe. Tranig langsam, langweilig und gelangweilt, die Triller während des pp im Schneckentempo, während der Fortissimo-Stellen etwas rascher, quält sich die Musik vorbei. Man meint, der Pianist imitiere ein Kind, das mit erfrorenen Fingern die Appassionata vom Blatt spiele. Nur selten vergißt er dabei, daß er ja vergessen machen wollte, der genialische Glenn Gould zu sein.*

This critique is strongly based upon the commonly accepted reading of the Beethoven text as a passionate message which calls for temperament and stormy dynamics in performance, and not for analytical cool vivisection of such a vital piece of literature. In Kaiser's characterization, Gould's production is like a "child with frozen fingers in a sight-reading performance". Here, the different and aberrant performance is incorrect, even forbidden. It is a norm which the politically incorrect Gould has broken and thus made the sonata ridiculous; Kaiser even comments that the sonata "remains silent" when confronted with such a misreading.

The basic hypothesis behind such an outrageous indignation is that Kaiser knows what and when and how the sonata (which is personalized here) would have communicated, and that crazy Gould just destroyed that known and accepted messaging. Kaiser in fact evokes an installed performance grammar which requires a passionate forte seventh degree cascade towards the piano on the dominant in bars 14-16 of allegro assai. Instead, Gould descends like a noble, bored lady and snobbishly sits down on the boring dominant fermata. No passion whatsoever.

The same, even more dramatic viz. ridiculous deformation can be observed in Gould's performance of sonata op.106, "Hammerklavier". This case is even worse since one just thinks that Gould did not understand a single word of the text, he simply was too stupid for the performance task.

What happened? And why was Gould's Bach so much more successful? Evidently, Gould's microscopic performance method works for Bach, and not for Beethoven. Why does this microscopic view fascinate and illuminate Bach's work whereas it virtually kills Beethoven's sonatas? The point is that in Beethoven's work, there is an inbuilt performance grammar which is not engraved in the score but stems from the performance tradition as such, an oral tradition so to speak, an element of rhetoric communication which transcends written code. Instead, Gould reads the same code from the Bach and from the Beethoven scores, and effectively demonstrates, that there is a huge defect in Beethoven's written code; it is quite trivial, at least locally, the written script is simply boring.

Gould has effectively given a quasi-mathematical demonstration that the same performance strategy cannot be applied to Bach and to Beethoven, that the same analytical insight and the same rhetoric shaping yield completely different results for these composers. To me, this is a sensational lesson to teach a characteristic difference between Bach and Beethoven. This is very clear in the descending seventh passage on bar 14th bar, which runs on semiquavers after a triggering triplet of quavers at the end of bar 13. Gould effectively takes the double temporal rate of the semiquavers with respect to the quaver triplet, without any tempo increase, without any dynamic profiling, just letting us see the anatomy of this triadic descent structure. The common reading [1066] of this passage is that of an explosion:

> *Die Explosion (a tempo, Auftakt zu T.14) erfolgt im niedersausenden Dominant-Arpeggio und f-Sextakkord (T.15), wird aber sogleich abgedämpft durch einen C-Dur-Sextakkord, p, T.16.*

With Gould, there is no explosion, just the written text, cleanly played, but antagonistic to any such musical drama to which an explosion would testify. The common reading in fact classifies this sonata as a musical drama, and asks interpreters to integrate this semantic into their performance. Gould plays the "Appassionata" minus the commonly implied drama. The question here is whether this dramatic character is implicit in the score structure or whether it is an external determinant which has been added by historical standards—which Gould filters away to lay bare what he believes is a poor structural essence [384]. So the question arises whether the commonly accepted dramatic performance is an expression of Beethoven's work or of an added character. Let us therefore analyze the specific performative shape of the passage in question. To begin, its agogics is profiled against the temporal neighborhood, i.e., not only is the indication "a tempo" valid from the last three quavers of bar 13, but in bar 14, the resumed tempo is again increased. The dramatic performance contains an increase of tempo, and within that level, also an increase of tempo towards the middle of the descent. Further, the dynamics is not only the forte at the end of bar 13, but the target tones of each descending intervallic movement of the descent is played louder, maybe to a *ff* or *sf*. As a whole, this descent (with its added ascending tail in bar 15) is not only a musical structure, but more an explosive gesture whose very beginning goes to the top pitch, falls down and bounces back to the dominant fermata. This is not a written rationale, but it is a semantic unit which can easily be deduced if gestural semantics is to be included in the performance shaping.

So Gould's experiment would demonstrate that Beethoven requires gestural rationales beyond analytical ones. Meaning that Beethoven's compositions have a performative added value of gestural nature which is not (yet?) virulent in Bach's architectural music. Observe however that this gestural character is not on the level of the interpreter's gestures, it is a rationale in the performance grammar, a semiotic layer which is added to the score system. Summarizing, Gould's politically incorrect performance withdraws from the common dramatizing approach and gives us an insight to Beethoven which would not have occurred otherwise.

46.3 Transversal Ethnomusicology

Summary. Ethnomusicology is essentially confronted with the *synchronic normalization* problem: Given two (simultaneously existing) cultural areas, understanding one of them from the other's perspective means normalizing fundamental categories, e.g., the "score" concept, such that *comparison* of different ethnics becomes feasible. When dealing with performance of works which belong to distant epochs, the ethnomusicological problem of understanding a performance is restated as *diachronic normalization*: what is the common ground for understanding the performance of a work which was written in the spirit of another epoch? We discuss the efforts of Nicolas Harnoncourt in this direction of "transversal" ethnomusicology.

– Σ –

In its common understanding, ethnomusicology deals with a transformation problem between synchronic music cultures. Such a transformation may cause major problems to the contents and forms of music, for example because of incompatible notation or even incompatible modalities of communication, such as oral traditions, rebuilding instruments for each new musical event, embedding of music in more global forms of art, etc. But synchronic ethnomusicology has the undeniable advantage that a feedback process may help deal with such problems, and eventually solve them in the ideal case[2]. This advantage cannot be claimed for diachronic ethnomusicology.

What is this type of ethnomusicology? It deals with transformation of music cultures which are at a temporal distance instead of a geographic or social distance of contemporaneous cultures. For example, if we play an opera of the late Renaissance composer Claudio Monteverdi in the 20th century, this is a diachronic type of ethnomusicology. In fact, the cultures of Monteverdi's time and of the 20th century are very different, and the communication between them is restricted to the historical proliferation channel. The historical distance influences strongly the communication of forms and contents. It is by no means automatic that everything is transmitted without loss of information. Above all, the socio-cultural background of a musical composition is easily blurred by historical filters.

But also, the instrumental practice and technology impose dramatic deformations of what was reality in a historically distant context. Nicolas Harnoncourt has with great success restarted a dialog with historically distant traditions in the sense of understanding those conditions and not imposing ours to a diachronically distant culture. His approach is based upon the basic position that instrumental constraints are very important for shaping one's performance and expressivity. Restriction gives one a clear frame, a limited field of activity where a composition must unfold its semantics, and not an unlimited Wagner-tailored orchestra, where the quasi-infinity of instrumental power and colors competes with and actually substitutes the efforts for better expressivity.

All this intelligent effort does not solve the communication problem in the historical dimension: in contrast to synchronic ethnomusicology, diachronic communication is unidirectional: No real feedback from ancient times is possible, we only have the sources and must try to understand without really having answers to our experiments. Harnoncourt's experiments may be an attempt to transform our performance practice back to Monteverdi's, but it is not a demonstration of anything.

So performance is not only a hic et nunc affair, but also a process which is coupled to spatio-temporal distances of cultures. Within this transformational framework, there are different serious asymmetries of communication. In the synchronic direction, feedback can be dealt with—however in the limits of socio-cultural asymmetries for inverse ethnology. In the diachronic direction, the communication is intrinsically unidirectional, and performance remains a challenge for the adequacy of cultural transformation.

[2] However, the real case may be far from ideal. For example, if we want to initiate *inverse ethnology*, i.e., the review of our own music culture from the point of view of an 'exotic' culture, major obstacles will occur, from financial ones to the intolerance against another culture which tries to relativize the usual perspective of occidental supremacy.

47

Critical Fibers

To see a World in a Grain of Sand,
And a Heaven in a Wild Flower,
Hold Infinity in the palm of your hand,
And Eternity in an hour.
William Blake (1757–1827)

Summary. Stemma theory offers a model for a critical understanding of performance *as* a complex process between interpretation and the rhetorics of expressive performance grammar. We make the model and its limitations explicit.

– Σ –

47.1 The Stemma Model of Critique

Summary. Modeling performance through stemma theory allows us to define inverse images of performances within a well-defined variety of stemmatic situations leading to the given performance. A critical fiber is such a variety. A critique is a choice of a point within this fiber. We discuss criteria to make such a choice, i.e., to select one critique among all possible critiques *within* the stemma model.

– Σ –

In Section 32.4 we discussed the four global aspects of performance, in particular the fourth point: stemmatic deployment of performance. In that discussion, we stressed the fact that stemmata are not just a learning process but much more a logical unfolding of deformation strategies. If we take this point of view for granted[1], music critique should deal with the stemmata which could possibly lead to a given performance, since they would unveil the anatomy and genealogy of a performance, and this must be the central issue of a critique which merits that name.

Probably the very construction of a stemma with its genealogical factors from cell hierarchies to performance operators and ramification architecture would help a critic to shape his/her style and criteria for judging performances. But beyond that, the general framework of stemma theory is far too generic for any concrete goal. What is the concrete goal?

Principle 27 *Try to construct a stemma that produces a given performance from the existing score!*

[1] Other models of performance genealogy are at hand, but we do not know of any such model that is technically as explicit as the stemma model.

G. Mazzola, *The Topos of Music II: Performance*, Computational Music Science,
https://doi.org/10.1007/978-3-319-64444-8_16

In this form, the problem has trivial solutions if we admit that any performance field is admitted without further articulation and operator constraints. The (non-trivial!) construction of a performance field from experimental performance data (e.g., on a MIDI file record) has been implemented by Stefan Müller, see Section 47.3, and can be used to construct a depth-one stemma whose operator just produces this performance field on a monolithic LPS. Such a solution is however not what a critical understanding of performance would preconize since this is nothing more than the brute 'sampling' of a performance transformation, and would not lay bare any of the semantically valid rationales.

The crucial point is that one should impose constraints on the admitted tools for stemma construction. This would lead to the more reasonable

Principle 28 *Given a defined set of constraints regarding the construction tools for stemmata, try to construct a stemma that produces a given performance from the existing score!*

In this form, the principle becomes a challenge for critics since it thematizes the strategies a critic could imagine to be used by a specific artist. Only after such a strategic preset can the reconstruction of a possible generating stemma be seriously tackled. The lesson to be learned from such an 'exercise' is that in criticism, one should learn to reflect very cautiously the conditions under which critical judgments are made.

The critical business now splits into two subtasks: comparative criticism of one and the same performance, and comparative criticism of a number of different performances of the given score. The first means that we are given a fixed performance and compare different reconstructions of backing stemmata: Which one is acceptable, which one is simpler but still adequate, in what respect could two criticisms be considered as being isomorphic, etc. The second one means that the phenomenological difference of performances is lifted to a genealogical difference of backing stemmata, and this one is the core activity of a music critic. However, it presupposes that usually, the constraints on the stemma reconstruction potential are the same, a condition for comparability which is not automatic.

47.2 Fibers for Locally Linear Grammars

Summary. Locally linear grammars are a special approach to stemma varieties, see Section 39.8. We give a first description of critical fibers in this context. They turn out to be varieties in the sense of algebraic geometry and are called grammatical varieties.

– Σ –

The general inverse problem of reconstructing a stemma for a given performance of a defined score is not solvable for two reasons: we first do not have a general formalism and second, even a reasonable general formalism would imply wild[2] mathematical classification problems.

We therefore want to consider a more tractable situation: the locally linear performance grammars which were introduced in Section 39.8. We refer to that section and keep its notation, recalling in particular from definition 110 that a locally $\mathbb{R}$-linear grammar is a family of $\mathbb{R}$-linear representations of the stemma quiver T which are parametrized by $\mathbb{R}$-vector spaces B_x of finite dimension s_x, $x \in V(T)$, the vertex set of T, the parametrization being given by affine maps $\varphi_x : B_x \to End(A_x)$.

Within this very precise setup, where the stemma and the locally linear grammar are fixed, we may ask for the structure of the fibers lying above the performances which are defined on the stemma's leaves by the given locally linear grammar.

Technically speaking, we proceed as follows: For each final vertex $z \in V(T)$ let $m_{z0} = r, m_{z1}, m_{z2}, \ldots$ $m_{zn_z} = z$, be the ordered sequence of elements of $M_z(T)$. For the root r and for each final vertex $z \in V(T)$ let us fix arbitrary $e_r \in A_r$, $e_z \in A_z$, respectively. The main task of this *inverse performance theory* is to study the set of solutions of the system of equalities

[2] In algebra, a classification problem is said to be wild, if its solution would imply the classification of any module category.

$$e_z = f_{m_{zn_z}}(f_{m_{zn_z-1}}(\dots f_{m_{z1}}(e_r, (c^r_{y1,m_{z1}}), (b_{y_1})), \dots), (c^{m_{zn_z-1}}_{y_{n_z},m_{zn_z}}), (b_{y_{n_z}})) \tag{47.1}$$

where for $k = 1, \dots n_z$ the vertexes y_k lie in the set $D_{m_{zk-1}}(T)$.

For the explicit calculation, we select a basis $v_{x1}, \dots v_{xs_x}$ of the vector space B_x. This means that for every vertex $x \in V(T)$ we may consider the linear operators $O_{xi} := \varphi^0_x(v_{xi}) \in End(A_x), i = 1, \dots s_x$ as a part of our data. Whence, if we identify B_x with $\mathbb{C}^{s_x}$ through the basis $v_{x1}, \dots v_{xs_x}$, we can define the homomorphism (39.50) as a triaffine map

$$f_x : A_m \times \mathbb{C}^{\#D_m(T)} \times \prod_{y \in D_m(T)} \mathbb{C}^{s_y}_y \to A_x$$

by associating to $(a_m, (c^m_{y,x})_{y \in D_m(T)}, (b_{yj})_{y \in D_m(T), j=1,\dots s_y})$ the sum

$$\sum_{y \in D_m(T)} \sum_{j=1}^{s_y} c^m_{y,x} b_{yj} i_{y,x}(O_{yj}(r_{m,y}(a_m))) + \sum_{y \in D_m(T)} c^m_{y,x} i_{y,x}(\varphi^t_y(r_{m,y}(a_m))).$$

The equations (47.1) become:

$$\begin{aligned} e_z = {} & \\ & \sum c^{m_{zn_z-1}}_{y_{n_z},m_{zn_z}} b_{y_{n_z} j_{n_z}} \cdots c^r_{m_{y_1},z1} b_{y_1 j_1} m_{z,y_1,\dots y_{n_z}, j_1,\dots j_{n_z}} + \\ & \dots + \\ & \sum c^{m_{zn_z-1}}_{y_{n_z},m_{zn_z}} \cdots c^r_{y_1,m_{z1}} m_{z,y_1,\dots y_{n_z}} \end{aligned} \tag{47.2}$$

where the sum is over all $y_1 \in D_r(T), y_2 \in D_{m_{z1}}(T), \dots y_{n_z} \in D_{m_{zn_z-1}}(T)$ and all $j_k = 1, \dots s_{y_k}$, for $k = 1, \dots n_z$. The leading vector summand of this linear combination equals

$$\begin{aligned} & m_{z,y_1,\dots y_{n_z}, j_1,\dots j_{n_z}} = \\ & i_{y_{n_z},m_{zn_z}}(O_{y_{n_z} j_{n_z}}(r_{m_{n_z-1},y_{n_z}}(\dots i_{y_1,m_{z1}}(O_{y_1 j_1}(r_{r,y_1}(e_r)))\dots))). \end{aligned} \tag{47.3}$$

The general vector summand refers to a choice of endomorphisms $X_{\dots} = O_{y_{n_z} j_{n_z}}$ or $X_{\dots} = \varphi^t_{y_{n_z}}$ etc. and is equal to

$$i_{y_{n_z},m_{zn_z}}(X_{\dots}(r_{m_{n_z-1},y_{n_z}}(\dots i_{y_1,m_{z1}}(X_{\dots}(r_{r,y_1}(e_r)))\dots))). \tag{47.4}$$

In order to describe the solution variety we can interpret (47.2) as a system of linear equations with parameters,

$$e_z = \sum_{y \in D_{m_{z,n_z-1}}(T)} c^{m_{n_z-1}}_{y,z} l_{yz}. \tag{47.5}$$

Observe that e_z, l_{yz} are vectors in A_z, so the solutions $c^{m_{n_z-1}}_{y,z}$ of equations (47.5) result from simultaneous solutions in the vectors' coordinates if we are given a basis of each A_z. The coefficients l_{yz} equal linear combinations of the vector summands (47.3) and (47.4) with coefficients which are monomials in the remaining variables $c^{\cdot}_{\cdot,\cdot}$ and the $b_{\cdot,\cdot}$.

If we now assume that all variables $c^{m_{n_z-1}}_{zy}$ are independent, we see that for any two different $z \in D_{m_{n_z-1}}(T)$, the solutions of the equations (47.5) do not interact with each other. This means that the solutions of the system (47.5) for $z \in D_{m_{n_z-1}}(T)$ is either empty (if the value e_z is not in the image of the linear map (47.5)) or equals

$$\prod_{z \in D_{m_{n_z-1}}(T)} L_z \subseteq \prod_{z \in D_{m_{n_z-1}}(T)} \mathbb{C}^{\#D_{m_{n_z-1}}(T)}$$

where L_z are linear subspaces of $\mathbb{C}^{\#D_{m_{n_z-1}}(T)}$. Their codimension can be read from the matrix of the coefficients. Since there is no other condition on the other variables $c^{m_{zk}}_{y,x}$ we obtain altogether the following result:

Theorem 35 *The solution space of* (47.2) *is a linear fibration over some appropriate affine space.*

Corollary 22 *The dimension of the non-empty fibers of this fibration is generically minimal, and it increases along some finite union of proper algebraic subschemes (the loci where the minors of the coefficients' matrices vanish). This means that the non-empty fibers $Fib(e_.)$, i.e., the solution spaces defined by* (47.1) *over a given output set $e_. = (e_z)_z$ are all generically isomorphic, i.e., isomorphic when restricted to appropriate open subschemes. However, the configuration of the specialization subschemes is not evident and depends on the particular vector summands. Also, the condition for non-empty fibers is not evident in general.*

How can we interpret this result in musicological terms? A quantitative measurement of a performance can be done by recording the values of the parameters that characterize a note, i.e., loudness, duration, and so on. In the last paragraph we have denoted this data by the vectors e_z. Equation (47.2) explains that in order to produce given values of the parameters from the weight functions, one has to find suitable values for all $c^{m_{zk}}_{y,x}$. It would be nice if the choice of these coefficients could be small, because this would mean that in order to produce a given performance one does not have a lot of freedom in the choice of system parameters. However from this model this is not the case. In fact, Corollary 22 tells us that for a fixed performance one has either no one or else infinitely many possibilities to choose system parameters which produce the given performance. Moreover, we learn that all performances with non-empty fibers $Fib(e_.)$ have open dense subsets $U(e_.) \subset Fib(e_.)$ which are all isomorphic with each other, so these generic subsets $U(e_.)$ are qualitatively equivalent, see also Figure 47.1. *In other words, the non-empty fibers only differ on their special loci apart from generic open subschemes $U(e_.)$.*

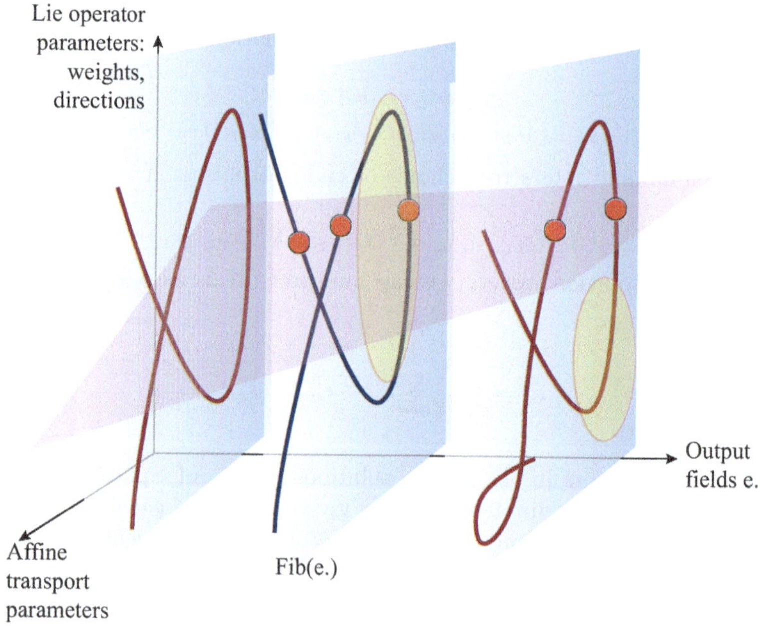

Fig. 47.1. All performances with non-empty fibers $Fib(e_.)$ have open dense subsets $U(e_.) \subset Fib(e_.)$ (elliptic regions) which are all isomorphic with each other

In musical terms, a fiber $Fib(e_.)$ could be called a *critical fiber* because its points are the possible background parameters—in the present stemmatic model—which lead to the given performance output data $e_.$, such as local tempi, articulations, dynamics, detunings, etc. So the fiber really includes the possible ways of understanding why a performer is playing his actual performance. In fact, finding out which parameters the interpreter could have used is (or should be) the core activity of a music critic. Having generically isomorphic fibers means that in any two critical fibers of two given performances, there are "dominant" open sets of "criticisms" (i.e., points in the fiber!) which are isomorphic with each other. This does not mean that the criticisms are the same for the two given performances, but that their structural contexts can be identified. Again, this does not mean that the relevant criticisms may be identified—on the contrary: maybe, these isomorphic contexts just describe the criticisms which are what everybody could say if no supplementary information about the specific performance culture of the interpreter is known.

So we should not discard this model as insignificant, instead we need to look at it in the right way. In fact, it is like pretending to explain the geometry of smooth plane curves, just using lines. The theory of lines can be useful for local questions, but not for global ones. The same thing happens here. The great flexibility of this model enables us to adapt it to the questions that one tries to answer, but they can not be too general. Indeed, one can select subschemes that are appropriate for the study of a particular problem. For example, if one is interested in comparing several performers under the point of view of the local/global way of playing, one could try to restrict the research to the level of the daughters of the root, and to use some more of the structures of the vector space A_x (if any). Another question that could be asked is that of the final/causal way of playing. This requires one to impose mutual dependence conditions upon the variables $c^{m_{zk}}_{y,x}$ for varying $z \in D_{m_{zk}}$ (see 47.4).

47.3 Algorithmic Extraction of Performance Fields

Summary. The algorithmic extraction of performance fields is a first step for systematic calculations in inverse performance theory. The extracted fields can further be used for visualization. We describe an approach how performance fields can be calculated from given scores and performances and present the tool that implements the theory.

– Σ –

In this section, we address the question of inverse performance theory which deals with the reconstruction of a performance field for a given performance on a determined score, including an implementation in the RUBATO® framework, named EspressoRUBETTE®. This question generalizes the well-known problem of constructing a tempo curve from a measured performance to form space $S = EHLD...$ of parameters such as onset E, pitch H, loudness L, etc. Such an attempt is first of all an interpolation task where a continuous performance field must be reconstructed from a discrete data set of performed notes. As such it is subjected to (a) the problem of matching symbolic and performance events, (b) ambiguities in the local definition of field vectors, (c) algorithmic constraints for real-time objectives, and (d) visualization options for performance fields.

47.3.1 The Infinitesimal View on Expression

The concept of expression is ambiguous as far as the content of the expression and its reality layer are not a priori clear, see Chapter 36 for the details. If we aim at analyzing expression, this does not regard the psychological perception of a performance by humans. The psychological aspect is a legitimate one, but it touches a category which relates the performed music to human categorizations in terms of emotional response. Such a perspective is, for example, dealt with in [480] or in [584]. In contrast, our point of view is expression as a rhetorically shaped transfer of structural score contents by means of the deformation mapping of symbolic data into a physical parameter space. The psychological implications are not the subject of this perspective, it is a purely mathematical description of this mapping, not of the emotional correlates.

The theory of performance fields is derived from the general hypothesis that performance is a smooth (continuously differentiable) isomorphism $\wp = R \to RP$ on a frame neighborhood R of the given local composition $C \subset S$. This is of course a strong hypothesis, but it is, at least locally on the given composition, a reasonable one. In our inverse problem, we are not given $\wp$, but only its restriction $\wp|_C$ to the given local composition. Accordingly, we shall not really construct the performance field $\mathfrak{S}$ associated with the unknown map $\wp$, but a discrete performance field, defined on the points of C, which is determined by the restriction $\wp|_C$.

We shall now construct such discrete fields, their interpolation on a neighborhood of C, as well as their visualization by means of color fields. It will also be possible to calculate difference fields in order to compare two performances of the same local composition.

47.3.2 Real-Time Processing of Expressive Performance

An implementation of the performance field theory should be able to operate in real-time, especially for interactive applications, where immediate feedback, either visible or audible, is desired. As we shall see, the complete calculation of the performance fields can be split up in dedicated, communicating modules for specific tasks. Particularly important for performance is the extraction of tasks that can be processed in advance. Figure 47.2 gives an overview of the modules and the flow of control (as shown by the vertical

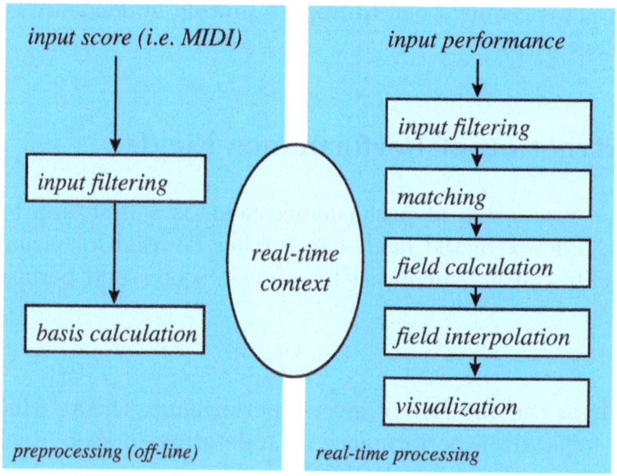

Fig. 47.2. Overview of the modules and the flow of control (as shown by the vertical arrows) in our implementation of the EspressoRUBETTE®.

arrows) in our implementation. Modules are notified by events when new data for processing is ready. The modules themselves are *stateless*, they share their information with other modules in the *real-time context*, a data structure which contains all relevant information for the whole process, thus minimizing the risk of inconsistency. Of course, asynchronous accesses to the context have to be synchronized using locks or a similar synchronization technique.

For increased flexibility and efficiency all modules accept lists of events, therefore making off-line and real-time processing structurally identical. For example, the input filtering modules for the score and for the performance are the same. The former accepts the score as a whole and the latter processes individual events as they are received in real-time. Further, modules can be prioritized and be put to sleep if there is not enough processing power to support all present modules temporarily.

The following items give short descriptions for the modules shown in Figure 47.2. Details will be presented in the subsequent sections. It is important to see that the described architecture allows the definition of additional modules as needed. This mostly depends on application requirements. Also some application might not need certain already defined modules, i.e., field interpolation in a computer accompaniment system.

Input filtering. This module translates incoming note events to the representation defined in the real-time context. It also processes structural information, such as different voices, tempo changes, etc. The input filtering module must be implemented for any external representation (e.g., MIDI or RUBATO®s denotator format).

Basis calculation. The calculation of the bases depends only on the input score, not on the performance and can thus be performed off-line. For each event an appropriate basis has to be calculated. Typically this is a time-consuming process.

Matching. The incoming performance events have to be matched to the corresponding score events. As we shall see in the designated section, this is a non-trivial task and has developed to a research field on its own.

Field calculation. The individual field vectors for each note are be calculated based on the precalculated basis and the given match.

Field interpolation. The field vectors calculated by the former step are typically not aligned on a grid. However, for visualization a 2D or 3D grid-like field, with field vectors defined anywhere in this grid is desired. The interpolation step allows the definition of such a grid and performs the translation from the note field to the interpolated field.

Visualization. Finally the calculated field is ready for visualization. Here, many user defined parameters such as scaling, color-specification, ranges, etc., have to be taken into account.

47.3.3 Score-Performance Matching

A lot of research effort has been put into score-performance matching techniques. *Score-following*, the real-time matching and tracking of soloists performing a given score was first published by Dannenberg [240] and Vercoe [1083]. Puckette [867] presented the methods used on the IRCAM Signal Processing Workstation (ISPW). Heijink et al. [456] have given an evaluation of different approaches to score-performance matching.

Literature has typically differentiated between two types of algorithms: For *real-time algorithms*, mostly used in real-time accompaniment software, good performance had higher priority than matching quality. Off-line algorithms, where calculation time is less important, were mostly used for in-depth analysis applications requiring a high level of matching quality. We however experienced that with todays processing power high quality matching can be performed in real-time, particularly when the algorithms are well suited for extensive preprocessing of the given score. Mathematically, the matching problem is complex and depends upon the difference which one allows between score and performance. For example, if chords remain chords and all notes are played exactly once, the problem is trivial. But normal performance includes more or less strong arpeggiation of chords, omissions of notes or playing additional notes by error or by ambiguous definition of the notes, such as is common for trills and other ornaments. We have implemented an algorithm which will not be described in detail here since it is not our principal subject. The algorithm is a kind of matching along a 'wave front of notes which are defined by the temporal unfolding of the performance and thereby fits in the real-time constraints. We nevertheless should sketch the principle ideas behind our algorithm.

Usually, matching is thought bottom-up in that the performance map of the whole piece is constructed from the performance map $X \mapsto \wp(X)$ on the single elements X. We rather tried a top-down strategy, i.e., to rebuild the element images from maps on sets of specific coverings I, J, respectively of the local composition C and its performance D. Typically, one considers the covering of C by hyperplane sections in each parameter (for example onset slices). On D, a covering J is defined which is a more fuzzy version of I, for example neighborhoods of hyperplane sections (for example ε-neighborhoods in the onset dimension). If $\wp$ exists, different constraints can be imposed on the induced map on the coverings: First, $\wp$ induces a map $n_0(\wp) : I \to J$ such that $\wp(U) \subseteq n_0(\wp)(U)$ for all covering elements U in I. This yields a map $n(\wp) : n(I) \to n(J)$ of the simplicial nerves, and thus conditions on the map on the covering sets. Second, the sets of these coverings are linearly ordered[3] by $U \prec V$ iff either $U \subset V$ or both $U - V$ and $V - U$ are non-empty and $min(U-V) < min(V-U)$. In this ordering, we require that $U \preceq V \Rightarrow n_0(\wp)(U) \leqslant n_0(\wp)(V)$. Third, if one defines a distance $d(U,V)$ between the covering sets (for example the elastic shape distance from motif theory, see Section 22.2.1.3), one requires that $d(U, n_0\wp(U)) < \varepsilon$ for a given positive distance limit ε. With these constraints one may define the map, and then recover $\wp$ if every point X in C may be seen as the intersection of all covering sets of I which contain the point. This is evidently the case for the hyperplane sections described above.

[3] This is the usual linear ordering of powerset denotators as defined in Section 6.8.

Following these observations, an implementation typically makes use of structural properties of a musical score and a corresponding performance. Further, dynamic programming techniques help in coping with the real-time problem: Multiple possible solutions are created, maintained, and discarded as the matching process is running.

47.3.4 Performance Field Calculation

Let us consider the score space S, the performance space P, the performance transformation $\wp : S \to P$ and the constant vector field $\Delta(x) = \Delta = (1, ..., 1)$ for all $x \in P$. Recall from Section 33.2.2, equation (33.11) that the performance field is the inverse image of Δ and evaluates to

$$\mathbf{Z}(X) = J(\wp)^{-1}(X)\Delta$$

where $J(\wp)$ is the Jacobian matrix

$$J(\wp)(X) = \left(\frac{\partial x_i}{\partial X_j}\right)\Bigg|_{X_j=E,H,L,D,...}^{x_i=e,h,l,d,...} = \begin{pmatrix} \frac{\partial e}{\partial E} & \frac{\partial e}{\partial H} & \cdots \\ \frac{\partial h}{\partial E} & \frac{\partial h}{\partial H} & \cdots \\ \vdots & \vdots & \ddots \end{pmatrix}(X)$$

at X. In order to calculate the field vectors in an element X of the given local composition C in S, we have to determine $J(\wp)$. Now, assume that we are given a matrix U_X of not necessarily orthogonal basis vectors based in X. $J(\wp)$ can be rewritten as:

$$J(\wp) = J(\wp)U_X U_X^{-1} = V_X U_X^{-1}$$

where the basis matrix V_X is the image of the basis matrix U_X. Then:

$$\mathbf{Z}(X) = U_X V_X^{-1}\Delta = det(V_X)^{-1}U_X Adj(V_X)\Delta.$$

The last term is used to identify three cases:

1. V_X is regular, thus $\mathbf{Z}(X)$ is defined,
2. $det(V_X) = 0, Adj(V_X) \neq 0$: only the direction of the vector is given, not its length, and
3. $det(V_X) = 0, Adj(V_X) = 0$: no information at all is given.

While we are now able to calculate the field vectors, the question of how to find the appropriate bases is still open.

47.3.4.1 Obtaining the Bases

The only information available for basis calculations are difference vectors of the given score notes. Basically any difference vectors could be considered as basis vectors, but due to the following restrictions, the candidates have to be selected carefully: First, only notes in a small neighborhood of X should be considered. This *principle of locality* ensures that the basis consists of notes that are in the local musical context.

The second restriction is of a mathematical nature: we have seen that the transformed basis has to be regular in order to be able to calculate the field vector. Because the performance is allowed to have arbitrary deviations from the score, there is no general solution to this problem. What can be done is to decrease the possibility that the transformation of the basis U_X yields to a non-regular basis matrix: This can be accomplished by making U_X as orthogonal as possible. Thus, the selection of the basis vectors is based on the following two criteria:

1. *Locality:* $|det(U_X)|$ is minimal, and
2. *Orthogonality:* $|det(U_X^{Norm})|$ is maximal (where U_X^{Norm} is the matrix of normalized basis vectors).

Note that the two criteria are to some extent in competition, so they have to be weighted and combined. Consequently, a basis-calculation algorithm has to select bases by searching for

$$min\left(w_{loc}|det(U_X)| + w_{orth}\frac{1}{det(U_X^{Norm})}\right)$$

with w_{loc} and w_{orth} being positive pre- or user-defined weight values.

Unfortunately, there is still one case that has to be dealt with: the case where it is not possible to find a regular basis matrix U_X in a small neighborhood of X. This may occur if all notes have the same loudness, if the basis has to be calculated for an isolated chord, where all onsets are equal, or for repeated notes with the same pitch. The only option left here is to construct orthogonal basis vectors that ensure that the basis remains regular.

Finally, the pseudo-code for a basis-calculation algorithm can be given:

```
for(each Note X in Score S) {
  List neighbors = S.getNeighborList(X, maxDist);
  List basisVectors = emptyList;
  for(each Note N in neighbors)
    basisVectors.add(N - X);
  List bases = getCandidates(basisVectors);
  X.basisCost = infinity;
  for(each Basis B in bases) {
    float basisCost = wLoc * abs(B.det()) + wOrth / abs(B.norm().det());
    if(basisCost < X.basisCost) {
      X.basisCost = basisCost;
      X.B = B;
  }
}
```

The function `getCandidates()`, whose pseudo-code was omitted here, generates a list of bases containing the permutations of the basis candidates, and also adding constructed basis vectors if necessary. The above algorithm can be optimized by generating the permutations on the fly, the best expected ones first. In that case, the candidate list can be sorted by increasing distance, and the distance is used to cancel the loop, as soon as it is known that a lower `basisCost` can not be reached anymore.

47.3.5 Visualization

One of the most straightforward applications of a calculated performance field is its visualization. Vector field visualization has been successfully used in many science and engineering domains, i.e., in gas and fluid dynamics. Thus, many different techniques and their corresponding implementations are available. Common to all those methods is that they should be *accurate*, *fast*, and display the field in an *intuitive way*. See [167] for an advanced method that is suited for 2D as well as for 3D visualization. This section shows how the calculated field vectors need to be processed in order to make them available to such standard visualization methods.

So far, we have dealt with a score C, consisting of a set of notes, the corresponding performance D, and the associated set of calculated field vectors F. The points in those sets reside in an n-dimensional space, n being the number of symbolic sound parameters. For visualization, n will normally be too large, so as a first step it has to be decided which parameters are to be used for visualization. For instance, we may choose onset E as the horizontal axis and pitch H as the vertical axis in a 2D setup. The remaining sound parameters are omitted. Further, the desired field vector components have to be selected, for example E in horizontal direction and D in vertical direction for a tempo-articulation field.

47.3.5.1 Field Interpolation

Typically, when dealing with vector fields, the field vectors are arranged in a grid of a given resolution. In contrast, our setup implies that the score points reside at arbitrary locations, making it impossible to use standard vector field visualization methods. Thus, a conversion from the calculated field vectors to vectors located on a grid is necessary. This can be accomplished through *interpolation.*

At first sight, a triangulation of the given set could be considered, making it easy to calculate the interpolated grid vectors in the resulting triangles. However, since the different symbolic sound parameters have different meaning, triangulation is not well suited here: interpolation should occur in a musically meaningful way. Therefore, it makes sense to perform interpolation in a defined recursive order. For instance, when interpolating an ED field, first the D axis of the grid is considered and then the E axis. More precisely, one draws hyperplanes $\mathrm{H}_1, \mathrm{H}_2, ..., \mathrm{H}_k$ perpendicular to the n-th axis in the symbolic parameter space S such that every point of the given composition C sits in one such hyperplane. By recursion, we suppose that the interpolation is available for the first $n-1$ coordinates. To get the interpolation value on an arbitrary point X, one draws the straight line through X and parallel to the n-th axis. This line cuts two neighboring hyperplanes in points P, Q. The values in P and Q are then interpolated by a cubic spline with zero slope in P and Q. For a detailed description see Section 32.3.2.1. What happens at the boundaries of the given set? Since no field vectors are available, boundary vectors have to be defined. When having a look at the theory of the former sections, it becomes clear that outside the boundaries a frame of diagonal vectors has to be placed. The distance between the boundaries and the frame is constant and has to be predefined.

47.3.6 The EspressoRUBETTE®: An Interactive Tool for Expression Extraction

The methods for algorithmic extraction of musical expressions have been implemented in a tool called EspressoRUBETTE®. The tool can run as a stand-alone Java application. The Swing and Java2D classes take care of the user interface, and the user can manipulate calculation and visualization parameters through a simple dialog panel.

As an alternative and more flexible approach, the software also implements the RUBETTE® interface and can thus be integrated into the Distributed RUBATO® framework, see Chapter 40.

47.3.6.1 Example 1: Tempo Field of a Chromatic Scale

Let us now give examples of calculated performance fields. Figure 47.3 shows a chromatic scale and its performance. In this case representation is close to the one of a piano roll: the horizontal axis represents onset, pitch is mapped to the vertical axis. The width of events corresponds to their duration. Note that the EspressoRUBETTE® allows arbitrary redefinition of those mappings. The top section shows the score containing the chromatic scale, twelve note events in increasing pitch order, all with the same duration. A hypothetical performance of the twelve events is shown in the bottom section. The first three events are played at the same speed as the original MIDI score. Then the performance is getting slower, and towards the end it is getting faster again. The last two notes are played faster than the MIDI score.

This situation is depicted by the calculated and visualized vectors in the top section: the first, the second and the eleventh vector are diagonal vectors, stating that the notes are played at the given tempo. The angle of the other vectors depends on the local tempo played at a given note. The middle section shows the corresponding interpolated field, at a resolution of 400 times 200 cells. Here, the slope of each vector has been mapped to a color, and its length is related to the brightness of a cell.

47.3.6.2 Example 2: Excerpt from Czerny's Piano School

The second performance is a real-world example, namely a performance field for an exercise from Carl Czerny [228], as recorded on an MIDI file. Figure 47.4 shows the first two measures of the exercise, axes and note representations are as in the previous example. The upper half shows the performance field on the score

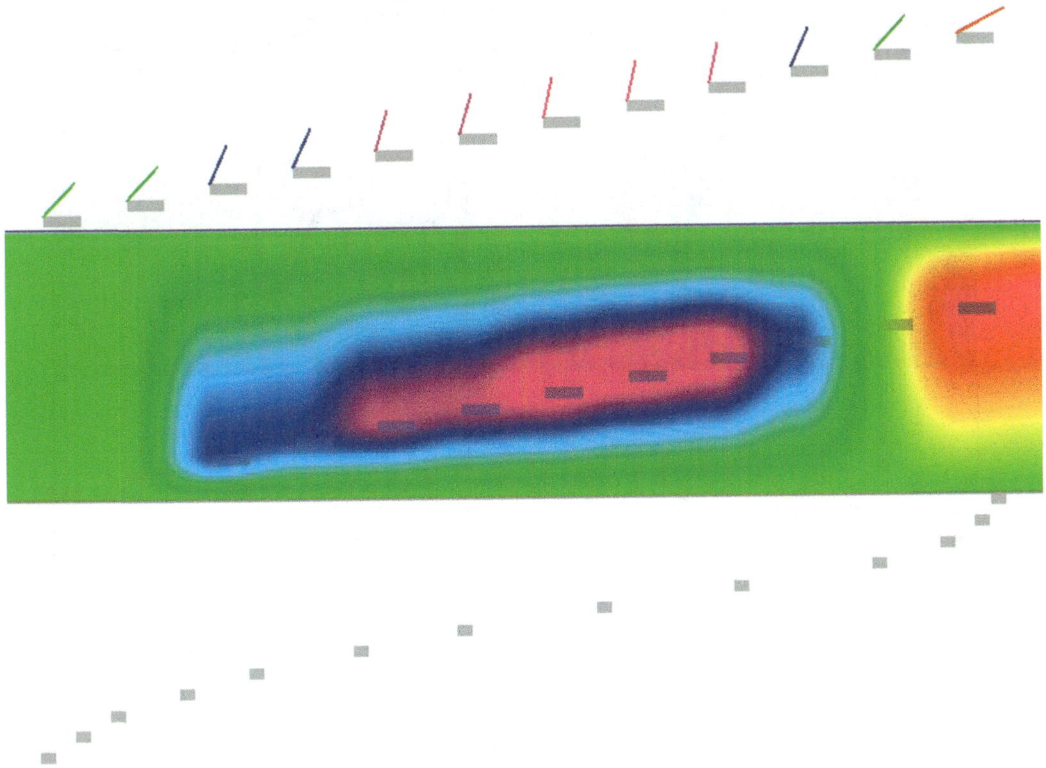

Fig. 47.3. A chromatic scale and its performance: Above the field vectors, middle: the color encoded and interpolated field vector, below the performed scale.

space $S = EH$, the lower half shows the physical space $S = eh$ of the performed piece (physical parameters being written in small letters). The exercise shows the Chopin rubato, i.e., the right hand plays the melody slightly shifted in time against the firm left hand chords in such a way that synchronization is recovered at the end of measures. The field shows the E- and D-components of the four-dimensional $EHLD$-performance field, as encoded by colors. One recognizes that the left hand notes are quite near in color to the green color which encodes the diagonal unit field for mechanical, unshaped performance. We see that there is a right hand rubato effect in the middle of each measure, and significantly more in the second measure. The cyclic coloring effects are due to a multiple covering of the color circle in order to make the fine slope differences of the performance field more visible.

Remark 22 The results - calculated and interpolated performance fields - contain explicit expressive information and are available for visualization or for other performance analysis tools. The algorithms are not restricted to specific sound parameters, and the method can thus be used for extensive expressive analysis. Currently, basis calculation imposes the biggest limitation. In some cases, the calculated basis of a note does not correspond to its musical context, resulting in field vectors that are - while being mathematically correct - hard to understand. Here, ongoing research will definitely deliver better results.

A promising field of further research is also the insight that performance fields are not restricted to musical data. In medical applications and in computer-aided anthropology, the growth information of human bones and organs can be extracted in a similar manner - in which case we may talk about *Nature's Performance*.

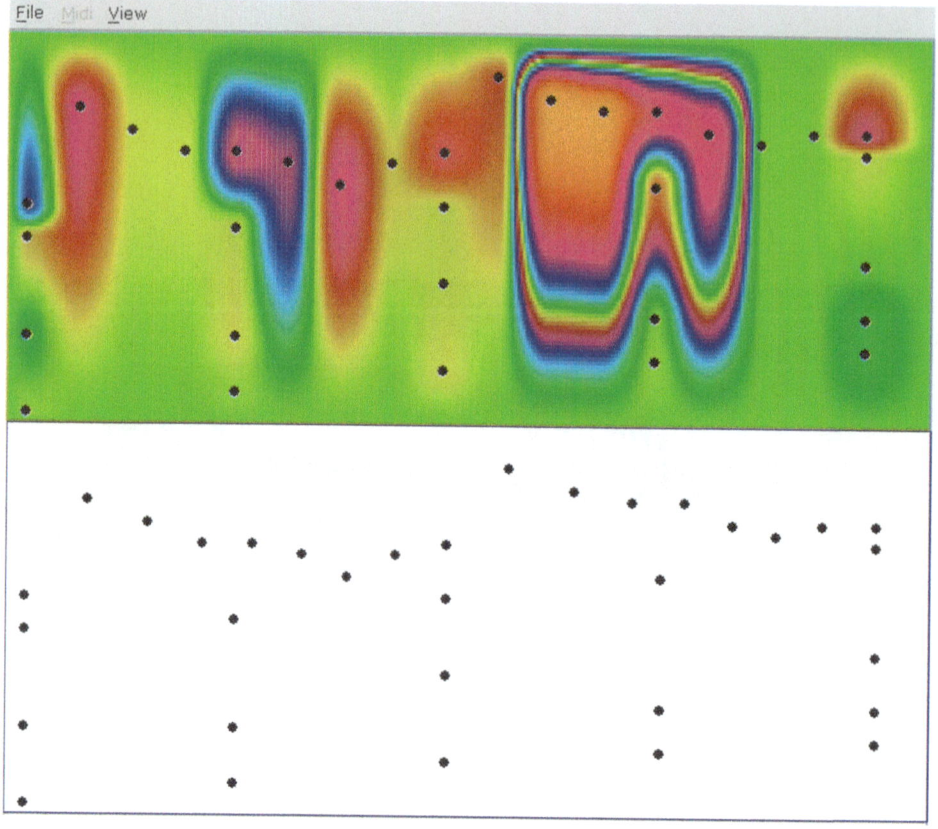

Fig. 47.4. A performance field for an exercise from Carl Czerny's piano school.

47.4 Local Sections

Summary. We study canonical sections, i.e., selectors of points in grammatical varieties. The subject deals with problems of how to choose a determined critique out of an entire variety of possible critiques.

– Σ –

The generic isomorphism of critical fibers as shown in the above Corollary 22 suggests that we should look for musically motivated restrictions on the admitted system parameters in order to obtain more specific information. In this section, we propose a model which is based upon the causality and finality of locally linear grammars and gives a manageable approach to the structure of causality and finality in the transition parameters $c^m_{x,y}$ from a daughter x of a mother m to its sister y (notation inherited from Section 47.2).

The idea is to parametrize the transition parameters by a small set of shaping parameters which give the system of all $c^m = (c^m_{x,y})_{D_m(T)}$ a causal-final coherence. Suppose that the daughters in $D_m(T)$ are linearly ordered with respect to time. For example, this may happen if these daughters are defined by onset intervals of a sequence of adjacent measures or periods. Then we may just enumerate these daughters $x_1 < x_2 < \ldots x_{d(m)}$ in ascending order and temporarily abbreviate $c^m_{x_i,x_j} = c_{i,j}$. In this notation, if $i < j$, this means that $c_{i,j}$ measures the *causal* influence of the prior LPS x_i on the later LPS x_j, while $i > j$ yields a measure $c_{i,j}$ for the *final* influence of the later LPS x_i on the prior LPS x_j. The diagonal values $c_{i,i}$ measure the "autocorrelation" of LPS x_i.

With this in mind, we want to give the value matrix c^m a simplified shape as follows: Suppose that the index set $1, 2, 3, \ldots d(m)$ is evenly distributed on the interval $[1, -1]$, i.e., $\lambda(i) = 1 - (i-1)2/(d(m)-1)$. We now define the function $deformLiGr$ on the square $[1, -1] \times [1, -1]$ on the real plane as follows:

$$
\begin{aligned}
&spl(x) = e^{-(0.4x)^2}\\
&LiGramm(causal, final, x) = \begin{cases} (1 + 0.1.causal.x^2)spl(x) \text{ if } 0 \leqslant x,\\ (1 + 0.1.final.x^2)spl(x) \text{ else,}\end{cases}\\
&LiGr(causal, final, x, y) = LiGramm(causal, final, (x + y)/2)\\
&deform(start, end, y) = 0.5((start - end)y + (start + end))\\
&deformLiGr(causalStart, causalEnd, finalStart, finalEnd, x, y) =\\
&LiGr(deform(causalStart, causalEnd, y), deform(finalStart, finalEnd, y), x, y). \qquad (47.6)
\end{aligned}
$$

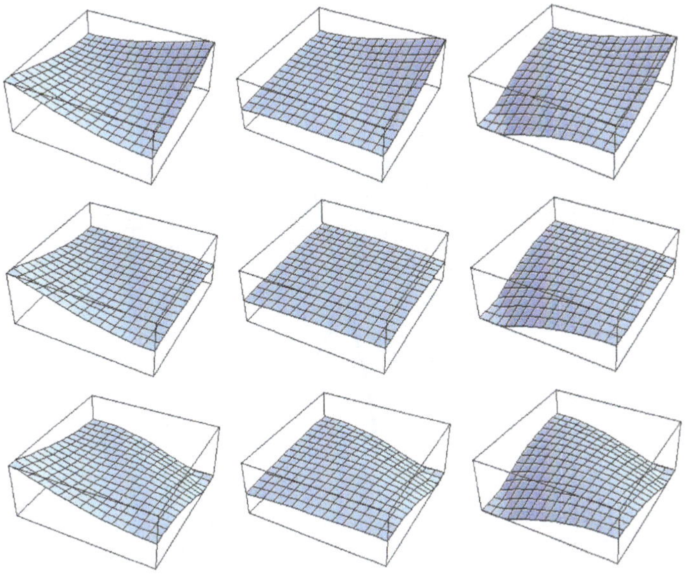

Fig. 47.5. Nine shapes of the function *deformLiGr* for the system parameter 4-tuples *causalStart, causalEnd, finalStart, finalEnd* (from left top to right bottom): (3,3,3,3), (3,3,1.5,1.5), (3,3,0,0),(1.5,1.5,3,3),(1.5,1.5,1.5,1.5),(1.5,1.5,0,0),(0,0,3,3),(0,0,1.5,1.5), (0,0,0,0). The causal extremum $(1, d(m))$ is to the left, the final extremum $(d(m), 1)$ is to the right of each surface. The horizontal diagonal is the autocorrelation area which has the constant value 1.

For each fixed pair x, y, the function *deformLiGr* is an affine function of the four system parameters *causalStart, causalEnd, finalStart, finalEnd*. The system parameters contain information on the strength of the causal and final correlations at start and end of the given time interval. Figure 47.5 gives an image of the function for nine different system parameter combination. The function is evaluated on each pair $x = \lambda(i), y = \lambda(j)$ and yields $c^m_{i,j}$ as an affine function of the four system parameters. We see that low parameter causal and final parameter values, respectively, give low correlations in causal and final direction, respectively. For example, the top right "flying carpet" is strong in final, but weak in causal direction.

This means that we are looking at an algebraic variety which is defined by the substitution of the generic transition coefficients $c^m_{i,j}$ by functions

$$c^m_{i,j}(causalStart^m, causalEnd^m, finalStart^m, finalEnd^m)$$

of four system parameters which are a function of the given mother m. If $m(T)$ is the total number of mothers of the stemma quiver T, this yields a total number of $4.m(T)$ *causal-final variables* which add to the analytical variables from the modules B_x.

To be clear, the shape of a "flying carpet" is a transcendental function involving exponential and quadratic components, but the causal-final system variables which define the actual shape of a carpet are only involved in an affine way.

47.4.1 Comparing Argerich and Horowitz

Summary. This section applies the comparative theory to a particular case of agogics performance in Robert Schumann's "Träumerei" by Martha Argerich and Vladimir Horowitz (Figure 47.6). It turns out that Argerich's agogical coherence is global compared to a rather local coherence with Horowitz.

– Σ –

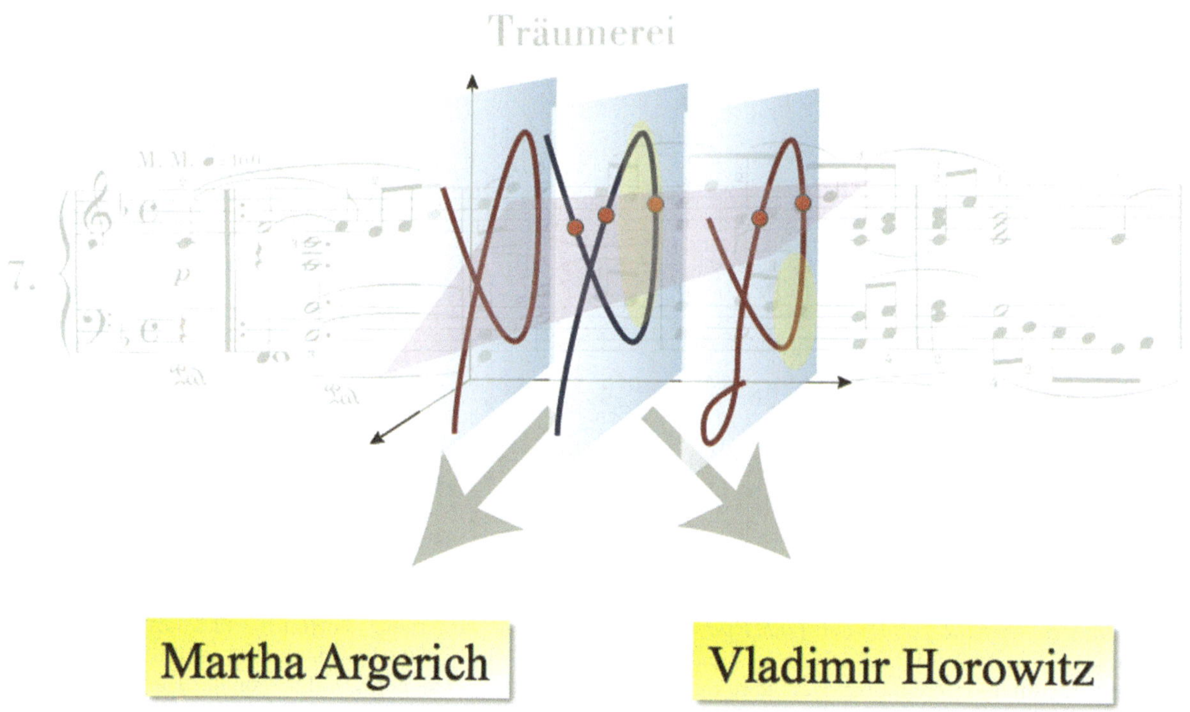

Fig. 47.6. The restriction of the generic fiber spaces to 'flying carpet' coherence domains yields characteristic differences between Argerich's and Horowitz' performances.

In this section, we shall apply the preceding approach to the inverse problem regarding tempo curves and their shaping by use of motivic weights and associated operators. The piece is Robert Schumann's Kinderszene 7 "Träumerei", the performances are those by Martha Argerich (ARG) and Vladimir Horowitz (HO1 in Repp's list), both measured by Bruno Repp [886] among a total number of 28 performances that were already discussed in Chapter 44.

For our example, we take the stemma on the onset space which is defined by the primary mother R_M frame that extends from the first onset through the end of the last event. This mother has four daughters $R_A, R_{A'}, R_B, R_{A''}$. Frame R_A is the time frame of part A from the very beginning to the end of measure 8. The second frame $R_{A'}$ is the time frame of the eight measures of the first period, the frame R_B is that of the third eight measure period, the last frame $R_{A''}$ is that of the reprise eight measure period to the end. Each of these daughters is the mother of eight measures, except the first, which has a first daughter including the upbeat quarter in the first full measure. Let us denote the daughters of R_A by $R_{A,1} \ldots R_{A,8}$, then those of $R_{A'}$ by $R_{A',1} \ldots R_{A',8}$, those of R_B by $R_{B,1} \ldots R_{B,8}$, and those of $R_{A''}$ by $R_{A'',1} \ldots R_{A'',8}$.

For the analytical parameters, we take a constant vector space $B_x = B = \langle W_1, W_2, \ldots W_t \rangle$ generated by a finite family $(W_i)_i$ of global weights on the onset frame R_M. For a given onset frame of our stemma, we have the linear map $\varphi_x : B \to End(A_x)$ into the vector space of C^1 tempo curves on the stemma frame x. It maps the weight W to the scaling transformation by the quantity $\varphi_x(W) = |x|^{-1} \int_x W$. This operation leaves constant fields constant, polynomial fields polynomial, etc., it conserves any reasonable special type of tempo curves. Observe that we really restrict global weights to local frames and do not recalculate local weights when averaging.

This being the case, we start by a global constant default tempo dT, then take its restrictions to the period daughters, then apply all the scaling transformations to the restricted tempi, then add up all the shifted contributions that are multiplied by the crosscorrelation coefficients from a "flying carpet" with causal-final variables

$$causalStart^M, causalEnd^M, finalStart^M, finalEnd^M$$

on the periods. Then we have the 32 restrictions to the single measures, the application of the scalings according to the averaged weight restrictions to the measures, and then ending by the cross correlations according to "flying carpets" for the four periods and eight measures each, with the total of 16 causal-final variables

$$\begin{gathered} causalStart^A, causalEnd^A, finalStart^A, finalEnd^A, \\ causalStart^{A'}, causalEnd^{A'}, finalStart^{A'}, finalEnd^{A'}, \\ causalStart^B, causalEnd^B, finalStart^B, finalEnd^B, \\ causalStart^{A''}, causalEnd^{A''}, finalStart^{A''}, finalEnd^{A''}. \end{gathered}$$

We have applied this system to the one and only motivic boiled-down weight

$$x^{P_{Träumerei}}_{MeloRubette, Param_{melodic}}$$

which was also used in the statistical analysis of Chapter 44. This yields a total of 21 variables and a total of 32 cubic polynomials (one for each measure) in these variables, whose solutions yield the variety lying above the average measured tempi on the 32 measures. Therefore, it cannot be expected that we really have zeros of all these equations.

With Mathematica® routines, we have therefore calculated local minima of these polynomial equations and found the following results [690, III] in terms of causal-final variable values:

Result 7 *Period level:*

- *In the inter-period coherence, Argerich is more final than Horowitz, whereas the causal level is more pronounced by Horowitz.*

Result 8 *Measure level:*

- Horowitz *plays the first period with pronounced causal and final coherence, whereas the causal coherence decreases to a very low level towards the end of the piece.*
- *The repetition A′ of the first period A shows a 'relaxation of coherence' which may be justified by the repetitive situation.*
- *The development section B slightly increases the final character.*
- *The recapitulation seems to be quite 'tired': the causal character is very low, the final character is decreased.*
- *For* Argerich, *the first period has a less coherent ambitus than with Horowitz.*
- *In contrast to Horowitz, the final coherence of Argerich increases as the piece goes on.*

- *The development and the recapitulation are pronouncedly final. The development and the recapitulation shows a consciousness of the end of the piece which is absent with Horowitz.*
- *In other words, Argerich's recapitulation is 'prospective' and not 'retrospective'.*

These calculations are however only locally relevant, and a global solution space with more subtle estimations should also be calculated. Nonetheless, an identical algorithm is applied to two samples and therefore, comparing these inverse performances is admitted.

Part XIII

Operationalization of Poiesis

Check for updates

48

Unfolding Geometry and Logic in Time

Keeping time, time, time,
In a sort of Runic rhyme,
To the tintinnabulation that so musically wells
From the bells, bells, bells, bells.
Edgar Allan Poe: The Bells (1849)

Summary. Musical poiesis in composition (and performance) is intimately related to a projection of abstract objects into time. We discuss the logical and geometric aspects of this mapping process. The subject is crucial to the entire art of music since music is involved in the creation of autonomous time beyond the physical "tyranny" of real time. This enforces a review of Michael Leyton's theory of time as a philosophical category which is *derived* from spatial symmetry transformations [607]. We also discuss the role of unfolding insight within the syntagmatic discourse of music.

– Σ –

After the development and analysis of the various aspects of musical structure and process in this book, it should no longer astonish that the creation of music involves an incredibly complex semiotic, communicative, and reality-critical construction. Far beyond a simple representation and performance of elementary sound events, music is a narration of strong logical and geometric categories, or, at least: without such an intense existentiality, music would never reach the status of a valid antiworld which takes us to an autonomous time and space. This is valid all the more than straightforward common signification processes fail, and meaning has to be built without external references.

So the question is legitimate: What is the story that compositional narration is likely to convey to the listener? This question is critical in two regards: first, the narrative discourse without common, external content is not likely to be a remake of the narrative discourse in ordinary language. Second: The absence of an ordinary content (the story) makes it a questionable point whether narration in music is a reasonable category at all, or whether it is rather a metaphor which one instantiates *à défaut de mieux*, and which one should avoid rather than abuse.

Of course, the musical performance is embedded in the physical time-line, just like the ordinary story which is being told in a piece of literature. But in literature, this time surface transports a hierarchy of time strata which owe their existence to told and telling times of a time-sensitive reality, be it a description of events and movements in a fictitious world, be it the story teller's discursive telling time. In music, this is not possible. Except physical time, there are no timely contents that are referred to and represented, including a beginning, middle, and ending part, as Aristotle has described narration in his *poetics*. This means that if narration does exist in music, it must not only invent an autonomous content layer, but in addition, it has to create the time hierarchy that organizes the narrative stream. In this short chapter, we want to give some remarks on how such a creative process and its contents could be conceived.

G. Mazzola, *The Topos of Music II: Performance*, Computational Music Science,
https://doi.org/10.1007/978-3-319-64444-8_17

48.1 Performance of Logic and Geometry

Summary. In contrast to linguistic discourse, musical syntax does not, a priori, tell a story which has its own time dimension. We contend that the intrinsic story in music is about *performing logic*. This includes—in particular—that musical time means above all syntagmatic time, and not the material time shared by the parametrized music events. We relate the subject to Algirdas Julien Greimas' theory of narrativity [390].

– Σ –

Although there is no common storyboard in music, it does nevertheless realize a narrative discourse. We shall see in chapter 53 that the discursive dialog among four humanist persons (as represented by four instruments from the family of violins) is a characteristic feature of the string quartet. This is all the more remarkable that string quartet music is quite the contrary of a program music, it has always been the art form of "absolute music" where an abstract "musical idea" is processed.

So what are these musical personalities talking about? In common musicology, one would argue that they are organizing musical ideas. We suggest that more precisely, they are constructing a poetical work along the principles of Jakobson's poetical function: projecting the paradigmatic axis into the syntagmatic axis, see Section 11.6.1 for this function. Such a paradigm could be a pitch or chord class, a tonality, a motif or a melody, a contrapuntal line such as cantus firmus or discant, a rhythm, or an instrumental color, for example. The projection of such paradigms into the syntagmatic axis means (1) selecting representative instances of such paradigms and (2) arranging these instances along the syntagmatic contiguity. So the paradigms in absentia are syntagmatically represented in praesentia. This may be a contiguity in time, such as is the case for a succession of tonalities, or else a contiguity in pitch, such as we have discussed for contrapuntal voices, or a contiguity in a sound color space for a multi-instrumental projection.

The point here is that these syntagmatic relations are not linear, i.e., the contiguity can extend to different dimensions. And it can also take place on relatively abstract levels in the sense that, for example, a tonality need not be the concrete score level, but a preliminary organizational level which is a generic scheme for the score realization. We shall also see in chapter 52 on the OpenMusic software that the abstract syntagms are completely natural in intelligent implementations of compositional strategies.

The syntagmatic arrangement of paradigmatic representatives creates the first instance of Greimas' fundamental categories of narration, i.e., succession. But we again have to stress that succession is not necessarily one in physical time, it is one in an abstract parameter space, or even in a space form of generic concepts. The concatenation of successive units is given a logical justification by the insertion of transformational process units, such as contrapuntal interval relations, contrapuntal or harmonic rules of progression, modulatory parts between adjacent tonalities, transformations of motives to their variations, for example. This enrichment is associated with Greimas' category of transformation which, together with the states of succession defines his "programme narratif"[1]. In the third stage of the narrative organization, the category of "mediation" is recovered in order to embed the narrative program in a global reasoning, i.e, the declaration of a purpose behind the narrative program. This can be, for example, the overall strategy for a modulatory plan as we have known it in the analysis of the modulatory landscape in Beethoven's op. 106, see Section 28.2.

Summarizing, the narrative organization is a performance of the logical strategies which dispatch paradigmatic units in the syntagm of their logical concatenation, the latter being explicated on successive transformations, whereas the whole expresses what one really could call "the musical idea". But the syntagm of this organization is by no means a linear one, and even less one in the physical time of a typical telling instance. And accordingly, the story being told is much more than a temporal succession of events, it is a logical construction of local and global compositions, of morphisms between such objects, of universal constructions in the corresponding categories, short: of logically and geometrically motivated predicate instances. This is why constraint programming has become an interesting field in computer aided composition, but see chapter 52 for details.

[1] "On appelle programme narratif (PN) la suite d'états et de transformations qui s'enchaînent sur la base d'une relation S-O [Sujet-Object] et de sa transformation."

At this point we have to a certain degree reached what Eduard Hanslick [438] calls "tönend bewegte Formen". So what is missing? Essentially, there are two major gaps at this stage:

- A narrative time concept and its organization is still missing.
- The unfolding of the syntagmatic logical display into physical performance time is outstanding.

We are now going to discuss these issues.

48.2 Constructing Time from Geometry

Summary. We sketch predicates of logical and geometric nature which are designed to set up syntagmatic time in music. In this context, Michael Leyton's theory of time is briefly reviewed.

– Σ –

As the present organism is a purely logical and geometric one, the only hope to generate narrative time is by means of mechanisms which turn logic and geometry into time. On the logical level, for example, the implication form $A \implies B$ induces a time category where the condition is antecedent to the implied statement. On the geometric level, there are several symmetry-related "time generators". In fact, a symmetry S can be viewed as a transformation, where objects x are moved around, i.e., $x \mapsto S(x)$, whence a time-stamped relation from x to $S(x)$.

A radical approach to time is proposed by Michael Leyton [607, 608]: In its cognitive *and* physical reality, the time concept is viewed as a derived one. Leyton's theory develops structures to reduce time ontology to a dynamical syntax of spatial symmetries. These symmetry groups are wreath products[2] $F \wr C$ of a control group C and a fiber group F. Leyton's approach allows us to understand music as a *natural* reconstruction method of time *via* syntax of symmetries. This syntax aligns symmetries and their broken variants as a ordering relation: first symmetry, then its broken version. Time is generated via symmetry breaking.

This implies that musical time is generated by such symmetry breaking relations. Together with the logical time construction, we obtain a variety of ordering relations which are induced by symmetry breaking processes. More generally speaking, time is generated by logical and geometrical succession relations. For example, if we recall the model for tonal modulation (Chapter 80) the modulation quantum is a global composition whose inner symmetries relate the antecedent to the successor tonality, but it is not possible to distinguish these tonalities in their logical role from the structure of this symmetric quantum. Leyton would say [608, symmetry and asymmetry principles p.41] that symmetry is without memory, since it is symmetry breaking that constructs memory. The time ordering of the involved tonalities is only generated by marking the trace of the modulation quantum in the target tonality: which defines in fact the modulation steps. This is also the Greimas transformation part relating the two units in the succession chain. Such a breaking of a symmetry in modulatory degrees creates time and bans the symmetry from the syntagmatic surface.

A good example of memory-less construction by unbroken symmetries is Schönberg's dodecaphonic method. It starts on an original dodecaphonic series and proposes a display of some of its (generically) 48 transformed versions under the known group $D_{12,12}$, (see the discussion following definition 22 in Section 8.1.1). There is no specification of the method regarding the instantiation of the involved symmetries, they are external to the composed syntagm, and there is no symmetry breaking that would create a time-line among the 48 variants in that orbit. So the narrative structure of this method is not specified. Maybe this drawback, which lives in a pronounced contrast to the excellent narrative sonata form, is one of the reasons for the failure of *pure* dodecaphonism, i.e., dodecaphonism that is not enriched by syntagmatic concepts for the narrative construction.

So let us assume that syntagmatic time is generated as an ordering relation between logical and/or geometric units. Such a time is not necessarily linear, there may be several competing time strings. For example, recall from counterpoint theory that "punctus contra punctum" relates to the horizontal interval

[2] See Appendix 103.

succession as well as to vertical cantus firmus vs. discantus succession (as the melodic implication). So we have two time strings which are simultaneously present in the "contra" process. In a modulatory process, the time-line of tonalities can be superimposed by a vertical time-line of melodic deduction from the harmonic basement, and a time-line induced by an instrumental hierarchy. Such abstract time-lines are the narrative bricks of the musical counter-world.

Suppose that we now dispose of such a narrative organization. Then, its unfolding is the following project: In performance, we have to unwind the logical and geometric time-lines in a space of performance time. We have discussed at length the time hierarchies and stemmata of performance in parts VIII and IX. Evidently, these tools must be used to unfold the abstract narrative organization in physical reality. Presently, such a theory is lacking, but the prerequisites are at hand. The main problem with regard to human cognitive capacities is to find an equilibrium between the complexity of the narrative instances of logical and geometric time-lines and the limitation of comprehension of such structures and processes in the physical performance space-time.

This means that we have to name principles that define the basic structures of a global composition with regard to its narrative communication. A global composition has a basic syntagm that is defined by its covering and, more specifically, its nerve. This is already present in Schönberg's harmonic strip between successive chords, which is in fact an elementary but prototypical syntagmatic junction (the transformational part between two adjacent succession units in Greimas' theory). So the composition's nerve is an excellent syntagmatic device in that it includes narrative paths (on the nerve's one-dimensional skeleton) of neighboring local charts in praesentia. For example, if we have the interpretation of a diatonic scale by the seven triadic chords, we obtain the harmonic (Möbius) strip as a global composition. The geometric relation of two neighboring degrees on the strip, I_C, III_C, say, can be given a narrative direction in the succession I^6_C, III_C where the common notes e, g of the inversion I^6_C and the third degree may be held such that we only have the time event of a downward movement $c \mapsto b$ to indicate the time-line from I^6_C to III_C. We contend that any unfolding of the syntagmatic logical and geometric time-lines should locally *on the physical time axis* produce a small linear storyboard, i.e, minimize the multi-dimensionality of the abstract syntagm, or at least try to stress its most important components. To put it the other way round: Understanding the abstract narrative structure of a composition's nerve amounts to realizing a sequence of paths through the nerve's one-dimensional skeleton such that eventually, all important parcours of the nerve's 'city map' have been exposed. The problem here is analogous to the problem of how the eye's path should look when it observes a painting in order to obtain a good understanding of what is seen, or else to the problem of how we should walk through a house in order to optimally understand it as an architectural organism. It is the problem of curvilinear reduction of high-dimensional objects, a problem which by the human language stream of words has been solved in a very special context and with the success of a very limited textual representation of the world.

48.3 Discourse and Insight

Summary. This final section locates the musical discourse as a means of gaining insight into complex multidimensional structures by syntagmatic sequentialization (linearization). This contrasts with the linguistic discourse, since the latter has rich semantic rootings which are absent in music, must therefore be balanced by explicit discoursivity.

– Σ –

The advantage of common language above music is the presence of external semantics, the language stream runs around any kind of complex entities that are *not* part of the language structure. This makes language an easy business since it does not include its objects, but only points to them. In contrast music cannot point outside (except for trivial special cases of onomatopoiesis), it has to build everything from an inner discourse. This is also its advantage and scope: to be able to invent worlds instead of pointing to them. In this sense it is an excellent exercise to learn what is really needed to create a 'world' in our imagination, i.e., a meaningful spiritual architecture without external roots. The investigation of the narrative discourse should teach us how we gain insight into things when telling their story, meaning: how to understand music while playing it.

49

Local and Global Strategies in Composition

Aus den Auswahlkriterien also entsteht die Dialektik der Abfolge oder Verknüpfung von Lokalstrukturen, wobei diese Auswahlkriterien bestimmend sind für die Eingliederung der Lokalstrukturen in die große allgemeine Struktur, die Form.
Pierre Boulez [138, II]

Summary. This chapter sketches the compositional process between paradigmatic selection and syntagmatic combination within the musical sign system. Apart from these semiotic perspectives, the process is characterized by a "dialectic" interrelation of local and global criteria. These features—well known from the general structure theory of global compositions—reappear in the special light of poiesis: The construction of a composition resembles the step-by-step completion of a puzzle of logical units, distributed in syntagmatic time, and selected to optimize association to already placed units. This activity is remunerative and fed back by a successive accumulation of poetical semantics.

– Σ –

We do not intend to describe the psychological path in the composer's mind, but the fundamental strategic steps, independently of the psychological or cognitive realization. We also do not impose this strategy on any composer, but want to sketch a possible system for composition which can also be implemented on the software level. We shall see that two implementations: *presto*® and OpenMusic realize quite a portion of these ideas.

Essentially, the poiesis of a musical composition is the construction of a global composition, i.e., a patchwork of local charts, together with gluing transformations. Moreover, the charts are selected from a set of paradigms and combined according to a set of syntagmatic rules. But the concrete making of a composition is not a one-step procedure, it results from a successive completion of an ensemble of charts. Ideally, one is

1. given a simplicial complex and would like to realize it as the nerve of a covering by local charts.
2. One also wants the charts to pertain to determined isomorphism classes or else being specializations of representatives of such classes.
3. One finally wants particular gluing transformations on determined parts of the charts.

Point two reflects the analytical approach using the paradigmatic theme of Ruwet and Nattiez, see Section 11.7. But it extends it in that the nerve is planned as a syntagmatic design pattern, further specialization and not only transformation of basic paradigms is allowed, and finally, the gluing transformations for the syntagmatic combination are thematized.

G. Mazzola, *The Topos of Music II: Performance*, Computational Music Science,
https://doi.org/10.1007/978-3-319-64444-8_18

49.1 Local Paradigmatic Instances

Summary. Local paradigmatic strategies can be split into transformational and topological procedures. In each case, we describe the basic options.

– Σ –

Recall from chapter 10 that paradigmatic relations can be either transformational or topological—or combinations thereof. Whereas the first is related to symmetries, the latter is related to the idea of variation. Usually, transformations will quite brutally change the auditory impression of a local composition and must be used with care, whereas topological variation is more adequate for instantaneous recognition of similarity.

49.1.1 Transformations

Summary. Starting from Schönberg's dodecaphonic method, we describe varieties of local compositions produced by a determined set of transformation groups, *the paradigmatic themes of poiesis*, acting on a selected set of local "germs".

– Σ –

This method starts from a sequence $S = (G_1, \ldots G_c)$ of local germs, for instance thematic units such as the singleton $S = (G_1)$ containing a dodecaphonic series, or a two-element set $S = (G_1, G_2)$ containing a main motif and a side motif of a sonata. Then, for each local germ G_i, we are given a transformation group P_i, the paradigmatic group of this germ, which describes the a priori allowed transformations. For instance, for dodecaphonic compositions, we have $P_1 = D_{12,12}$. It is however not necessary to realize the entire orbits $P_i.G_i$ of these group actions, a composition will usually select a small number of such transformed germs. In the construction of the germinal melody of the "Synthesis" composition in Section 11.6.3, see also Figure 11.16, we have 26 isomorphism classes of three-element motives and take one representative of each class, where the transformation group is the same for all classes, i.e., the full affine group on $\mathbb{Z}_{12}$.

The main point in this procedure is the instantiation of the transformations within the composition. In fact, it is very rarely sufficient to place just two objects $G_i, p.G_i$ in contiguous position in order to communicate the particular symmetry p which is responsible for the succession. It is part of the failure of dodecaphonism that its exponents never did care about instantiating p. A nice example of such an instantiation in Beethoven's op. 106 was given in Section 28.2.5.

49.1.2 Variations

Summary. Instead of transformation groups, a set of local "germs" can be deformed according to a set of similarity paradigms (e.g., gestalt topologies)

– Σ –

In this approach, there is no group action and the original germ G_i must be associated with one of its specializations by use of deformation procedures. For example, one might superimpose a force field on the given germ and then move its points around according to the local force action. Such a procedure has been implemented in the OrnaMagic module of the *presto*® software, see Section 50.3. This variational change of perspective is very precious for understanding a germ's potential, but in general will not give us back the original germ. The quantity of change is a function of the used gestalt topology whose choice depends on the composer's preferences.

49.2 Global Poetical Syntax

Summary. According to rules of contiguity and semantic "added value", a variety of transformed and/or deformed local germs is distributed along the syntagmatic space. We systematize the possible procedures according to horizontal and vertical poetical functions.

– Σ –

The construction of a global assembly of local charts which are provided by the local techniques is very delicate, see also Pierre Boulez' reflections [138, II]. For instance, the germinal melody in "Synthesis" was built from the generic motif class representative, then gluing it with a representative of the first different motif class that fits with the first representative on a subset of two points. But the possibility to find such a solution until every class was represented exactly once is not obvious, it is a constraint to the compositional material that could have ended in an incomplete solution. The normal procedure is a puzzle reconstruction: One begins on any interesting germ and adds other germs at any syntagmatic position (not necessarily adjacent), and successively densifies the display, thereby observing constraints on the nerve that one wants to construct. For example, in the sonata "L'essence du bleu" [708], the plan as described in [670] was to construct a global composition on a motivic zig-zag $\{c, c_\sharp, d, d_\sharp, e, d_\sharp, d, c_\sharp, c\}$ by use of nine three-element submotives in such a way that the nerve turns out to be a Möbius strip.

The construction of a nerve is not only complicated as a constraint problem, it is also a problem of narrative character: the intersection configurations must also be mediated on the concrete time axis, and in such a way that the transformations between associated units can be made explicit, e.g., the Möbius strip must be boiled down to the time axis such that its one-dimensional skeleton (we stick to the ideas from Section 48.2) is transgressed on reasonable paths.

Evidently, the main point of this global patchwork is the syntagmatic distribution of paradigmatic representatives in order to achieve an added poetical value in the sense of Jakobson's poetical function. This objective splits into two subtasks: horizontal and vertical poetical functions according to Jakobson and Posner, respectively.

49.2.1 Roman Jakobson's Horizontal Function

Summary. Distribution of a variety of transformed and/or deformed local compositions can follow Jakobson's correspondence [500] along the syntagmatic axis.

– Σ –

Jakobson's poetical function was introduced in Section 11.6.1. Counterpoint, harmonic syntax or dodecaphonic row distribution are examples of this mechanism. The most elementary realization of this function is the action of a translation group on rhythmical units in order to produce a cyclic character of our composition. This semantic enrichment creates a coherence of the compositional corpus which is basic to all other elements. It may, for example, connect motives, harmonies and similar units if they are distributed on the regular positions of the rhythmic basis. But it also creates a time circle that abolishes the straight physical time-line and rebuilds and "good" cyclic eternity (in Hegel's sense) which our finite life was to negate.

49.2.2 Roland Posner's Vertical Function

Summary. Jakobson's poetical function is orthogonal to Posner's vertical function which relates different signification levels in denotation and connotation. We discuss strategies which make use of this functionality in the production of poetical "added value".

– Σ –

Jakobson's poetical function relates signs of the same semiotic system, for example, two phonological signs such as "bad", "dad" which are positioned in syntagmatically equivalent places, in a rhyme, say. But this structure is enchained with the semantic sign system (by the double articulation of language, see [703]) where the phonological equivalence in the rhyming places produces a paradigmatic equivalence between the bad and dad, i.e., "dad is bad", a meaning which is neither inherent in "bad" nor in "dad". So the level of equivalence is shifted from the phonological to the semantic one. This shifting of equivalence in Jakobson's function illustrates the vertical function as proposed by Roland Posner [860].

In music, such a construction is not obvious since connotative semantic levels are not automatic, but must be constructed or simulated. This can be achieved by a small global composition which is attached to a local 'signal' composition whose equivalence to another local 'signal' composition on the horizontal Jakobson function induces an equivalence of the attached small global compositions. Such a small global composition can be a configuration consisting of a chord interpretation and a melodic unit, for example. So here, the connotative levels are constructed by successively enriched small global compositions within the total global composition.

49.3 Structure and Process

Summary. The poietic process is confronted with the resulting 'score' structure in order to recall a more general approach to the score concept.

– Σ –

We have stressed the global composition as a target of compositional strategies. However, the making of this composition can be more relevant than the result. In other words, the general score concept which we discussed in Section 33.3.2 may be specialized to the interior score known from jazz: Instead of a more or less fixed score structure in form of a global composition, the processes that lead to the effectively played material are stressed, while the resulting structure is quite secondary (albeit not irrelevant, as some fundamental critics of free jazz have argued) in that it is one possible variant of those processes, and another variant—as an exemplification of the operating processes—would do as well. The situation is comparable to algorithms for complex shapes, such as L-systems or fractals. They are defined as processes, and the resulting variants are all an expression of the same basic process type, as with biological phenotypical expression of the genotype[1].

It is a very interesting research area to classify process score types and to implement corresponding tools in music composition software. An example of such an approach (the composition software OpenMusic) is given in chapter 52. The following chapter 50 describes a software (the composition software *presto*®) which is mainly based on paradigmatic strategies. As a matter of fact, such strategies are very difficult to handle because a paradigm is, by definition, in absentia to the text, whereas syntagmatic structures are in praesentia, which means that a software that is built on the LEGO-like juxtaposition of bricks will have an easier acceptance than the paradigmatic one. However, if the bricks are no longer concrete musical material, but abstract units of processual character, the ease of the LEGO approach is no longer valid.

[1] *Phenotype* is the "outward, physical manifestation" of the organism, while *genotype* is the "internally coded, inheritable information" (written in the genetic code) carried by all living organisms.

50

The Paradigmatic Discourse on *presto*®

Damit können Sie mich eine ganze Nacht lang allein lassen!
Herbert von Karajan to Guerino Mazzola
on the occasion of the presentation of the
presto® prototype in Salzburg 1984

Summary. The *presto*® composition software was developed as a commercial implementation and operationalization of mathematical music theory. Several large compositions have been successfully realized on this software [101, 103, 680]. One of them will be discussed in chapter 51. We describe the overall architecture and functionality of *presto*®. This is mainly driven by a local/global paradigmatic perspective. The paradigmata of transformation and deformation are realized on (1) the level of modular affine transformations in the four-dimensional space of onset, pitch, duration, and loudness, and (2) the level of variational deformations. Both paradigmata are discussed and exemplified. We conclude the chapter with a remark on the problem of abstraction in paradigmatic composition, since composers tend to have major difficulties to get familiar with abstract paradigmatic structures.

– Σ –

Although the software was implemented in ANSI-C on now historic Atari®computers, we first describe its functional scheme since it represents a prototypical and still unique way of thinking paradigmatic composition strategies (to our knowledge, no similar strategy has been implemented on software to the present). The *presto*® software as well as the entire source code are GPL and can be downloaded—together with a manual, examples, and a concept presentation software `prestino.prg`—from the book's CD-ROM, see page xlvii.

50.1 The *presto*® Functional Scheme

Summary. We describe the functional scheme of *presto*®. The scheme is centered around the back bone of global score, score, and local score. This threefold architecture reflects an early stage of the global-local paradigm from mathematical music theory.

– Σ –

The functional architecture of *presto*® consists of a series of so-called scores, see Figure 50.1. The global score, the score, and the local score deal with geometric representation and editing of notes (mental note events) on global to local space levels. The recording score allows for mouse-driven and instrumental real-time input, the transformation score deals with graphical input of affine transformations, and the grid score allows for graphical input of ornamental grids. The input goes via recording via mouse or MIDI instrument, loading from MIDI or *presto*® data files, and painting with the mouse. Editing features split into affine

G. Mazzola, *The Topos of Music II: Performance*, Computational Music Science,
https://doi.org/10.1007/978-3-319-64444-8_19

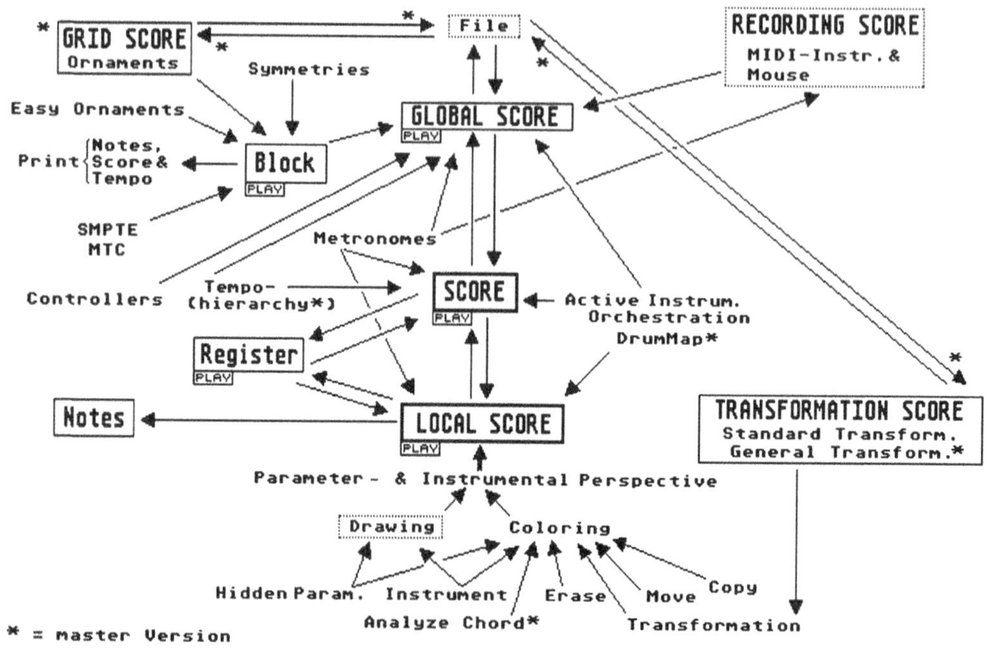

Fig. 50.1. The functional scheme of the *presto*® software. It is centered around the back bone of global score, score, and local score. The graphics of this chapter are all from the original *presto*® manual [680].

transformations on charts of notes, building of ornaments, all-parametric variations according to ornamental attraction and repulsion fields, instrumental and parametric coloring of local charts, Boolean combination operations on groups of notes, and performance editing, especially construction of complex tempo hierarchies. For the latter, please refer to our corresponding discussion in Section 38.2. Output is split into audio and SMPTE, saving is on MIDI and presto formats.

Figure 50.2 shows the main window of *presto*®. The rectangular score window in the middle shows onset (horizontal) and pitch (vertical, can also be set to loudness or duration), and Schumann's Kinderszene "Haschemann". The black subrectangle marks a block which may be copied, cleared, moved, etc. Below the score, the tempo curve is visible. Above the score, the total composition space of the global score (narrow rectangle) is seen, below the tempo curve, six registers for provisional local parts of the score are placed. On top, we see 16 icons for instrumental colors (MIDI program change data), checked means the instrument is active; while editing, one may work on any checked subset. The different score and register windows implement some chart types in the global composition strategy. One may perform any Boolean operations on such parts, e.g., on two registers. Evidently, modern object-oriented windowing techniques could vastly generalize this elementary implementation.

For a detailed editing of small composition parts, the local score window is available, see Figure 50.3.

The local score can be opened by double-clicking on a selected position on the score; after editing it can be merged to the score. A miniature view of the local score's content is visible on the right lower corner of the main window in Figure 50.2. The local global paradigm is realized on the local score via the feature of "coloring". On the local score, we view a four-dimensional cube, representing the four-dimensional discrete torus $\mathbb{Z}_{71}^4$, in any of the six relevant projections $Onset + Pitch, Onset + Loudness$, etc. onto two of the four parameters of pitch, onset, duration, and loudness. All parameters are integers modulo 71 and suitably calibrated: Pitch is a MIDI key number in the interval $[27, 97]$, loudness is a MIDI velocity value between velocity 0 and 127, equally distributed among 71 values, onset and duration are in integer units that can be set to be different fractions of a bar, depending on the time signature and the resolution preferences for each metrical unit.

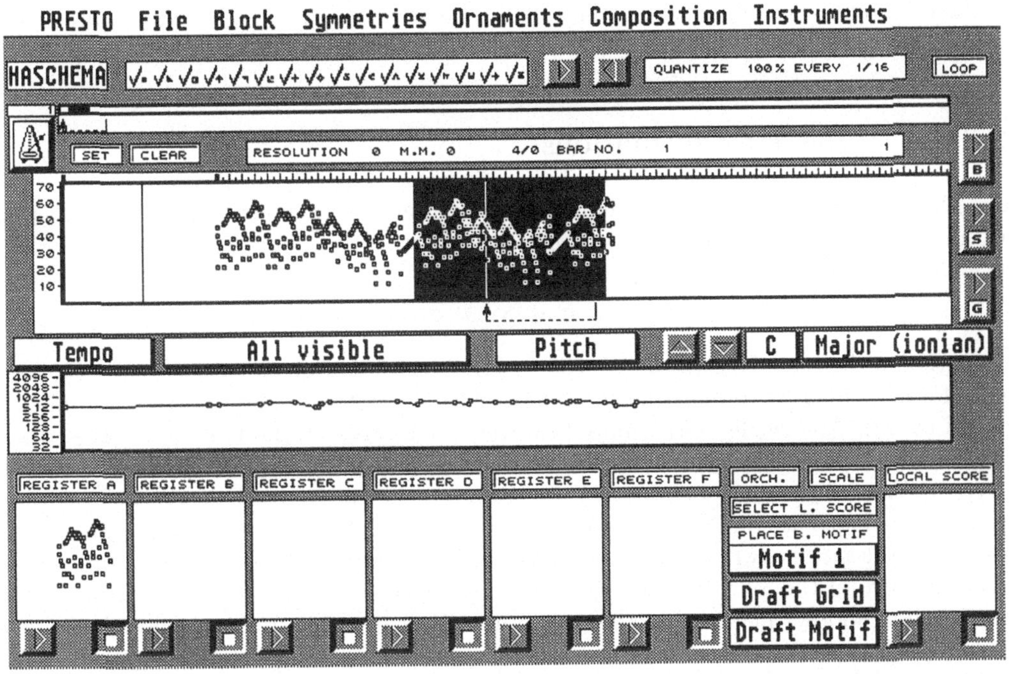

Fig. 50.2. The main window of *presto*®. The rectangular score window in the middle shows onset (horizontal) and pitch (vertical, can also be set to loudness or duration), and Schumann's Kinderszene "Haschemann".

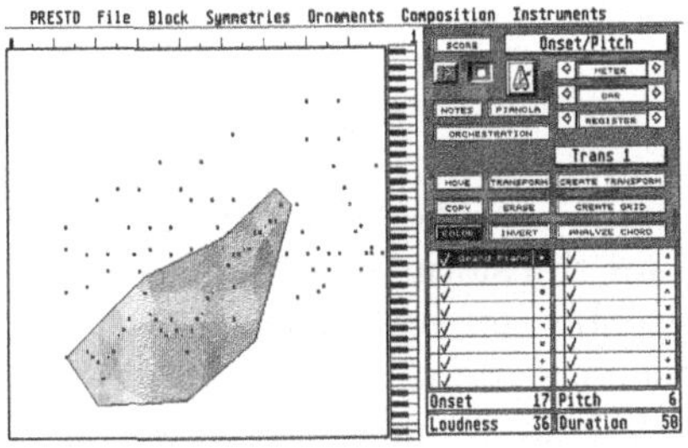

Fig. 50.3. The local score is the interface for detailed editing of small portions of the actual composition. The darkened polygon marks a region where any operation, such as copy, erase, transform, change the instrumental icon, or simply play can be performed. New events can be defined by mouse drawing actions.

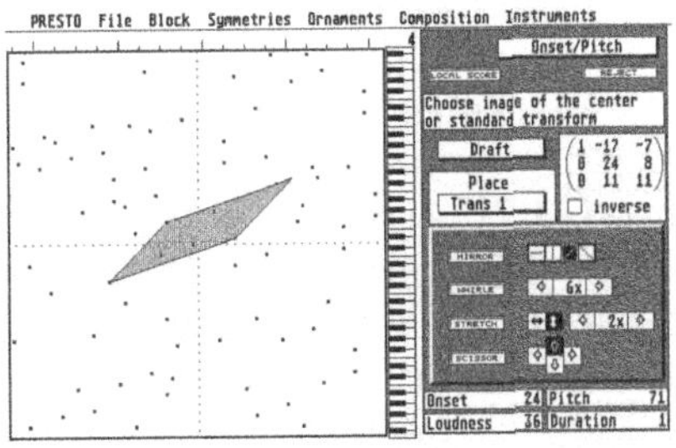

Fig. 50.4. The transformation score is the interface for detailed editing of affine transformations of the torus $\mathbb{Z}^2_{71}$. The user's editing acts on a number of standard transformations (buttons to the right, middle), matrix text fields, and a graphical editing option by direct drawing of affine images of the unit square.

The local-global paradigm is realized by the coloring feature: The user may define any closed polygon in a 2D projection, and thereby select all events that lie in that region with regard to the given projection. Such a coloring domain can be used to do many different things, such as moving around on $\mathbb{Z}^4_{71}$, copying, erasing, transforming, setting new instrumental icons, playing. By mouse-driven drawing actions, new events can be inserted. The parameter plane can be changed at any time. To the right, we see the orchestration (only piano in Figure 50.3, see below for complex orchestras), where any instrumental icon can be unchecked for instrumentally specified editing options.

Of course selection of a toroidal representation and manipulation (for example for shifting operations, where events exit to the right and reenter from the left) is questionable, but once you are dealing with affine operations, such finiteness decisions have to be taken. The advantage of this selection will become clear in the next Section 50.2.

50.2 Modular Affine Transformations

Summary. This section describes the mathematical framework of modular affine transformations in $\mathbb{Z}_{71}^2$, their four-dimensional extension and the combination options for such transformations, their action on selected local charts, and the graphical input features.

– Σ –

The selection of the torus $\mathbb{Z}_{71}^4$ was motivated by small number of pixels in those early screens, together with the need for icons in order to represent instrumental 'colors'[1]. Now, the selection of a prime number was not by case. In fact, the choice offers us a finite field where non-zero determinants of matrices automatically yield invertible matrices. So a large number of transformations becomes invertible, and therefore a large number of compositional processes become non-destructive. The software offers a transformation interface where any affine endomorphism in $\mathbb{Z}_{71}^4 @ \mathbb{Z}_{71}^4$ can be defined graphically as well as numerically see Figure 50.4. We first use the fact (see Section E.3.6, Theorem 61) that $\mathbb{Z}_{71}^4 @ \mathbb{Z}_{71}^4$ is generated by its transformations which only move around two of the four dimensions. So the user may transform his/her material on the selected plane, then switch to another plane for further transformations, etc. The user may directly define a product of any number of plane transformations (for different planes) by a feature which memorizes a list of plane transformations. We have a total of

$$71^{16}.71^4 = 10'596'610'576'391'421'032'662'867'140'133'202'401 \approx 1.05966 \times 10^{37}$$

elements in $\mathbb{Z}_{71}^4 @ \mathbb{Z}_{71}^4$, and the number of

$$10'445'260'466'832'483'579'436'191'905'936'640'000 \approx 1.04453 \times 10^{37}$$

elements in $\overrightarrow{GL}(4, 71)$ according to the formulas in Appendix C.3.5. The transformation score in Figure 50.4 shows the basic input choices: matrix coefficients, standard transformations, and graphical editing by definition of the images of the three points: origin, head of horizontal unit vector, head of vertical unit vector. The parallelogram image of the unit square is visualized on the local score.

The geometric advantages of the prime number 71 are abundant, among others, we have these facts: Since the multiplicative group $71^\times$ is cyclic of order 70, i.e., isomorphic to $\mathbb{Z}_{70} \xrightarrow{\sim} \mathbb{Z}_2 \times \mathbb{Z}_5 \times \mathbb{Z}_7$, we have fifth and seventh roots of unity in $\overrightarrow{GL}(4, 71)$. We also have $SO_2(71) \xrightarrow{\sim} \mathbb{Z}_{72}$ (see Appendix C.3.5), and therefore, a generator D of the special orthogonal group. Such a generator is

$$D = \begin{pmatrix} 30 & 33 \\ 38 & 30 \end{pmatrix}$$

which means that we may view D as a rotation by $360° : 72 = 5°$. In the software, we have implemented the rotation by the triple, i.e., $D^3 \sim 15°$-rotation. Of course, such rotations in the modular torus can transform harmless compositions into very wild looking variants, but we have always experienced that the transformed version maintains certain regularities that were present in the original form.

50.3 Ornaments and Variations

Summary. Modular affine transformations are used to build ornaments by translation grids of "cells" of notes. Such ornaments are used directly as periodic note sets, for instance in drum patterns. They may also

[1] Because of low resolution, color was not a good commercial option in 1988

be used as "background" ornaments whose points act as centers of attraction or repulsion for other notes. The latter method is a multidimensional generalization of the well-known variational technique, in particular of the alteration of pitch sets (scales, chords, motives, etc.).

– Σ –

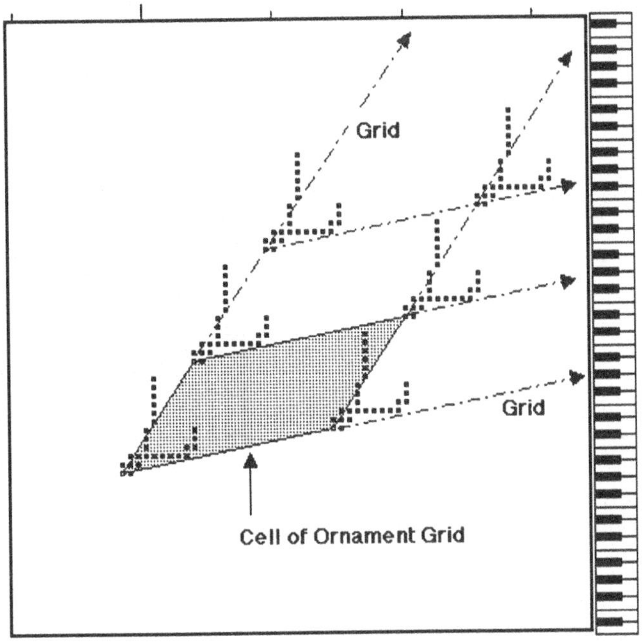

Fig. 50.5. An ornament is defined by a motif, together with a translation grid and a range in each direction of the defining grid vectors. The grid cell is the parallelogram spanned by the generating vectors.

The software's module for ornaments is termed OrnaMagic. The idea is this: The user first defines a motif M, either a small one on the local score, or an arbitrary large one on the score. M is just a local composition on the four-dimensional $EHLD$ space of the software. Next, two (usually linearly independent) translation vectors g_h, g_v define the grid, i.e., the group $\langle e^{g_h}, e^{g_v}\rangle$. This group operates on the motif M and yields a translated motif $M_{i,j} = e^{i \cdot g_h + j \cdot g_v}.M$ for each integer pair (i, j), see Figure 50.5.

The user defines a two-dimensional ornament of a special ornament window, the grid score, see Figure 50.6. This gives us a 'grid of translations' $\bigcup_{a \leqslant i \leqslant b, c \leqslant j \leqslant d} M_{i,j}$. A second method allows the user to define also larger grid vectors on the score level, but the principle is the same.

In the composition "Synthesis" to be described in the subsequent chapter 51, we recognize a number of superimposed ornaments of drum sounds; here in the second movement, see Figure 50.7. The compositional principles will be described in Section 51.3.

Apart from this explicit usage of ornaments, there is a second truly paradigmatic, more precisely: topological usage by *two-dimensional alterations*. Here is the general procedure. We have two local compositions, L, G, where G plays the role of the driving grid, whereas L is the composition which we want to deform. Usually, G is defined by an ornamental construction as described above, but this is not mandatory, i.e., one may also take a large motif M and just apply the grid group on the zero range ($a = b = c = d = 0$).

Given these data, the alteration works as follows. Each event x of L is shifted towards or from G according to a procedure which starts by the calculation of a 'nearest point' $G(x)$ to x in G. This point is found starting on x, and then moving along a spiral outward around x, until the first point of G is hit; this is $G(x)$. Following this first algorithm, several alteration strategies are offered. First, one may choose a degree

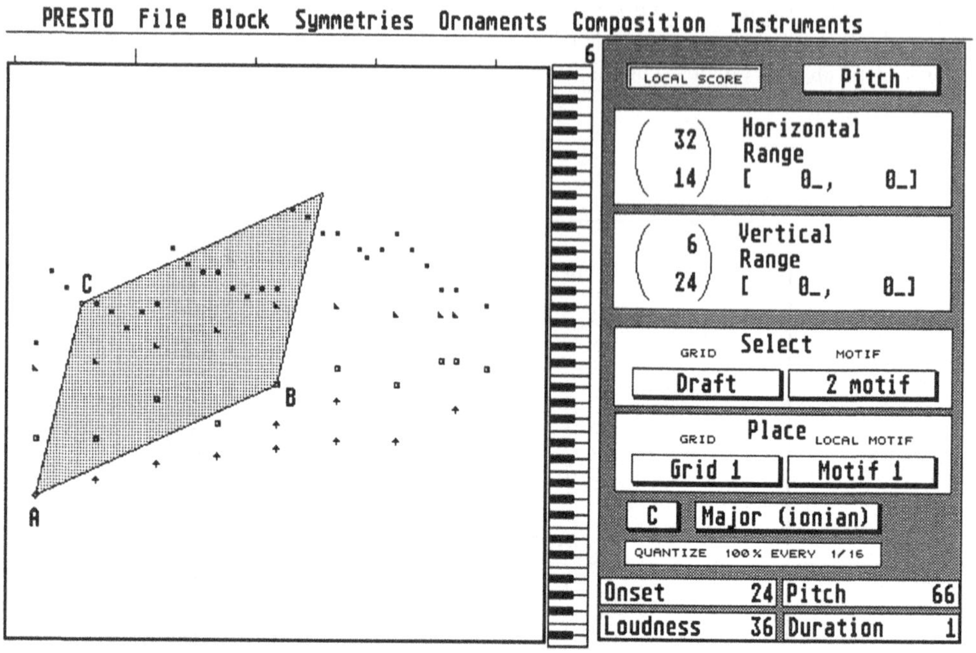

Fig. 50.6. The grid score is the interface for defining an ornament via its grid vectors (so-called "horizontal" g_h and "vertical" g_v) and the range $a \leqslant i \leqslant b, c \leqslant j \leqslant d$, i.e., the interval of horizontal/vertical translation $e^{i \cdot g_h + j \cdot g_v}$ to be effected on the motif M.

of deformation, say $y\%$, a positive or negative real number. This means that the vector $G(x) - x$ is stretched to $y.(G(x) - x)$, and that the alteration of x is set to $x + y.(G(x) - x)$. However, this is not the last word in alteration, one may also define its direction. Typically, this is vertical (conserving onset time), or horizontal (conserving pitch), etc. But the vector $G(x) - x$ is neither one nor the other. So we have to introduce the projection $p_t(G(x) - x)$ of the difference $G(x) - x$ to the direction t. This yields the final alteration

$$A_{t,y,G}(x) = p_t(y(G(x) - x)) + x$$

of x according to the system variables t, y, G.

This is all implemented in the OrnaMagic module of *presto*®. Moreover, given a block with onset limits $u < v$, the user can define a successively increased alteration by setting the percentage to $y(u), y(v)$ and altering from $A_{t,y(u),G}$ to $A_{t,y(v),G}$ as onset moves from u to v. Such procedures have been used in the composition "Mystery Child" in [103]. This piece can also be heard from the book's CD-ROM, see page xlvii. The effect resembles a morphing operator which lets the background ornament G act on L with successive variation of its "alteration force field". As a special case, we have the classical *pitch alteration* that is driven by a background tonality G, as well as the onset alteration, i.e., better known as *groove effect* on sequencers, defined by an onset grid that is derived from a rhythmical onset configuration. These special effects are implemented as "easy alteration variants" in *presto*®.

Example 58 A special application of pitch alteration is *tonal inversion*: If we want, for example, to apply tonal alteration of the C-major scale around e (leaving e fixed, exchanging d and f, c and g, b and a), we may first apply the real inversion U_e and then apply the tonal alteration downwards with respect to C-major.

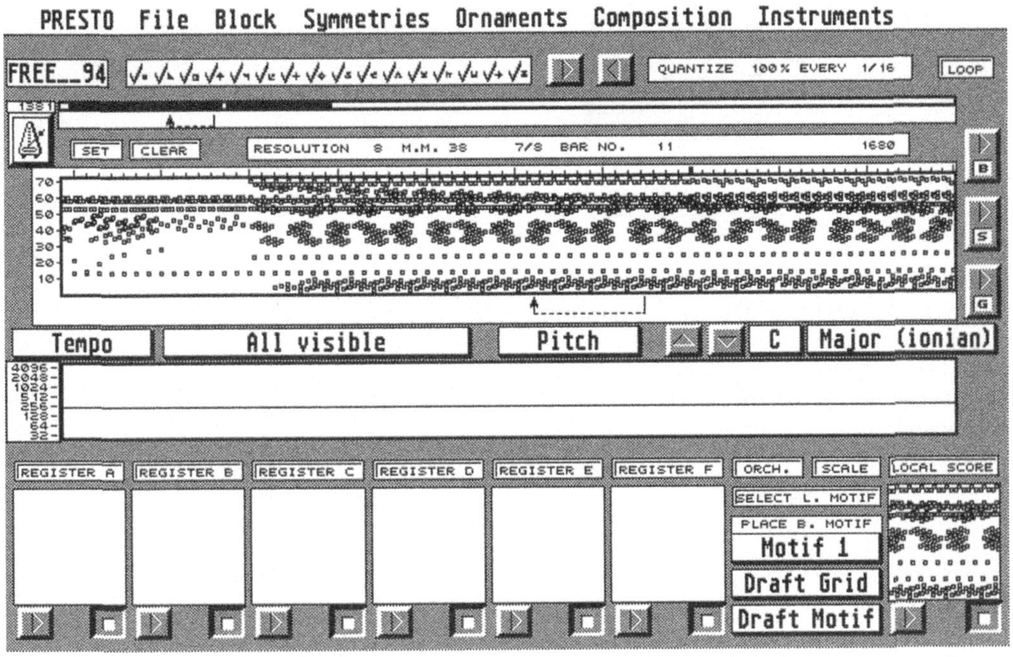

Fig. 50.7. Drum patterns in the composition "Synthesis".

50.4 Problems of Abstraction

Summary. Whereas syntagmatic contiguity is a relation *in praesentia*, paradigmatic associativity is a relation *in absentia*. This difference has heavy consequences on the ease of managing paradigmatic composition tools. As a consequence, composers rarely transcend straightforward paradigmatics and prefer sticking to syntactical composition software, such as MAX, or common sequencer software.

– Σ –

The composition software *presto*® was designed as a tool for everybody, even for the non-experts in musical notation. The publicity and the product concept tried to invoke the statement "Beethoven's creativity at everyone's reach". The argument was that drawing music (see Figure 50.8) is an interface that could give everybody access to the otherwise abstract and heavily codified music. Also, the simultaneous presentation in geometric coordinates of $EHLD$ space, together with the 2D projections, and the presentation in notes (see Figure 50.9) was thought to be a big advantage against classical approaches to composition.

Now, at the center of this concept are evidently the TransforMaster (symmetries), the OrnaMagic (ornaments), and the AgoLogic (tempo curve hierarchies) modules. The first two of them are paradigmatic tools par excellence. They are powerful, but require strong abstraction capabilities. The *absentia* of the paradigm enforces a selection process of representatives, and a memorization of their relation to the paradigm. This reality switch is a major obstacle to a breakthrough of such a software concept. It seems to be a major task of future composition software development to build concepts that enable seamless transitions from abstract to concrete composition layers. In the limit, one should have very abstract objects, such as self-addressed compositions or even more generally addressed objects (why not play a functorial global composition, how would one do that?) at one's fingertips and handle them completely naturally as if they were normal notes or chords, and in fact: they *are* completely normal, only we did not learn to sonify them in appropriate shape. In this regard, Tom Johnson's compositorial approach [510] is one of the most promising in trying to morph mathematical objects into musical events.

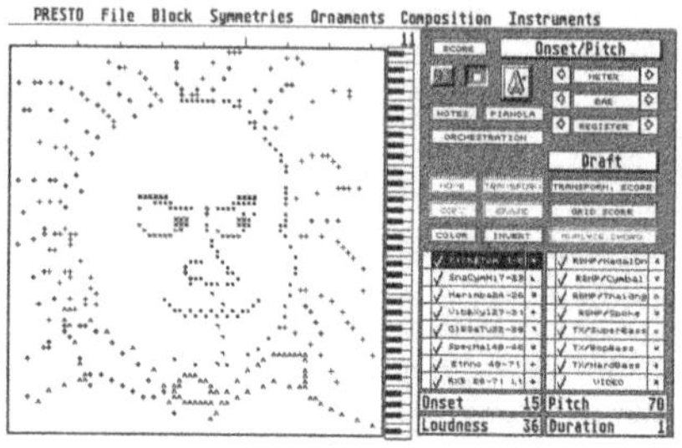

Fig. 50.8. The *presto*® concept tried to merge Beethoven's creativity and ease of a graphically interactive interface for musical composition. The local score here shows a drawing of Beethoven in a geometric space, and using instrumental icons.

Fig. 50.9. The geometric representation on the local score can also be viewed in classical score signs. One can immediately understand the geometric shape, whereas the common notation yields a completely cryptic object.

Check for updates

51

Case Study I: "Synthesis" by Guerino Mazzola

Habe die CD nun gründlich mir angehört und
frage mich, woher kommt bloss
die Kraft, die so was
Schönes schafft.
Jazz Saxophonist Werner Lüdi on "Synthesis" [629]

Summary. "Synthesis" is a composition for piano, percussion and e-bass. Its global and local organization was driven by classification of local compositions and modulation theory on one hand, and by the *presto®* software tool on the other. We describe the overall organization and the four movements.

– Σ –

"Synthesis" is documented on CD [681]. It is the result of a composition grant of the city of Zurich and was composed and recorded in 1990 using the *presto®* software, together with this hardware: one Atari®Mega ST4 computer, the synthesizers Roland R-8M (drums and percussion), Yamaha RX5 (drums, percussion, and special sounds), and Yamaha TX802 (bass), and a Steinway grand for the piano part played by the author.

The drum, percussion, and bass parts of the composition were written on *presto®* in four months and then completed by the piano part in an additional two months.

The music critics did not recognize on the CD (where no trace of the computerized music was given) that the whole composition's percussive and bass parts were synthetic. This is what Gérard Assayag rightly calls a *Turing test for the viability of computer-aided composition*, in particular since in the field of jazz this technology is thought to be an impossible tool.

The original presto-files (extender `.sto`) as well as the audio-files of "SYNTHESIS" are available on the book's CD-ROM, see page xlvii.

51.1 The Overall Organization

Summary. We discuss the overall organization: its material, composition principles in the four movements, and the instrumentation strategy.

– Σ –

The overall organization is a four-fold view on a generic material: the 26 classes of three-element motives in $OnPiMod_{12,12}$ (see Appendix O.3 for the class list). The multiplicity of views is given by classical forms:

- the sonata form,
- the cycle of variations,
- the scherzo, and
- a finale.

G. Mazzola, *The Topos of Music II: Performance*, Computational Music Science,
https://doi.org/10.1007/978-3-319-64444-8_20

51.1.1 The Material: 26 Classes of Three-Element Motives

Summary. The entire composition is based on the 26 isomorphism classes of three-element motives in $OnPiMod_{12,12}$ (see Section 11.3.8) and its specialization tree. We make precise the different usage modes which have been realized in "Synthesis".

– Σ –

After the inspiring analysis of the Schubert–Stolberg work in 11.6.2, the usage of the 26 classes was recommended. The classes were used in different contexts. We refer to Section 11.3.8 for the theory and to Appendix O.3 for the 26 classes, in particular Figures O.1, O.2 for representatives and the specialization Hasse diagram, see Section 12.2.2 for specialization.

To begin with, representatives of all classes were patched together to build the germinal melody already described in Section 11.6.3, see especially Figure 11.16. This melody appears explicitly in the introduction to movement four, see Figure 51.1. The 26 motives also appear as percussive "phonemes" in the third movement,

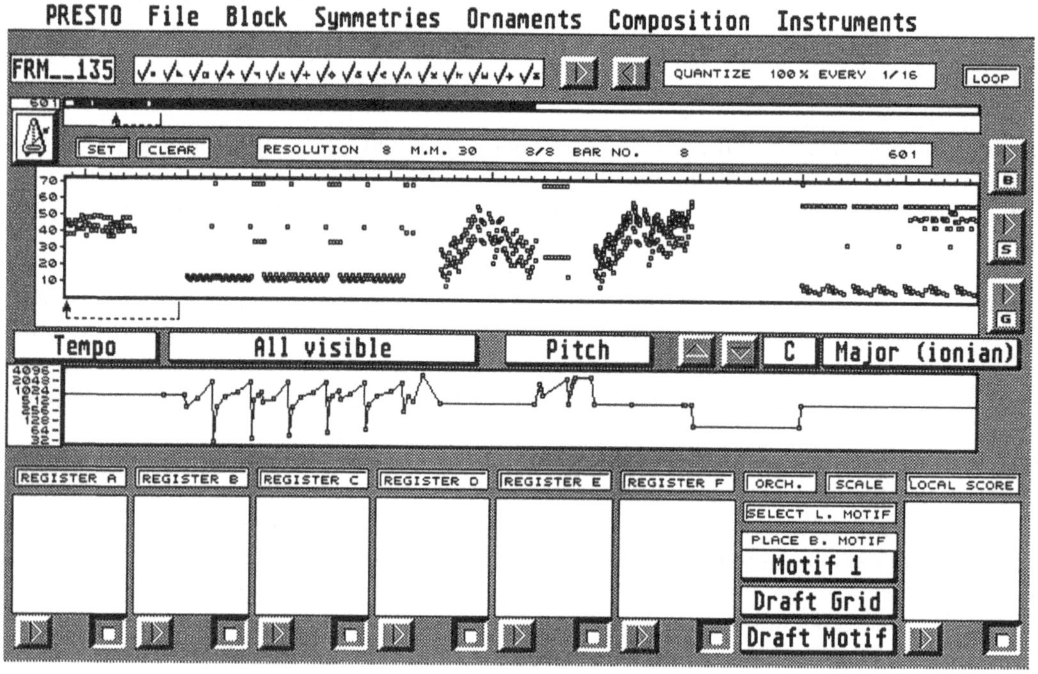

Fig. 51.1. The beginning percussion rolls in movement four of "Synthesis", as extracted from *presto*®'s main window, together with the repeatedly accelerating tempo curve. The germinal melody comes twice just before the piano intro solo.

as already discussed in Section 11.6.3. In the second movement, the germinal melody is altered according to ornamental deformation techniques and yields a sequence of characteristic melodies which are played by the bass and harmonically ornamented by the piano. Already the very first percussive motif in movement one (just after the piano solo intro) *is* the germinal melody, however played with pitches being encoded by percussive sounds. And in movement four, these motives are used to define ornaments and fractal refinements thereof. More schematically, this strategy is shown in Figure 51.2.

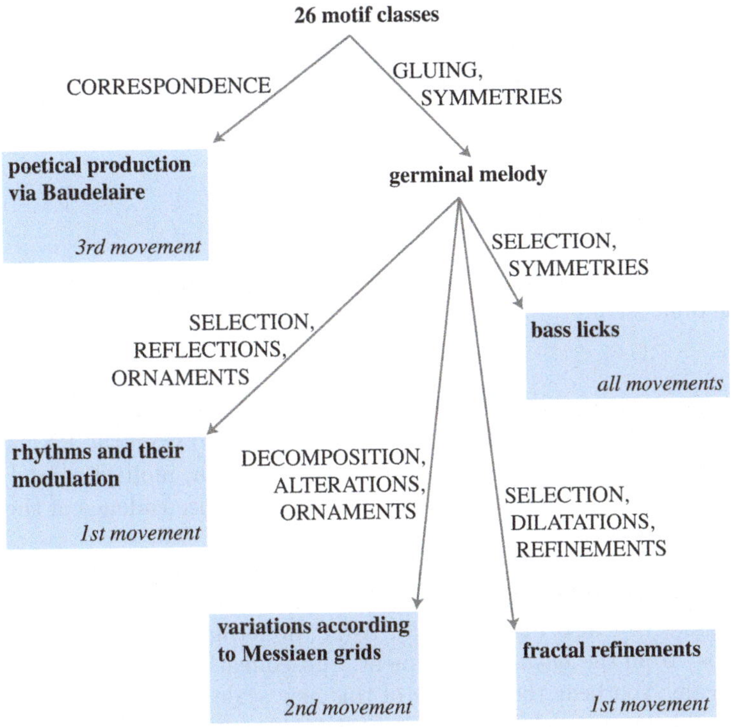

Fig. 51.2. The overall strategy in "Synthesis" is driven by exploiting the 26 classes of three-element motives.

51.1.2 Principles of the Four Movements and Instrumentation

Summary. The four movements include a sonata form, a variational sequence, a scherzo and a finale. The instrumentation issue is intimately related to the structural aspects. Each movement bears its specific constraints on the role of instrumentation.

– Σ –

The principle of a four movement concert such as are encountered in the classical concert form is realized as follows, each movement with a total of 122 instruments:

First movement: Earthquake/Full Force, duration: 10:46. It is devoted to the Greek element of earth. It is a sonata form which exposes the ideas, followed by modulatory development, a reprise and ending by a coda.
Instrumentation: Yamaha RX5, TX802, Roland R-8M (PCM cards: Jazz, Jazzbrush, Ethnic Percussion).

Second movement: Liquid Colours/Sea of Faces, duration: 14:05. It is devoted to the Greek element of water. It is a cycle of variations. These variations are taken from the germinal melody and are concatenated in a paradigmatic way.
Instrumentation: Yamaha RX5, TX802, Roland R-8M (PCM cards: Jazzbrush, Ethnic Percussion, Mallets).

Third movement: Poem of Wind/Fly! Fly! Fly!, duration: 09:27. It is devoted to the Greek element of wind. It is a scherzo. The motivic alphabet of the 26 classes is produced in poetical arrangement of rhythmic patterns.
Instrumentation: Yamaha RX5, TX802, Roland R-8M (PCM cards: Jazz, Contemporary Percussion, Mallets).

Fourth movement: Burning Spears/Interstellar Space, duration: 10:28. It is devoted to the Greek element of fire. It is a finale. Self-similar repetitions and refinements of the germinal structure successively densify the musical material.
Instrumentation: Yamaha RX5, TX802, Roland R-8M (PCM cards: Jazz, Jazzbrush, Contemporary Percussion)

51.2 1st Movement: Sonata Form

Summary. The first movement "Earthquake" is a sonata form which uses the modulation theorem (see Section 27.1.4) in time dimension.

– Σ –

This movement is a sonata form for rhythmical structures. After the piano intro, the rhythmical germinal melody is played on *toms* (pitches are associated with different toms) from the RX5 synthesizer. During the exposition, the germinal subject is enriched by drum and percussion, multiplied and profiled by bass lines. The flight of the piano over the rhythmical carpet indicates an intense cadence of the exposition, ending on a fermata.

The development consists of three rhythmical modulations (see Sections 28.3.1 and 28.3.2). This delicate process includes four different rhythmical macro-scales, which each consists of seven rhythmical three-tone motives from the 26 classes (see also Figure 28.13). After the fermata, the first scale is built and cadenced by regular "falling drops" from the *ohkawa* instrument. Subsequently, its character is neutralized and transformed into a second scale. You hear the cadence of this new scale as being again marked by the ohkawa's regular "falling drops", this time enriched by a rain of light piano pearls.

After a further fermata, the second scale is altered into a third one, whose cadence coincides with an intermediate climax of the piano. When the piano finally recedes, the third scale is neutralized until a march-like turning point introduces the fourth scale, whose cadence terminates the development by another fermata.

The reprise follows in a slightly altered instrumentation. The finale starts after a short fermata. It is recognized on a heavy rock rhythm which is played in reduced tempo on the *kicks*.

51.3 2nd Movement: Variations

Summary. The second movement "Sea of Faces" follows a syntax of melodic variations of a fundamental melody which entails a particular "harmolodic" color in the vein of Ornette Coleman's music/theory.

– Σ –

This adagio is a labyrinthic wandering of *piano, bass, barafons, glockenspiel, sanzas, gender, tube bells* etc. through seven melodic variations of the germinal melody. Drums and percussion also obey these deformation forces, each harmonic, rhythmical, and melodic change is a specific expression of these forces.

Scales and rhythms are known to be periodic structures in pitch and onset. By the choice of two scales from the first three Messiaen scales, one for pitch, one for onset, we generate an ornament, a "harmonical-rhythmical" scale.

After certain rotations of such ornaments, the harmonic and rhythmical Messiaen components in onset and pitch are completely mixed. Here, the deformation forces of our germinal theme become manifest: Each tone event is displaced according to the ornament alteration algorithm described in 50.3.

The deformation uses the following grids (see Figure 51.3): We take the Messiaen scales M_1, M_2, M_3 as described in Example 11 of Section 8.1.1. For each grid, we take a pair (M_i, M_j), $i, j = 1, 2, 3$ and build the cartesian product $M_i \times M_j$ of two such scales, one in pitch, the other in onset direction, and each of them with a large range such that it acts on the germinal melody as if it were infinitely extended. The action is

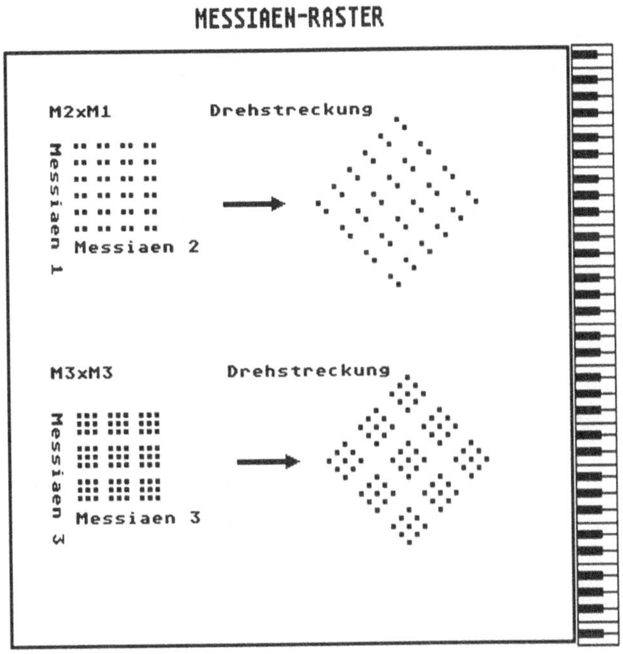

Fig. 51.3. The original graphics for the grid construction. To the left two Messiaen grids, to the right their transformation via a rotation-dilatation which is used for rhythmical connections. The Messiaen grids are cartesian products of the Messiaen scales and rhythms.

taken to be 100% without any directional constraints. This yields nine variations of the germinal melody, as shown in Figure 51.4.

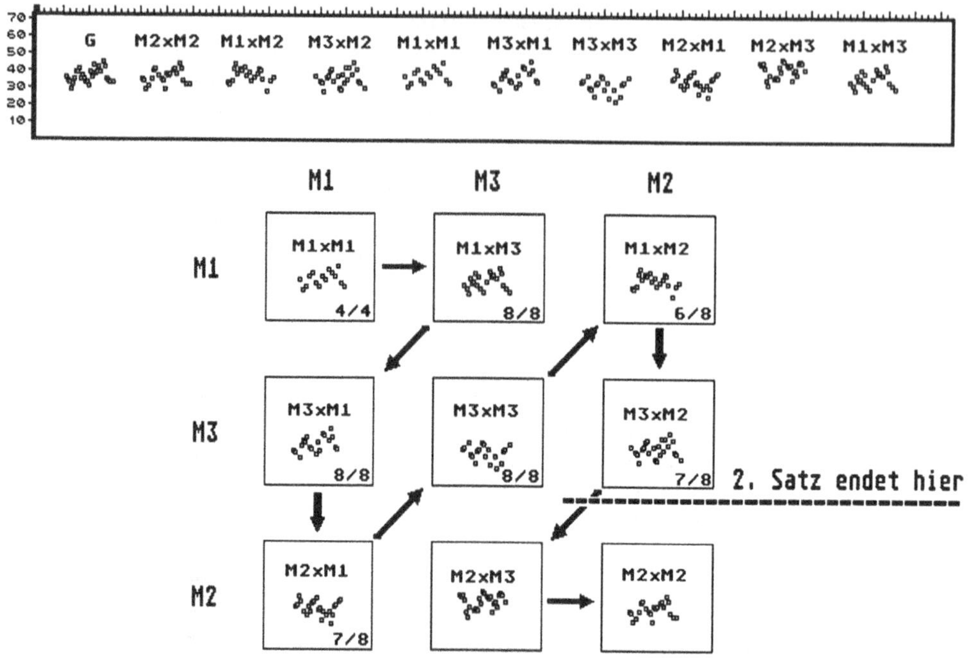

Fig. 51.4. The original graphics for the nine variations of the germinal melody G according to the Cartesian product grids $M_i \times M_j$ (upper part). The variations are played in the order shown below, however ending on variation $M_i \times M_j$, for reasons of the length limits of the movement

In the second movement, seven of the nine variations are used in the order shown in Figure 51.4, lower part. The bass plays these variations in a very extended slow gesture, and the piano ornaments these melodic lines by a harmonization work.

Each couple of such variations is connected by a rhythmically complex intermediate structure, see Figure 51.5. The intermediate parts terminating in the $M_i \times M_j$-variation are constructed as follows: We take the Messiaen grid $M_i \times M_j$ and operate a dilatation-rotation by 45° on this object, see Figure 51.3, right part, for two such transformations. From these transformed local compositions (or from a very similar skew transformation) we build an interpretation by a partition into five subcompositions which are again transformed as visualized in Figure 51.6. These charts are then taken as motives of different ornaments which are generated by the grids as shown in Figure 51.7. The upper part of Figure 51.7 shows the superposition of the five ornaments generated from Messiaen grid $M_3 \times M_3$. This rhythmical construction is also seen in the right and lower part of Figure 51.5.

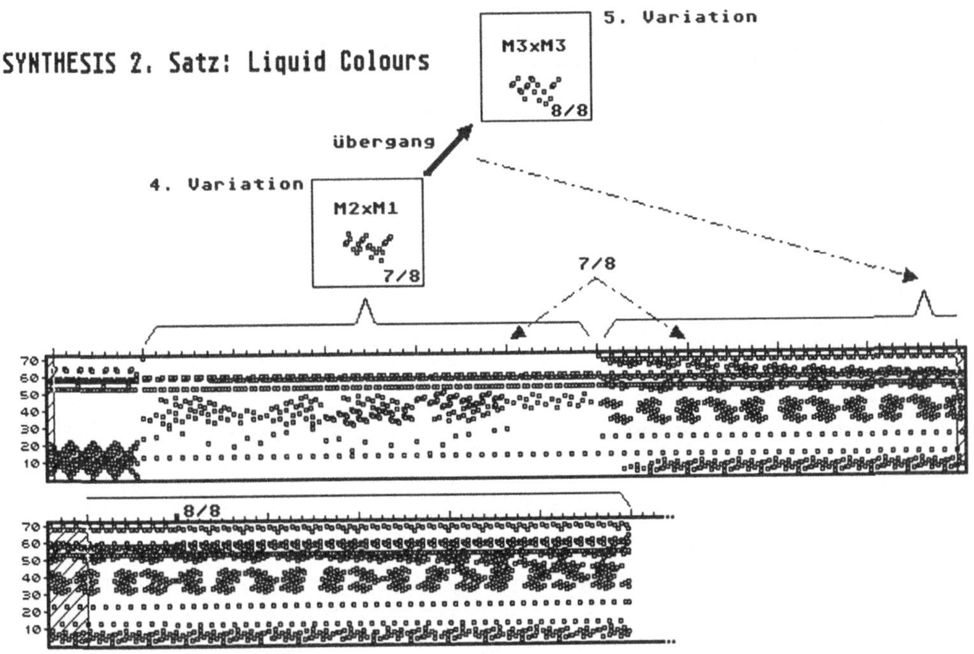

Fig. 51.5. The original graphics for the intermediate structure between two successive variations, here connecting the $M_2 \times M_1$-variation to the $M_3 \times M_3$-variation. The connecting structure is a rhythmical construction that is deduced from the generating Messiaen grid $M_3 \times M_3$.

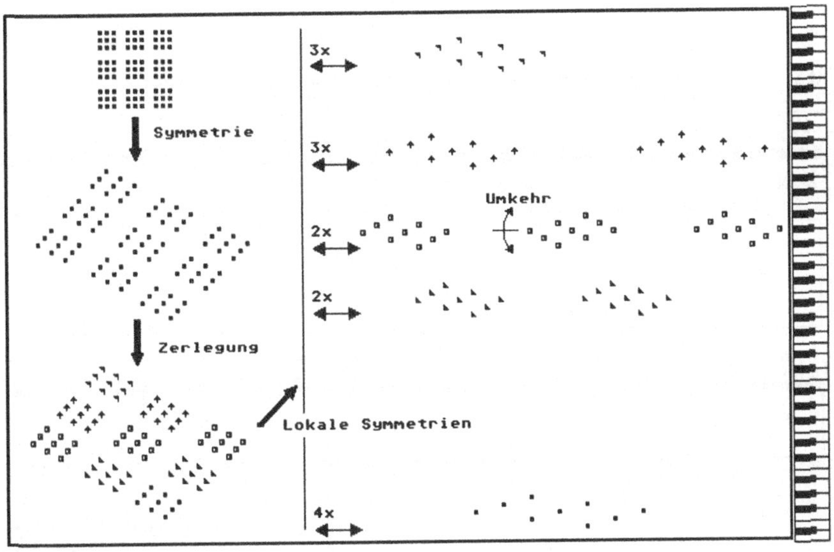

Fig. 51.6. The original graphics for the interpretation of the rotated Messiaen grid and the symmetries applied to the partition's components.

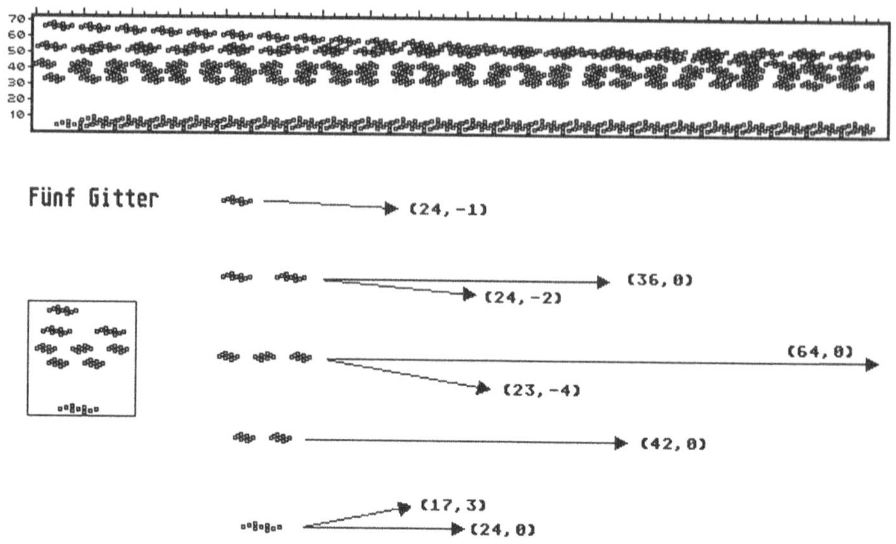

Fig. 51.7. The original graphics for the ornaments that are generated by charts of a rotated Messiaen grid, following different grid translation vectors.

51.4 3rd Movement: Scherzo

Summary. The scherzo "Poem of Wind" is based on a transcription of Charles Baudelaire's last poem "La mort des artistes" in "Les fleurs du mal".

– Σ –

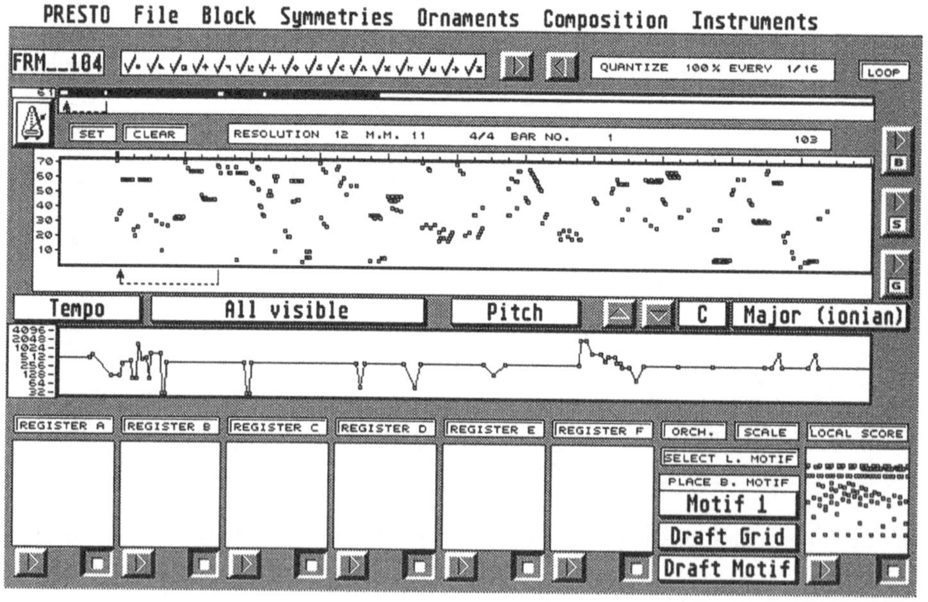

Fig. 51.8. The beginning of the third movement with its breakneck changes in agogics.

This movement is in complete contrast to the quiet flowing of that adagio. It resembles Ernst Jandl's concrete poesy. A poem of wind: without firm ground or romantic sky. Here, the piano builds an expressive dialog with a pointillistic percussion.

The first two strophes of Charles Baudelaire's "La mort des artistes"is the last poem in his famous "Les fleurs du mal". The transformation of this poem was already discussed in Section 11.6.3. This procedure transforms every word into a sequence of rhythmical motives, we hear it as a coherent sound sequence which is separated from its successor by a short fermata. At the end of each verse, longer interruptions are inserted.

From this raw material the final composition is constructed by an extremely refined agogical architecture, changing from breakneck accelerandi to stagnating ritardandi. Here, the third movement makes extensive use of the *presto*®-AgoLogic for tempo curves. The single motives are also enriched by echo-like variations and repetitions, such that their color engraves itself on ones mind.

The answer of the piano to this witty poesy is the albatross in the air, which flies, falls and plays with Cecil Taylor's florescence as if it were an old tale. Just one not too many salty swift—and goodbye.

51.5 4th Movement: Fractal Syntax

Summary. The fourth movement "Burning Spears" is constructed in a completely non-syntactic way from fractal principles.

– Σ –

This finale is a steadily evolving rhythmical organism whose form could be compared to Alexander Scriabin's ecstatic sonata "Vers la flamme".

The movement gets off by a drum intro, followed by a bass intro which announces fanfare-like that something is to happen. It receives a promising answer from the low piano keys until a samba-like rhythm makes the final run.

This starting rhythm is again derived from the germinal melody. You can hear its essential elements from the drum intro. After two roll groups, the samba reappears, but now with augmented time values, whereas the now opened gaps are filled up by new micro rhythms (see Figure 51.9 where the original motives are seen in the low and high regions, and Figure 51.10, where the stretched motives are now also visible in these regions). This process of intensification is cadenced until a cutting bass line terminates the developmental part. The piano dances like an entranced sorcerer over this rhythmical lava.

In the reprise, the samba figure reappears, but with a higher tempo. It ends in a strong pulse which leads to a further time augmentation after a bass cadence. This time, the black dancing extension figure is refined by a cackling entwinment by *wood blocks* which is joined by a fast dialog with the piano.

The following earthy rhythm is introduced by a bass drive as if we should be warned that we are now changing the civilizations. Here too, the construction is an augmentation of rhythm and an intensification. What we are hearing is in fact a fractal repetition of self-similar time structures.

The last and extremely fast part expresses the force of a burning spear, which is thrown into the sound sky by a delirating dervish dancing on an exploding volcano... So the musical principle of this finale is not architectural, but a self-renewing process.

52.3 Maquettes: Objects in Time

Summary. The *maquette* is an OpenMusic *meta-object* aiming at representing, in a same object, patches (musical process) and containers (musical material). It is a new concept of score where the static description of musical structures and the definition of dynamic computational processes seamlessly coexist. The user may go back and forth between these two metaphors by considering the maquette as a score (in traditional or graphical notation) or as a set of interconnected processes. As external objects like MIDIFiles or SoundFiles may also be imported, maquettes offer an original environment for music creation.

– Σ –

The maquette is an original concept in OpenMusic which allows us to solve the problem of combining the design of high level hierarchical musical structures, the arrangement of musical material in time, and the specification of musical algorithms. Just like other meta-objects, maquettes may appear as a view or may be opened in a maquette editor, which is basically a 2-dimensional surface with time flowing along the x-axis. This surface contains several blocs that we call *temporal boxes*. In the maquettes the hierarchical imbrication of musical structures and their temporal order can be represented in an explicit visual way. In the musical sketch presented in Figure 52.21, made by the composer Mikhail Malt, *temporal boxes* have been disposed on the maquette surface.

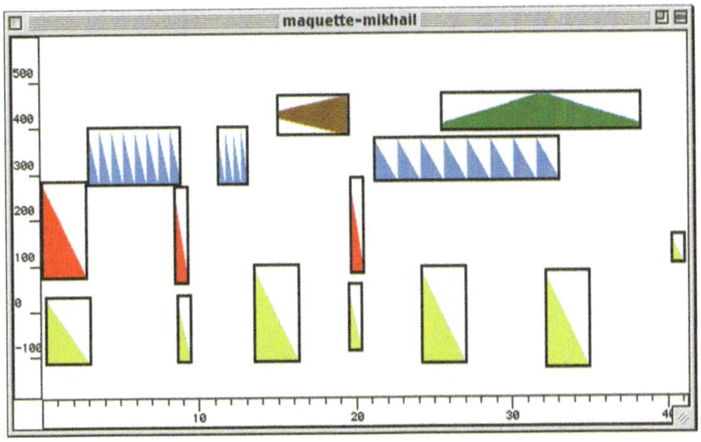

Fig. 52.21. Temporal boxes in a maquette.

Horizontal temporal box positions correspond to onset values in 'absolute' (physical) time. Durations and intensities are given by their horizontal and vertical extensions. Pictures have been associated to the boxes in order to give an elementary musical semiotics. Thus, triangles correspond to chords whose resonance decrease quickly. Multiple triangles are associated with chords *ostinato*. Other figures are triangles whose intensity follows the geometrical contour of the picture.

In Figure 52.22, connections between temporal boxes are shown. They represent a different kind of musical information. We can see that temporal boxes are inferred one from another by functional relations. For instance, the ostinatos are linked to the first chord. This level of information bears paradigmatic content, because analogies between any part of the structure can be derived.

If we open the editor of the third ostinato (Figure 52.23), we can see that the chord coming from the first box (which is represented by the box called *input*) is transposed by an augmented-fourth (18 half-tons or 1800 MIDI-cents), then repeated six times and finally sent to a *factory* that builds an instance of the class chord-sequence. Elements of the musical material take place syntagmatically in the final result.

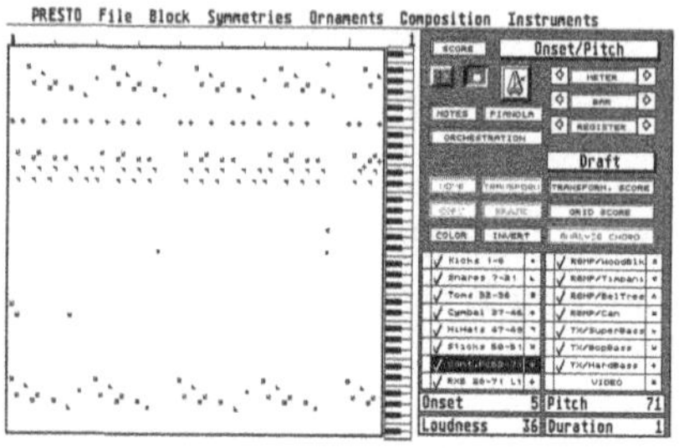

Fig. 51.9. The original motives are seen in the low and high regions of the local score.

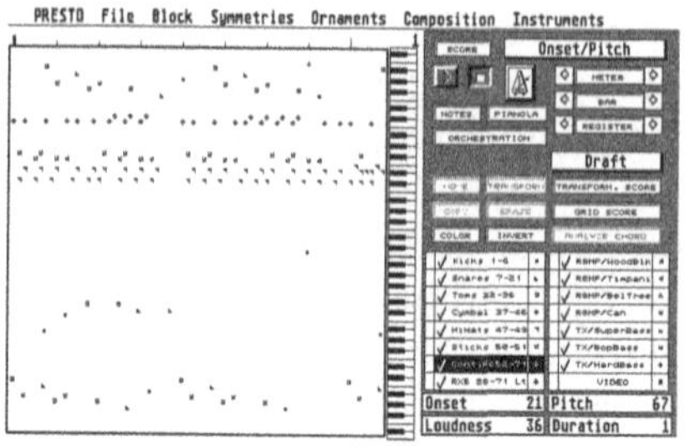

Fig. 51.10. The stretched motives are also visible in the low and high regions.

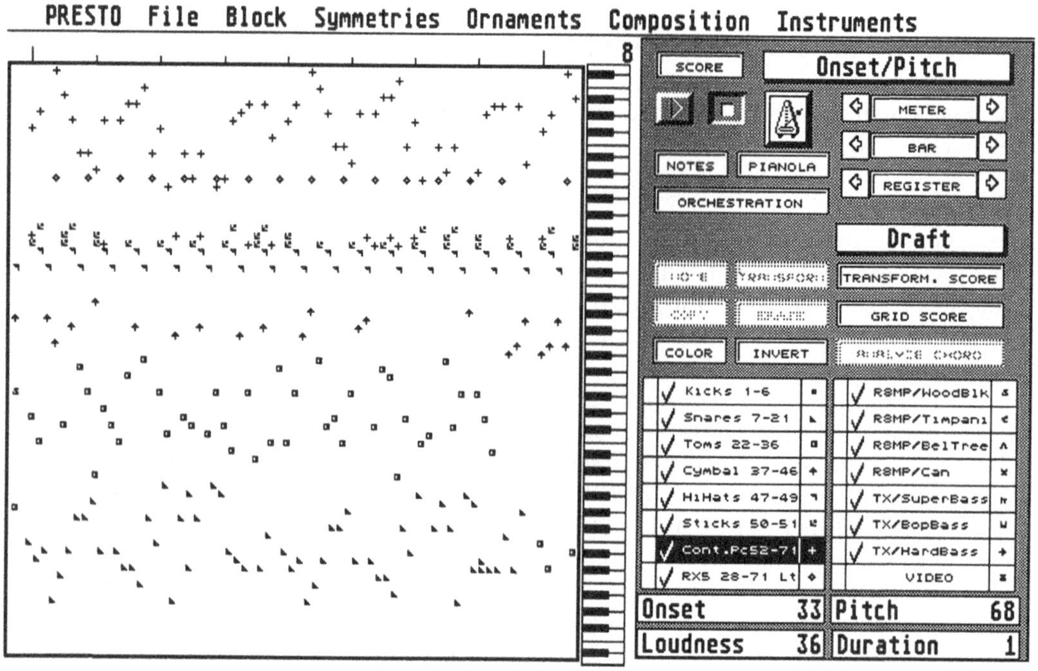

Fig. 51.11. The end of the fourth movement shows a particularly dense and polyrhythmic percussive part.

52

Object-Oriented Programming in OpenMusic

Le problème que l'ordinateur pose au compositeur n'est pas d'abord d'ordre spécifiquement musical, mais avant tout culturel et philosophique. Il implique une refonte complète des rapports de l'abstrait et du concrete. Le musicien, le musicologue, l'auditeur doivent bien se rendre à l'evidence de ces changements sans aller pour autant jusqu'à la maladie de l'adaptation.

Hugues Dufourt [282]

Summary. OpenMusic is a visual programming language for music composers. It was designed and implemented by the Musical Representation Team at Ircam-Centre Georges Pompidou. OpenMusic is based and implemented on CLOS (Common Lisp Object System) [1014]. It shows several original features, such as reflexivity, meta-programming capacities, handling of the duality between musical and computational time, and provides a framework of predefined musical objects for handling sound, MIDI and musical notation. OpenMusic combines different technics of programming, e.g., functional programming, constraint programming and object-oriented programming. We will focus in this chapter on the last one. Object-oriented programming is crucially connected with the categorical approach. Category theory helps formalize in an original way concepts like inheritance, methods, classes, etc. More details on this relation can be found in [613] or in Section 9.4.2 of this book.

– Σ –

Although one may consider music composition to be an important issue in any computer music research or development, the term Computer Assisted Composition (CAC) has taken a specialized meaning during the past years. As opposed to the generation and processing of audio signal, by means of DSP hardware or software technologies, CAC systems such as OpenMusic focus on the formal structure of music. The software technology is rather based on symbolic computation, where the typical data structures (trees, graphs, sets, collections, associative memory, etc.) and algorithms (often issued from discrete mathematics) are suited to handle the complex structures involved in a compositional process. The great diversity of esthetic, technical, formal (or anti-formal) models, implies that one cannot conceive an environment of CAC as a fixed application which provides a fixed collection of generative procedures and musical transformations. On the contrary, we conceive an environment as a programming language, helping each composer to constitute his personal universe. Of course, there is no sense in providing a traditional programming language, the control of which requires a great technical expertise. For this reason, our purpose is to build a programming language conceived especially for composers. This leads us to think about the various existing models of programming, intuitive graphical interfaces (which enable control of this programming) and the internal as well as external representations of musical structures that will be built and transformed when using this programming. Therefore the main goal is to implement a language containing the concept of notation of the result (a musical score) as well as the concept of notation of the process leading to this result (visual program).

G. Mazzola, *The Topos of Music II: Performance*, Computational Music Science,
https://doi.org/10.1007/978-3-319-64444-8_21

52.1 Object-Oriented Language

Summary. In this section we describe the main entities of our object language. Such language entities are usually called *meta-objects*. We describe each meta-object in a graphical and formal way. Basic calculus for object-oriented programming inherits the approach imposed by the precursory language Simula and its successors. The attempts to formalize this family of object models used the concept of parametric polymorphism [178]. More recently, languages based on multiple-dispatching (methods that dispatch on a product of types rather than a single type) such as CLOS could be formalized using concepts of overloading or ad-hoc polymorphism, found in *λ&-calculus* [179]. OpenMusic, which is based on CLOS, may be formally described in this way. Although we do not give here a real formalization of OpenMusic, we use the *λ&-calculus* to give a general idea of each meta-object. From a visual point of view, the meta-objects of our calculus are represented as graphical entities called frames.

– *Σ* –

Object-oriented programming is based on simple concepts. A program can be seen as a set of entities called *objects*. An object is made of data (*slots*) and operations applied on it (*methods*). Objects communicate in a specific way, usually called *message passing*. Most of the object-oriented languages implement the notion of *class* in order to abstract similar objects. New classes can be created from existing classes using the mechanism of *inheritance*. Inheritance allows the extension or partial modification of a class. If a class A inherits from a class B, A is called the *subclass* of B and B is called the *superclass* of A. Object-oriented programming offers in a natural way a dynamic management of resources, which means that we can create new objects at any time. The mechanism of object creation from a class is called *instantiation*. The new created object is called an *instance* of the class.

Meta-objects are represented either as composed frames or simple frames. Several frames (i.e., different point of views) may be produced for the same object. The simple frames which represent an object are called *views*. They generally appear as icons. The composed frames representing an object are called *editors*. For more details about formal and graphical description of OpenMusic see [13].

52.1.1 Patches

A *patch* is a meta-object specific to OpenMusic. It reifies the notion of program. A patch is the place where objects will be interconnected in order to specify musical algorithms. From a formal point of view, a patch can be seen as a *λ-function*. Patches are composed by boxes (icons) and connections between them. Boxes represent functional calls while connections represent functional composition. Figure 52.1 shows the *view* and the *editor* of a patch implementing an algorithm for the expression $x^2 + 2$. It can be formalized as the lambda function $\lambda x(x \cdot x + 2)$.

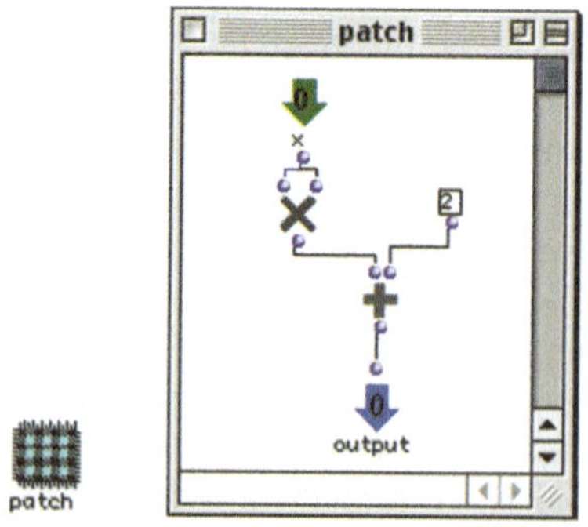

Fig. 52.1. View and editor of a patch.

52.1.2 Objects

In the *λ&-calculus*, objects are represented as registers. A register can be seen as a set of labeled fields $l = v$ where l is called the label and v is called the value. For instance, an object representing the note C3 can be written as: $note = \langle pitch = C, octave = 3\rangle$. Figure 52.2 shows the *view* and the *editor* of this note object.

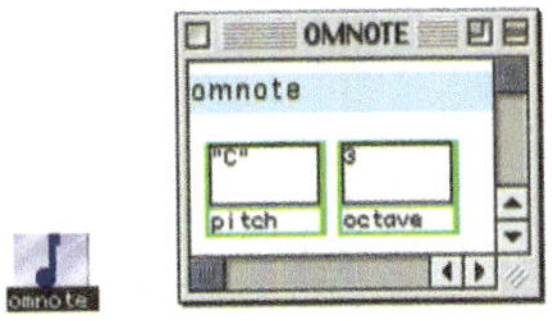

Fig. 52.2. View and editor of an object.

The next two rules define the field selection and the field writing, respectively:

$$\langle l_1 = v_1, \ldots, l_n = v_n\rangle \circ l_i \mapsto v_i,$$
$$\langle l_1 = v_1, \ldots, l_i = v_i, \ldots, l_n = v_n\rangle[l_i \mid v] \mapsto \langle l_1 = v_1, \ldots, l_i = v, \ldots, l_n = v_n\rangle.$$

52.1.3 Classes

If objects are seen as registers, classes will then be seen as generators of registers. The editor for a class is an ordered collection of *views* representing *slots*. Slots contain information about their name, their type (a class icon), a default value and a flag that indicates if the slot is public or private. *View* and *editor* for the class of the note object defined in the previous section are shown in Figure 52.3.

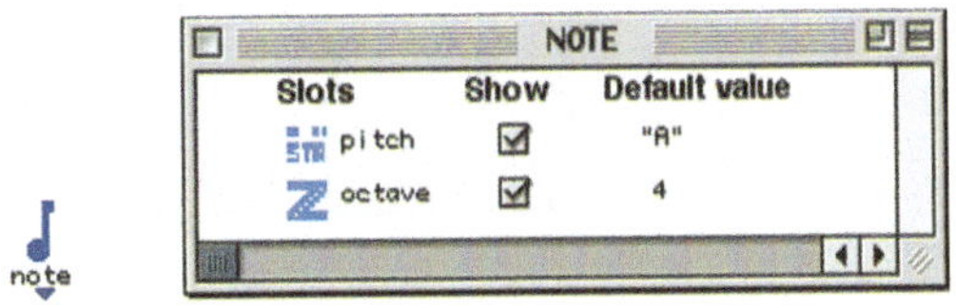

Fig. 52.3. View and editor for a class.

52.1.4 Methods

Methods are simple functions (*λ-abstractions*) where arguments are typed by classes. The editor in Figure 52.4 shows a method with two inputs *self* and *num* of type *Note* and *Integer* respectively.

The body of this method increments the value of the slot *octave* of *self* by *num* and returns the note object. This method is formalized by the expression:

$$\lambda self{:}note\ \lambda num{:}integer\ self[octave \mid self \circ octave + num].$$

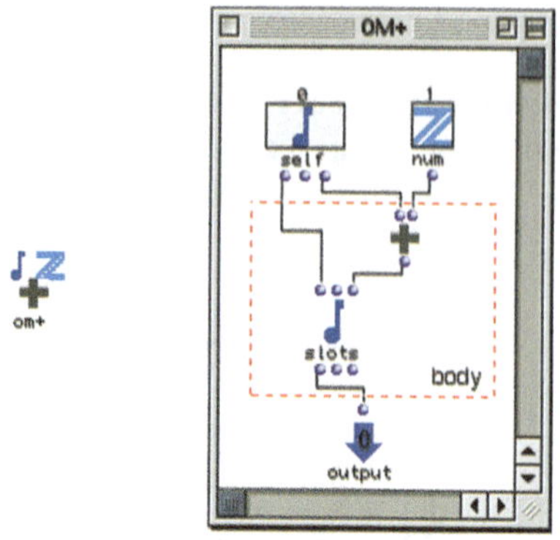

Fig. 52.4. View and editor for a method.

52.1.5 Generic Functions

A generic function is a collection of methods ($\epsilon\&M_1\&...\&M_m$) where ϵ is the empty method. The type of a generic function containing methods M_i of type $U_i \to V_i$, $1 \leqslant i \leqslant m$, is $\{U_1 \to V_1, \ldots, U_m \to V_m\}$. However, not any set of methods can be seen as a generic function. A set of methods with types $U_i \to V_i$ is a generic function iff the two following conditions are satisfied for all i,j:

$$U_i \leqslant U_j \Rightarrow V_i \leqslant V_j,$$
$$U \textit{ is maximal in } LB(U_i, U_j) \Rightarrow \exists U_k, U_k = U$$

where $LB(U, V)$ indicates the set of common lower bounds[1] V of types U. Figure 52.5 shows a generic function composed of two methods: the first one is the method described in the previous section; the second one is specialized for inputs of type *Number*.

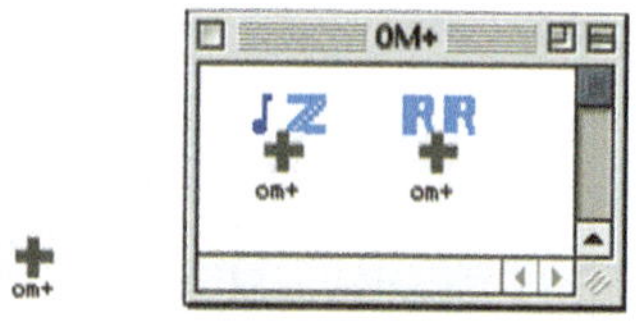

Fig. 52.5. View and editor for a generic function.

52.1.6 Message Passing

Message passing is achieved by generic function invocation. We distinguish the application of a simple function from the application of a generic function, which will be indicated by the operator $\bullet$. The application of a generic function G to arguments N of type U consists of two steps: selection of a method M_j among the methods of G and normal application of M_j to N .

$$(\epsilon\&M_1\&...\&M_m) \bullet N \mapsto M_j N.$$

Note that U may not be contained in the set U_i of input types of the generic function. In this case we select the method M_j satisfying : $U_j = min_{i=1...m}\{U_i \mid U \leqslant U_i\}$. The *CPL* of a class is a linearization of its superclasses. The details of the linearization are not crucial for our purposes, see [1014].

[1] The ordering $\leqslant$ is defined by the Class Precedence List (*CPL*) of the class U.

52.1.7 Inheritance

The inheritance mechanism is defined by the subtyping and the mechanism of method selection. Subtyping is defined by the following rules:

$$\frac{U_2 \leqslant U_1 \qquad V_1 \leqslant V_2}{U_1 \to V_1 \leqslant U_2 \to V_2} \qquad\qquad \frac{\forall i \in I,\ \exists j \in J,\ S_j \to T_j \leqslant U_i \to V_i}{\{S_j \to T_j\}_{j \in J} \leqslant \{U_i \to V_i\}_{i \in I}}$$

The rule on the left is the usual contravariant-covariant rule for arrow types. The other one states that an overloaded type is smaller than or equal to another overloaded type if for every branch in the latter, there is a branch in the former smaller than or equal to this one [179].

Graphically, a class inherits from another one when there exists an arrow from the superclass to the subclass, as is shown in Figure 52.6. The class *rnote* extends the class *note* by adding a new slot called *rhythm* of type *string* with default value *quarter*.

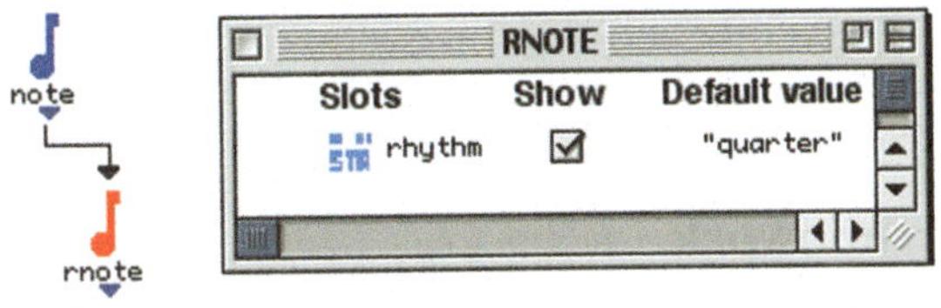

Fig. 52.6. Graphic inheritance.

52.1.8 Boxes and Evaluation

Boxes are placed in patches and allow other meta-object to be involved in the calculus. Boxes are composed of an icon and an ordered set of *inlets* and *outlets*. There are different types of boxes depending on the referenced meta-object. The user may create boxes in a patch by dragging meta-objects into it. Figure 52.7 shows different types of boxes and their references.

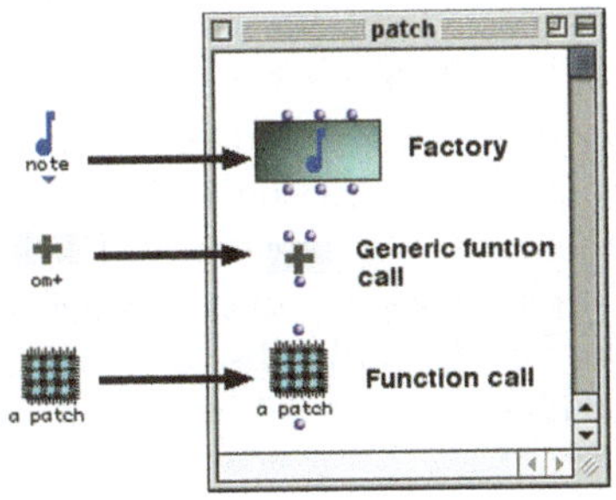

Fig. 52.7. Boxes in a patch.

Inlets in functional call and generic function call boxes correspond to the function arguments. For factory boxes, inlets and outlets refer to the slots of the class. Boxes and connections in a patch can be seen as a graph of functional compositions. By clicking on the output of any box, an evaluation from the corresponding point into the graph is induced. A box evaluation can give rise to other box evaluations creating a chain corresponding to the execution of a program.

52.1.9 Instantiation

Users can create instances of a class with the aid of *factory boxes* (boxes built from a class). A factory contains a number of inputs corresponding to the public slots of the class. There are as many outputs as inputs. When evaluating an output, a new instance is created. The values returned by outlets are, from left to right, the instance itself, then the current value of its public slots. An instance can be visualized graphically as a box which can eventually be connected to other boxes (see Figure 52.8).

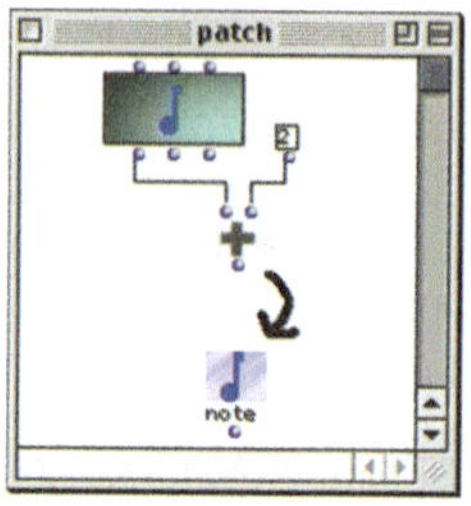

Fig. 52.8. Graphical instantiation.

52.2 Musical Object Framework

Summary. A framework is a set of reusable classes that one can use as building blocks for a specific software [358]. This section describes the musical OpenMusic framework.

– Σ –

OpenMusic offers a set of predefined classes and generic functions for musical representation and manipulation. This framework can be extended by using inheritance or by defining new methods in generic functions or by writing new generic functions. There are graphical editors for these definitions. In this section we will focus on the structure of musical objects (classes) rather than on their behavior (generic functions). The object-oriented concept of *encapsulation* can be defined as the separation between the internal representation and the interface of an object. These two aspects will be described in the next two sections.

52.2.1 Internal Representation

In OpenMusic any musical structure is a *container* which embeds other musical structures. All containers have a temporal extension (e) and an offset (o). e and o are expressed as a multiple of a rational unit $v = 1/u, u \in \mathbb{N}_+$, the set of positive integers, v is considered a fraction of the quarter-note. The value u can be redefined at each level of the embedded structure. From the temporal point of view, we have entities coming from different rational time scales. In order to express them in a homogeneous way, we have implemented a hierarchical unit system, which will be discussed briefly by showing an example. Figure 52.9 shows a music fragment and an integer hierarchy describing it.

Exercise 82 *Define the corresponding form whose denotators are these rhythmical hierarchies.*

We can calculate the extension of each element e_i by the recursive formula $d(e_i) = \frac{e_i}{S} d(F(e_i))$, where S is the sum of e_i's brothers including itself, and $F(e_i)$ is e_i's father. The deepest triplet in our example has an extension given by: $\frac{1}{3}\frac{2}{5}\frac{2}{4}\frac{4}{1} = \frac{4}{15}$. For this reason, we set u to 15 (the quarter note is equivalent to 15 units). In this scale the new extension of the triplet is 4. Following the idea of a variable scale adapted to each embedded level in a musical structure the rhythm of Figure 52.9 can be represented as shown in Figure 52.10.

Some basic operations for container manipulation are described below:

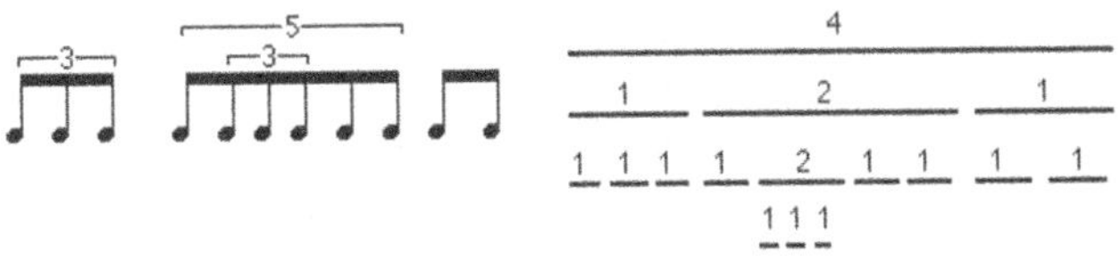

Fig. 52.9. Hierarchical representation of a rhythm.

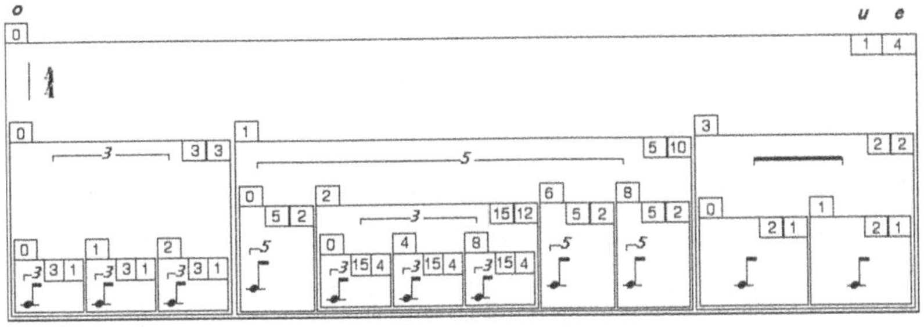

Fig. 52.10. Temporal organization of musical instances.

- *NewContainer*. Create an empty container (u equal 1 by default).
- *AddTo*$(c1, c2, at)$. Set the container $c2$ into the container $c1$ at the position at and calculate new values for u, e and o attributes.
- *RemoveFrom*$(c1, c2)$. Remove the container $c2$ from the container $c1$.
- *QReduce*$(c, [n])$. Computes $u' = ppcm(u_i)$, where u_i are units for all subcontainers of c. The attributes u of the container c is set to u'. New values for parameters u_i, o_i and e_i are calculated too. This operation allows comparison between all parts of a container. If a value n is given, then $u' = ppcm(u_i, n)$.
- *QNormalize*(c). Sets u value for c and each of its subcontainers to optimal, i.e., the smallest possible integer number, considering the offset and extent values that have to be expressed as integer multiples of $1/u$.

The *QNormalize* operation sets all extensions to integer values. If we want to compare all values we need to choose for the whole measure one unit u equal to $ppcm(2, 3, 15) = 30$. In this case the new values for u are 30 for the last eighth note and 30 for the triplet, see Figure 52.11. The choice of using only integers for our representation allows us an easier translation to music notation and a coexistence of objects described in hierarchical time and in continuous time. Moreover, in this way we avoid reversibility problems coming from the translation between reals and integers (continuous and discrete time).

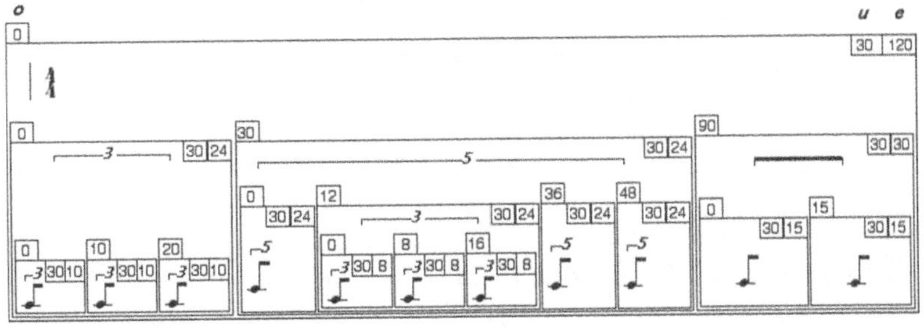

Fig. 52.11. A normalized container.

This model of containers and their operations represents in a natural way traditional musical structures like polyphony, measures, chords, etc. We will see in the next section how these containers are passed to the composer in an object context.

52.2.2 Interface

A summary of the predefined musical classes available in OM is given in Figure 52.12.

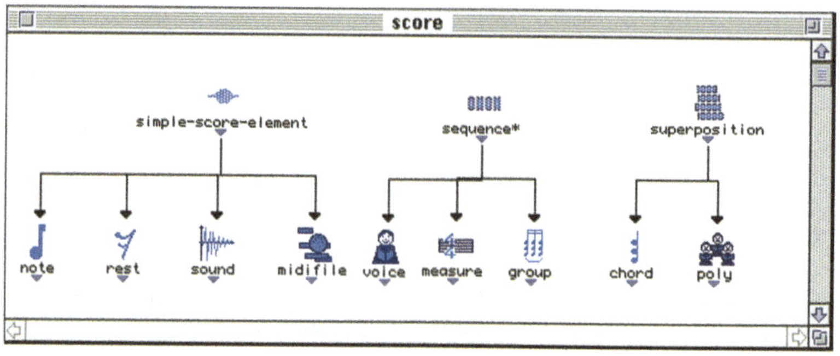

Fig. 52.12. Musical framework.

There are three main classes of musical objects:

- *superposition*, chords and polyphonies are made of other objects placed in parallel.
- *sequence*, objects like voices, measures, etc are composed by other objects one after the other.
- *simple-score-element*, these objects are terminals (empty containers).

There is a predefined set of generic functions with methods for all musical classes. They are used to apply transformations to musical instances (i.e., transposition, inversion, etc.) or simply to play or visualize them. Editors of musical instances have been replaced by musical notation editors (Figure 52.2 is replaced by Figure 52.13). We use the paradigm *Model-View-Controller*, where the model is the musical instance, the view is its representation on the screen as a score and the controller is a user interface allowing us to change slot values.

Fig. 52.13. View and editor of a note instance.

As we have seen in the previous section, musical entities can be represented as containers. However it is not suitable for composers to have to build musical objects from containers. In order to create a musical instance we must provide to the composer a symbolic abstraction of the *real* musical object. For instance, *MIDIcents* are a good way to code pitch values. On the contrary, problems arise when the rhythmic content is taken into account. We propose in the next section a representation of rhythmic information adapted to the composer.

52.2.2.1 Rhythmic Trees

Rhythmic Trees (RTs) are the base of the rhythmic representation in OpenMusic. They must be understood as an alternative description of symbolic rhythmic structures using traditional music notation, and an external specification for container objects.

- *Syntax* An RT is defined as a pair $(D\ S)$ where D is an integer ratio (> 0) and S is a list of n elements. Each element in S can be either an integer or a RT. Here is an example corresponding to this syntax: (2 ((1 (1 1 1 1)) (1 (1 1 1 1))).

  Remark 23 Solving the above exercise 82, we have here a nice example of a doubly circular form. To begin with, given a form F, a list form over F is the circular form

  $$List(F) \underset{Id}{\longrightarrow} \mathbf{Colimit}(Item(F), Terminal)$$

  with the terminal form $Terminal \underset{Id}{\longrightarrow} \mathbf{Simple}(\mathbb{Z})$ in order to formalize the end of a finite list and to indicate its length with the integer of the coordinator module $\mathbb{Z}$. The form $Item(F)$ is a Cartesian product

  $$Item(F) \underset{Id}{\longrightarrow} \mathbf{Limit}(F, List(F)).$$

  Building on this form, we define the form RT by

  $$RT \underset{Id}{\longrightarrow} \mathbf{Limit}(DurationValue, List(RT))$$

  and a simple duration form

  $$DurationValue \underset{Id}{\longrightarrow} \mathbf{Simple}(\mathbb{Q}).$$

  It is useful to visualize this doubly circular form in its construction graph as drawn in Figure 6.5.

- *Semantics* For a given $RT = (D, S)$, D expresses a duration and S defines a group of proportions of D. For instance, by taking as unity the quarter note we have for $RT = (1, (1, 1, 1, 1))$ the rhythm shown in Figure 52.14.

Fig. 52.14. Rhythm for $RT = (1, (1, 1, 1, 1))$.

RTs allow us to represent, in a homogenous way, different types of musical objects. Polyphonies, voices, measures groups, etc. are expressed as RTs. When the value D is at the measure level, we express it in whole note units. For example, the RT for the next rhythm will be $(3/4, (2, 1))$.

Fig. 52.15. $RT = (3/4, (2, 1))$.

As it was defined, S represents a sequence of proportions of D. The example in Figure 52.16 shows the case for $RT = (4/4, (1, 2, 1))$.

Until now, we have presented RTs where S was a list of integer elements, representing notes or beats. For an $RT = (4/4, (1, (2, (1, 1, 1)), (1, (1, 1, 1))))$, more complex elements appear as shown in Figure 52.17.

Fig. 52.16. $RT = (4/4, (1, 2, 1))$.

Fig. 52.17. $RT = (4/4, (1, (2, (1, 1, 1)), (1, (1, 1, 1))))$.

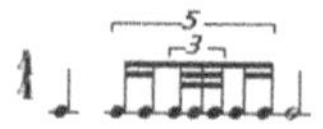

Fig. 52.18. $RT = (4/4, (1, (1, (1, 1, (1, (1, 1, 1)), 1, 1))2))$.

The RTs contained in S represent what we call *(rhythmical) groups*. In general, groups are graphically represented as beamed notes. As well as measures, groups can contain either notes or other groups. Here is an example for the $RT = (4/4, (1, (1, (1, 1, (1, (1, 1, 1)), 1, 1)), 2))$:

We extend the RT syntax in order to integrate rests and ties which will be represented respectively by negative numbers and floats[2]. For

$$RT = (4/4, ((1, (1, 1)), (2, (1.0, 1, -2, 1)), (1, (1.0, -1, 1))))$$

we obtain the following rhythm:

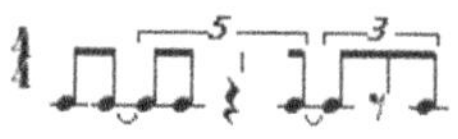

Fig. 52.19. $RT = (4/4, ((1, (1, 1)), (2, (1.0, 1, -2, 1)), (1, (1.0, -1, 1))))$.

RTs may encode the metric intention in a rhythm. The two rhythms in the next figure are encoded by different RTs $RT1 = (2/4, (2, 2, 2, 2))$ and $RT2 = (2/4, ((1, (1, 1)), (1, (1, 1))))$.

Fig. 52.20. RTs $RT1 = (2/4, (2, 2, 2, 2))$ and $RT2 = (2/4, ((1, (1, 1)), (1, (1, 1))))$.

The symbolic representation is essentially based on the hierarchical musical content. This allows an overall readability of the hierarchical structure and a symbolic format, which can be controlled by algorithms and other transformations such as inversion, recursion, etc. This is made possible by the fact that the format is the abstraction of the object itself. RTs are automatically encoded into optimal container structures.

[2] Perhaps, a better distinction of domains for ties would be to take imaginary rational numbers $i.x/y$ instead of floats, since rational numbers are not really different from floating point numbers. A more musical solution would be to take the form $TimeSig$ defined in formula (6.92) instead of $DurationValue$ for normal duration values (see also Appendix A.2.1), to reserve $DurationValue$ for ties, and to replace $DurationValue$ by the colimit of $DurationValue$ and $TimeSig$ in the above definition.

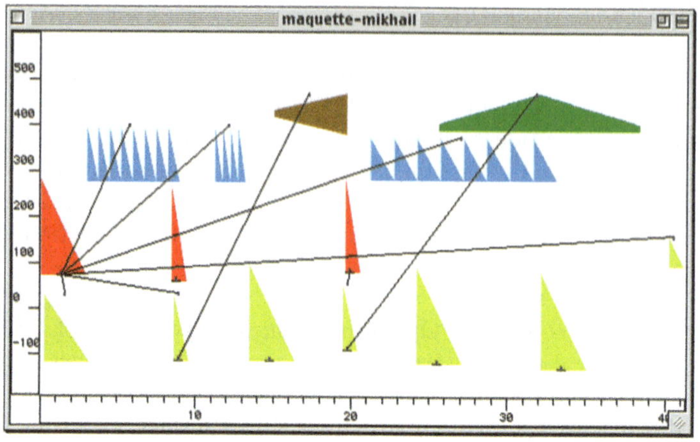

Fig. 52.22. Functional relation of temporal boxes.

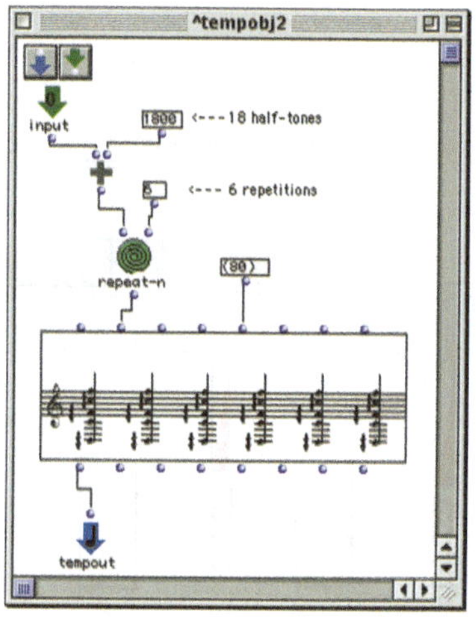

Fig. 52.23. Patch calculating a temporal box.

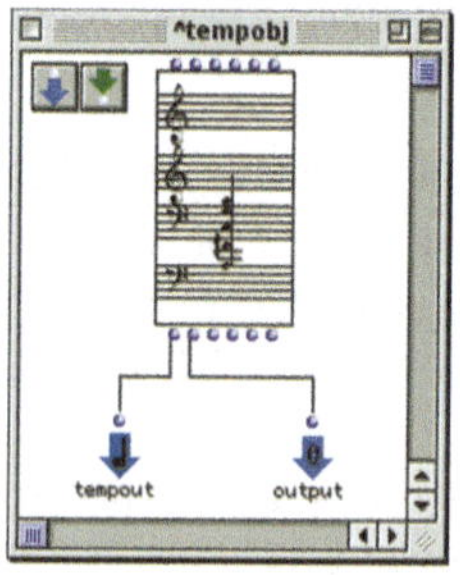

Fig. 52.24. Patch with musical material for a temporal box.

Let us take now the first chord that is provided as input for the ostinato. If we open its editor (Figure 52.24), we can observe that the generating algorithm is reduced to a given data. This is the fourth level of information which concerns the basic music material representation.

Finally, we can toggle the visual representation of boxes by showing the traditional representation of the musical result, see Figure 52.25.

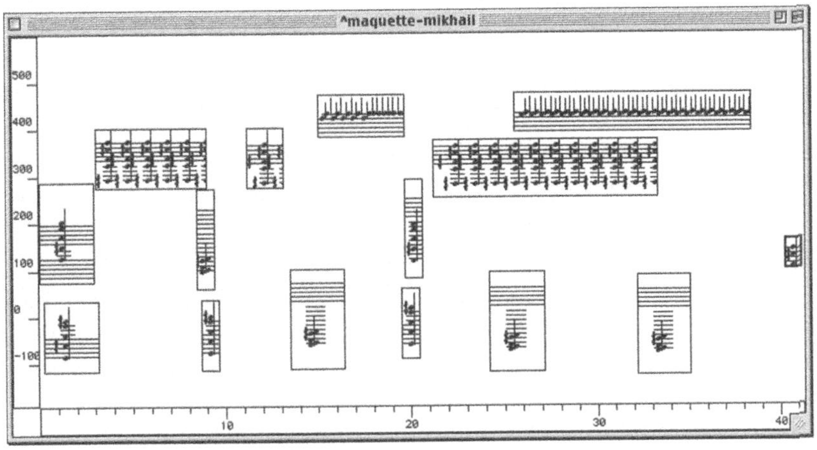

Fig. 52.25. Musical notation of a maquette.

In the previous example, we tried to show how an OpenMusic object scheme enables us to structure the musical information at different levels:

- the static level of the form, allowing us to create visual semiotic markers;
- the dynamic and paradigmatic level of the form (i.e., its functional relations between the temporal boxes);
- the syntactical level, it is the calculus building the musical discourse inside the temporal boxes;
- the material level.

These four levels of information are obviously interconnected. The most important advantage in the maquette concept is to offer a visualization of this interaction and at the same time, interactive control of it. This produces a source of experimentation:

- **Recombination at the form level**: Temporal boxes are moved and stretched in time without changing the other three levels.
- **Modification of functional relations**: We do not change the position of the blocks but their causal relation.
- **Syntax modification**: Algorithms which build the material can be changed according to the compositional goals.
- **Change of the material**: The color of the piece may change while keeping all the formal organization.

When combining these procedures, sophisticated musical experiments can be carried out.

52.4 Meta-object Protocol

Summary. Meta-object protocols (MOP) provide an alternative framework that opens the language implementation to user's intervention. OpenMusic was implemented using meta-programming technics. We have extended CLOS meta-objects (methods, classes, generic functions, etc.) by adding visual counterparts. In the same way we extended CLOS as such by using the technique of meta-programming; the user can make extensions of the OpenMusic language thanks to the visual MOP. In this section we describe OpenMusic's graphical MOP, as well as musical applications.

– Σ –

Originally the MOP was conceived for solving problems in the design and implementation of CLOS [532]. Usually the internal architecture of a programming language is not interesting for the programmer but only for language designers. For object-oriented languages, a MOP of the language is an interface to the language, presented as a framework. Classes, generic functions, patches, maquettes and other entities discussed in the previous sections are simply instances of special classes. These instances are called *meta-objects* and the classes *meta-object classes*. The set of these classes, as a framework, constitute the MOP. In addition, a language may support a MOP only if it has two characteristics: *reflection* and *reification*. The reflection is the ability of a program to inspect and modify its state at run time. In order to achieve this purpose it is necessary to have a mechanism to represent the program's data and state, this mechanism is called reification. A MOP is composed by a *static* and a *dynamic* part. The static part is given by the hierarchy of meta-object classes. The dynamic part is made of a set of generic functions that can be applied to meta-objects in order to control their behavior. Examples of such functions are:

- *Get-Elements*. We will consider that a meta-object is composed by a set of meta-objects together with a relation between them. For example, a class is composed by an ordered list of slots, a generic function consists of a set of methods, a patch contains a list of boxes, etc. This function returns the list of elements of a meta-object.
- *Get-View* and *Get-Editor*. It returns the two possible graphical representations of a meta-object.
- *Open-Editor*. It shows the editor for a meta-object.
- *Add-Element* and *Remove-Element*. They allow editing of any basic object.
- *Box-Value*. It enables us to trigger the evaluation of a visual expression in a patch.

Figure 52.26 shows the graphical representation of the dynamic and static parts of the OpenMusic MOP. One may be surprised by the simplicity of the protocol. In general, a protocol which is too much specified is not very modular. It means that changes must be made in several places, and as consequence, modifications are not very reliable. On the other hand, a poor specification of the protocol makes it very difficult to find the place where modification should be included.

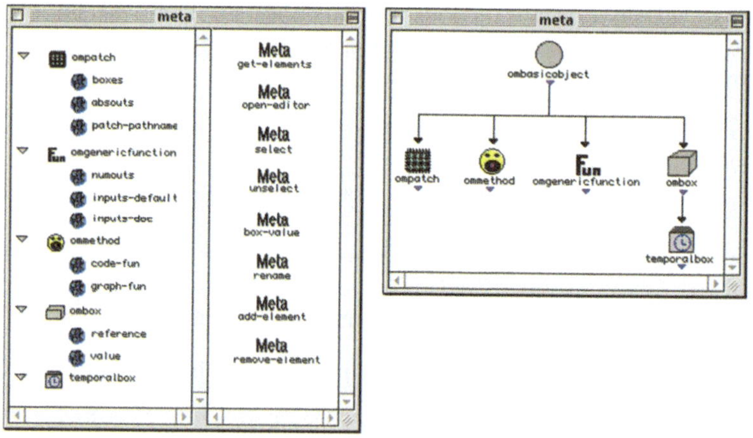

Fig. 52.26. Dynamic and static part of the OpenMusic MOP.

The main tools for the meta-programmer are: subclassing inside the static class part and redefining functions in the dynamic protocol part. It is clear that the user can change some features of the language, but the most interesting point is that he can extend the language without loss of compatibility with old programs. Programmers can profit from customizing the language semantics. The following example in Figure 52.27 shows the redefinition of the generic protocol function *Box-Value* which is called at each evaluation of a box. Graphical redefinition of these two methods changes the behavior of the language by introducing a visual trace of programs. The first method, which is defined with qualifier *before*, graphically selects any currently evaluated box before its execution. The second method, with qualifier *after*, deselects the box after its evaluation. These two simple modifications allow us to extend the language by tracing graphically and in an orthogonal way the functional composition.

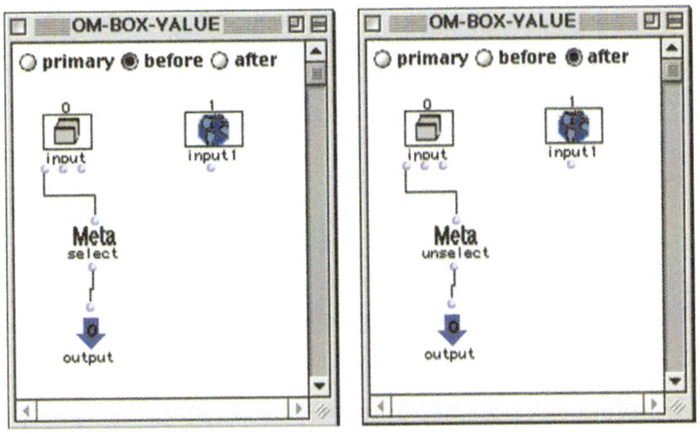

Fig. 52.27. Changing a generic function of the MOP.

52.4.1 Reification of Temporal Boxes

Finally, this section shows how we have extended the temporal boxes in maquettes by applying meta-programming concepts. Indeed, thanks to the graphical reification of temporal boxes, users have access to the temporal boxes at the same level as other objects like notes, chords, etc. Figure 52.28 shows a first example of the new possibilities. By opening the editor of the temporal box, we can see a new box *self* (pointed by one arrow) that represents the temporal box itself. The three public slots—available as outlets—are from left to right: the meta-object itself, its offset and its extension. The process associated to this temporal box builds a major chord whose pitches are transposed taking into account the box's time offset in the maquette (which means that the start time of the box is directly proportional to the transposition interval).

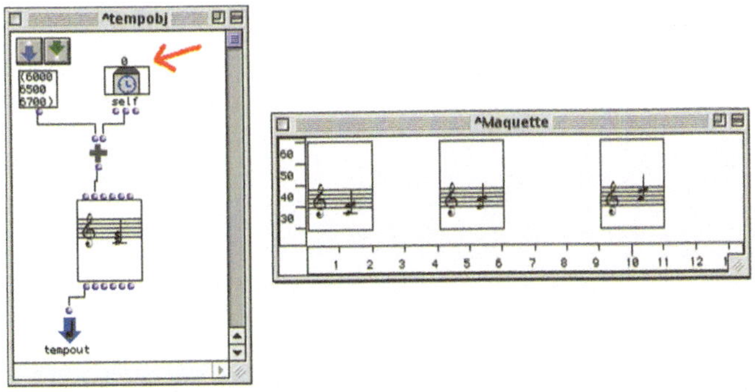

Fig. 52.28. Mixing calculations and temporal relations.

As has been shown above, temporal boxes may send and receive data by using connections. By reifying temporal boxes, we allow them to have access to other temporal boxes belonging to the maquette. In Figure 52.29 the upper temporal box is sent as a whole to the other one, which sets its own offset to the same value as the previous. One evaluation of the maquette will produce two chords and will align them in time.

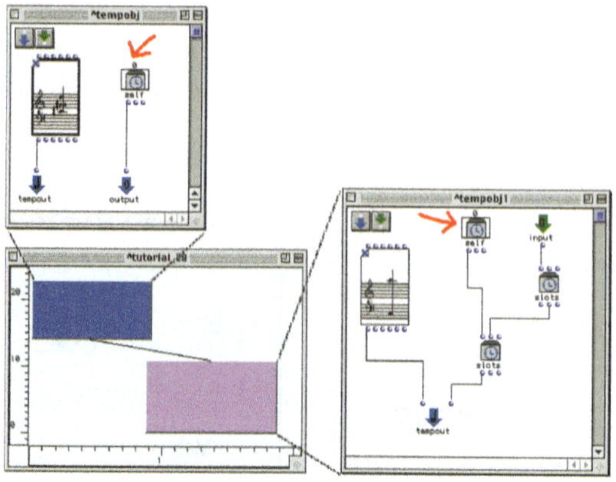

Fig. 52.29. Mixing calculus and temporal relations again.

We stress two principal temporal box relations within maquettes: a temporal relation, given by the horizontal position, and a causal relation, established by functional connection between boxes. Examples in Figures 52.28 and 52.29 expose how we have combined these relations by using the MOP. In the first example, the result of the calculation is affected by the position of the box. Informally, we can say that time changes calculation. In the second example, the evaluation of the functional composition of the boxes changes their position in time. In this case, the calculation changes the time organization.

We finish this chapter with a musical example of composition using OpenMusic.

52.5 A Musical Example

Summary. *Encore*, By Jean-Luc Hervé for two ensembles, live electronics, and two MIDI-controlled pianos (commissioned by IRCAM-EIC). Created April 9, 2000 by Ensemble Intercontemporain, Conductor : Patrick Davin. Frédéric Voisin was musical assistant and designed in OpenMusic the maquette described here.

– Σ –

This piece is concluded by a cadence for two mechanical pianos (actually two MIDI-controlled pianos) followed by a short orchestral finale. The cadence is actually an OpenMusic maquette that is played through MIDI .

The piece is based on the concept of instrumental gesture. A gesture is a small musical unit with a typical energy profile, such as a glissando, a strong note preceded by a group of grace notes, a repeated note, etc. The gestures performed by the instruments may be continuous, such as a glissandi.

The piano cadence is prepared by a densification of orchestral gestures, soon imitated by the piano. At the end of the cadence, the instruments enter back again and the cadence progressively sinks into the orchestral mass.

The idea behind the cadence is to accumulate and superpose an enormous quantity of discrete gestures (played by the piano) into a well-defined architecture so that the total construction sounds as a quasi continuous sculpted sound shape. This is why it has been realized as a maquette: it was in fact impossible to write it manually in detail. Rather, the architecture is specified by the hierarchical inclusion of maquettes within maquettes (three levels).

At the deeper level, one finds the elementary gestures. Elementary gestures are temporal boxes that contain all the algorithmics necessary to generate them.

Groups of gestures inside a submaquette are linked by two kinds of relations. Firstly, temporal logic relations force them to begin and end simultaneously, even if they are moved or stretched in time by the composer. Secondly, they have inputs that are fed by links coming from a special block. This block is not a true temporal block in the sense that it is here only for computing harmonic material from which the gestures are colored. But this block does not generate any in-time music by itself. We will not detail the algorithmics behind gesture generation, we only want to make precise the relation between the computation of music material and the time information. Whenever the composer stretches a temporal block in order to experiment with duration, the note durations inside the block are not extended proportionally, as the default maquette behavior would enforce it. Rather, the composer's intention is to keep the same density of notes within a different duration. This is achieved by using the 'extent' outlet of the 'self' reflexive box inside a temporal block. By connecting this outlet to the right place in the visual algorithm, a gesture possessing the same overall profile is generated by inserting more notes picked out from the harmonic reservoirs.

The whole process is illustrated in Figures 52.30 to 52.36.

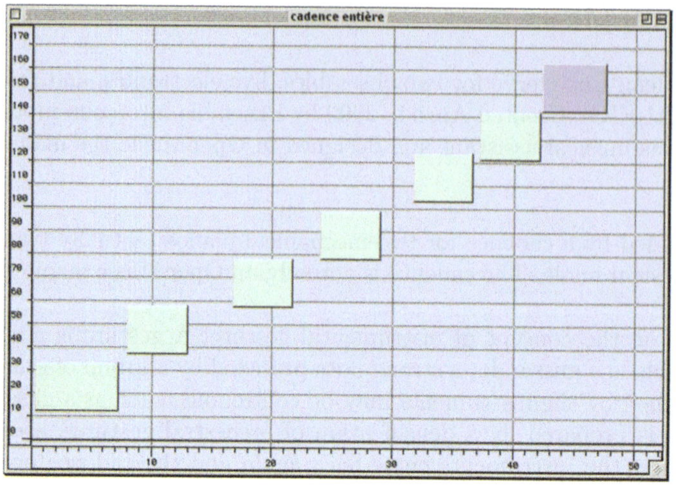

Fig. 52.30. The piano cadence maquette of "Encore".

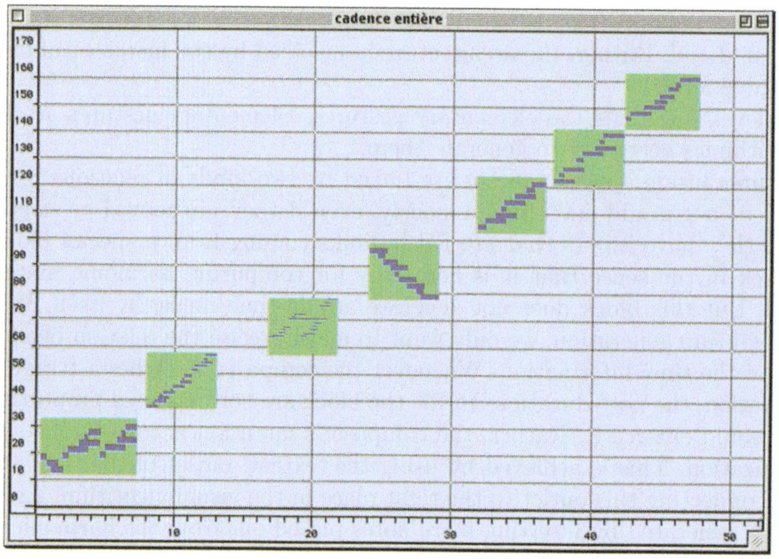

Fig. 52.31. The second maquette level revealed.

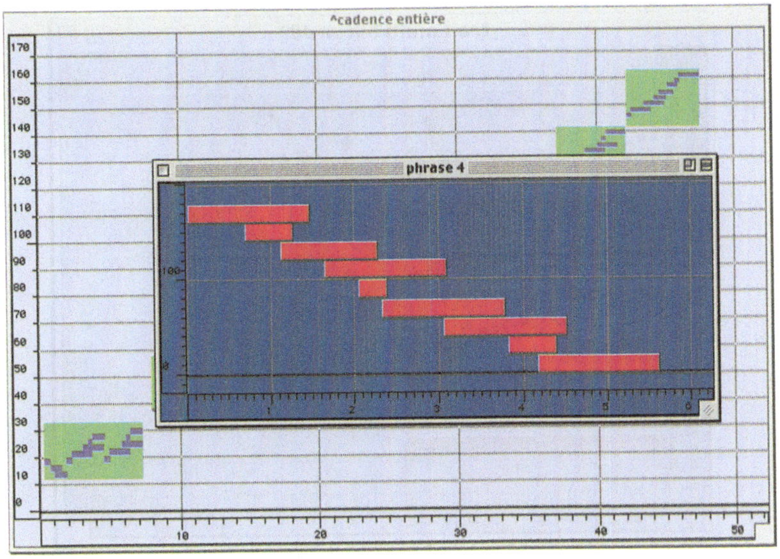

Fig. 52.32. The fourth submaquette (starting left) opened.

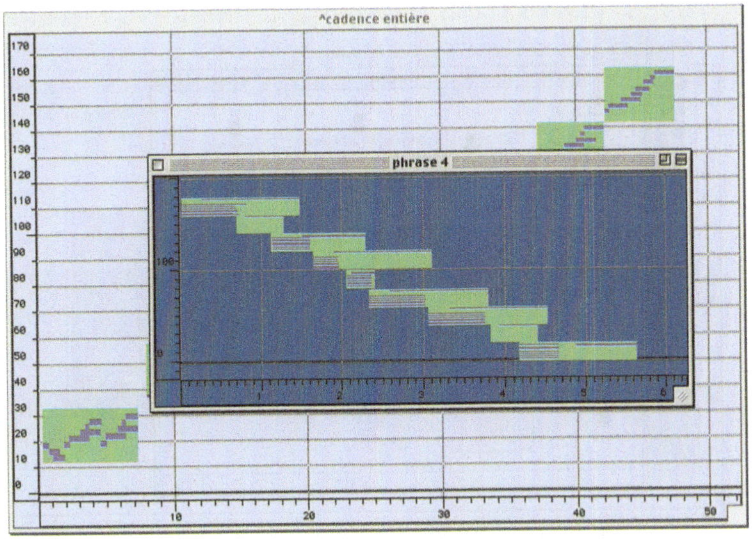

Fig. 52.33. The third maquette level revealed.

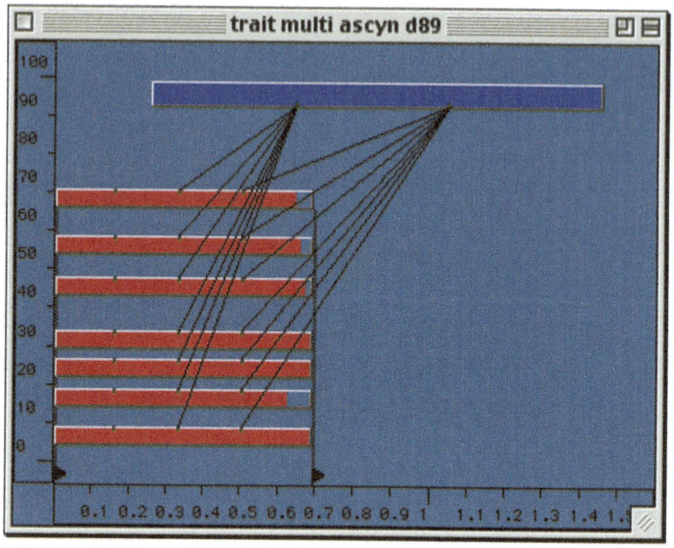

Fig. 52.34. The fifth block (starting left) opened. The block on the top generates the harmonic material. The seventh block underneath are gestures. The vertical lines are temporal logic constraints.

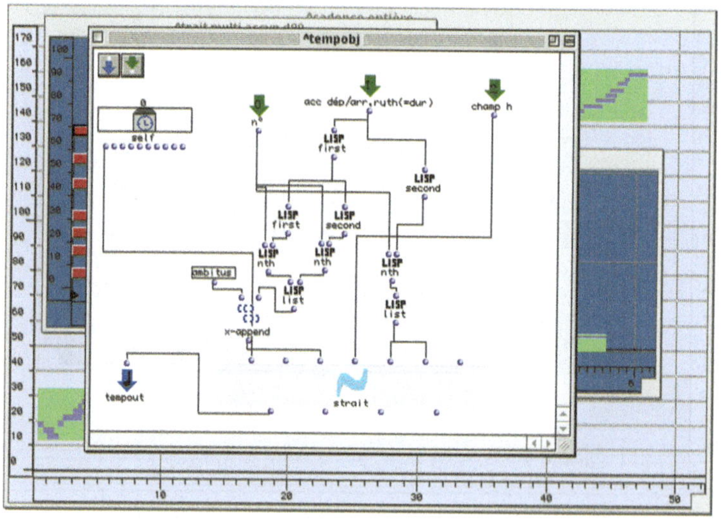

Fig. 52.35. A gesture opened. The algorithmics generate in-time music material. The link coming from the 'self' box informs the algorithm about time conditions.

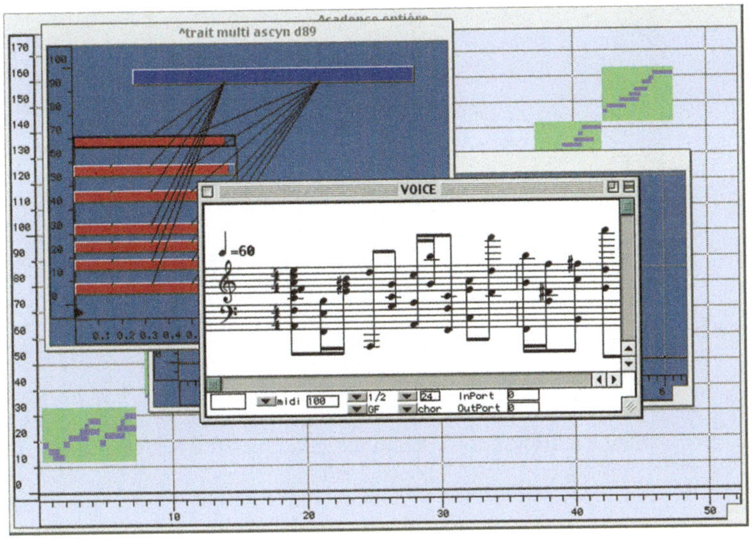

Fig. 52.36. A gesture opened in music notation mode. Manual modifications can be achieved at that time.

Part XIV

String Quartet Theory

53

Historical and Theoretical Prerequisites

Bei der nämlichen Gelegenheit fragte ich Haydn, warum er nie ein Violinquintett geschrieben habe, und erhielt die lakonische Antwort, er habe mit vier Stimmen genug gehabt.
Ferdinand Ries [908, p.287]

Summary. This chapter introduces the best evolved theoretical part of instrumentation: string quartet theory. It starts with a short historic synopsis and then reviews Ludwig Finscher's work [320] on string quartet theory. We then focus on the technical core subject: the violin family as an instrumentation paradigm for the string quartet. The chapter concludes with a general discussion of semantics of sound colors.

– Σ –

The scope of this part is the elaboration of a systematic foundation of the distinguished role of the string quartet at the end of the eighteenth century. This part is a synthesis of methods and results from modulation theory (chapter 80), classification (chapter 15), and counterpoint theory (part VII), combined with knowledge about the nature of sound parameters (see also Appendix A). It is not astonishing that precisely the classical string quartet—one of the absolute highlights of instrumental music—needs a complex theoretical background for its comprehension. But since we deal with a systematically as well as historically founded phenomenon, it is adequate to prepend some remarks on the theory and history of the string quartet, remarks which we shall orient towards Ludwig Finscher's pioneering work [320].

This art form, whose instruments include two violins, one viola and one violoncello, can be characterized in the following way [737, p.409]: "In the string quartet, individuality and character of the single players are combined to a harmonic whole, where each one finds himself with an added profile. This reaches from a single complex chord which—in contrast to the piano—is not played and nuanced by the hand of one player, but by four players, to the entire performance. In the quality as a whole lies the specificity of the string quartet. In its harmonically contrasting togetherness, this chamber music corresponds to the ideal world view and to the high humanistic vision of the classical epoch."

53.1 History

Summary. The history of the string quartet is exceptionally short since this instrumental species almost instantly appeared around 1760 with the works of Luigi Boccherini and Joseph Haydn. It almost instantly installed a *leading* stream of sophisticated instrumental expression. We discuss the historical background of this phenomenon.

– Σ –

G. Mazzola, *The Topos of Music II: Performance*, Computational Music Science,
https://doi.org/10.1007/978-3-319-64444-8_22

The *prehistory of the string quartet* is more complicated than that of any other instrumental art form of the eighteenth century. It cannot be causally deduced from any single one of the threads of tradition from where it comes. To a certain degree it is the creative act, the invention out of a moment of the delicate historic equilibrium, the *kairos* in the sense of ancient Greek thinking. The prehistory dates only from about 1720 to 1760 when *Luigi Boccherini* and *Joseph Haydn* independently invented the string quartet. In 1761, Boccherini wrote his first quartets in northern Italy, they were published 1767–68 in Paris under the name of "quatuor concertant". Probably Haydn had written quartet "divertimenti" already in the 1750s in Vienna, they were however only well-known in 1760. The sparse regional, instrumental, and stylistic rootedness in the string quartet's prehistory, from which this new art form has quite spontaneously emerged, provokes the question whether *beyond historical rationales* a more *systematic understanding* could better enlighten the 'string quartet phenomenon'. The problem is to question *this precise date* (1760) of the rise of *this precise instrumental art form* (the string quartet) in the context of the European music from the systematic point of view. In this question Dahlhaus [235, p.105,p.119] is fairly right in stating that

> *(...) erst die systematische Konstruktion den Blick dafür öffnet, welche Tatsachen einer Geschichte angehören, die zu erzählen lohnend scheint. (...) Daß etwa das Ausmaß in dem die Besetzung von Instrumentalmusik im 18. Jahrhundert gattungsprägend wurde, mit dem Grad ästhetischer Autonomie, mit der Herausbildung musikalischen 'Formdenkens' und mit der Festigung der Institution des öffentlichen Konzerts eng zusammenhing, ist keineswegs nur eine geschichtliche Tatsache, die sich empirisch feststellen läßt, sondern erscheint auch als Sinnzusammenhang, der sich einer phänomenologischen — also 'systematischen' — Analyse erschließt.*

53.2 Theory of the String Quartet Following Ludwig Finscher

Summary. In his habilitation thesis [320] Ludwig Finscher investigated the theory of the string quartet and exhibited three perspectives: the texture of four parts, the topos of conversation among cultivated humanists, and the family of violins. We discuss this threefold theory.

– *Σ* –

If we have decided to turn the string quartet into the subject of a formally valid investigation, this is because the theoretical reflections on the string quartet have reached a scientifically valid status. Finscher remarks [320, p.279]:

> *Das Streichquartett ist die einzige Gattung der neueren Instrumentalmusik, die eine solche an einem einzigen künstlerischen Modell entwickelte, vergleichsweise genau und detailliert ausformulierte und als allgemeinverbindlich akzeptierte Theorie ausgebildet hat.*

This theory is based on two fundamentals:

- the four part texture;
- the topos of a conversation of four humanistically educated persons.

53.2.1 Four Part Texture

Summary. The texture of four parts is a basic structural prerequisite for the string quartet theory. We review its implications.

– *Σ* –

The four part texture was the ideal type of structured polyphony which was oriented on the counterpoint with its long tradition. This is the formal, or better: formalized element of string quartet theory. We have to take it in the full conceptual ambiguity, i.e., on the one hand the texture "note against note" in its linear temporal progression in the sense of classical counterpoint. On the other, it is a texture of vertical units—charts in the terminology of global compositions—as an expression of harmonic relations. In the radically harmonic thinking, which is realized in the work of August Kollmann [541] from 1796, it is even possible to reverse the tradition in that the following thesis is proposed: Counterpoint should not start from the intervallic two-part texture, but from the four-part texture, since "a complete harmony" is four-part and not two- or three-part. It seems that the formation of theories was already influenced by Haydn's success with his famous "Russian" string quartets from 1782.

It is interesting to observe that this ideal type of an instrumental art form was fixed to exactly four voices, not one more. Especially with Haydn one could imagine that he could have added a fifth voice to "enrich the texture". But it is reported that he 'failed' on several occasions with this 'experiment'. Ries reports 1838 [320, p.287]:

> *Bei der nämlichen Gelegenheit fragte ich Haydn, warum er nie ein Violinquintett geschrieben habe, und erhielt die lakonische Antwort, er habe immer mit vier Stimmen genug gehabt. Man hatte mir nämlich gesagt, es seien drei Quintette von Haydn begehrt worden, die er aber nie hätte komponieren können, weil er sich in den Quartettstil so hineingeschrieben habe, daß er die fünfte Stimme nicht finden könne; er habe angefangen, es sei aber aus einem Versuche am Ende ein Quartett, aus dem anderen eine Sonate geworden.*

Presently, Haydn's argumentation that he has "enough" with the four voices, cannot be understood. We come back to this point at the end of our discussion in Section 55.3.

53.2.2 The Topos of Conversation Among Four Humanists

Summary. String quartet tradition is intimately related to non-verbal humanistic conversation. This gives the species a *rhetoric* characteristic which has important consequences for the instrumentation problem.

– Σ –

On the one hand, this is a topos which stems from the analogy of contrapuntal texture to the conversation and argumentation of humans. For example, Mattheson [320, p.285] says on imitation:

> *...daß eine Stimme die andere gleichsam gesprächsweise unterhalte.*

This topos must be cautiously distinguished from the well-known topos of a "Klangrede"[1], i.e., from the similarity of musical expression or semantics to the common language. In the case of the string quartet, the more important thing than speaking is the dialog, a fact that becomes more evident in the French expression "quatuor dialogué" for "string quartet" (in fact the invention of a publisher).

The association of a discourse to the string quartet was initiated by the musician Johann Friedrich Reichhardt in 1777 [320, p.287]:

> *Bei dem Quartett habe ich die Idee eines Gesprächs unter vier Personen gehabt.*

Like Haydn, Reichhardt also views the number of four as being the upper limit for a good dialog. He tries to add a fifth person to the quartet. But he fails:

> *Die fünfte Person ist hier ebensowenig zur Mannigfaltigkeit des Gesprächs noth-wendig, als zur Vollstimmigkeit der Harmonie; und in jenem verwirrt sie nur und bringt Undeutlickeiten in's Stück.*

The same happens to Schumann [320, p.289] in a discussion about a viola that was added in a quintet:

> *Man sollte kaum glauben, wie die einzige hinzugekommene Bratsche die Wirkung der Saiteninstrumente, wie sie sich im Quartett äußert, auf einmal verändert, wie der Charakter des Quintetts ein ganz anderer ist, als der des Quartetts. Die Mitteltinten haben mehr Kraft und Leben; die einzelnen Instrumente wirken mehr als Massen zusammen; hat man im Quartett vier einzelne Menschen gehört, so glaubt man jetzt eine Versammlung vor sich zu haben.*

---
[1] German for "sound speech", however, difficult to translate.

The quartet discourse as a dialog is very well suited to communicate understanding within music. This is the sense of the dialog. Goethe stresses this aspect in an enlightening comment on a concert by Niccolò Paganini [320, p.288], [373]:

> *Mir fehlte zu Dem, was man Genuß nennt und was bei mir immer zwischen Sinnlichkeit und Verstand schwebt, eine Basis zu dieser Flammen- und Wolkensäule. Wäre ich in Berlin, so würde ich die Möserschen Quartettabende selten versäumen. Dieser Art Exhibitionen waren mir von je her von der Instrumentalmusik das Ver-ständlichste: man hört vier vernünftige Leute sich untereinander unterhalten, glaubt ihren Diskursen etwas abzugewinnen und die Eigentümlichkeiten der Instrumente kennen zu lernen. Für diesmal fehlte mir in Geist und Ohr ein solches Fundament; ich hörte nur etwas Meteorisches und wußte mir weiter davon keine Rechenschaft zu geben.*

The connection between the dialogical discourse and the communication of understanding which Goethe indicates means a valuation of the dialog, its qualification in function of communication of understanding. This valuation in turn is related to the instance of varying competition by Karl Popper [858]: "The value of a dialog depends above all from the manifold of competing opinions." But this *manifold of opinions*, always related to a given subject, is nothing else than a variation of points of view, of the perspectives of the participants. Therefore the *dialogical principle* of the string quartet turns out to be an instance of the *Yoneda philosophy* discussed in Section 9.3: *Understanding or classification, respectively, by a variation of the point of view.* It is not astonishing that this phrase is thoroughly indebted to humanism with which the string quartet is deeply associated.

Result 9 *Summarizing, one root of the string quartet, the four part texture, appears as a form which, in the sense of Hanslick, is a carrier of musical spirit. This carrier in turn is made accessible to our understanding by the second root: the dialogical discourse of four violin personalities, by means of a variation of perspectives.*

At this point of our discussion the question arises, why the string quartet and the end of the eighteenth century are related to each other, more precisely: *What is the connection between the four-ness within the family of violins (and its above all violins, not violinists who speak!) and the paradigm of four part texture at Haydn's time?*—This question leads to a mathematically tractable apparatus in the manifold of violin sounds in defined parameter spaces.

53.2.3 The Family of Violins

Summary. We discuss the exceptional role of the family of violins in the building process of the string quartet species. In Finscher's work, other instrumental families are compared to the violins, we comment on the results of this study.

– Σ –

The formation of the string quartet would not have been thinkable without the collaboration of the homogeneous sound of the instruments of the violin family, which in the eighteenth century was discovered as an ideal type of four-part music and also perfected on the artisanal level after the violin's creation from local variants of string instruments in the fifteenth century.

The high quality of violins is guaranteed by a technical standard which through its fineness, sometimes also imponderableness such as the choice of woods and the varnish covering, contradicts any normalization. The individual sound color of every good violin is characteristic to the family. But the family of violins is also strongly differentiated from other families of string instruments. Finscher describes [320, p.124/125] the characteristic of violins as compared to the gambas, which were the preferred solo instruments in the seventeenth century, as follows:

> *Die Violinen hatten gegenüber den Gamben jedoch noch eine weitere zukunftsträchtige Eigenart: Sie gliederten den Tonraum, der ehemals in Analogie zu den menschlichen Stimmengattungen gebildet, nun aber in der Tiefe wie in der Höhe längst kräftig erweitert worden war, klarer und sinnfälliger,*

mit deutlicherer Individualisierung ihrer jeweiligen Tonbereiche. (...) Für das klassische Streichquartett, das die Beweglichkeit, den Lagenwechsel, den Kontrast- und Farbreichtum des symphonischen Streichersatzes mit der größtmöglichen Annäherung an eine streng auskomponierte Vierstimigkeiit zu verbinden suchte, bot sich das vierstimmige Ensemble aus Gliedern der Violinfamilie als das ideale Instrument an.

It will be a task of the following chapter to access the specificities of the sound of the violin family via a mathematical description in the frame of parameter spaces.

54

Estimation of Resolution Parameters

Si tous les instrumets jouent l'accord staccato,
je supprime ce moyen naturel d'analyse,
et la perception ne peut plus discerner,
à l'interieur du bloc sonore,
de quelle combinaison il s'agit.
L'identification dépend, dans ce cas,
de la présence ou de l'absence des
éléments essentiels pour la perception.
Pierre Boulez [139, p.547]

Summary. This chapter is a technical account of the variety of sound parameters which intervene for the family of violins.

– Σ –

Based on the preceding historical, systematic, and philosophical reflection we argue that the classification theory of global compositions (Chapters 15 through 17) is an excellent candidate to be applied to the theory of the string quartet. From classification theory it follows that the resolution of a global composition is a formalization and optimization of the process of understanding this composition. We shall equally apply this approach to string quartets, which means that we have to think of the composition as being a global composition G^I, whose resolution Δ_{G^I} one wants to represent. In Chapter 55, we shall specify the global composition associated with a string quartet composition. Our approach to the resolution is essentially realized via a variation of the perspective of the resolution parameters, i.e., by a differentiated change of their dichotomy into essential and accessory parameters. Within the string quartet, this dynamics of weighting of performance parameters is a subtle tool for communication of understanding and articulation as a vehicle of dialog. It is as if you walk around an object—which is the resolution in our case—and observe it from one, and then from another perspective.

If we now envisage the resolution of a string quartet composition, we first have to investigate the possibility of parametrization by concrete physical performance parameters as they intervene on the arsenal of the violin family.

54.1 Parameter Spaces for Violins

Summary. We describe the denotator types for violins, including generic geometric parameters, sound color parameters, and technical parameters, in particular those related to vibrato.

– Σ –

G. Mazzola, *The Topos of Music II: Performance*, Computational Music Science,
https://doi.org/10.1007/978-3-319-64444-8_23

As we already did in classification theory, we want to work over $\mathbb{Q}$, which is a reasonable postulate for string quartet theory since we are dealing with physical parameters.

Here, we envisage a fundamental problem of mathematics since we have to deal with a huge number of independent numerical parameters. If, for example, a sound color has to be represented by means of amplitudes and phases of the Fourier representation, and by the envelope, we easily add up hundreds of numerical parameters. Even though locally, the structure of music thinking may concern a small number of parameters, the techniques of classification theory, by their requirement of points in general position, enforce a number of geometric degrees of freedom that might possibly be far larger. We shall have to work with charts in modules $\mathbb{Q}^n$ with large dimension n. This should however not prevent us from *visualizing essential aspects of our reflections in three space* $\mathbb{Q}^3$. The only point here is to build mathematically representative analogies in three space.

What is the shape of the physical parameter space of a violin sound? To begin with, we dispose of parameters which we call *geometric*[1], i.e.,

- Onset
- Duration
- Pitch

The amplitude (loudness) is omitted in this presentation since the poietic violin parameters are coupled to the way of generation of the sound color aspect.

The other attributes of the violin sound span the *color space.* To begin with, this aspect includes the following ingredients:

- Envelope
- Amplitude
- Fourier spectrum

What is the range of variability of these color aspects of a violin sound? Regarding this question, we refer to [734]. For example, we consider the amplitude spectrum for pitch g and $g_\sharp$ of a "Guarneri del Gesù" as compared to an F horn, see Figure 54.1. One recognizes that the spectra between g and $g_\sharp$ are significantly more different for the "Guarneri" than for the F horn. In the representation of the amplitude spectrum as a vector in the color space, one may say that the change from g to $g_\sharp$ for the F horn is essentially a dilatation, whereas the corresponding change for the "Guarneri" violin also includes a change of direction (Figure 54.2).

In other words: The g and $g_\sharp$ spectral vectors of the "Guarneri" span a plane in the color space, whereas the corresponding vectors for the F horn lie on a line. This statement evidently is not meant in a strictly physical sense, but in the sense of valence theory (see Appendix B.2): For the F horn, the g- and $g_\sharp$-spectral vectors are indistinguishable in the auditory perception from a pair of linearly dependent vectors[2].

This instrumental difference is justified by the fact that for winds, the sound color is essentially built from an air column, but not from the material. In contrast, for the violins, sound color is an essential function of the material, i.e., of the corpus' resonance properties. Moreover, the spectrum depends on the string on which a fixed pitch is intonated (to this end, the spectral envelopes for different strings are compared [734]).

All this turns the sound color of violins substantially into a function of the individual construction: from the material through the artisanal manufacturing to the individual history of the particular instrument. This variability is out of the question with wind instruments. For the same reasons, string pianos are inferior to violins. Figure 54.3 shows the principal configuration of the spectral vector for three violins as opposed to three such vectors for string pianos.

To these instrumental parameter properties, central techniques of instrumental practice of sound shaping are added, techniques which for several other instruments do not even exist. The string player has the following possibilities to vary parameters:

[1] See Appendix A.1.2.1.

[2] This discussion is somewhat speculative since precise measurements should be made and relations to valence theory should be investigated in a more quantitative way. However, the means for such an investigation depend on the insight in the basic problem setup.

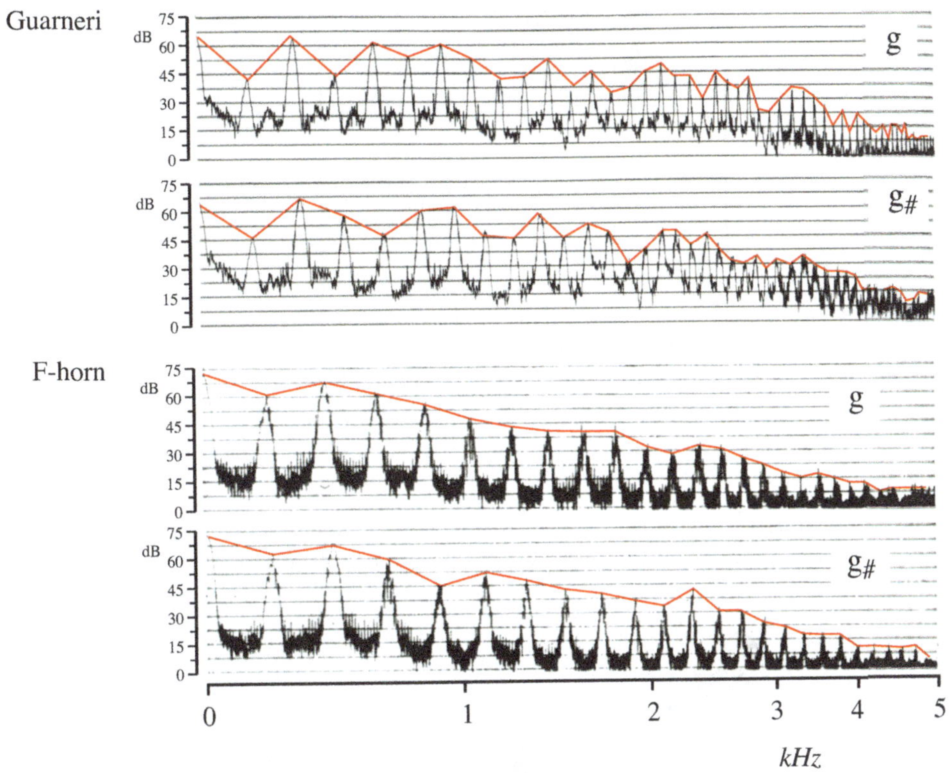

Fig. 54.1. We consider the amplitude spectrum for pitch g and $g_\sharp$ of a "Guarneri del Gesù" (top) as compared to a F horn (bottom). One recognizes that the spectra from g to $g_\sharp$ are significantly more different for the "Guarneri" than for the F horn.

- Bow pressure
- Bow velocity
- Contact point of bow and string
- Bow angle

Further, the string player may shape his/her *vibrato* according to the following points of view:

- Delay time with respect to the tone's onset
- Modulation frequency (frequency of the finger's movement)
- Pitch modulation (Range of finger movement on the string)
- Amplitude modulation (Contact point of the finger-tip)

Compared to the color attributes envelope, amplitude, and spectrum, these *vibrato parameters* are new. They enable the violinist to realize sound in still larger spaces. Together with the four *bow parameters*, the four vibrato parameters define an additional eight-dimensional space. These eight parameters which emerge in contrast to the instrumental parameters are called *technical parameters*. Figure 54.4 shows the effect of bow pressure and contact point variation while bow velocity and sound color remain constant.

54.2 Estimation

Summary. This section is devoted to a theorem giving an estimation of the maximal possible chart dimension $ch(n)$ within a global composition, which can be produced by an orchestra consisting of n individual strings

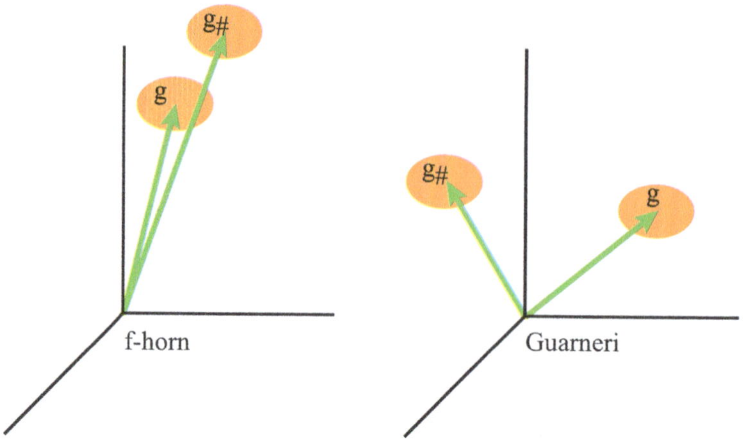

Fig. 54.2. In the representation of the amplitude spectra as vectors (represented in 3D space here), the "Guarneri" vector differs beyond the valence limit when comparing the g sound to the $g_\sharp$ sound.

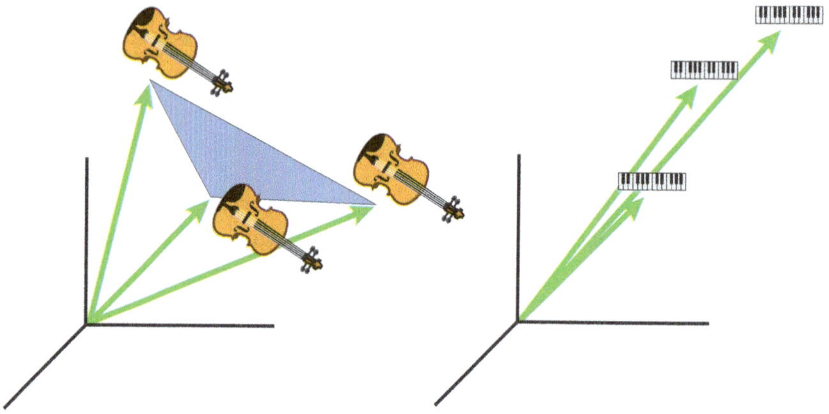

Fig. 54.3. The instrumental sound parameters of violins are more variable individually than these parameters are for pianos. Here the amplitude spectra of three violins against three pianos are visualized schematically in 3D space. In contrast to the piano spectra, those of violins lie in general position.

(in the violin family). We make use of the resolution theory for classification of global compositions (see Chapter 15.2).

– Σ –

In the last step of our parameter analysis, we deal with differentiation within the technical parameters. Let us recall that we are searching for parameter spaces, i.e., coordinate functions which are suited for charts of global compositions. But this signifies that one has to distinguish parameters which can be varied independently of each other in short parts of the composition, in fact on charts, from those parameters which may very well vary from player to player or from situation to situation within a larger composition, but which are relatively constant on local regions. The latter, which may strongly depend on the player's personality, include:

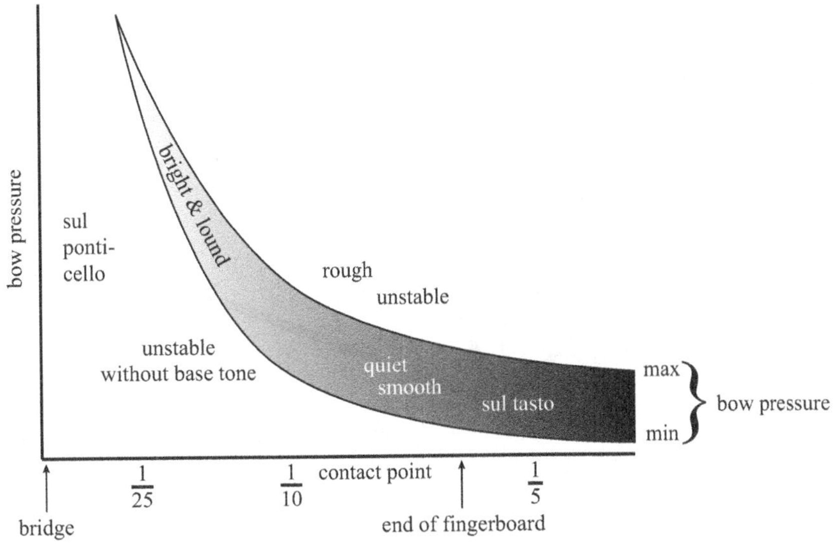

Fig. 54.4. The effect of a variation of bow pressure and contact point while bow velocity and sound color remain constant.

- the vibrato parameters. Even for a professional violinist they can scarcely be separated from the personality and are difficult to control;
- The bow angle and contact point are relatively inert parameters, therefore not suited for extremely local purposes.

The local variability is therefore distributed on two dimensions:

- Bow pressure which above all acts on the amplitude. (This is the reason why we did not add amplitude to the geometric parameters here: it is only an aspect of the action of bow pressure!)
- Bow velocity which above all acts on the dynamics of partials.

Evidently the bow parameters are coupled with each other in their action; we only indicated the main actions.

Summarizing, we have found three types of color parameters:

1. *Instrumental parameters* (Violin type, choice of strings, performance conditions),
2. *global technical parameters* (vibrato, bow angle, contact point), which are a strong function of the individual player,
3. *local technical parameters* (bow pressure, bow velocity), which can be steered quite independently of each other.

Therefore, the *local* technical parameters define a plane in the color space. Since these are under general control and can be steered objectively, we may assume that these planes $H_1, H_2, \ldots$ for player 1, player 2, ... are one and the same player-independent plane H up to an individual translation (Figure 54.5). Observe that our hypothesis is made in the context of the widespread string quartet at the end of the eighteenth century, and that in fact the signification of the string quartet is a pronounced reality of music sociology with regard to normal music practice in bourgeois saloons. It would be unrealistic to model our theory upon elite ensembles in this context.

Based on the above observations about the individuality of colors and the violinists' personalities, we may set forth the following hypothesis in the spirit of string quartet theory:

Assumption 3 *The n "instrumental vectors" $v_1, v_2, \ldots$ which are spanned by the instrumental and global technical parameters can be chosen to be linearly independent of each other and of the common plane H.*

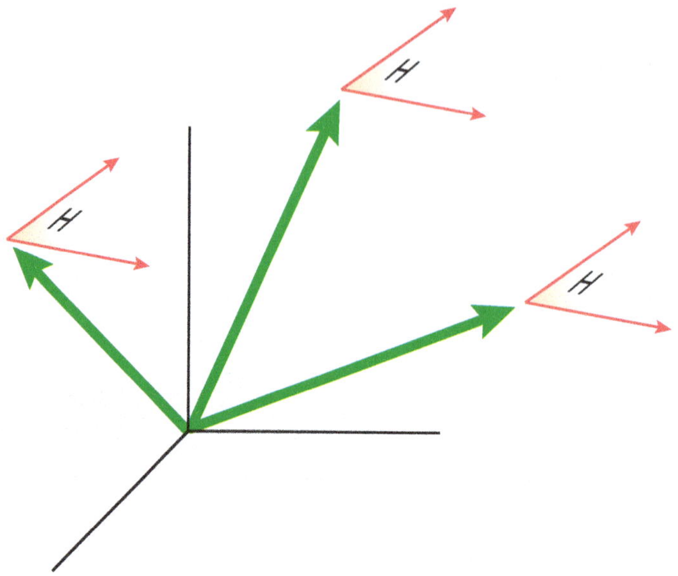

Fig. 54.5. In the color space of violins a plane H is appended to the heads of individual vectors v_s of instrumental and global (inert) color parameters and yields $H_s = v_s + H$ for player s.

This means that the submodule is spanned by $v_1, \ldots v_n$ and H has dimension $n + 2$. In other words: If H is spanned by two vectors h, k, the zero vector and the heads of the vectors $v_1, \ldots v_n, h, k$ are in general position[3].

If we add the three-dimensional module $G \xrightarrow{\sim} \mathbb{Q}^3$ of geometric parameters to the color space (direct sum), we obtain a *total parameter space*, where the n instrumental vectors are positioned, of which to each is attached a $2 + 3 = 5$-dimensional space $H + G$ of local parameters, see Figure 54.6.

In view of classification theory our question is how many points in the total parameter space can maximally be distributed in general position on the n affine subspaces $H + G + v_s, s = 1, \ldots n$. So every such point has the form $x = v_s + h + g, h \in H, g \in G$, see Figure 54.6. It can be shown (Appendix E.2.1, Theorem 55) that there are maximally $n+5$ points, where 5 means the dimension of $G+H$. The musicological meaning of this result for classification theory is this:

Theorem 36 *With n string players from the violin family, charts of global compositions in local parameters can be defined with maximally $n + 5$ points in general position.*

[3] See Appendix E.3.4.

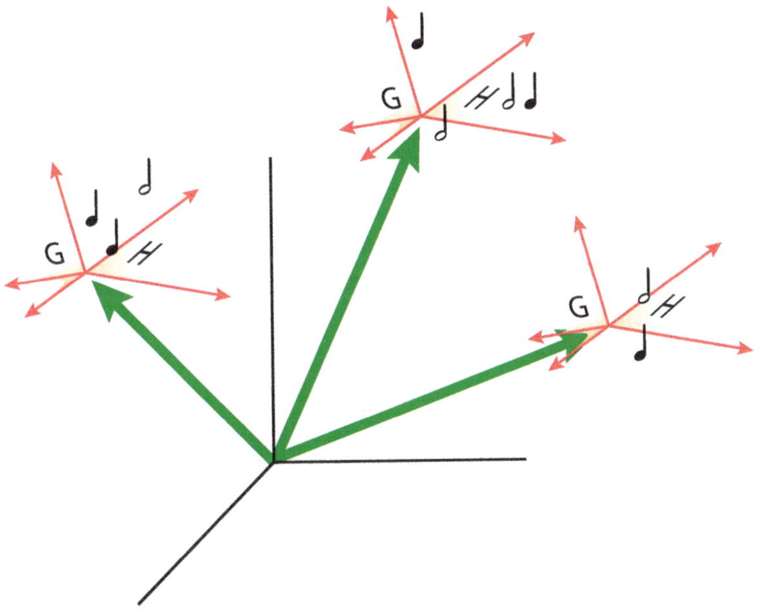

Fig. 54.6. Viewed as points for the n instruments of the violin family, the sounds are distributed on n affine subspaces $H+G+v_s, s=1,\ldots n$, in the total attribute space. Here G is the three-dimensional space of sound geometry (pitch, onset, duration), H is the plane of the local technical parameters bow pressure and bow velocity, and v_s is the sth instrument vector. One may assume that the $\mathbb{Q}$-module spanned by $G, H, v_s, s=1,\ldots n$ has dimension $n+5$.

55

The Case of Counterpoint and Harmony

Au fur et à mesure que l'orchestre s'agrandit,
que le rôle de l'instrument devient,
je ne dirais pas flou, mais ductile, multiple,
les formes, elles aussi, s'amplifient.
Pierre Boulez [139, p.544]

Summary. This final chapter on string quartet theory deals with the analytical conditions on global compositions which are powerful enough to comprehend the structural richness of central European music in the epoch of Boccherini and Haydn. These structures are—essentially—counterpoint and harmony. As a germ for a systematic theory of instrumentation we propose an estimation of maximal necessary chart dimension for Fuxian counterpoint and traditional harmony (including cadence and modulation).

– Σ –

As the last member of our model for the string quartet, we need the initially announced information to solve the question of how a string quartet should be defined as a global composition. To this end we imagine that a score, written by Haydn, say, is given as a local composition. From this one would like to construct an adequate interpretation (in the technical sense of interpretable global compositions). And it is here where the historical moment comes into the game. The interpretation should be such that the European structural music thinking in the four part texture at the end of the eighteenth century is expressed in its essential features. In our presentation of the roots of the string quartet theory, the structural basis was first presented: the four part texture with its polysemic meaning as a contrapuntal as well as harmonic setup. These two main components are therefore to be investigated for the construction of an atlas.

55.1 Counterpoint

Summary. In this section, we calculate the upper limit of chart dimensions for counterpoint. We make use of the counterpoint model exposed in part VII.

– Σ –

In this section we shall refer to the discussion of the core theory of counterpoint, "note against note", as it was presented in Chapters 29 through 31. For intervals which are represented by arrows in counterpoint theory, we need two-element charts (admitting that what is physically played are not arrows, but their heads and tails). For the cantus firmus and the discant steps we also need two-element charts, one for each. For the consideration of a progression from interval to interval, one needs four-element charts. It is advantageous to include also the succession of two interval steps, for example regarding hidden/composed tritones, we need six-element charts. This is however *not* the statement of the rules of counterpoint! As we know, this would be much more complex. But we have defined the "cartographical" setup, and that is what we need. Therefore:

G. Mazzola, *The Topos of Music II: Performance*, Computational Music Science,
https://doi.org/10.1007/978-3-319-64444-8_24

Result 10 *The classical contrapuntal texture as it was codified by Johann Joseph Fux in 1725, the early days of the prehistory of the string quartet, requires in its core structure maximally 6-element charts. All more complex configurations can be reduced to this core structure: Thereby charts are glued together, but not enlarged.*

55.2 Harmony

Summary. In this section, we calculate the upper limit of chart dimensions for harmony. We make use of the cadence (Chapter 26) and modulation (Chapter 80) models.

– Σ –

The degree theory of chords as vertical structures interprets a chord as being a subset of a triadic covering (see Chapter 25), i.e., by three-element charts.

A cadence (see Chapter 26) can be thought within the scheme consisting of three degrees (typically: IV-V-I), where the minimal cadential sets (here: IV,V) are completed by the first degree or else by the tonic note of the given tonality. This produces an interpretation of maximal nine tones per chart, consisting of one, three or nine points.

The most complex situation in the harmony at the end of the eighteenth century is crystallized in the modulation process which we have formalized in the modulation model that is based on triadic degrees, see Chapter 80. In a modulation which is presented as a sequence of neutral degree, modulation degree, cadence degree, we may recognize these three charts, with three elements each. Moreover, like with a cadence, the entire process is collected in a nine-element chart that contains the three triadic degrees. The cadence as such, which we described above as a nine-element interpretation, intersects our modulation in the cadence degree and concludes it as a process. Therefore:

Result 11 *Harmony has a maximum of nine points for the relevant local charts.*

55.3 Effective Selection

Summary. As a result of the instrumental parameter estimation made in Chapter 54 and the structural parameter estimation from this chapter, we obtain a global theorem estimating the *minimal number four of string instruments* (in the violin family) which is needed to express analytical music structures in the compositions of the classical epoch of Boccherini and Haydn. This theorem makes essential use of the resolution theory for global compositions, as exposed in Chapter 15.2.

– Σ –

Finally, we are in a state of indicating the minimal number of instruments from the violin family in order to provide the resolution of an interpretation of a score under the structure as preconized by harmony and counterpoint at the end of the eighteenth century by a sufficient number of free parameters.

In Sections 55.1 and 55.2 we have seen that a maximal number of nine points per chart are present in such an interpretation. Since in a resolution, all charts have their points in general position, the number n of instruments must, according to Theorem 36 at the end of Section 54.2, suffice the inequality

$$n + 5 \geqslant 9. \tag{55.1}$$

We therefore need at least four instruments for these scopes. As announced, our model of the string quartet has been deduced from the information about parameters, modulation theory, counterpoint, and the classification technique. From this point of view, it is not astonishing that classical string quartet composers such as Haydn did not see any sense in the accumulation of instruments: For a purely economical point of view, they were superfluous. Four string players were perfectly sufficient in order to provide the textural structure with a profiled representation in its resolution.

Part XXIV

References and Index

References

1. Abel V and Reiss P: MUTABOR II - Software Manual. Mutabor Soft, Darmstadt 1991
2. Abraham R and Marsden J: Foundations of Mechanics. Benjamin, New York et al. 1967
3. Abramowitz M and I A Stegun: Handbook of mathematical functions. With formulas, graphs and mathematical tables. Dover, 1965
4. Ackermann Ph: Computer und Musik. Springer, Wien and New York 1991
5. Aczel P: Non-well-founded Sets. No. 14 in CSLI Lecture Notes. Center for the Study of Language and Information, Stanford 1988
6. Ackermann Ph: Developing Object-Oriented Multimedia Software. dpunkt, Heidelberg 1996
7. Adorno Th W: Fragment über Musik und Sprache. Stuttgart, Jahresring 1956
8. Adorno Th W: Zu einer Theorie der musikalischen Reproduktion. Suhrkamp, Frankfurt/M. 2001
9. Adorno Th W: Der getreue Korrepetitor (1963). Gesammelte Schriften, Bd. 15, Suhrkamp, Frankfurt am Main 1976
10. Agawu V K: Playing with Signs. Princeton University Press, Princeton 1991
11. Agmon E: A Mathematical model of the diatonic system. JMT, 33, 1-25, 1989
12. Agmon E: Coherent Tone-Systems: a study in the theory of diatonicism. JMT, 40(1), 39-59, 1996
13. Agon A: OpenMusic: Un langage visuel pour la composition musicale assistée par ordinateur. PhD Dissertation, Université Paris VI, Paris 1998
14. Agustín-Aquino O A: Counterpoint in $2k$-tone equal temperament. Journal of Mathematics and Music, Vol 3, nr. 3, pp. 153-164, 2009
15. Agustín-Aquino O: Extensiones Microtonales de Contrapunto. PhD Thesis, Facultad de Ciencias, UNAM, Mexico City, 2012
16. Agustin-Aquino, O A, J Junod, G Mazzola: Computational Counterpoint Worlds. Springer Series Computational Music Science, Heidelberg 2014
17. Akmajian A et al.: Linguistics. MIT Press, Cambridge MA 1995
18. Alain: Alain Citation on Almada's painting in the Gulbenkian Foundation Center. Lisbon 1968
19. Albert R S and M A Runco: A History of Research on Creativity. In: R J Sternberg (Ed.): Handbook of Creativity. Cambridge University Press, New York 1999
20. Aleinikov A G: On Creative Pedagogy. Higher Education Bulletin, 12, 29-34, 1989
21. D'Alembert J Le Rond: Einleitung zur Enzyklopädie (1751). (German Translation) Fischer, Frankfurt/Main 1989
22. Alforhs L V: Complex Analysis. McGraw-Hill, New York et al. 1966
23. Alpaydin R and G Mazzola: A Harmonic Analysis Network. Musicae Scientiae, Vol. 19(2) 191–213, 2015
24. Alunni Ch: Diagrammes & catégories comme prolégomènes à la question: Qu'est-ce que s'orienter diagrammatiquement dans la pensée? In: Batt, N (éd): Penser par le diagramme. Presses Universitaires de Vincennes no.22-2004, Vincennes 2004
25. Alunni Ch: Relativités et puissances spectrales chez Gaston Bachelard. Revue de synthèse, 1999, 120-1, p. 73-110, 1999
26. Alunni Ch: Diagrammes et catégories comme prolégomènes à la question : qu'est-ce que s'orienter diagrammatiquement dans la pensée ? In: Théorie Littérature Enseignement no 22 - Penser par le diagramme de Gilles Deleuze à Gilles Châtelet, p. 83-93, 2004

G. Mazzola, *The Topos of Music II: Performance*, Computational Music Science,
https://doi.org/10.1007/978-3-319-64444-8

27. Alunni Ch: Le Lemme de Yoneda : enjeux pour une conjecture philosophique ? (variations sous la forme pro-lemmatique, mais en prose), in Andreatta M, F Nicolas and C Alunni (Eds.): A la lumière des mathématiques et à l'ombre de la philosophie. Coll. Musique/Sciences, ENS, Delatour, Ircam, Centre Pompidou, p. 195-211, 2012
28. American Mathematical Society (AMS): Mathematical Subject Classification 2010, 47 pages http://www.ams.org/msc/pdfs/classifications2010.pdf
29. Amuedo J: Computational Description of Extended Tonality. Master's Thesis, U Southern California, Los Angeles 1995
30. Andreatta M: Group-theoretical Methods Applied to Music. Independent Study Dissertation, University of Sussex, Sussex 1997
31. Andreatta M: Méthodes algébriques en musique et musicologie du XXème siècle : aspects théoriques, analytiques et compositionnels. PhD thesis, Ircam/EHESS 2003
32. Andreatta M, A Ehresmann, R Guitart, G Mazzola: Towards a Categorical Theory of Creativity for Music, Discourse, and Cognition, in J. Yust, J. Wild, and J.A. Burgoyne (Eds.): MCM 2013, LNAI 7937, pp. 19-37, 2013
33. Andreatta M: La théorie mathématique de la musique de Guerino Mazzola et les canons rythmiques. Mémoire DEA, EHESS Paris IV & IRCAM, Paris 1999
34. Andreatta M and G Mazzola: From a Categorical Point of View: K-nets as Limit Denotators. Perspectives of New Music, volume 44, number 2 (Summer 2006)
35. http://www.staps.uhp-nancy.fr/bernard
36. http://www.staps.uhp-nancy.fr/bernard/revues_corps.htm
37. Ansermet E: Die Grundlagen der Musik im menschlichen Bewusstsein. 4th Ed. Piper, München and Zürich 1986
38. Apfel E: Diskant und Kontrapunkt in der Musiktheorie des 12. bis 15. Jahrhunderts. Heinrichshofen, Wilhelmshafen 1982
39. Apollonius: Les coniques d'Apollonius de Perge, trad. Paul Ver Eecke. Blanchard, Paris, 1959
40. Aristoteles: Topik (Organon V, 345 b.C.). Rolfes E (German transl.), Meiner, Hamburg 1992
41. Artin M: Algebra. Birkhäuser, Basel et al. 1993
42. Artin M, Grothendieck A, Verdier J L: Théorie des Topos et Cohomologie Etale des Schémas (Tomes 1,2,3). Springer Lecture Notes in Mathematics 269, 270, 305, Springer, New York et al. 1972-1973
43. Askenfelt A et al.: Musical Performance. A Synthesis-by-Rule Approach. Computer Music J. 7/1, 1983
44. Assayag G: Du calcul secret au calcul visuel. In: Delalande F and Vinet H (Eds.): Interface homme-machine et création musicale. Hermes, Paris 1999
45. Assayag G, G Bloch, M Chemillier: OMax-Ofon. Sound and Music Computing (SMC), Marseille 2006
46. Assayag G and S Dubnov: Using Factor Oracles for Machine Improvisation. Soft Computing 8, pp. 1432-7643, September 2004
47. Assmann J: Schöpfungsmythen und Kreativitätskonzepte im Alten Ägypten. In: R M Holm-Hadulla (Ed.): Kreativität. Springer, Heidelberg 2000
48. Auroux S: La sémiotique des encyclopédistes. Payot, Paris 1979
49. Babbitt M: Some Aspects of Twelve-Tone Composition. In: Hays W (Ed.): The Score and IMA Magazine, 12, 53-61, 1955 (reprinted in "Twentieth Century Views of Music History", 364-371, Scribner, New York 1972)
50. Babbitt M: Twelve-Tone Invariants as Compositional Determinants. Musical Quarterly, 46, 245-259, 1960
51. Babbitt M: Set Structure as a Compositional Determinant. JMT, 5(2), 72-94, 1961
52. Babbitt M: Twelve-Tone Rhythmic Structure and the Electronic Medium. PNM, 1(1), 49-79, 1962
53. Babbitt M: The Structure and Function of Music Theory. College Music Symposium, Vol.5, 1965 (reprinted in Boretz and Cone, 1972, 10-21)
54. Babbitt M: Words about Music. Dembski S and Straus J N (Eds.), University of Wisconsin Press, Madison 1987
55. Bach J S: Krebskanon. In: *Musikalisches Opfer*, 70. (BWV 1079), Neue Gesamtausg. sämtl. Werke Ser. VIII, Bd.1, Bärenreiter, Kassel 1978
56. Bach J S: Choral No.6, In: *Himmelfahrtsoratorium* (BWV 11), Neue Gesamtausg. sämtl. Werke Ser. II, Bd.8, Bärenreiter, Kassel 1978
57. Bachelard G: La Valeur inductive de la Relativité. 257 p., Vrin, Paris 1929
58. Bacon F: De dignitate et augmentis scientiarum. 1st Ed. 1605, improved 1623
59. Bacon F: Novum Organum, trad. Malherbe et Pousseur. coll. épiméthée, PUF, Paris, 1986
60. Badiou A: Court traité d'ontologie transitoire. Seuil, Paris, 1998

61. Badiou A, F Tarby: La philosophie et l'événement. Entretiens. Suivis d'une courte introduction à la philosophie d'Alain Badiou. Germina, 2010
62. Badiou A: La République de Platon. Pluriel, 2012
63. Badiou A: Le Séminaire – Parménide. L'être 1 - Figure ontologique. 1985-1986. Fayard, 2014
64. Badiou A: Le Séminaire – Heidegger. L'être 3 - Figure du retrait. 1986-1987. Fayard, 2015
65. Badiou A: Éloge des mathématiques. Flammarion, 2015
66. Baer J and J C Kaufman: Creativity Research in English-Speaking Countries. J C Kaufman and R J Sternberg (Eds.): The International Handbook of Creativity. Cambridge University Press, Cambridge, pp. 10-38, 2006
67. Bagchee S N: Understanding Raga Music. BPI, New Delhi 1996
68. Bahle J: Der musikalische Schaffensprozess. Psychologie der schöpferischen Erlebnis- und Antriebsformen. S Hirzel, Leipzig 1936
69. Baker J, Beach D, Bernard J: Music Theory in Concept and Practice. Eastman Studies in Music, University of Rochester Press, 1997
70. Bakhtin M: Toward a Philosophy of the Act. Translated by Vadim Liapunov. University of Texas Press, Austin 1993
71. Balzano G: The group-theoretic description of 12-fold and microtonal pitch systems. CMJ, 4, 66-84, 1980
72. Bandemer H and Gottwald H: Fuzzy Sets, Fuzzy Logic, Fuzzy Methods. Wiley, New York et al. 1995
73. Banter H: Akkord-Lexikon. Schott, Mainz 1982
74. Barbin E: Histoire et enseignement des mathématiques : pourquoi ? comment ?. Bulletin de l'Association Mathématique du Québec, XXXVII-1, 20-25
75. Barbin E: La révolution mathématique du XVIIème siècle. Ellipses, 2006
76. Barbin E et R Guitart: La pulsation entre les conceptions optiques, algébriques, articulées et projectives, des ovales cartésiennes, in: L'Océan indien au carrefour des mathématiques arabes, chinoises, européennes et indiennes. IUFM de La Réunion, 1998, pp. 359-394
77. Barbin E et R Guitart: Algèbre des fonctions elliptiques et géométrie des ovales cartésiennes. Revue d'histoire des mathématiques, SMF, n7, 2001, pp. 161-205
78. Barbin E: Gaston Tarry et la doctrine des combinaisons. Labyrinthes et carrés magiques, conférence au colloque *Les travaux combinatoires entre 1870 et 1914 et leur actualité*, Guéret, 30 Sept. - 2 Oct. 2015
79. Barbin E et J-P Cléro (Eds.): Les mathématiques et l'expérience : ce qu'en ont dit les philosophes et les mathématiciens. Hermann, Paris 2015
80. Barbin E: La révolution mathématique du XVIIème siècle. Ellipses, 2006
81. Bareil H: Analyse de "La pulsation mathématique", *A.P.M.E.P.* 427, March-April 2000
82. Barlow K: Über die Rationalisierung einer harmonisch irrationalen Tonhöhenmenge. Preprint, Köln 1985.
83. Barrow J and F Tipler: The Anthropic Cosmological Principle: Oxford University Press, New York 1986
84. Bars I: Extra Dimensions in Space and Time (Multiversal Journeys). Springer, Heidelberg 2009
85. Bartenieff I: The Roots of Laban Theory: Aesthetics and Beyond. In: Bartenieff et al.: Four Adaptations of Effort Theory. Dance Notation Bureau, New York 1970
86. Barthes R: Eléments de sémiologie. Communications 4/1964
87. Barwise J and L Moss: Hypersets. Mathematical Intelligencer, 13(4):3141, 1991
88. Basten A: Personal e-mail communciaction to G.M. November 22, 1996
89. Bastiani A (= A Ehresmann): Théorie des ensembles. CDU, 328 p., Paris, 1970
90. Bätschmann O: Einführung in die kunstgeschichtliche Hermeneutik. Wissenschaftliche Buchgemeinschaft Darmstadt, Darmstadt 1986
91. Batt N (éd): Penser par le diagramme. Presses Universitaires de Vincennes no.22-2004, Vincennes 2004
92. Bauer M: Die Lieder Franz Schuberts. Breitkopf und Härtel, Leipzig 1915
93. Bazelow A and F Brickel: A Partition Problem Posed by Milton Babbitt. PNM, 14(2), 15(1), 280-293, 1976
94. Becker M: Nineteenth Century Foundations of Creativity Research. Creativity Research Journal, Vol. 8, pp. 219-229, 1995
95. Bednarik R G: A Figurine from the African Acheulian. Current Anthropology, Vol. 44:3, pp. 405-413, 2003
96. Beethoven L van: *Grosse Sonate für das Hammerklavier* op.106 (1817-1818). Ed. Peters, Leipzig 1975
97. Békésy G von: Experiments in Hearing. McGraw-Hill, New York 1960
98. Bell J L: From absolute to local mathematics. *Synthese* 69, 409-426, 1986
99. Bellal A: Il faut sortir des mathématiques (humour) (see also [437]), `http/::environ-energie.org/category/humourpastiches/`
100. Benveniste É: La nature des pronoms. In: Benveniste É: Problèmes de linguistique générale, I. Gallimard, Paris 1966
101. Beran J: *Cirri*. Centaur Records, 1991

102. Beran J: Statistics in musicology. Chapman & Hall (CRC Press), New York 2003
103. Beran J and G Mazzola: *Immaculate Concept.* SToA music 1002.92, Zürich 1992
104. Beran J and G Mazzola G: Rational composition of performance. In Auhagen W, R Kopiez (Eds.): Regulation of creative processes in music, Staatliches Institut für Musikforschung (Berlin), pp. 37-67, Lang Verlag, Frankfurt/New York, 1997
105. Beran J and G Mazzola: Analyzing Musical Structure and Performance—a Statistical Approach. Statistical Science. Vol. 14, No. 1, 47-79, 1999
106. Beran J and G Mazzola: Visualizing the Relationship Between Two Time Series by Hierarchical Smoothing. Journal of Computational and Graphical Statistics, Vol. 8, No. 2, 213-238, 1999
107. Beran J and G Mazzola: Timing Microstructure in Schumann's "Träumerei" as an Expression of Harmony, Rhythm, and Motivic Structure in Music Performance. Computers and Mathematics with Applications, Vol. 39, Issue 5/6, 99-130, 2000
108. Beran J and G Mazzola: Musical composition and performance - statistical decomposition and interpretation. Student, Vol. 4, No.1, 13-42, 2001
109. Beran J: Maximum likelihood estimation of the differencing parameter for invertible short and long-memory ARIMA models. J.R. Statist. Soc. B, 57, No.4, 659-672, 1995
110. Beran J, R Goswitz, G Mazzola, P Mazzola: On the relationship between tempo and quantitative metric, melodic and harmonic information in Chopin's Prélude op. 28, No. 4: a statistical analysis of 30 performances. Journal of Mathematics and Music, Vol 8., No. 3, p. 225-248, 2014
111. Beranek L L: Acoustics, 1954 -1993, The Acoustical Society of America, ISBN: 0-88318-494-X
112. Berg A: Die musikalische Impotenz der Neuen Ästhetik Hans Pfitzners. Musikblätter des Anbruchs. Jg 02, S. 400, Wien 1920
113. Berger M: Geometry I, II. Springer, Berlin et al. 1987
114. Bernard J: Chord, Collection, and Set in Twentieth-Century Theory. In: Baker J et al. [69], 11-52
115. Béziau J-Y: La Théorie des Ensembles et la Théorie des Catégories : Présentation de deux Sœurs Ennemies du Point de Vue de leurs Relations avec les Fondements des mathématiques. Boletin de la Asociación Matemática Venezolana, Vol. IX, No 1, pp. 45-52, 2002
116. Bilkey D and K M Douglas: Amusia Is Associated with Deficits in Spatial Processing. Nature Neuroscience 10, pp. 915 - 921, 2007
117. Biok H-R: Zur Intonationsbeurteilung kontextbezogener sukzessiver Intervalle. Bosse, Regensburg 1975
118. Biot J-B: Essai de géométrie analytique, 8 ième éd. (1ère éd. 1802), Bachelier, Paris 1834
119. Birkhoff G: Universal Algebra. In: Rota G-C and J.S. Oliveira J S (Eds.): Collected papers of G. Birkhoff, pp. 111-115, Birkhäuser, Basel
120. Blind Boy Fuller: `ttp://www.allaboutbluesmusic.com/blind-boy-fuller/` 2015
121. Block I: Oscar Peterson Lays Down the Story of a Life Lived in the Swing of Things. The Gazette (Montreal), August 31, 2002
122. Blumenthal L M: Distance Geometry. Chelsea 1953
123. Boden M A: Conceptual Spaces. In [753], pp. 235-244
124. Boden M A: The Creative Mind: Myths and Mechanisms. Routledge, London 2004
125. Boden M A: Creativity & Art. Oxford University Press, Oxford 2010
126. Blumenthal L M: Distance Geometry. Chelsea 1953
127. Blood A, J Zatorre, P Bermudez, A C Evans: Emotional response to pleasant and unpleasant music correlates with activity in paralimbic brain regions. Nature Neuroscience, Vol. 2, No. 4, 382-387, April 1999
128. Boi L: Le problème mathématique de l'espace. Une quête de l'intelligible. Springer, 1995
129. Boirel R: Comment résoudre aisément les problèmes de mathématiques. Initiation au dynamisme opératoire mathématique. éd. Culture humaine, Sermaise, 1964
130. Boissière A: Geste, interprétation, invention selon Pierre Boulez. Revue DEMéter, Dec. 2002, Univ. Lille-3
131. Borceux F: Handbook of Categorical Algebra: Volume 1. Cambridge University Press, Cambridge 1994
132. Borceux, F: Handbook of Categorical Algebra: Volume 2. Cambridge University Press, Cambridge 1995
133. Borgo D: Sync or Swarm. Continuum International Publishing Group 2006
134. Boretz B and E T Cone: Perspectives on Contemporary Music Theory. W.W. Norton and Company, New York 1972
135. Bose R C and S S Shrikhande: On the construction of sets of mutually orthogonal Latin squares and the falsity of a conjecture of Euler. Trans. Amer. Math. Soc. 95, 191-209, 1960
136. Bose R C, S S Shrikhande and E T Parker: Further results on the construction of mutually orthogonal Latin squares and the falsity of Euler's conjecture. Canadian J. of Math. 12, 189-203, 1960

137. Bott J, J Crowley, J LaViola Jr: Exploring 3D Gestural Interfaces for Music Creation in Video Games. CFDG 2009 April 2630, 2009, Orlando 2009
138. Boulez P: Musikdenken heute I,II; Darmstädter Beiträge V, VI. Schott, Mainz 1963, 1985
139. Boulez P: Le timbre et l'écriture, le timbre et le langage. In: Bourgeois Chr (Ed.): Le timbre, métaphore pour la composition. IRCAM, Collection Musique/Passé/Présent, Paris 1991
140. Boulez P: Jalons (dix ans d'enseignement au Collège de France). Bourgeois, Paris 1989
141. Boulez P: L'Ecriture du geste. Bourgeois, Paris 2002
142. Boulez P: Structures, premier livre. UE, London 1953
143. Boulez P: Structures, deuxièxime livre. UE, London 1967
144. Bourgeois Chr (Ed.): Le timbre, métaphore pour la composition. IRCAM, Collection Musique/Passé/Présent, Paris 1991
145. Bourbaki N: Eléments de Mathématique, Algèbre, Ch.1-9. Hermann, Paris 1970-1973
146. Bourbaki N: Eléments de Mathématique, Algèbre Commutative, Ch.1-7. Hermann, Paris 1961-65
147. Bourbaki N: Eléments de Mathématique, Topologie Générale, Ch.1-4. Hermann, Paris 1971
148. Bourbaki N: Foundations of mathematics for the working mathematician. Journal of Symbolic Logic, 14, pp. 1-8, 1949
149. Boyer C B: Newton as an Originator of Polar Coordinates. American Mathematical Monthly 56: 73-78, 1949
150. Braem P B and T Braem. A Pilot Study of the Expressive Gestures Used by Classical Orchestra Conductors. In: Emmorey K and H Lane (Eds.): The signs of Language Revisited. Lawrence Erlbaum Associates, 2000
151. Brändle L: Die "Wesentlichen Manieren" (Ornamente in der Musik). Oesterreichischer Bundesverlag, Wien 1987
152. Brandt C and C Roemer: Standardized Chord Symbol Notation: Roerick Music, Sherman Oaks, CA 1976
153. Brecht B et al.: Conductor Follower. In: ICMA (Ed.): Proceedings of the ICMC 95, San Francisco 1995
154. Brockwell P J and R A Davis: Time Series: Theory and Methods. Springer, New York 1987
155. Brofeldt H: Piano Music for the Left Hand Alone. `http://www.left-hand-brofeldt.dk` Non-dated
156. Bruhn H: Harmonielehre als Grammatik der Musik. Psychologie Verlags Union, München et al. 1988
157. Bruter C-P: Comprendre les mathématiques. Les 10 notions fondamentales. Ed. Odile Jacob, Paris, 1996
158. Bulwer J: Chironomia. Thomas Harper, London 1644
159. Buteau Ch and G Mazzola: From Contour Similarity to Motivic Topologies. Musicae Scientiae 2, pp. 125-149, 2000
160. Buteau Ch and G Mazzola: From Contour Similarity to Motivic Topologies. Musicae Scientiae, Vol. IV, No. 2, 125-149, 2000.
161. Busch H R: Leonhard Eulers Beitrag zur Musiktheorie. Bosse, Regensburg 1970
162. Buser P and M Imbert: Audition. Hermann, Paris 1987
163. Buteau Ch: Motivic Topologies and their Signification in Musical Motivic Analysis. Master's Thesis, U Laval/Québec 1998
164. Buteau Ch: Reciprocity Between Presence and Content Functions on a Gestalt Composition Space. Tatra Mt. Math. Publ. 23, 17-45, 2001
165. Butsch C and H Fischer: Seashore-Test für musikalische Begabung. Huber, Bern/Stuttgart 1966
166. Byers W: How Mathematicians Think. Using Ambiguity, Contradiction and Paradox to Create Mathematics. Princeton U. Press, 2007
167. Cabral B and L Leedom: Imaging vector fields using line integral convolution. Computer Graphics 27 (SIGGRAPH'93 Proceedings), 263-272, 1993
168. Cadoz C and M M Wanderley: In: Gesture - Music. Trends in Gestural Control of Music, M M Wanderley and M Battier eds., Ircam - Centre Pompidou, 2000
169. Cadoz C: Le geste canal de communication homme-machine. La communication 'instrumentale'. Sciences Informatiques, numéro spécial: Interface homme-machine. 13(1): 31-61, 1994
170. Cage J: ASLSP. `http://www.john-cage.halberstadt.de`
171. Calvet O et al.: Modal synthesis: compilation of mechanical sub-structure and acoustical sub-systems. In: Arnold S, Hair G, ICMA (Eds.): Proceedings of the 1990 International Computer Music Conference. San Francisco 1990
172. Campos D G: Peirce on the Role of Poietic Creation in Mathematical reasoning, trans. Charles S. Peirce Society: A Quaterly J. in American Philosophy, Vol. 43, No 3, pp. 470-489, Summer 2007
173. Franco Ferrara's conducting technique, Personal conversation with maestro C. Caruso at State Music Conservatory of Palermo, 2014
174. Carnot L: Mémoire sur la Relation qui existe entre les distances respectives de cinq points quelconques dans l'espace, suivi d'un Essai sur la théorie des transversales. 1806

175. Castellengo M and N H Bernardoni: Interplay between harmonics and formants in singing: when vowels become music. Proceedings of ISMA, International Conference on Noise and Vibration Engineering 2014
176. Camuri A et al.: Toward a cognitive model for the representation and reasoning on music and multimedia knowledge. In: Haus G and Pighi I (Eds.): X Colloquio di Informatica Musicale. AIMI, LIM-Dsi, Milano 1993
177. Carey N and D Clampitt: Aspects of well-formed scales. MTS, 11, 187-206, 1989
178. Cardelli L and P Wegner: On Understanding types, data Abstraction and polymorphism. Computing Surveys, Vol. 17,4, 1985
179. Castagna G: Foundations of Object-oriented programming. ETAPS, Lisbon 1998
180. Castine P: Set Theory Objects. Lang, Frankfurt/Main et al. 1994
181. Cavaillès J: Méthode axiomatique et formalisme, p. 178, 1938. In: Oeuvres complètes de Philosophie des sciences. Hermann, 1994
182. Cayley A: On a Theorem in the Geometry of Position. Camb. Math. Jour., t. II, pp. 267-271, 1841
183. Celibidache Video: `http://www.youtube.com/watch?v=_BAsS5SmiNU`
184. Celibidache at Wikipedia: `http://en.wikipedia.org/wiki/Sergiu_Celibidache`
185. CERN. Internet Information via http://www.cern.ch 1995
186. Certeau M de: L'étranger ou l'union dans la différence. 1ère éd. 1969, Le Seuil, 2005
187. Châtelet G: Arithmétique et algèbre modernes, II, PUF, Paris 1956
188. Chastel A: Le geste dans l'art. Liana Levi, Paris 2001
189. Châtelet G: Les enjeux du mobile. Seuil, Paris 1993
190. Châtelet G: Figuring Space. Kluwer 2000 (translation of the French original *Les enjeux du mobile*, [189])
191. Chatterjee S and B Price: Regression Analysis by Example. Wiley, 2nd ed., New York 1995
192. Chauviré C: Peirce et la signification. PUF, Paris 1995
193. Chauviré C: L'œil mathématique. Essai sur la philosophie mathématique de Peirce. éd. Kimé, 2008
194. Cherlin M: Memory and Rhetorical Trope in Schoenberg's String Trio. Journal of the American Musicological Society, 51/3, pp. 559-602, fall 1998
195. Cherry D & E Blackwell: Mu (Parts I and II). Actuel 1969
196. Choi I: Cognitive Engineering of Gestural Primitives for Multi-Modal Interaction in a Virtual Environment. In: Proceedings of the 1998 IEEE International Conference on Systems, Man and Cybernetics (SMC'98), 1998
197. Chomsky N and M Halle: The Sound Pattern of English. Harper and Row, New York 1968
198. Chowning J: The Synthesis of Complex Audio Spectra by Means of Frequency Modulation. Journal of the Audio Engineering Society 21 (7), 1985
199. Chowning J: The synthesis of complex audio spectra by means of frequency modulation. Journal of the Audio Engineering Society, 21, 1973.
200. Clarke E: Imitating and Evaluating Real and Transformed Musical Performances. Music Perception 10/3, 317-341, 1993
201. Clayton M and G Fatone, L Leante, and M Rahaim: Imagery, Melody, and Gesture in Cross-Cultural Perspective. In: King E: New Perspectives on Music and Gesture (2016), Routledge 2011
202. Clough J and Myerson G: Variety and multiplicity in diatonic systems. JMT, 29, 249-270, 1985
203. Clough J and J Douthett: Maximally even sets. JMT, 35, 93-173, 1991
204. Clough J: Diatonic Interval Cycles and Hierarchical Structure. PNM, 32(1), 228-253, 1994
205. Clynes M: Sentics. The Touch of Emotions. Anchor Doubleday, New York 1977
206. Clynes M: Secrets of life in music. In: Analytica, Studies in the description and analysis of music in honour of Ingmar Bengtsson, Royal Swedish Academy of Music, No 47 pp. 3-15, 1985
207. Colbourn C J and J H Dinitz: Handbook of Combinatorial Designs. Chapman & Hall/CRC, 2007
208. Cont A, S Dubnov and G Assayag: On the Information Geometry of Audio Streams with Applications to Automatic Structure Discovery. IEEE Transactions on Audio, Speech and Language Processing 2009
209. Cohn R: Introduction to Neo-Riemannian Theory: A Survey and a Historical Perspective. JMT 42(2), 167-180, 1998
210. Conen H: Formel-Komposition. Schott, Mainz et al. 1991
211. Cook P: `http://www.cs.princeton.edu/ prc/SingingSynth.html`
212. Cook N: Playing God: Creativity, Analysis, and Aesthetic Inclusion. In: I Deliège and G A Wiggins (Eds.): Musical Creativity: Multidisciplinary Research in Theory and Practice. Psychology Press, Hove/New York, pp. 9-24, 2006
213. Coolidge J L: The origin of polar coordinates. American Mathematical Monthly, 59 : 78-85, 1952
214. Cooley J W and J W Tuckey: An Algorithm for the Machine Calculation of Complex Series. Math. Comput. 19, 297301 1965
215. Cope D: Virtual Music: Computer Synthesis of Musical Style. MIT Press, Cambridge, MA 2001

216. Cope D: Computer Models of Musical Creativity. MIT Press, Cambridge, MA 2005
217. Corbin A, J-J Courtine, G Viagarello (Eds.): Histoire du corps, T3, Les mutations du regard. Le XX siècle. Seuil, Paris 2006
218. Cox A: Hearing, Feeling, Grasping Gestures. In: Gritten A and E King (Eds.): Music and Gesture. Aldershot: Ashgate, 2006
219. Cooper K N D et al.: Handwritten Music-Manuscript Recognition. In: ICMA (Ed.): Proceedings of the ICMC 96, San Francisco 1996
220. Couasnon B et al.: Using a Grammar For a Reliable Full Score Recognition System. In: ICMA (Ed.): Proceedings of the ICMC 95, San Francisco 1995
221. Creutzfeldt O D: Cortex Cerebri. Springer, Berlin et al. 1983
222. Cros S C H: Théorie de l'homme intellectuel et moral, vol.2. Bachelier, 1842
223. Csíkszentmihályi M: Beyond Boredom and Anxiety. Jossey-Bass, San Francisco 1975
224. Csíkszentmihályi M: Society, Culture, and Person: A Systems View of Creativity. In: R J Sternberg (Ed.): The Nature of Creativity: Contemporary Psychological Perspectives. Cambridge University Press, New York, pp. 325-339, 1988
225. Csíkszentmihályi M: The Psychology of Optimal Experience. Harper and Row, New York 1990
226. Csíkszentmihályi M: Creativity: Flow and the Psychology of Discovery and Invention. HarperCollins, New York 1996
227. Cusick S: Feminist Theory, Music Theory, and the Mind/Body Problem. Perspectives of New Music 32/1 (Winter 1994): 8-27
228. Czerny C: Pianoforte Schule. 1840
229. Dacey J S and K Lennon: Understanding Creativity: The Interplay of Biological, Psychological, and Social Factors. Josey-Bass, San Francisco, 1998
230. Dahlhaus C: Zur Theorie des klassischen Kontrapunkts. Kirchenmusikalisches Jb 45, 1961
231. Dahlhaus C: Über den Begriff der tonalen Funktion. In: Vogel M (Ed.): Beiträge zur Musiktheorie des 19. Jahrhunderts. Bosse, Regensburg 1966
232. Dahlhaus C and H H Eggebrecht: Was ist Musik? Heinrichshofen, Wilhelmshaven et al. 1985
233. Dahlhaus C: Untersuchung über die Entstehung der harmonischen Tonalität. Bärenreiter, Kassel et al. 1967
234. Dahlhaus C et al.: Neues Handbuch der Musikwissenschaft, Bd. 1-13: Athenaion and Laaber, Laaber 1980-1993
235. Dahlhaus C and G Mayer: Musiksoziologische Reflexionen. In: Dahlhaus, C and de la Motte-Haber H (Eds.): Neues Handbuch der Musikwissenschaft, Bd. 10: Systematische Musikwissenschaft. Laaber, Laaber 1982
236. Dahlhaus C: Ludwig van Beethoven und seine Zeit. Laaber, Laaber 1987
237. D'Alembert J Le Rond: Einleitung zur 'Enzyklopädie'. Fischer, Frankfurt/Main 1989
238. D'Alembert, J Le Rond: Einleitung zur Enzyklopädie (1751). (German Translation) Fischer, Frankfurt/Main 1989
239. Dallos P: The active cochlea. J. Neurosci. Dec;12(12):4575-85, 1992
240. Dannenberg R B: An on-line algorithm for real-time accompaniment. In: ICMA (Ed.): Proceedings of the ICMC 84, San Francisco 1984
241. Dannenberg R B et al.: Automatic Ensemble Performance. In: ICMA (Ed.): Proceedings of the ICMC 94, San Francisco 1994
242. Alighieri, D: La Divina Commedia (1321). CreateSpace Independent Publishing Platform 2014
243. Danuser H et al. (Eds.): Neues Handbuch der Musikwissenschaft, Bd. 11: Interpretation. Laaber, Laaber 1992
244. Darwin Ch: The Origin of Species. London 1859
245. Daval R and G-T Guilbaud: Le raisonnement mathématique. PUF, 1945
246. de Bruijn N G: Pólya's Theory of Counting. In: Beckenbach E F (Ed.): Applied Combinatorial Mathematics, Ch.5. Wiley, New York 1964
247. de Bruijn N G: On the number of partition patterns of a set. Nederl. Akad. Wetensch. Proc. Ser. A 82 = Indag. Math. 41, 1979
248. Debussy C: *Préludes*, Livre I (1907-1910). Henle, München 1986
249. Dechelle F et al.: The Ircam Real-Time Platform and Applications. In: ICMA (Ed.): Proceedings of the ICMC 95, San Francisco 1995
250. Delalande F: La gestique de Gould. In: Guertin G (Ed.): Glenn Gould — pluriel. Corteau, Verdun 1988
251. Degazio B: A Computer-Based Editor For Lerdahl and Jackendoff's Rhythmic Structures. In: ICMA (Ed.): Proceedings of the ICMC 96, San Francisco 1996
252. de la Motte D: Harmonielehre. Bärenreiter/dtv, Kassel 1976
253. de la Motte-Haber H and H Emons: Filmmusik. Eine systematische Beschreibung. München, Hanser 1980
254. de la Motte-Haber H: Handbuch der Musikpsychologie. Laaber-Verlag, 2.Ed., Laaber 1996

255. de la Motte-Haber H: Rationalität und Affekt. In: Götze H and Wille R (Eds.): Musik und Mathematik. Springer, Berlin et al. 1985
256. de la Motte-Haber H: Musikalische Hermeneutik und empirische Forschung. In: Dahlhaus, C and de la Motte-Haber H (Eds.): Neues Handbuch der Musikwissenschaft, Bd. 10: Systematische Musikwissenschaft. Laaber, Laaber 1982
257. de la Motte-Haber H: Die Umwandlung der Interpretationsparameter in Struktureigenschaften. In: "Das Paradox musikalischer Interpretation", Symposion zum 80. Geburtstag von K. von Fischer, Univ. Zürich 1993
258. Deleuze G: Francis Bacon. La logique de la sensation. Éditions de la Différence, Paris 1981
259. Demazure M and P Gabriel: Groupes Algébriques. Masson & Cie./North-Holland, Paris/Amsterdam 1970
260. Dennett, D C: Quining qualia. In Marcel A and Bisiach E (Eds.): Consciousness in Contemporary Science. Oxford University Press, 1988
261. Desain P and Honing H: The Quantization of Musical Time: A Connectionist Approach. Computer Music Journal 13 (3), 56-66, 1989
262. Descartes R: Musicae Compendium. Herausgegeben und ins Deutsche übertragen als "Leitfaden der Musik" von J. Brockt, Wiss. Buchgesellschaft, Darmstadt 1978
263. Descartes R: Principia philosophiae (Principles of Philosophy). Translation with explanatory notes by Valentine Rodger and Reese P. Miller (Reprint ed.). Dordrecht: Reidel
264. Descartes R: Œuvres. Vrin, Paris, 1989. See also in English: The Geometry of René Descartes, transl. by E Smith and M L Latham, Dover Publ., 1954
265. Detienne M: Les Maîtres de Vérité dans la Grèce archaïque. 1ère éd. 1967, Le livre de poche/Références, no 611, Paris, 2012
266. De Tranquelléon: Sur l'intersection de deux cônes. Nouvelles Annales de mathématiques, 2ème série, tome III, pp. 539-540, 1864
267. Deyoung L: Pitch Order and Duration Order in Boulez' Structure 1a. PNM 16/2: 27-34, 1978
268. Dieudonné J: Foundations of Modern Analysis. Academic Press, New York et. al. 1960
269. Dieudonné J: La conception des mathématiques chez Valéry. In: Robinson-Valéry J: Fonctions de l'esprit. Treize savants redécouvrent Paul Valéry. Hermann, p. 183-191, Paris 1983
270. Dieudonné J: Mathematics – Music of Reason. Springer, 1992
271. Discogs: Horace Parlan. `http://www.discogs.com/artist/253065-Horace-Parlan` Not-dated
272. Döhl F: Webern - Weberns Beitrag zur Stilwende der Neuen Musik. Katzbichler, München et. al. 1976
273. Donald M: Mimesis and the Executive Suite: missing links in language evolution. In: J R Hurford, M Studdert-Kennedy, Ch Knight (Eds.): Approaches to the evolution of language, Cambridge U Press 1998
274. Dominguez M, D Clampitt and Th Noll: Well-formed Scales, Maximally Even Sets, and Christoffel Words. In: Proceedings of the MCM 2007, Springer, Heidelberg et al. 2009
275. Douady R: Jeux de cadres et dialectique outil-objet dans l'enseignement des mathématiques. Une réalisation dans le cursus primaire. Thèse, Paris 7, 1984
276. Douthett J and R Steinbach: Parsimonious Graphs: A Study in Parsimony, Contextual Transformations, and Modes of Limited Transposition. Journal of Music Theory 42.2, pp. 241-263, 1998
277. Doyle P, J Conway, J Gilman, B Thurston: Geometry and the Imagination in Minneapolis, version 0.91 dated 12 April 1994
278. Drake R M: Drake Musical Aptitude Tests. Science Research Associates, Chicago 1954
279. Dreiding A et al.: Classification of Mobile Molecules by Category Theory. In: Symmetries and Properties of Non-Rigid Molecules. Studies in Physical and Theoretical Chemistry, 23, 1983
280. Drossos C A: Sets, Categories and Structuralism. In: Sica, G. (Ed.): What is Category Theory?, Polimetrica, Monza 2005
281. Duffy M and S Tsuda: Shield Your Eyes Sonic Anarchy. The Sonic Alchemists, Minneapolis 2010
282. Dufourt H: Les difficultés d'une prise de conscience théorique. In: Le compositeur et l'ordinateur. 6-12, Ircam, Centre Georges Pompidou, Paris 1981
283. Duvignau K, B Gaume, J-L Nespoulous: Proximité sémantique et stratégies palliatives chez le jeune enfant et l'aphasique. Revue Parole, numéro spécial: 'Handicap langagier et recherches cognitives: Apports mutuels', J-L Nespoulous and J Virbel (Eds.), UMH Belgique, vol. 31-32: 219-255, 2004
284. Eberle G: "Absolute Harmonie" und "Ultrachromatik". In: Kolleritsch O (Ed.): Alexander Skrjabin. Universal Edition, Graz 1980
285. Ebisch S J H, M G Perrucci, A Ferretti, C Del Gratta, G Luca Romani, V Gallese: The Sense of Touch: Embodied Simulation in a Visuotactile Mirroring Mechanism for Observed Animate or Inanimate Touch. Journal of Cognitive Neuroscience 20:1-13, 2008
286. Eco U: Kunst und Schönheit im Mittelalter. Hanser, Wien 1991

287. Eco U: The Search for the Perfect Language. Wiley-Blackwell 1997
288. Eco U: Die Suche nach der vollkommenen Sprache. Beck, München 1994
289. Eco U: Le Signe. biblio essais n4159, éd. Labor 1988
290. Edelsbrunner H and J L Harer: Computational Topology. AMS 2010
291. Eggebrecht H H: Interpretation. In: "Das Paradox musikalischer Interpretation", Symposion zum 80. Geburtstag von K. von Fischer, Univ. Zürich 1993
292. Eggebrecht H H: Musik im Abendland. Piper, München and Zürich 1996
293. Ehrenfels Chr von: Über Gestaltqualitäten. Vierteljahresschrift für wissenschaftliche Philosophie XIV, 1890
294. Ehresmann C: Œuvres complètes et commentées. Part. IV-2 : esquisses et structures monoïdales fermées. ed. par Andrée Ch. Ehresmann, Amiens, 1983
295. Ehresmann A C, J-P Vanbremeersch: Memory Evolutive Systems. Hierarchy, Emergence, Cognition. Elsevier Amsterdam, 2007
296. Ehresmann A and J Gomez-Ramirez: Conciliating Neuroscience and Phenomenology via Category Theory. Progress in Biophysics and Molecular Biology, Vol. 19, issue 2, Elsevier, 2015
297. Eitz C: Das mathematisch reine Tonsystem. Leipzig 1891
298. Eimert H: Grundlagen der musikalischen Reihentechnik. Universal Edition, Wien 1964
299. Eisenbud D: Commutative Algebra with a View Toward Algebraic Geometry. Springer New York et al. 1996
300. Eisenbud D, and J Harris: The Geometry of Schemes. Springer, New York 2000
301. Engström B: Stereocilia of sensory cells in normal and hearing impaired ears. Scand. Audiol. Suppl. 19, 1-34, 1983
302. Engquist B and W Schmid (Eds.): Mathematics Unlimited — 2001 and Beyond. Springer, 2001
303. Ericson K: Heinz Holliger. Spurensuche eines Grenzgängers. Peter Lang, Bern 2004
304. Erlewine S: Movin' & Groovin' - Horace Parlan. `http://www.allmusic.com/album/movin-groovin-mw0000381525` 2015
305. Essl K: Strukturgeneratoren. Beiträge zur elektronischen Musik 5, IEM, Graz 1996
306. Euclid: Euclid's Elements. Translated with introduction and commentary by Sir Th L Heath, vol. 1, 2 and 3, Dover, 1956
307. Euler L: Tentamen novae theoriae musicae (1739). In: Opera Omnia, Ser.III, Vol.1 (Ed. Bernoulli, E et al.). Teubner, Stuttgart 1926
308. Euler L: Conjecture sur la raison de quelques dissonances générales reçues dans la musique (1764). In: Opera Omnia, Ser.III, Vol.1 (Ed. Bernoulli, E et al.). Teubner, Stuttgart 1926
309. Euler L: De harmoniae veris principiis per speculum musicum representatis (1773). In: Opera Omnia, Ser.III, Vol.1 (Ed. Bernoulli, E et al.). Teubner, Stuttgart 1926
310. Euler L: Recherches sur une nouvelle espèce de quarrés magiques. Opera Omnia, Ser. I, Vol. 7: 291-392, 1782
311. Fauré G: Requiem in D minor, Op. 48. Edited by P. Legge 2000
312. Feldman J et al.: Force Dynamics of Tempo Change in Music. Music Perception, *10*, 1992
313. Felver Chr: Cecil Taylor - All The Notes. A Chris Felver Movie, 2003
314. Ferretti R and G Mazzola: Algebraic Varieties of Musical Performances. Tatra Mt. Math. Publ. 23, 59-69, 2001
315. Feulner J et al.: MELONET: Neural Networks that Learn Harmony-Based Melodic Variations. In: ICMA (Ed.): Proceedings of the ICMC 94, San Francisco 1994
316. Fichte J G: Doctrine de la science. Nova methodo (1799), Livre de poche 4621, LGF, 2000
317. Fichtner R: Die verborgene Geometrie in Raffaels "Schule von Athen". Oldenburg 1984
318. Fink E: Grundlagen der Quantenmechanik. Akademische Verlagsgesellschaft, Leipzig 1968
319. Finke R A, Th B Ward, S M Smith: Creative Cognition: Theory, Research, and Applications. MIT Press, Cambridge, MA 1992
320. Finscher L: Studien zur Geschichte des Streichquartetts. Bärenreiter, Kassel 1974
321. Finsler P: Über die Grundlegung der Mengenlehre. Erster Teil. Die Mengen und ihre Axiome. Math. Z. 25, 683-713, 1926
322. Finsler P: Aufsätze zur Mengenlehre. Unger G (Ed.), Wiss. Buchgesellschaft, Darmstadt 1975
323. Fisher G and J Lochhead: Analyzing from the Body. In: Theory and Practice, Journal of the Music Theory Society of New York State, Vol. 27
324. Fleischer A: Eine Analyse theoretischer Konzepte der Harmonielehre mit Hilfe des Computers. Magisterarbeit, MWS, HU Berlin 1996
325. Fleischer A, G Mazzola, Th Noll: Zur Konzeption der Software RUBATO für musikalische Analyse und Performance. Musiktheorie, Heft 4, 314-325, 2000
326. Florida R: The Rise of the Creative Class. Basic Books, 2003

327. Floris E: La rupture cartésienne et la naissance d'une philosophie de la culture dans les œuvres juvéniles de J.-B. Vico. Thèse de 3ème cycle, dir. P. Ricœur, Université Paris X-Nanterre, 22 juin 1974
328. Flusser V: Gesten — Versuch einer Phänomenologie. Fischer, Frankfurt am Main 1994
329. Forster M: Technik modaler Komposition bei Olivier Messiaen. Hänssler, Neuhausen-Stuttgart 1976
330. Forte A: Structure of Atonal Music: Practical Aspects of a Computer-Oriented Research Project. In: Musicology and the Computer. Musicology 1966-2000. A Practical Program. Three Symposia. American Musicological Society, NY 1970
331. Forte A: Structure of Atonal Music. Yale University Press, New Haven 1973
332. Forte A: La Set-complex theory: Elevons les enjeux! Analyse musicale, 4e trimestre, 80-86, 1989
333. Foster W C: Singing Redefined. Recital Publications 1998
334. Frank H: RUBATO® Broadcast. ORF2: Modern Times, Jan. 10, 1997
335. Freedman D Z and P van Nieuwenhuizen: Supergravitation und die Einheit der Naturgesetze. In: Dosch H G (Ed.): Teilchen, Felder und Symmetrien. Spektrum der Wissenschaft, Heidelberg 1984
336. Friberg A: Generative Rules for Music Performance: A Formal Description of a Rule System. Computer Music Journal, Vol. 15, No. 2, 1991
337. Friberg A et al.: Performance Rules for Computer-Controlled Contemporary Keyboard Music. Computer Music Journal, Vol. 15, No. 2, 1991
338. Friberg A et al.: Recent Musical Performance Research at KTH. In: Sundberg J, (Ed.): Generative Grammars for Music Performance. KTH, Stockholm 1994
339. Friberg A: A Quantitative Rule System for Musical Performance. KTH PhD-Thesis, Stockholm 1995
340. Fripertinger H: Enumeration in Musical Theory. Beiträge zur elektronischen Musik 1, Hochschule für Musik und Darstellende Kunst, Graz 1991
341. Fripertinger H: Die Abzähltheorie von Pólya. Diplomarbeit, Univ. Graz 1991
342. Fripertinger H: Endliche Gruppenaktionen in Funktionenmengen—Das Lemma von Burnside—Repräsentantenkonstruktionen—Anwendungen in der Musiktheorie. Doctoral Thesis, Univ. Graz 1993
343. Fripertinger H: Untersuchungen über die Anzahl verschiedener Intervalle, Akkorde, Tonreihen und anderer musikalischer Objekte in n-Ton Musik. Magisterarbeit, Hochschule für Musik und Darstellende Kunst, Graz 1993
344. Fripertinger H: Anwendungen der Kombinatorik unter Gruppenaktionen zur Bestimmung der Anzahl "wesentlich" verschiedener Intervalle, Chorde, Tonreihen usw. Referat an der Univ. Innsbruck, Math. Institut d. Karl-Franzens-Univ., Graz 1996
345. Fripertinger H: Enumeration of Mosaics. Discrete Mathematics, 199, 49-60, 1999
346. Fripertinger H: Enumeration of Non-isomorphic Canons. Tatra Mt. Math. Publ. 23, 47-57, 2001
347. Frova A: Fisica nella musica. Zanichelli, Bologna 1999
348. Fussi F and S J Ghiotti: Il canto fra terminologia e didattica, 2015 `http://www.voceartistica.it/it-IT/index-/?Item=ghiotti`
349. Fux J J: Gradus ad Parnassum (1725). Dt. und kommentiert von L. Mitzler, Leipzig 1742
350. Gabriel P: Personal Communication. Zürich 1979
351. Gabriel P: Des catégorie abéliennes. Bull. Soc. Math. France 90, 1962
352. Gabriel P: Representations of Finite-Dimensional Algebras. In: Kostrikin, A.I.; Shafarevich, I.R. (Eds.): Encyclopaedia of Mathematical Sciences, Vol. 73, Springer Berlin 1992
353. Gabrielsson A: Music Performance. In: Deutsch D (Ed.): The Psychology of Music (2nd ed.). Academic Press, New York
354. Gabrielsson A: Expressive Intention and Performance. In: Steinberg R (Ed.): Music and the Mind Machine. Springer, Berlin et al. 1995
355. Gaither C C and A E Cavazos-Gaither: Mathematically Speaking. A Dictionary of Quotations. IOP Publishing Ltd, 1998
356. Galenson D W: Old Masters and Young Geniuses: The Two Life Cycles of Creativity. Princeton University Press, Princeton, NJ 2006
357. Gamer C: Some combinatorial resources of equal-tempered systems. JMT, 11, 32-59, 1967
358. Gamma E, R Helm, R Johnson, J Vlissides: Design Patterns, Elements of Reusable Object-Oriented Software. Addison-Wesley, Reading, MA et al., 1994
359. Gardner H: Kreative Intelligenz: Was wir mit Mozart, Freud, Woolf und Gandhi gemeinsam haben (Translated by Andreas Simon). Campus Verlag, Frankfurt/New York 1999
360. Gasquet S: L'illusion mathématique. Le malentendu des maths scolaires. Syros, 1997
361. Geisser H, G Mazzola, S Onuma: Imaginary Time. Video, Vimeo 2015

362. Gentilucci M & and M C Corballis: From manual gesture to speech: A gradual transition. Neuroscience & Biobehavioral Reviews Vol. 30, Issue 7, pp 949-960 2006
363. Gerwin Th: IDEAMA, Zentrum fur Kunst und Medientechnologie, Karlsruhe 1996
364. Geweke J: A comparison of tests of independence of two covariance-stationary time series. J. Am. Statist. Assoc., 76, 363-373, 1981
365. Giannitrapani D: The Electrophysiology of Intellectual Functions. Karger, Basel 1985
366. Gilson E: Introduction aux arts du beau. Vrin, Paris 1963
367. Ginzburg C: Leçon de Montaigne. Sud-Ouest, 17 octobre 2015
368. Ginzburg C: L'estrangement. In à distance, Neuf essais sur le point de vue en histoire. Traduction française par Pierre Antoine Fabre, éditions Gallimard NRF, Paris 2001
369. Giocanti S: L'art sceptique de l'estrangement dans les *Essais* de Montaigne, in *L'estrangement, Retour sur un thème de Carlo Ginzburg* [573, p. 19-35]
370. Godement R: Topologie algébrique et théorie des faisceaux. Hermann, Paris 1964
371. Godøy R I and M Leman (Eds.): Musical Gestures—Sound, Movement, and Meaning. Routledge, New York and London 2010
372. Goeller S: Object Oriented Rendering of Complex Abstract Data. PhD. thesis, Universität Zürich, 2004
373. Goethe J W von: Brief an Zelter. 9. Nov. 1829
374. Goffman E: The Presentation of Self in Everyday Life. University of Edinburgh, Edinburgh 1956
375. Goffman E: Behavior in Public Places. The Free Press, New York 1963
376. Goldblatt R: Topoi. North-Holland, Amsterdam et al. 1984
377. Goldin-Meadows S: Hearing Gesture: How Our Hands Help Us Think. Harvard University Press 2003
378. Goldstein J L: An Optimum Processor Theory for the Central Formation of the Pitch of Complex Tones. J. Acoust. Soc. Am. 54, 1496 1973
379. Gomes Teixeira F: Traité des courbes spéciales remarquables planes et gauches. 3 vol., (Coïmbre, Impr. de l'Université, 1908-1915), éd. Jacques Gabay, 1995. Traduction de *Tratado de las curvas especiales notables*, Ac. Sc. Madrid, 1899
380. Interview with Emilia Gómez: "PHENICX is a new way of experiencing a classical music concert", `http://www.upf.edu/enoticies/en/entrevistes/1232.html#.VbfU7s63QUU`
381. Gorenstein D: Classifying the finite simple groups. Bull. A.M.S. 14, 1-98, 1986
382. Gottschewski H: Tempohierarchien. Musiktheorie, Heft 2, 1993
383. Götze H and R Wille R (Eds.): Musik und Mathematik. Springer, Berlin et al. 1985
384. Gould G: The Glenn Gould Reader. Alfred A. Knopf, New York 1984
385. Goupillaud P, A Grossmann, J Morlet: Cycle-octave and related transforms in seismic signal analysis. Geoexploration, 23, 85-102, 1984-1985
386. Grabusow N: Vielfalt akustischer Grundlagen der Tonarten und Zusammenklänge - Theorie der Polybasiertheit. Musiksektion des Staatsverlags, Moskau 1929
387. Graeser W: Bachs "Kunst der Fuge". In: Bach-Jahrbuch 1924
388. Graeser W: Der Körpersinn. Beck, München 1927
389. Gramain A: Géométrie élémentaire. Hermann, 1997
390. Greimas A J: Les actants, les acteurs et les figures. In: Chabrol C (Ed.): Sémiotique narrative et textuelle. Larousse, Paris 1974
391. Greub W: Linear Algebra. Springer, Berlin et al. 1967
392. Grisey G: Tempus ex machina: a Composer's Reflections on Musical Time. Contemporary Music Review 2, no. 1:238-275, 1987
393. Grondin J: L'herméneutique. Coll. Que sais-je ? no 3758, PUF, Paris 2006
394. Gross D: A Set of Computer Programs to Aid in Music Analysis. Ph. Diss. Indiana Univ. 1975
395. Grothendieck A and J Dieudonné: Eléments de Géométrie Algébrique I. Springer, Berlin et al. 1971
396. Grothendieck A and J Dieudonné: Eléments de Géométrie Algébrique I-IV. Publ. Math IHES no. 4, 8, 11, 17, 20, 24, 28, 32, Bures-sur-Yvette 1960-1967
397. Grothendieck A: Correspondence with G. Mazzola. April 1, 1990
398. Grothendieck A: Récoltes et Semailles. Université Univ. Sci. et Tech. Languedoc et CNRS, Montpellier, 1st edn. 1985
399. Guevara R C L et al.: A Modal Distribution Approach to Piano Analysis and Synthesis. In: ICMA (Ed.): Proceedings of the ICMC 96, San Francisco 1996
400. Guillaume P: La psychologie de la forme. Journal de Psychologie, XXII, p. 768-800, 1925
401. Guillaume P: La psychologie de la forme. Flammarion, 1979

402. Guitart R: Sur l'ébauche des structures. Proc. 3d Congress of Bulgarian math., summaries part II, p. 354, Varna, Sept. 6-15, 1972
403. Guitart R: Sur le foncteur diagrammes, CTGD XIV-2, pp. 181-182, 1973
404. Guitart R: Remarques sur les machines et les structures, CTGD XV-2, pp. 113-144, 1974
405. Guitart R: Remarques sur les machines et les structures. CTGD XV-2, pp. 113-144, 1974
406. Guitart R and L Van den Bril: Décompositions et lax-complétions. CTGD XVIII-2, pp. 333-407, 1977
407. Guitart R: Des machines aux bimodules. polycopié, Univ. Paris 7, 24 p., 1978
408. Guitart R: Relations et carrés exacts. Ann. Sc. Math. Qué., vol. IV-2, p. 103-125, juillet 1980
409. Guitart R and C Lair: Calcul syntaxique des modèles et calcul des formules internes, *Diagrammes*, vol. 4, déc. 1980, 106 p.
410. Guitart R: On the geometry of computations, I and II. Cahiers Top. Géo. Diff. Cat. XXVII,4, 107-136, 1986; XXIX,4, p. 297-326, 1988
411. Guitart R: Mathématiques. 1ère année. Polycopié U. Paris 7; 27 chap., 432 p. 1992-1993
412. Guitart R: La pulsation mathématique. L'Harmattan, Paris, 1999
413. Guitart R: évidence et étrangeté. PUF, Paris, 2000
414. Guitart R: Sur les places du sujet et de l'objet dans la pulsation mathématique. Questions éducatives, revue du Centre de Recherche en éducation de l'Université Jean Monnet de Saint-Étienne, no. 22-23, pp. 49-81, décembre 2002
415. Guitart R: Toute théorie est algébrique et topologique. Cahiers Top. Géo. Diff. Cat. vol. XLIX-2 (2008), p. 83-128. 2008 (first partial version: "Toute théorie est algébrique", in Journée Mathématique en l'Honneur d'Albert Burroni, vendredi 20 septembre 2002, Université Paris 7, Institut de Mathématiques de Jussieu, Prépublication 368, p. 79-102, avril 2004
416. Guitart R and Ch Ehresmann: Au carrefour des structures locales et algébriques. Colloque International Ch Ehresmann : 100 ans, Amiens 7-9 octobre 2005; Cahiers Top. Géo. Diff. Cat., vol. XLVI-3, p. 172-175, 2005
417. Guitart R: L'évidemment des objets et le dehors comme substance. Colloque Le lemme de Yoneda, "Enjeux mathématiques et philosophie", ENS, Paris 18 juin 2007
418. Guitart R: An anabelian definition of abelian homology. Cahiers Top Géo Diff Cat, XXXXVIII, 4, pp. 261-269, 2007
419. Guitart R: Mathematical Pulsation at the root of invention. Anais do Congresso Htem 4, Rio de Janeiro, de 5 a 9 de maio de 2008
420. Guitart R: Sur la représentation de l'herméneutique comme commerce gratuit des interprétations, conférence au *CERCI*, Université de Nantes, 7 mars 2009
421. Guitart R: Cohomological Emergence of Sense in Discourses (As Living Systems Following Ehresmann and Vanbremeersch). Axiomathes, 19(3), 245-270, 2009
422. Guitart R: Les coordonnées curvilignes de Gabriel Lamé, représentations des situations physiques et nouveaux objets mathématiques, 12 p. In: Actes du Colloque International Gabriel Lamé. Les pérégrinations d'un ingénieur au XIXème siècle, 15-16-17 janvier 2009, Nantes. SABIX, no 44, p. 119-129, octobre 2009
423. Guitart R: Sense and Signs in the Mathematical Invention of Coordination (in the Perspective of Arrows, Relations and Sketches). Talk at the colloquium *Semiotic Approaches to Mathematics, the History of Mathematics and Mathematics Education (SemMHistEd)*, 3rd Meeting Aristotle University of Thessaloniki, July 16-17, 2009
424. Guitart R: Modélisation qualitative catégoricienne : modèles, signes et formes. In: Andreatta M, F Nicolas and Ch Alunni (Eds.): à la lumière des mathématiques et à l'ombre de la philosophie, Coll. Musique/Sciences, ENS, Delatour, Ircam, Centre Pompidou, pp. 133-147, Paris 2012
425. Guitart R: Bachelard et la pulsation mathématique. Revue de synthèse, tome 136, 6ème série, n. 1-2, p. 33-74, 2015
426. Guitart R: Note sur deux problèmes pour une épistémologie transitive des mathématiques, *Revue de synthèse*, tome 136, 6ème série, n. 1-2, p. 237-279, 2015
427. Guitart R: Nietzsche face à l'expérience mathématique, in: Barbin E and J-P Cléro éds, *Les mathématiques et l'expérience : ce qu'en ont dit les philosophes et les mathématiciens*, Hermann, p. 247-278, 2015
428. Guitart R: History in mathematics according to André Weil. ESU7, Copenhagen, July 2014
429. Guitart R: L'évidement et la nécessité de la contingence. Conférence au Colloque *Les fins de cure : la poursuite du désêtre*, 7/8, AECF Lille, novembre 2015
430. Gurlitt et al. (Eds.): Riemann Musiklexikon/Sachteil. Schott, Mainz 1967
431. Gusdorf G: La révolution galiléenne, tomes I et II. Payot, Paris, 1969
432. Hába A: Neue Harmonielehre. (Reprint from 1927) Universal Edition, Wien 1978
433. Habermann R: Elementary Applied PDEs, Prentice Hall 1983

434. Hadamard J: An Essay on the Psychology of Invention in the Mathematical Field. Princeton University Press, 1945, Dover, 1954
435. Halsey D and E Hewitt: Eine gruppentheoretische Methode in der Musiktheorie. Jahresber. d. Dt. Math.-Vereinigung 80, 1978
436. Handschin J: Der Toncharakter. Zürich 1948
437. Hansen P: Pourquoi il faut sortir des mathématiques. Mercredi 1er avril 2015 (see also [99]). `http://www.energie-crise.fr/spip.php?article176`
438. Hanslick E: Vom Musikalisch-Schönen. Breitkopf und Härtel (1854), Wiesbaden 1980
439. Hardy G H: A Mathematician's Apology. Cambridge U. Press, 1967
440. Hashimoto K: D-Brane. Springer, Heidelberg et al. 2012
441. Hardt M: Zur Zahlenpoetik Dantes. In: Baum R. and Hirdt W (Eds.): Dante Alighieri 1985. Stauffenburg, Tübingen 1985
442. Harkness R: Songs of Seoul. University of California Press 2014
443. Harris C and A R Brinkman: A unified set of software tools for computer-assisted set-theoretic and serial analysis of contemporary music. Proc. ICMC 1986, ICA, San Francisco 1986
444. Hartshorne R: Algebraic Geometry. Springer, New York et al. 1977
445. Hashimoto S and H Sawada: Musical Performance Control Using Gesture: Towards Kansei Technology for Art. In: Kopiez R and Auhagen W (Eds.): Controlling Creative Processes in Music. Peter Lang, Frankfurt am Main et al. 1998
446. Hatten R: Interpreting Musical Gestures, Topics, and Tropes. Indiana University Press 2004
447. Hawking S: A Brief History of Time: From the Big Bang to Black Holes. Bantam Books, New York 1988
448. Haugh L D: Checking for independence of two covariance stationary time series: a univariate residual cross correlation approach. J. Am. Statist. Assoc., 71, 378-385, 1976
449. Haus G et al.: Stazione di Lavoro Musicale Intelligente. In: Haus G and Pighi I (Eds.): X Colloquio di Informatica Musicale. AIMI, LIM-Dsi, Milano 1993
450. Hayes J R: Cognitive Processes in Creativity. In: J A Glover, R R Roning, C R Reynolds (Eds.): Handbook of Creativity. Plenum, New York 1989
451. Hebb D O: Essay on mind. Hillsdale, New Jersey, Lawrence Erlbaum Associates, 1980
452. Hegel G W F: Wissenschaft der Logik I (1812). Felix Meiner, Hamburg 1963
453. Hegel G W F: Wissenschaft der Logik, Erster Band, Erstes Buch. Schrag, Nürnberg 1812
Online `http://www.deutschestextarchiv.de/book/show/hegel_logik0101_1812`
Modern German online version:
`http://www.zeno.org/Philosophie/M/Hegel,+Georg+Wilhelm+Friedrich/Wissenschaft+der+Logik`
454. Hegel G W F: Hegel's Science of Logic.
Online `http://www.marxists.org/reference/archive/hegel/works/hl/hlconten.htm`
455. Heiberg J L and H Menge (Eds.): Euclidis opera omnia. 8 vol. & supplement, in Greek. Teubner, Leipzig 1883-1916.
456. Heijink H, P Desain, H Honing, L Windsor: Make me a match: An evaluation of different approaches to score-performance matching. Computer Music Journal, 24(1), 43-56, 2000
457. Helmholtz H von: Die Lehre von den Tonempfindungen als physiologische Grundlage der Musik (1863). Nachdr. Darmstadt 1968
458. Henck H: Karlheinz Stockhausens Klavierstück IX. Verlag für systematische Musikwissenschaft, Bonn-Bad Godesberg 1978
459. Hentoff N: Liner notes to Coltrane's last album "Expression". Impulse AS-9120, New York 1967
460. Hecquet S and A Prokhoris: Fabriques de la danse. PUF, Paris 2007
461. Herbort H J: Keine Ausweispflicht für cis. Die Zeit Nr. 43, 21. Oktober 1988
462. Hermann J: Consideratio curvarum in punctum positione datum projectarum, et de affectionibus earum inde pendentibus. Commentarii Academiae Petropolitanae, IV, 37-46, 1729
463. Hesse H: Das Glasperlenspiel (1943). Fretz und Wasmuth, Zürich 1943
464. Hesse H: Materialien zu H. Hesses "Das Glasperlenspiel", Vol. I. : Vom Wesen und von der Herkunft des Glasperlenspiels (1934). Suhrkamp, Frankfurt am Main 1981
465. Heussenstamm G: Norton Manual of Music Notation. Norton & Comp., New York 1987
466. Hichert J: Verallgemeinerung des Kontrapunkttheorems für die Hierarchie aller starken Dichotomien in temperierter Stimmung. Diplomarbeit, TU Ilmenau 1993
467. Hindemith P: Unterweisung im Tonsatz. Schott, Mainz 1940
468. Hilbert D and S Cohn-Vossen: Geometry and the Imagination. AMS Chelsea Pub. (transl. by Paul Nemenyi from *Anschauliche Geometrie*, 1932), 1952

469. Hiller L and P Ruiz: Synthesizing sounds by solving the wave equation for vibration objects. J. of the Audio Engineering Soc. 19: 463-470, 542-551, 1971
470. Hilton P J and S Wylie: Homology Theory. Cambridge University Press, London et al. 1967
471. Hjelmslev L: La Stratification du Langage. Minuit, Paris 1954
472. Hjelmslev L: Prolégomènes à une théorie du langage. Minuit, Paris 1968-71
473. Hjelmslev L: Nouveaux essays. PUF, Paris 1985
474. Hofstädter D: Gödel, Escher, Bach. New York: Basic Books, New York 1979
475. Hofstadter D R: Staring Emmy Straight in the Eye—and Doing my Best not to Flinch. In: D Cope (Ed.): Virtual Music: Computer Synthesis of Musical Style. MIT Press, Cambridge, MA 2001
476. Hodgson J: Mastering Movement: the life and work of Rudolf Laban. Routledge, New York 2001
477. Holden S P: Right-Handed and Right On. The New York Times, No. 3, June 2, 1999
478. Holm-Hadulla R M: Kreativität. Konzept und Lebensstil. Vandenhoeck & Ruprecht, Göttingen 2005
479. Hong Y: Testing for independence between two covariance stationary time series. Biometrika, 83, No.3, 615-626, 1996
480. Honing H: Expresso, a strong and small editor for expression. In: ICMA (Ed.): Proceedings of the ICMC 92, San Francisco 1992
481. Hooft G 't: Symmetrien in der Physik der Elementarteilchen. In: Dosch H G (Ed.): Teilchen, Felder und Symmetrien. Spektrum der Wissenschaft, Heidelberg 1984
482. Hörnl D et al.: Learning Musical Structure and Style by Recognition, Prediction and Evolution. In: ICMA (Ed.): Proceedings of the ICMC 96, San Francisco 1996
483. Horry Y: A Graphical User Interface for MIDI Signal Generation and Sound Synthesis. In: ICMA (Ed.): Proceedings of the ICMC 94, San Francisco 1994
484. Hosken D: An Introduction to Music Technology. Routledge New York 2011
485. Howe H: Some combinatorial properties of Pitch-Structures. PNM, 4(1), 45-61, 1965
486. Howe M J A: Prodigies and Creativity. In: R J Sternberg (Ed.): Handbook of Creativity. Cambridge University Press, New York 1999
487. Hu S-T: Mathematical Theory of Switching Circuits and Automata. University of California Press, Berkeley and Los Angeles 1968
488. Hudak P et al.: Haskore Music Notation - An Algebra of Music. J. Functional Programming Vol. 6 (3) 1996
489. Hudspeth A J and D P Corey: Sensitivity, Polarity, and Conductance Change in the Response of Vertebrate Hair Cells to Controlled Mechanical Stimuli. Proc. Nat. Acad. Sci. Am. 74(6), 2407-2411, 1977
490. Hull P et al.: The Music Business and Recording Industry. Third Edition. Routledge, New York and London 2011
491. Humphreys J E: Introduction to Lie Algebras and Representation Theory. Springer, New York et al. 1972
492. Hung R et al.: The Analysis and Resynthesis of Sustained Musical Signals in the Time Domain. In: ICMA (Ed.): Proceedings of the ICMC 96, San Francisco 1996
493. Hunziker E and G Mazzola: Ansichten eines Hirns. Birkhäuser, Basel 1990
494. Husmann H: Einführung in die Musikwissenschaft. Heinrichshofen, Wilhelmshaven 1975
495. Ilchmann A: Kritik der Übergänge zu den ersten Kategorien in Hegels Wissenschaft der Logik. Hegel-Studien, Bd. 27, Bouvier, Bonn 1992
496. Jackendoff R and F Lerdahl: A Generative Theory of Tonal Music. MIT Press, Cambridge MA, 1983
497. Jackson J D: Classical Electrodynamics. Wiley, New York 1998
498. Jairazbhoy N A: What Happened to Indian Music Theory? Indo-Occidentalism? Ethnomusicology 52 (3): 34977, 2008
499. Jakobson R and M Halle: Fundamentals of Language. Mouton, Le Hague 1957
500. Jakobson R: Linguistics and Poetics. In: Seboek, TA (Ed.): Style in Language. Wiley, New York 1960
501. Jakobson R: Language in relation to other communication systems. In: Linguaggi nella società e nella tecnica. Edizioni die Communità, Milano 1960
502. Jakobson R: Six leçons sur le son et le sens. Ed. de Minuit, Paris 1976
503. Jakobson R: Hölderlin, Klee, Brecht. Suhrkamp, Frankfurt/Main 1976
504. Jakobson R and K Pomorska: Poesie und Grammatik. Dialoge. Suhrkamp, Frankfurt/Main 1982
505. Jakobson R: Semiotik - Ausgewählte Texte 1919-1982. Holenstein E (Ed.), Suhrkamp, Frankfurt/Main 1988
506. Jauss H R: Rückschau auf die Rezeptionstheorie—Ad usum Musicae Scientiae. In: Danuser H and Krummacher F (Hsg.): Rezeptionsästhetik und Rezeptionsgeschichte in der Musikwissenschaft. Laaber, Laaber 1991
507. Jeans J H: Science and Music. 1938, reprinted by Dover, 1968
508. Jeppesen K: Kontrapunkt. Breitkopf und Härtel, Wiesbaden 1952
509. John F: Partial Differential Equations. Springer, Heidelberg et al. 1978

510. Johnson T: See his web page at http://www.tom.johnson.org
511. Johnson-Laird P N: Freedom and Constraint in Creativity. In: R J Sternberg (Ed.): The Nature of Creativity: Contemporary Psychological Perspectives. Cambridge University Press, New York, pp. 202-219, 1988
512. Junod J: Counterpoint Worlds and Morphisms. Dissertation, University of Zurich, Zurich 2010
513. Julia G: Sur l'itération des fonctions rationnelles. J. de Math. Pure et Appl. 8, 1918
514. Jurek T: The Complete Blue Note Horace Parlan Sessions. AllMusic 2015. `http://www.allmusic.com/album/the-complete-blue-note-horace-parlan-sessions-mw0001007267`
515. Kagel M: Translation - Rotation. Die Reihe Bd.7, Universal Edition, Wien 1960
516. Kahle W: Taschenatlas der Anatomie, Bd.3: Nervensystem und Sinnesorgane. Thieme/dtv, Stuttgart 1979
517. Kaiser J: Beethovens 32 Klaviersonaten und ihre Interpreten. Fischer, Frankfurt/Main 1979
518. Kan D: Adjoint functors. Transactions of the American Mathematical Society, 87: 294-329, 1958
519. Kant I: Kritik der reinen Vernunft. Meiner, Hamburg, 2nd edn. 1957
520. Kantor J-M (Ed.): Ou en sont les mathématiques?. Vuibert/SMF, 2002
521. Karg-Elert S: Polaristische Klang- und Tonalitätslehre (1931). Out of print, cf.: Schenk P.: Karg-Elerts polaristische Harmonielehre. In: Vogel, M (Ed.) Beiträge zur Musiktheorie des 19. Jahrhunderts. Bosse, Regensburg 1966
522. Kasner E and J Newman: Mathematics and the imagination. Simon and Schuster, New York, 1940
523. Katayose H et al.: Demonstration of Gesture Sensors for the Shakuhachi. In: ICMA (Ed.): Proceedings of the ICMC 94, San Francisco 1994
524. Katsman R: Anthropoetic Gesture: A Key to Milorad Pavić's Poetics (Landscape Painted With Tea). Academic Electronic Journal in Slavic Studies No. 16, Univ. Toronto, 2006
525. Katsman R: Gestures Accompanying Torah Learning/recital among Yemenite Jews. Gesture 7 (1): 119, 2007
526. Kayser H: Akróasis. Die Lehre von der Harmonik der Welt. Schwabe, Basel 1946
527. Kelley J L: General Topology. Van Nostrand, Princeton et al. 1955
528. Kempe A B: A memoir on the theory of mathematical form. Philosophical Transactions of the Royal Society of London, 177, 1-70, 1886
529. Kendon A: How Gestures Can Become like Words. Crosscultural Perspectives in Nonverbal Communication. 1988
530. Kendon A: Gesture: Visible Action as Utterance. Cambridge University Press, Cambridge 2004
531. Keyser C J: The Meaning of Mathematics. Scripta Mathematica, 1933, p. 15-28
532. Kiczales G, J Rivieres, D G Bobrow: The Art of the Metaobject Protocol. The MIT Press, Boston 1991
533. Kinderman W: Artaria 195. University of Illinois Press, Urbana and Chicago 2003
534. King E: New Perspectives on Music and Gesture. Routledge 2016
535. Kirsch E: Wesen und Aufbau der Lehre von den harmonischen Funktionen. Leipzig 1928
536. Klemm M: Symmetrien von Ornamenten und Kristallen. Springer, Berlin et al. 1982
537. Klumpenhouwer H: Deep Structure in K-net Analysis with special reference to Webern's opus 16,4. Unpublished manuscript
538. Klumpenhouwer H: A Generalized Model of Voice-Leading for Atonal Music. Ph.D. Thesis, Harvard University 1991
539. Knapp J: Franz Liszt. Berlin 1909
540. Köhler E: Brief an Guerino Mazzola. Hamburg 1988
541. Kollmann A: An Essay on Musical Harmony. London 1796
542. Komparu K: The Noh Theatre. Weatherhill/Tankosha, New York et al. 1983
543. Koenig Th: Robert Schumanns *Kinderszenen* op.15. In: Metzger H-K und Riehn R (Hrg.): Robert Schumann II. edition text+kritik, München 1982
544. Kononenko N O: Ukrainian minstrels and the blind shall sing. Armonk, N.Y.: M E Sharpe, 1998
545. Kopiez R and J Langner: Entwurf einer neuen Methode der Performanceanalyse auf Grundlage einer Theorie oszillierender Systeme. In: Behne K-E and de la Motte H (Eds.): J.buch der D. Ges. für Musikpsychologie. **12** (1995), Wilhelmshaven 1996
546. Kopiez R: Aspekte der Performanceforschung. In: de la Motte H: Handbuch der Musikpsychologie. Laaber, Laaber 1996
547. Kopiez R: Mensch - Musik - Maschine. Musica, 50 (1), 1996
548. Kopiez R: "The most wanted song/The most unwanted song" – Klangfarbe als wahrnehmungsästhetische Kategorie. Musicology Conference, Halle 1998
549. Kopiez R, J Langner, Ch Stoffel: Realtime analysis of dynamic shaping. Talk at the 6th International Conference on Music Perception and Cognition (ICMPC), Keele, England, August 5-10, 2000

550. Kopiez R, J Langner and B Feiten: Perception ad Representation of Multiple Tempo Hierarchies in Musical Performance ad Composition: Perspectives from a New Theoretical Approach. In: Kopiez R and W Auhagen (Eds.): Controlling Creative Processes in Music. Peter Lang, Frankfurt am Main et al. 1998
551. Körner T W: Fourier Analysis. Cambridge University Press 1988, Cambridge
552. Kostelanetz R: John Cage. Praeger, New York 1968
553. Kouneiher J, D Flament, P Nabonnand, J-J Szczeciniarz (Eds.): Geometrie au XXème siècle, 1930-2000. Histoire et horizons. Hermann, Paris 2005
554. Kouzes R T et al.: Collaboratories: Doing Science on the Internet. Computer, August 1996
555. Kozbelt A, R A Beghetto, M A Runco: Theories of Creativity. In: J C Kaufman and R J Sternberg (Eds.): The Cambridge Handbook of Creativity. Cambridge University Press, New York, pp. 20-47, 2010
556. Krömer R: La théorie des catégories : un outil de musicologie scientifique aux yeux de la critique philosophique. In: Andreatta M, F Nicolas et C Alunni (Eds.): Á la lumière des mathématiques et à l'ombre de la philosophie. Coll. Musique/Sciences, ENS, Delatour, Ircam, Centre Pompidou, p. 213-222, Paris 2012
557. Kronland-Martinet R: The Wavelet Transform for Analysis, Synthesis, and Processing of Speech and Music Sounds. Computer Music Journal, 12 (4), 1988
558. Kreiman J and D Sidtis: Foundations of Voice Studies. Wiley-Blackwell 2013
559. Kronland-Martinet R: The Wavelet Transform for Analysis, Synthesis, and Processing of Speech and Music Sounds. Computer Music Journal, 12 (4), 1988
560. Kronman U and J Sundberg: Is the Musical Ritard an Allusion to Physical Motion? In: Gabrielsson A (Ed.): Action and Perception in Rhythm and Meter. Bibl. of the Royal Swedish Acad. of Sci. 55, Stockholm 1987
561. Kuratowski K and A Mostowski: Set Theory. North Holland, Amsterdam 1968
562. Kühner H: Virtual Table. CG TOPICS 3/97
563. Laban R: Language of Movement: A guidebook to Choreutics, ed. Lisa Ullmann, Plays, Boston 1974
564. Laban R: Die Welt des Tänzers. W. Seifert, Stuttgart 1920
565. Laban R: The Laban Sourcebook. McCaw D (Ed.), Routledge, New York 2011
566. Laban R: The Mastery of Movement, 4th ed. Macdonald & Evans, Plymouth 1980
567. Laban R: The Rhythm of Effort and Recovery I. LAMG Magazine, in Hodgson, Mastering Movement, November 1959
568. Lachenmann H: Mouvement (- vor der Erstarrung) für Ensemble. Breitkopf und Härtel, 1984
569. Lages Lima E: Fundamental Groups and Covering Spaces. Peters, Natick, MA 2003
570. Lakoff G and R Núñez: Where mathematics comes from: How the embodied mind brings mathematics into being. Basic Books, New York 2000
571. Lamé G: Examen des différentes méthodes employées pour résoudre les problèmes de géométrie (1903). Hermann, Paris 1918
572. Lamé G: Leçons sur les coordonnées curvilignes, et leurs diverses applications. Mallet-Bachelier, 1859
573. Landi S (Ed.): L'estrangement, Retour sur un thème de Carlo Ginzburg. Essais, revue interdisciplinaire d'humanités, No Hors série, école doctorale Montaigne-Humanités, Univ. Michel de Montaigne, Bordeaux 3, 2013
574. Landi S (Ed.): Avant-propos. In: L'estrangement, Retour sur un thème de Carlo Ginzburg [573, p. 9-17]
575. Landry, E and J-P Marquis: Categories in Context: Historical, Foundational, and Philosophical. Philosophia Mathematica 13.1: 1-43, 2005
576. Lang S: Introduction to Differentiable Manifolds. Interscience, New York et al. 1962
577. Lang S: Elliptic Functions. Addison-Wesley, Reading, Mass. 1973
578. Lang S: $SL_2(\mathbb{R})$. Addison-Wesley, Reading, Mass. et al. 1975
579. Langley P, H A Simon, G L Bradshaw, J M Zytkow: Scientific Discovery: Computational Explorations of the Creative Process. MIT Press, Cambridge, MA, 1987
580. Langner G: Evidence for Neuronal Periodicity Detection in the Auditory System of Guinea Fowl: Implications for Pitch Analysis in the Time Domain. Exp. Brain Res. 52, 333-355 1983
581. Langner G: Periodicity coding in the auditory system. Hear Res. Jul;60(2):115-42, 1992
582. Langer S: Feeling and Form, Routledge and Kegan Paul, London 1953
583. Lawson R A: The First Century of Blues: One Hundred Years of Hearing and Interpreting the Music and the Musicians. Southern Cultures, pp. 39-61, 2007
584. Langner J, R Kopiez, Ch Stoffel, M Wilz: Realtime analysis of dynamic shaping. In: Woods C et al. (Eds.): Proceedings of the Sixth International Conference on Music Perception and Cognition, Keele, UK, 2000
585. Lawvere F W: An elementary theory of the category of sets. Proc. Natl. Acad. Sci. 52, 1506-1511, 1964
586. Lawvere F W and R Rosebrugh: Sets for mathematics, Cambridge U. Press, 2003

587. Lazier R: Making Schoenberg Dance. In: Sound Moves, Proceedings of the Int. Conf. on Music and Dance, Roehampton University 2005
588. Leach J L: Towards a Universal Algorithmic System for Composition of Music and Audio-Visual Works. In: ICMA (Ed.): Proceedings of the ICMC 96, San Francisco 1996
589. Ledderose L: Kreativität und Schrift in China. In: R M Holm-Hadulla (Ed.): Kreativität. Springer, Heidelberg 2000
590. Lee, J M: Introduction to Topological Manifolds. Springer, Heidelberg et al. 2000
591. Legrand M: Revue critique de "La pulsation mathématique". Repères-IREM, 39, avril 2000
592. Le Guin E: Boccherini's Body. University of California Press, Berkeley and Los Angeles 2006
593. Leman M: Schema Theory. Springer, Berlin et al. 1996
594. Lerdahl F: Timbral Hierarchies. Contemporary Music Review, vol.2, no.1, 1987
595. Lerdahl F: Cognitive Constraints on Compositional Systems. In: ed. J Sloboda (Ed.): Generative Processes in Music. Oxford University Press, Oxford 1988. Reprinted in Contemporary Music Review 6, no. 2 (1992):97-121
596. Levy M: Intonation in North Indian Music. Biblia Impex, 1982
597. Lévi-Strauss C: Le cru et le cuit: Mythologies I. Plon, Paris 1964
598. Lewin D: Re: Intervallic Relations between Two Collections of Notes. JMT, 3(2), 298-301, 1959.
599. Lewin D: Klumpenhouwer Networks and Some Isographies that Involve Them. Music Theory Spectrum 12/1, 1990
600. Lewin D: The Intervallic Content of a Collection of Notes, Intervallic Relations between a Collection of Notes and Its Complement: An Application to Schoenberg's Hexachordal Pieces. JMT, 4(1), 98-101, 1960
601. Lewin D: Forte's Interval Vector, My Interval Function, and Regener's Common-Note Function. JMT, 21(2), 194-237, 1977
602. Lewin D: A Formal Theory of Generalized Tonal Functions. JMT 26(1), 32-60, 1982
603. Lewin D: On Formal Intervals between Time-Spans. Music Perception, 1(4), 414-423, 1984
604. Lewin D: Musical Form and Transformation: 4 Analytic Essays. Yale University Press, New Haven and London 1993
605. Lewin D: Generalized Musical Intervals and Transformations (1987). Cambridge University Press 2007
606. Lewis C I: Mind and Word Order. Dover, New York 1956
607. Leyton M: Symmetry, Causality, Mind. MIT Press, Cambridge/MA and London 1992
608. Leyton M: A Generative Theory of Shape. Springer, Berlin et al. 2001
609. Lichtenhahn E: Romantik: Aussen- und Innenseiten der Musik. In: "Das Paradox musikalischer Interpretation", Symposion zum 80. Geburtstag von K. von Fischer, Univ. Zürich 1993
610. Lidov D: Mind and Body in Music. Semiotica 66/1 (1987), pp. 77-78
611. Ligeti G: Pierre Boulez: Entscheidung und Automatik in der Structure Ia. Die Reihe IV, UE, Wien 1958
612. Ligeti G and G Neuweiler: Motorische Intelligenz. Wagenbach, Berlin 2007
613. Lippe E and A H M ter Hofstede: A Category Theory Approach to Conceptual Data Modeling. Informatique Theorique et Applications, Vol 30, 1, 31-79, 1996
614. Liszka J: A general introduction to the Semeiotic of Charles S. Peirce. Bloomington M.A. : Harvard University Press, 1996
615. Liszt F: Mephisto-Walzer I. Ed. Peters, No. 7203, (Edited by E. Von Sauer) 1913-17
616. Locke J: An essay concerning human understanding (first publication 1690). Pennsylvania State University 1999
617. Loomis L H and S Sternberg: Advanced Calculus. Addison-Wesley, Reading, MA et al. 1968
618. Loria G: Perfectionnements, Evolution, Métamorphoses du concept de "coordonnées" Contribution à l'histoire de la géométrie analytique. Osiris, Vol. 8, pp. 218-288, 1948
619. Loria G: A. L. Cauchy in the History of Analytic Geometry. Scripta Mathematica, p. 123-128, 1933
620. Losada C: Isography and Structure in the Music of Boulez. Journal of Mathematics and Music 2008, Vol. 2, no. 3, pp. 135-155, 2008
621. Luck G, P Toiviainen, M Thompson: Perception of Expression in Conductors' Gestures: A Continuous Response Study. Music Perception, 20(1), 47-57, 2010
622. Louis A K et al.: Wavelets. Teubner, Stuttgart 1994
623. Lubet A: (Paralyzed on One) Sideman: Disability Studies Meets Jazz, through the Hands of Horace Parlan. Critical Studies in Improvisation/Études critiques en improvisation 6.2. 2010 `http://www.criticalimprov.com/article/view/1268`
624. Lubet A: Music, Disability, and Society. Temple University Press, Philadelphia 2011
625. Lubet A: Social Confluence and Citizenship: A View from the Intersection of Music and Disability. In: Hirschmann N J, and B Linker (Eds.): Civil Disabilities: Citizenship, Membership, and Belonging. University of Pennsylvania Press, Philadelphia 2015

626. Lubet A: Oscar Peterson's piano prostheses: Strategies of performance and publicity in the post-stroke phase of his career. Jazz Research Journal, 7(2), pp. 151-182, 2015
627. Lubet A: A Case of the Blues (misc). Unpublished, 2014
628. Lucas E: Récréations mathématiques. Gauthier-Villars, Paris, 1882
629. Lüdi W: Fax to G Mazzola. Malans 01/23/1991
630. Ludwig H: Marin Mersenne und seine Musiklehre. Olms, Hildesheim 1971
631. Lussy M: Traité de l'expression musicale. Paris 1874
632. Lutz R: Le poète dans son oeuvre. Seminar Jung, Univ. Zürich, 1980
633. Lyttleton H: The Best of Jazz. Robson Books, London 1998
634. Machado A: Chant XXIX, Proverbios y cantarès, Campos de castilla, (1917), Ed. Cátedra, Madrid, 1989, 2015
635. Mac Lane S: Homology. Springer, Heidelberg et al. 1963
636. Mac Lane S: The development of mathematical ideas by collision: The case of Categories and Topos Theory. In: Categorical Topology and its relation to analysis, algebra and combinatorics, 1989
637. Mac Lane S: Categories for the Working Mathematician. Springer, New York et al. 1971
638. Mac Lane S: Mathematics, Form and Function. Springer, New York 1986
639. Mac Lane S and Moerdijk I: Sheaves in Geometry and Logic. Springer, New York et al. 1994
640. MacLean P D: The triune brain, emotion, and scientific bias. In: Schmitt F O (Ed.): The Neurosciences: Second Study Program, 336-348, Rockefeller Univ. Press, New York 1970
641. Maeder R: Programming Mathematica. Addison-Wesely, Reading, Mass. 1991
642. Manin Y: Interview in the Notices of the American Mathematical Society, November 2009, p. 1268
643. Mannone M and G Mazzola: Hegel's Conceptual Group Action on Creative Dynamics in Music. Gli spazi della musica, University of Torino 2014
644. Mannone M and G Compagno: Characterization of the degree of Musical non-Markovianity. Unpublished
645. Mannone M and S Lo Franco and G Compagno: Comparison of non-Markovianity criteria in a qubit system under random external fields. Phys. Scr. T, 2013
646. Mannone M: Segmentation des séries temporelles pour l'orchestration automatique. IRCAM - UPMC, Paris 2013 `http://www.atiam.ircam.fr/Archives/Stages1213/MANNONE_Maria.pdf`
647. Mannone M: Dalla Musica all'Immagine, dall'Immagine alla Musica - Relazioni matematiche fra composizione musicale e arte figurativa. Compostampa 2011
648. Mannone M and G Mazzola: Hypergestures in Complex Time: Creative Performance Between Symbolic and Physical Reality. Proceedings of the MCM 2015 conference, Springer, Heidelberg 2015
649. Mannone M and G Mazzola: Hypergestures in Complex Time: Creative Performance Between Symbolic and Physical Reality. Talk at MCM conference 2015, Queen Mary U, London 2015
650. Mansouri B: Normes internationales élargies et dépassement du calcul classique de la valeur. Thèse, Doctorat ès Sciences de gestion, Université de Bordeaux, 321 p., 3 juillet 2015
651. Marek C: Lehre des Klavierspiels. Atlantis 1977
652. Marino G, M-H Serra, J-M Raczinski: The UPIC system: Origins and innovations. Perspectives of New Music, 31(1):258–69, 1993
653. Marquis J-P: Category theory and the foundations of mathematics: philosophical excavations, *Synthese*, 103, pp. 229-254, 1995
654. Marrou H-I: De la connaissance historique. 1954, nouv. éd. coll. Points/Histoire, H21, Paris, 1975
655. Martinet A: Eléments de Linguistique Générale. Colin, Paris 1960
656. Martin B: CeDell Davis: The Last Man Standing. Huffington Post, 18 February 2015 `http://www.huffingtonpost.com/barrett-martin/cedell-davis-the-last-man_b_6696710.html`
657. Martino D: The Source Set and Its Aggregate Formations. JMT, 5(2), 224-273, 1961
658. De Martino E: La terra del rimorso. Bellaterra edizioni, 1999
659. Marx A B: Lehre von der musikalischen Komposition. Breitkopf & Härtel, Leipzig 1837-1847
660. Mason R M: Enumeration of Synthetic Musical Scales (...). J. of Music Theory 14, 1970
661. Massey W S: A Basic Course in Algebraic Topology. Springer, New York et al. 1991
662. Mathiesen Th J: Transmitting Text and Graphics in Online Database: The Thesaurus Musicarum Latinarum Model. In: Hewlett W B and Selfridge-Field E (Eds.): Computing in Musicology Vol.9, CCARH, Menlo Park 1993-94
663. Matisoff S: The legend of Semimaru, blind musician of Japan. Columbia University Press, New York 1978
664. Mattheson J: Der vollkommene Kapellmeister. Hamburg 1739
665. Maxwell H J: An Expert System for Harmonizing Analysis of Tonal Music. In: Balaban M et al. (Eds.): Understanding Music with AI: Perspectives on Music Cognition, MIT Press. Cambridge, MA 1992
666. May J P: A Concise Course in Algebraic Topology. The University of Chicago Press, Chicago 2007

667. Mazzola G: *Akroasis—Beethoven's Hammerklavier-Sonate in Drehung (für Cecil Taylor)*. LP, recorded August 30, 1979, WERGO/Schott, Mainz 1980
668. Mazzola G: Musique et théorie des groupes. Conference at Institut de Mathématique, U Genève, January 31, 1980
669. Mazzola G: Die gruppentheoretische Methode in der Musik. Lecture Notes, Notices by H. Gross, SS 1981, Mathematisches Institut der Universität, Zürich 1981
670. Mazzola G: Gruppen und Kategorien in der Musik. Heldermann, Berlin 1985
671. Mazzola G et al.: Rasterbild-Bildraster. Springer, Berlin et al. 1986
672. Mazzola G: Die Rolle des Symmetriedenkens für die Entwicklungsgeschichte der europäischen Musik. In: Symmetrie, Katalogband Vol.1 zur Symmetrieausstellung, Mathildenhöhe, Darmstadt, 1986
673. Mazzola G: Obertöne oder Symmetrie: Was ist Harmonie?. In: Herf F R (Ed.): Mikrotöne, Helbling, Innsbruck 1986
674. Mazzola G: Mathematische Betrachtungen in der Musik I,II. Lecture Notes, Univ. Zürich 1986/87
675. Mazzola G: Der Kontrapunkt und die K/D-Dichotomie. Manuscript, University of Zürich 1987
676. Mazzola G: Die Wahl der Zahl - eine systematische Betrachtung zum Streichquartett. In: Dissonanz 17, 1988
677. Mazzola G and G R Hofmann: Der Music Designer MD-Z71 - Hardware und Software für die Mathematische Musiktheorie. In: Petsche H (Ed.): Musik - Gehirn - Spiel, Beiträge zum 4. Herbert-von-Karajan-Symposion. Birkhäuser, Basel 1989
678. Mazzola G et al.: A Symmetry-Oriented Mathematical Model of Classical Counterpoint and Related Neurophysiological Investigations by Depth-EEG. In: Hargittai I (Ed.): Symmetry II, CAMWA, Pergamon, New York 1989
679. Mazzola G et al.: Hirnelektrische Vorgänge im limbischen System bei konsonanten und dissonanten Klängen. In: Petsche H (Ed.): Musik - Gehirn - Spiel, Beiträge zum 4. Herbert-von-Karajan-Symposion. Birkhäuser, Basel 1989
680. Mazzola G: *presto* Software Manual. SToA music, Zürich 1989-1994
681. Mazzola G: *Synthesis*. SToA 1001.90, Zürich 1990
682. Mazzola G: Geometrie der Töne. Birkhäuser, Basel 1990
683. Mazzola G and Muzzulini D: Tempo- und Stimmungsfelder: Perspektiven künftiger Musikcomputer. In: Hesse H P (Ed.): Mikrotöne III. Edition Helbling, Innsbruck 1990
684. Mazzola G and Muzzulini D: Deduktion des Quintparallelenverbots aus der Konsonanz-Dissonanz-Dichotomie. Accepted for publication in: Musiktheorie, Laaber 1990, but unpublished (!)
685. Mazzola G: Mathematische Musiktheorie: Status quo 1990. Jber. d.Dt. Math.-Verein. 93, 6-29, 1991
686. Mazzola G: Mathematical Music Theory—An Informal Survey. Edizioni Cerfim, Locarno 1993
687. Mazzola G: RUBATO at SMAC KTH, Stockholm 1993
688. Mazzola G and O Zahorka: Tempo Curves Revisited: Hierarchies of Performance Fields. Computer Music Journal **18**, No. 1, 1994
689. Mazzola G and O Zahorka: The RUBATO Performance Workstation on NeXTSTEP. In: ICMA (Ed.): Proceedings of the ICMC 94, San Francisco 1994
690. Mazzola G and O Zahorka: Geometry and Logic of Musical Performance I, II, III. SNSF Research Reports (469pp.), Universität Zürich, Zürich 1993-1995
691. Mazzola G et al.: Analysis and Performance of a Dream. In: Sundberg J (Ed.): Proceedings of the 1995 Symposium on Musical Performance. KTH, Stockholm 1995
692. Mazzola G et al.: The RUBATO Platform. In: Hewlett W B and E Selfridge-Field (Eds.): Computing in Musicology 10. CCARH, Menlo Park 1995
693. Mazzola G and O Zahorka: The PrediBase Data Base System of RUBATO on NEXTSTEP. In: Selfridge-Field E (Ed.): Handbook of Musical Codes. CCARH, Menlo Park 1995
694. Mazzola G: Inverse Performance Theory. In: ICMA (Ed.): Proceedings of the ICMC 95, San Francisco 1995
695. Mazzola G et al.: Neuronal Response in Limbic and Neocortical Structures During Perception of Consonances and Dissonances. In: Steinberg R (Ed.): Music and the Mind Machine. Springer, Berlin et al. 1995
696. Mazzola G and O Zahorka: RUBATO und der Einsatz von Hypermedien in der Musikforschung. Zeitschrift des Deutsches Bibliotheksinstituts, Berlin Jan. 1996
697. Mazzola G and O Zahorka: Topologien gestalteter Motive in Kompositionen. In: Auhagen W et al. (Eds.): Festschrift zum 65. Geburtstag J.P. Fricke. Köln 1995 `http://www.uni-koeln.de/phil-fak/muwi/publ/fs_fricke`
698. Mazzola G: Towards Big Science. Geometry and Logic of Music and its Technology. In: Enders B and Knolle N (Eds.): Symposionsband Klangart '95, Rasch, Osnabrück 1998

699. Mazzola G et al.: The RUBATO Homepage. http://www.rubato.org, Univ. Zűrich, since 1996
700. Mazzola G: Objective C and Category Theory. Seminar Notes, Multimedia Lab, CS Department, U Zurich, Zurich 1996
701. Mazzola G et al.: *Orbit.* Music & Arts CD-1015, Berkeley 1997
702. Mazzola G and J Beran: Rational Composition of Performance. In: Kopiez R and W Auhagen (Eds.): Proceedings of the Conference "Controlling Creative Processes in Music". Lang, Frankfurt and New York 1998
703. Mazzola G: Semiotic Aspects of Musicology: Semiotics of Music. In: Posner R et al. (Eds.): A Handbook on the Sign-Theoretic Foundations of Nature and Culture. Walter de Gruyter, Berlin and New York 1998
704. Mazzola G: Semiotic Aspects of Musicology: Semiotics of Music. - Art. 152, In: Posner et al. (eds): Semiotics (Handbook, Vol.III). W. de Gruyter, Berlin 1999
705. Mazzola G: Music@EncycloSpace. In: Enders B and J Stange-Elbe (Eds.): Musik im virtuellen Raum (Proceedings of the Klangart congress'98). Rasch, Osnabrück 2000
706. Mazzola G: Humanities@EncycloSpace. FER-Studie Nr.XX. Schweizerischer Wissenschaftrat, Bern 1998 http://www.swr.ch
707. Mazzola G: Die schöne Gefangene – Metaphorik und Komplexität in der Musikwissenschaft. In: Kopiez R et al. (Eds.): Musikwissenschaft zwischen Kunst, Aesthetik und Experiment. Königshausen & Neumann, Würzburg 1998
708. Mazzola G: L'Essence du Bleu (sonate pour piano). Acanthus, Rüttenen 2002
709. Mazzola G: Les mathématiques de l'interprétation musicale. Séminaire MaMuX, IRCAM, Paris, May 18, 2002
710. Mazzola G, G Milmeister, J Weissmann: Comprehensive Mathematics for Computer Scientists, Vols. I, II; 700 pp. Springer, Heidelberg et al. 2004
711. Mazzola G: The Topos Geometry of Musical Logic. Proceedings of the Fourth Diderot Symposium of the European Math. Soc., Springer, Heidelberg et al. 2002
712. Mazzola G: Degenerative Theory of Tonal Music. In: Proceedings of the Klangart congress'99. Universität Osnabrück 2002
713. Mazzola G: Classifying Algebraic Schemes for Musical Manifolds. Tatra Mt. Math. Publ. 23, 71-90, 2001
714. Mazzola G et al.: The Topos of Music. Birkhäuser, Basel 2002
715. Mazzola G, G Milmeister, K Morsy, F Thalmann: Functors for music: The Rubato Composer System. In: Digital Art Weeks Proceedings, ETH Zürich, 2006
716. Mazzola G: Global Networks in Computer Science? Invited Talk at the Computer Science Department Colloquium, ETH Zurich, January 2006
717. Mazzola G: Elemente der Musikinformatik. Birkhäuser, Basel et al. 2006
718. Mazzola G: La vérité du beau dans la musique. Delatour/IRCAM, Paris 2007
719. Mazzola G and M Andreatta: From a Categorical Point of View: K-nets as Limit Denotators. Perspectives of New Music, Vol. 44, No. 2, 2006
720. Mazzola G and M Andreatta: Diagrams, Gestures, and Formulas in Music. Journal of Mathematics and Music 2007, Vol. 1, no. 1, 2007
721. Mazzola G and P B Cherlin: Flow, Gesture, and Spaces in Free Jazz—Towards a Theory of Collaboration. Springer Series Computational Music Science, Heidelberg et al. 2009
722. Mazzola G et al.: Topos Theory for a Creative Analysis of Boulez's Structures. In: Naimpally S, G Di Maio (Eds.): Quaderni di Matematica, Theory And Applications of Proximity, Nearness and Uniformity, Vol. 23, 2009
723. Mazzola G: Categorical Gestures, the Diamond Conjecture, Lewin's Question, and the Hammerklavier Sonata. Journal of Mathematics and Music Vol. 3, no. 1, 2009
724. Mazzola G, C Losada, F Thalmann, S Tsuda: Topos Theory for a Creative Analysis of Boulez's Structures. In: Somashekhar Naimpally, Giuseppe Di Maio (Eds.): Quaderni di Matematica, Theory And Applications of Proximity, Nearness and Uniformity, Vol 23, 2009
725. Mazzola G: Musical Performance. Springer, Heidelberg et al. 2011
726. Mazzola G, J Park, F Thalmann: Musical Creativity. Springer, Heidelberg 2011
727. Mazzola G: Singular Homology on Hypergestures. Journal of Mathematics and Music Vol. 6, Nr. 1, 2012
728. Mazzola G: Manifolds and Singular Homology of Compositions, Networks, and Gestures: Classification, Connecting Functors, Examples. Unpublished
729. Mazzola G and F Thalmann: The BigBang Rubette: Gestural Music Comoposition With Rubato Composer. Proceedings of the ICMC 2008, ICMA, Ann Arbor 2008
730. Mazzola G and F Thalmann: Musical Composition and Gestural Diagrams. In Agon C et al. (Eds.): Proceedings of the Third Conference on Mathematics and Computation in Music (MCM), Springer, Heidelberg, 2011

731. Mazzola G, M Mannone, Y Pang: Cool Math for Hot Music. Springer, Heidelberg 2016
732. Mesnage M: La Set-Complex Theory: de quels enjeux s'agit-il? Analyse musicale, 4e trimestre, 87-90, 1989
733. Messiaen O: Technique de mon langage musical. Leduc, Paris 1944
734. Meyer J: Akustik und musikalische Aufführungspraxis. Verlag Das Musikinstrument, Frankfurt/Main 1980
735. Meyer-Eppler W: Grundlagen und Anwendungen der Informationstheorie. Springer, Berlin 1959
736. Meyer-Eppler W: Zur Systematik der elektrischen Klangtransformationen. In: Darmstädter Beiträge III, Schott, Mainz 1960
737. Michels U: dtv-Atlas zur Musik I,II. dtv/Bärenreiter, München/Kassel 1977, 1985
738. Miklaszewski K: A case Study of a Pianist Preparing a Musical Performance. Psychology of Music, 17, 95-109, 1989
739. Milmeister G: The Rubato Composer Music Software: Component-Based Implementation of a Functorial Concept Architecture. Springer Series Computational Music Science, Heidelberg 2009
740. McKay B D and I M Wanless: On the number of Latin squares. arXiv:0909.2101v1 [math.CO] 11 Sep 2009
741. McNeill D: Gesture and Thought. University of Chicago Press, Chicago 2005
742. McPartland M: Marian McPartland's Piano Jazz. National Public Radio, 9 November 1997 `http://www.npr.org/player/v2/mediaPlayer.html?action=1&t=1&islist=false&id=17622315&m=90469170`
743. Mead V H: More than Mere Movement: Dalcroze Eurhythmics. Music Educators Journal, Feb. 1986
744. Mendes R S et al.: Universal patterns in sound amplitudes of songs and musical genres. Phys. Rev. E, Vol. 83, pp. 017101 ff., 2011
745. Menez M: L'expérience mathématique. In: Dorier J-L, S Coutat (Eds.): Enseignement des mathématiques et contrat social : enjeux et défis pour le 21ème siècle — Actes du Colloque EMF20123 (GT4, pp. 632-640)
746. Meredith D, G Wiggins, K Lemstrom: Pattern induction and matching in polyphonic music and other multidimensional datasets. In: Callaos N et al. (Eds.): Proceedings of the 5th World Multiconference on Systemics, Cybernetics and Informatics. Orlando 2001
747. Meredith D, K Lemstrom, G Wiggins: Algorithms for discovering repeated patterns in multidimensional representation of polyphonic music. Journal of New Music Research, Vol 4, 2002
748. Merker B H: Layered Constraints on the Multiple Creativities of Music. In: I Deliège and G A Wiggins (Eds.): Musical Creativity: Multidisciplinary Research in Theory and Practice. Psychology Press, Hove/New York, pp. 25-41, 2006
749. Merleau-Ponty M: The Phenomenology of Perception in The Body: Classic and Contemporary Readings, (ed. Donn Welton), Blackwell, Oxford 1999
750. Merleau-Ponty M: Phenomenology of Perception. Trans. Colin Smith, Routledge & Kegan Paul, London 1978 [1962]
751. Merleau-Ponty M: Phénoménologie de la perception. Gallimard 1945
752. Merleau-Ponty M: The prose of the world, translation by J O'Neill. Northwestern University Press, Evanston 1973 (La prose du monde, Gallimard, 1969)
753. Meusburger P, J Funke, E Wunder (Eds.): Milieus of Creativity: An Interdisciplinary Approach to Spatiality of Creativity. Springer, 2009
754. Miller M: A Jazz Legend Handles his Own Legacy with Care. Globe and Mail (Canada), 23 February: C3. 1999
755. Misch C and Wille R: Stimmungslogiken auf MUTABOR: Eine Programmiersprache. In: Herf F R (Ed.): Mikrotöne II. Edition Helbling, Innsbruck 1988
756. `http://forumnet.ircam.fr/701.html`
757. Mokkapati V: Modal Action: An Indo-European Perpective on Gregorian Chant. Phd Thesis, Wesleyan University 1996
758. Molino J: Fait Musical et Sémiologie de la Musique. Musique en Jeu 17 1975
759. Monti G: Théorie des catégories et méthode conditionnelle appliquées aux comptes en normes internationales élargies. Thèse, Doctorat ès Sciences de gestion, Université de Bordeaux 2017
760. Montiel-Hernandez M: El denotador: Su estructura, construcción y papel en la teoría matemática de la musica. Master's thesis, UNAM, Mexico City, 1999
761. Moog R: MIDI, Musical Instrument Digital Interface. Audio Eng. Soc. 34, Nr.5, 1986
762. Moon T: A Master of Improv is Renewed by Melody. Philadelphia Inquirer, 29 September: H02, 2002
763. Moore C-L: The Harmonic Structure of Movement, Music and Dance According to Rudolf Laban. Edwin Meller Press, Lewiston 2009
764. Moritz R E: On Mathematics. A collection of witty, profound, amusing passages about mathematics and mathematicians. Copyright 1914 Robert Edouard Moritz, 410 p., Dover 1958
765. Morris R D: Composition with Pitch-Classes. Yale University Press, New Haven et al. 1987
766. Morris R D: Compositional Spaces and Other Territories. PNM 33, 328-358, 1995

767. Morris R D: K, Kh, and Beyond. In: Baker J et al. [69], 275-306
768. Morse Ph M and H Feshbach: Methods of Theoretical Physics, Part I. McGraw-Hill, New York et al. 1953
769. Mozart W A: *Die Zauberflöte.* (Klavierauszug) Ed. Peters 1932
770. Müller W: Darstellungstheorie von endlichen Gruppen. Teubner, Stuttgart 1980
771. Müller S: Pianist's Hands: Synthesis of Musical Gestures. PhD Thesis, University of Zurich, Zurich 2004
772. Müller S and G Mazzola: Constraint-based Shaping of Gestural Performance. Proceedings of the ICMC 03, ICMA, Ann Arbor 2003
773. Müller S and G Mazzola: The Extraction of Expressive Shaping in Performance. Computer Music Journal, Vol. 27(1), MIT Press, 2003
774. Mumford D: Lectures on Curves on an Algebraic Surface. Princeton University Press, Princeton 1966
775. Mumford D and K Suominen: Introduction to the theory of moduli. In: Oort F (Ed.): Algebraic Geometry Oslo 1970. Wolters-Noordhoff, Groningen 1972
776. Murenzi R: Wavelets. Combes J M, Grossman A, Tchmitchian P (Eds.), Springer Berlin et al., 1988
777. MusicKit Online-Documentation. Version 4.0 1994
778. Muzzulini D: Konsonanz und Dissonanz in Musiktheorie und Psychoakustik. Lizenziatsarbeit, MWS Univ. Zürich 1990
779. Muzzulini D: Musical Modulation by Symmetries. J. for Music Theory, Vol. 39, No. 2, pp. 311-328, 1995
780. Muzzulini D: Tempo Modifications and Spline Functions. NF-Report 1993, Univ. Zürich 1993
781. Narayan S S, Temchin A N, Recio A, Ruggero M A: Frequency tuning of basilar membrane and auditory nerve fibers in the same cochleae. Science. 282(5395):1882-4, Dec 4 1998
782. Nattiez J-J: Fondements d'une Sémiologie de la Musique. Edition 10/18 Paris 1975
783. Negro G Del: The Passeggiata and Popular Culture in an Italian Town: Folklore and the Performance of Modernity. Montréal; McGill-Queen's University Press 2004
784. Neumaier W: Was ist ein Tonsystem? Lang, Frankfurt/Main et al. 1986
785. Neuman D A: A House of Music: The Hindustani Musician and the Crafting of Traditions. PhD Thesis, Columbia University 2004
786. Neuwirth G: Josquin Desprez, "Erzählen von Zahlen". In: Musik-Konzepte 26/27, edition text+kritik, München 1982
787. Newlove J and J Dalby: Laban for All. Routledge, New York 2004
788. NeXTSTEP Online-Documentation; Version 3.3. NeXt Inc., Redwood City 1995
789. Nicolas F: René Guitart : la pulsation d'une intellectualité mathématique étrangement évidente. Conférence à la *Journée en l'honneur de René Guitart pour ses 65 ans*, 9 novembre 2012, Université Paris 7 Denis Diderot, 8 p, inédit. See a summary in: `http://www-lmpa.univ-littoral.fr/ sic`
790. Nicolas F: Le Monde-Musique III. Le musicien et son intellectualité musicale. Aedam Musicae, 384 p., juin 2015
791. Nicotra E: Introduzione alla tecnica della direzione d'orchestra. Edizioni Curci 2007
792. Nieberle R: IRCAM Signal Processing Workstation. Keyboards, Dec. 1992
793. `http://www.nime.org`
794. Noether E: Hyperkomplexe Grössen und Darstellungstheorie. Math. Zeitschr., Vol. XXX 1929
795. Noland C: Agency and Embodiment: Performing Gestures/Producing Culture. Harvard University Press, Cambridge 2009
796. Noll J: Musik-Programmierung. Addison-Wesley, Bonn 1994
797. Noll Th: Morphologische Grundlagen der abendländischen Harmonik. Doctoral Thesis, TU Berlin 1995
798. Noll Th: Fractal Depth Structure of Tonal Harmony. In: ICMA (Ed.): Proceedings of the ICMC 95, San Francisco 1995
799. Noll Th: `http://www.cs.tu-berlin.de/ noll/ChordDictionary.sea.hqx`, TU Berlin 1996
800. Noll Th: Harmonische Morpheme. Musikometrika 8, 7-32, 1997
801. Noll Th: The Consonance/Dissonance-Dichotomy Considered from a Morphological Point of View. In: Zannos I (Ed.): Music and Signs. ASCO Publ., Bratislava 1999
802. Noll Th: Morphologische Grundlagen der abendländischen Harmonik. Doctoral Thesis, TU Berlin 1995
803. Noll Th: Wie blättert man in einem hypermedialen Gebärdenlexikon? In: Schmauser C, Th Noll (Eds.): Körperbewegungen und ihre Bedeutungen. Berlin Verlag Arno Spitz, Berlin 1998
804. Noll Th et al.: Computer-aided transformational analysis with tone sieves. Contribution au congrés: SMC 2006
805. Nunn D et al.: Acoustic Quanta. In: ICMA (Ed.): Proceedings of the ICMC 96, San Francisco 1996
806. Oettingen A von: Das duale Harmoniesystem. Leipzig 1913
807. Okopenko A: Lexikon-Roman. Residenz Verlag, Salzburg 1970
808. Opcode: MAX. http://www.opcode.com/products/max/, Opcode Systems Inc. 1997

809. Marco Orsini conducting Le Nozze di Figaro 2014 https://www.youtube.com/watch?v=M7QuLXy_61U
810. Osgood C E et al.: The Nature and Measurement of Meaning. Psychological Bulletin 49, 1952
811. Osher S and R Fedkiw: Level Set Methods and Dynamic Implicit Surfaces. Springer, 2003
812. Owens Th: Bebop: The Music and the Players. Oxford University Press, New York 1995
813. Pakkan M and V Akman: Hypersets. Information Sciences: An International Journal, 85(1):4361, 1996
814. Pallasmaa J: The Thinking Hand. Wiley 2009
815. Palmer R: Liner notes, Oscar Peterson: A Night in Vienna. DVD. Verve 02498 6253-4, 2004
816. Pappas T: The Music of Reason: Experience the Beauty of Mathematics Through Quotations. Wide World Publishing/ Tetra, 1995
817. Parker W: Conversations. Rogueart, 2011
818. Parlan H: The Maestro. CD. SteepleChase SCCD 31167, 1995
819. Parlan H: Horace Parlan by Horace Parlan. DVD. Image ID1505R5DVD, 2000
820. Parlan H: In Copenhagen. CD. Storyville 101 8521, 2008
821. Parlan H: Musically Yours. CD SteepleChase SCS 31141, 2010
822. Parmentier M: Connaissances, vérités et idées mathématiques dans l'*Essai sur l'entendement humain* de Locke. In: Barbin E et J-P Cléro (éds): Les mathématiques et l'expérience : ce qu'en ont dit les philosophes et les mathématiciens, p. 95-114, Hermann, Paris 2015
823. Parncutt R: The Perception of Pulse in Musical Rhythm. In: Gabrielsson A (Ed.): Action and Perception in Rhythm and Music. Royal Swedish Adademy of Music, No.55
824. Parncutt R: Recording Piano Fingering in Live Performance. In: Enders B, Knolle N (Eds.): KlangArt-Kongress 1995, Rasch, Osnabrück 1998
825. Parncutt R et al.: Interdependence of Right and Left Hands in Sight-read, Written, and Rehearsed Fingerings. Proc. Euro. Sco. Cog. Sci. Music, Uppsala 1997
826. Parncutt R: Modeling Piano Performance: Physics and Cognition of a Virtual Pianist. In: ICMA (Ed.): Proceedings of the ICMC 97, San Francisco 1997
827. Patras F: La pensée mathématique contemporaine. PUF, Paris, 2001
828. Patras F: La possibilité des nombres. PUF, Paris, octobre 2014
829. Peck R: Klein-Bottle Tonnetze. MTO, Vol. 9, no. 3, August 2003
830. Penrose R: The Road to Reality. Vintage, London 2002
831. Perle G: Serial Composition and Atonality: An Introduction to the Music of Schoenberg, Berg and Webern. 5th ed., revised, University of California Press, Berkeley 1981
832. Peterson O: An Oscar Peterson Christmas. CD, Telarc 515956K, 1995
833. Peterson O: Oscar in Paris. Telarc 2CD-83414, 1997
834. Peterson O: A Tribute to Oscar Peterson: Live at the Town Hall. Telarc 2CD-83401
835. Peterson O: A Summer Night in Munich. CD. Telarc CD-83450, 1999
836. Peterson O: A Night in Vienna. DVD, Verve 02498 6253-4, 2004
837. Peterson O, R Brown, M Jackson: The Very Tall Band: Live at the Blue Note. CD. Telarc CD-83443, 1999
838. Peterson O, R Brown, M Jackson: What's Up? The Very Tall Band. CD. Telarc CD-83663, 2007
839. Peterson O and B Green: Oscar and Benny. CD. Telarc CD-83406, 1998
840. Peterson O, R Hargrove, R Moore: Oscar Peterson Meets Roy Hargrove and Ralph Moore. CD. Telarc, CD-83399, 1996
841. Peterson O, and M LeGrand: Trail of Dreams: A Canadian Suite. CD. Telarc CD-83500, 2000
842. Peterson O, and I Perlman: Side by Side. CD. Telarc, CD-83341, 1994
843. Petsche H et al.: EEG in Music Psychological Studies. In: Steinberg R (Ed.): Music and the Mind Machine. Springer, Berlin et al. 1995
844. Petsche H: Private correspondence. Vienna, March 2001
845. Piaget J: Morphismes et categories. Comparer et transformer, Delachaux et Niestlé, Neuchâtel, 1990
846. Pfeifer R and J Bongard: How the Body Shapes the Way We Think: A New View of Intelligence. MIT Press 2006
847. Peirce C S: Peirce, *Collected Papers*, vol. III, vol. IV The simplest mathematics, 1960
848. Pierce A: Deepening Musical Performance through Movement: The Theory and Practice of Embodied Interpretation (Musical Meaning and Interpretation). Indiana University Press, Bloomington 2010
849. Plato: Phaidron.
850. Plomp R and W Levelt: Tonal Consonance and Critical Bandwidth. J. Acoust. Soc. Am. 38, 548, 1965
851. Plotinus: The Six Enneads, in *Great Books of the Western World*, Vol. 17. Encyclopædia Britannica, Inc., Chicago, 1952
852. Poincaré H: La valeur de la science. Flammarion, Paris 1905

853. Pólya G: Mathematical Discovery. On Understanding, Learning and Teaching Problem Solving (Combined Edition). Wiley, New York, 1962, 1965, 1981
854. Pólya G: Comment poser et résoudre un problème. Dunod, 1965
855. Polyanin A D: Handbook of Linear Partial Differential Equations for Engineers and Scientists. Chapman & Hall/CRC, 2002; see also: `http://eqworld.ipmnet.ru`, section 3.2.
856. Pope S T: Music Notations and the Representation of Musical Structure and Knowledge. Perspectives of New Music, Spring-Summer 1986
857. Popelard M-D and A Wall: Des Faits et gestes. éditions Bréal, Rosny-sous-Bois 2003
858. Popper K R: Conjectures and Refutations. Routledge & Kegan Paul, London 1963
859. Popper K: The open society and its enemies. Vol. 2, Routledge & Kegan, London, 1945
860. Posner R: Strukturalismus in der Gedichtinterpretation. In: Blumensadth H (Ed.): Strukturalismus in der Literaturwissenschaft. Kiepenheuer & Witsch, Köln 1972
861. Prame E: Measurements of the Vibrato Rate of 10 Singers. In: Friberg A et al. (Eds.): Proceedings of the 1993 Stockholm Music Acoustic Conference. KTH, Stockholm 1994
862. Pribram K H: Brain and Perception. L Erlbaum, Hillsdale, NJ 1991
863. Priestley M B: Spectral Analysis of Time Series. Academic Press, London 1981
864. Prinz W: Modes of linkage between perception and action. In W Prinz & A-F Sanders (Eds.): cognition and motor processes (pp. 185–193), Springer, Berlin 1984
865. Promies W: Symmetrie in der Literatur. In: Symmetrie, Katalogband Vol.1 zur Symmetrieausstellung. Mathildenhöhe, Darmstadt, 1986
866. Promies W: Stolbergs Gedicht im Poesiegarten In: Symmetrie, Katalogband Vol.3 zur Symmetrieausstellung. Mathildenhöhe, Darmstadt, 1986
867. Puckette M and Lippe C: Score Following in Practice. In: ICMA (Ed.): Proceedings of the ICMC 92, San Francisco 1992
868. Pulkki V et al.: DSP Approach to Multichannel Audio Mixing. In: ICMA (Ed.): Proceedings of the ICMC 96, San Francisco 1996
869. Puttke M: Learning to Dance Means Learning to Think! In: Bläsing B, Puttke M, Schack Th. (Eds.): The Neurocognition of Dance. Psychology Press, Hove and New York 2010
870. Quinn I: Fuzzy Extensions to the Theory of Contour. Music Theory Spectrum. Vol. 19/2, 1997
871. Radl H: Versuch über die Modulationstheorie Mazzolas in reiner Stimmung. Diploma Thesis, U Augsburg, Augsburg 1998
872. Rahaim M: Musicking Bodies. Wesleyan U P, Middletown, CT 2012
873. Rahn J: Basic Atonal Theory. Longman, New York 1980
874. Rahn J: Review of D. Lewin's "Generalized Musical Intervals and Transformations". JMT, 31, 305-318, 1987
875. Raffman D: Language, Music, and Mind. MIT Press, Cambridge et al. 1993
876. Rameau J-Ph: Traité de l'Harmonie, Réduite à ses Principes Naturels. Paris 1722
877. Ramachandran V S: Mirror Neurons and imitation learning as the driving force behind the "great leap forward" in human evolution, 2015. See: `http://www.edge.org/3rd.culture/ramachandran/ramachandran p6.html`
878. Ramstein C: Analyse, Représentation et Traitement du Geste Instrumental. PhD thesis, Institut National Polytechnique de Grenoble, December 1991
879. Ratz E: Einführung in die musikalische Formenlehre. Universal Edition, Wien 1973
880. Rauchfleisch U: Musik schöpfen, Musik hören. Ein psychologischer Zugang. Vandenhoeck und Ruprecht, Göttingen and Zürich 1996
881. Rav Y: Why do we prove theorems?. Philosophia mathematica, 7 (1), p. 5-41, 1999
882. Read G: Music Notation. Crescendo Publ., Boston 1969
883. Reeves H: Patience dans l'azur. L'évolution cosmique. Seuil, Paris 1981
884. Reichardt J F: Vermischte Musikalien. Riga 1777
885. Reidmeister K: Elementare Begründung der Knotentheorie. Abh. Math. Sem. Univ. Hamburg, 5, 24-32, 1926
886. Repp B: Diversity and Commonality in Music Performance: An Analysis of Timing Microstructure in Schumann's "Träumerei". J. Acoustic Soc. Am. 92, 1992
887. Repp B: e-mail communication of tempo data. Haskins Laboratories, New Haven, June 2, 1995
888. Repp B: Patterns of note onset asynchronies in expressive piano performance. J. Acoustic Soc. Am. **100**, 1996
889. Repp B: Pedal Timing and Tempo in Expressive Piano Performance: A Preliminary Investigation. Psychology of Music **24**, 1996
890. Repp B: The Art of Inaccuracy: Why Pianists' Errors Are Difficult to Hear. Music Perception, *14*, 2 1997
891. Repp B: Expressive Timing in a Debussy Prélude: A Comparison of Student and Expert Pianists. Musicae Scientiae 1, 1997

892. Reti R: The Thematic Process in Music (1951). Greenwood Press, Westport, 2nd ed. 1978
893. Reti R, commented by Kopfermann M: Schumanns Kinderszenen: quasi Thema mit Variationen. In: Musik-Konzepte Sonderband Robert Schumann II, edition text + kritik, München 1982
894. Riguet J: Systèmes de coordonnées relationnels. *CRAS*, t. 236, p. 2369-2371, séance du 22 juin 1953
895. Riguet J: Systèmes de coordonnées relationnels II. Applications à la théorie des groupes de Kaloujnine. *CRAS* t. 238, p. 435-437, séance du 25 janvier 1954
896. Riguet J: Systèmes de coordonnées relationnels III. Fermetures et systèmes symétriques. *CRAS*, t. 238, p. 1763-1765, séance du 3 mai 1954
897. Rhode W S and L Robles: Evidence from Mössbauer Experiments for Nonlinear Vibration in the Cochlea. J. Acoust. Soc. Am. 48, 988, 1970
898. Richter P et al.: How Consistent are Changes in EEG Coherence Patterns Elicited by Music Perception? In: Steinberg R (Ed.): Music and the Mind Machine. Springer, Berlin et al. 1995
899. Riemann B: Über die Hypothesen, welche der Geometrie zugrunde liegen (1854). Gött. Abh. No.13 (published in 1867)
900. Riemann H: Musikalische Logik. Leipzig 1873
901. Riemann H: Musikalische Syntaxis. Leipzig 1877
902. Riemann H: Vereinfachte Harmonielehre oder die Lehre von den tonalen Funktionen der Akkorde. London 1893
903. Riemann H: Über Agogik. In: Präludien und Studien II. Leipzig 1900
904. Riemann H: System der musikalischen Rhythmik und Metrik. Breitkopf und Härtel, Leipzig 1903
905. Riemann H: Handbuch der Harmonielehre. Leipzig 6/1912
906. Riemann H: Grundriss der Musikwissenschaft. Leipzig 1928
907. Riemann H and H H Eggebrecht: Riemann Musiklexikon, Sachteil, 12. Auflage. Schott, Mainz 1967
908. Ries F: Biographische Notizen über L. van Beethoven (1838). New edition by Kalischer A Ch, 1906
909. Risset J-C: Real-time: Composition or performance? Reservations about real-time control in computer music and demonstration of a virtual piano partner. In: Kopiez R and W Auhagen (Eds.): Controlling creative processes in music. Peter Lang, Frankfurt/Main 1998
910. Rizzolati G and M Fabbri-Destro: Mirror neurons: from discovery to autism. Exp. Brain Res. 200:223-237, 2010
911. Roads C: The Computer Music Tutorial. MIT Press, Cambridge, Mass. and London 1998
912. Rodet X et al.: Xspect: a New X/Motif Signal Visualization, Analysis and Editing Program. In: ICMA (Ed.): Proceedings of the ICMC 96, San Francisco 1996
913. Rodet X: Synthesis and Processing of the Singing Voice. Proc.1st IEEE Benelux Workshop on Model based Processing and Coding of Audio (MPCA-2002), Leuven, Belgium, November 15, 2002
914. Roederer J G: Physikalische und psychoakustische Grundlagen der Musik. Springer, Berlin 1977
915. Rogers C R: Toward a theory of creativity, *ETC: A review of General Semantics*, Vol 11, 1954, 249-260
916. Rossi A: Die Geburt der modernen Wissenschaft in Europa. Beck, München 1997
917. Rostand F: Souci d'exactitude et scrupules des mathématiciens. Vrin, Paris 1960
918. Rüetschi U-J: Denotative Geographical Modelling—an attempt at modelling geographical information with the Denotator system. Diploma thesis, University of Zürich, 2001
919. Rufer J: Die Komposition mit zwölf Tönen. Bärenreiter, Kassel 1966
920. Runco M A and I Chand: Cognition and Creativity. Educational Psychology Review, Vol. 7:3, pp. 243-267, 1995
921. Ryan R A: Introduction to Tensor Products of Banach Spaces. Springer, Heidelberg 2002
922. Ruwet N: Langage, Musique, Poésie. Seuil, Paris 1972
923. *S-Plus*. MathSoft Inc., Seattle, Washington 1995
924. Sachs K-J: Der Contrapunctus im 14. und 15. Jahrhundert. AMW, Franz Steiner, Wiesbaden 1974
925. Sacks O W: Musicophilia: tales of music and the brain. (Rev. and expanded 1st ed.) Vintage Books, New York 2008
926. Saiber A and H S Turner (Eds.): Mathematics and the Imagination. Configurations, 17, 195 pp. 2009
927. Sandall R: Miraculously, Magically, the Years Fell Away. Daily Telegraph (London), 1 July: 21 2006
928. Sander E: Notre pensée progresse par analogies. La conscience. La Recherche Hors-série, no 15, pp. 54-57, octobre-novembre 2015
929. Sattler D E (Ed.): Friedrich Hölderlin – Einhundert Gedichte. Luchterhand, Frankfurt/Main 1989
930. Salzer F: Structural Hearing: Tonal Coherence in Music. (German: Strukturelles Hören. Noetzel, Wilhelmshaven 1960) Dover, New York 1962
931. Sarasúa A and E Guaus: Dynamics in Music Conducting: A computational Comparative Study Among Subjects. Proceedings of the International Conference on New Interfaces for Musical Expression, NIME'14, Goldsmiths London, UK, 2014

932. Sauvageot F: Pierre Deligne. *Images des mathématiques*, CNRS, 12 mai 2013 http://images.math.cnrs.fr/Pierre-Deligne.html
933. Saussure F de: Cours de Linguistique Générale. Payot, Paris 1922
934. Saussure F de: Linguistik und Semiotik - Notizen aus dem Nachlass. Fehr J (Ed.), Suhrkamp, Frankfurt/Main 1997
935. Savan D: An introduction to C.S. Peirce's Full System of Semeiotic. Toronto Semiotic Circle, Toronto 1988
936. Sawyer R K: Improvisation and the creative process: Dewey, Collingwood, and the aesthetics of spontaneity. The Journal of Aesthetics and Art Criticism, Vp;. 58, pp. 149-61, 2000
937. Sawyer R K: Group Creativity: Music, Theater, Collaboration. Lawrence Erlbaum Associates, Mahwah 2003
938. Scarborough D L et al.: Connectionist Models for Tonal Analysis. Computer Music Journal Vol. 13, 1989
939. Schenker H: Fünf Urlinien-Tafeln. Universal Edition, Wien 1932
940. Schenker H: Theorien und Phantasien III: Der freie Satz. Universal Edition, Wien 1935
941. Scherchen H: Manuale del direttore d'orchestra. Edizioni Curci 2009
942. Schewe K-D: Type concept in OODB Modelling and its Logical Implications. In Kawaguchi E et al. (Ed.), Information Modelling and Knowledge Bases XI, pp. 256–274. IOS Press, 2000
943. Schmalz R: Out of the mouth of mathematicians. A quotation book for philomaths. Mathematical Association of America, 1993
944. Schmidt J (Ed.): Celibicache. Book and Movie PARS (ISBN 3-9803265-1-9), München 1992
945. Schmidt-Biggemann W: Topica universalis. Meiner,Hamburg 1983
946. Schmitt, J-C: La raison des gestes dans l'occident médiéval. Gallimard, Paris 1990
947. Schneps L and P Lochak (Eds.): Geometric Galois Actions. Cambridge University Press, Cambridge, 1997
948. Schönberg A: Harmonielehre (1911). Universal Edition, Wien 1966
949. Schönberg A: Die Komposition mit zwölf Tönen. In: Style and Idea, New York 1950
950. Schönberg A: Fundamentals of Musical Composition. Edited by Gerald Strang, with an introduction by Leonard Stein. St. Martin's Press, New York 1967
951. Schopenhauer A: De la quadruple racine du principe de raison suffisante. éd. complète (1813-1847), Vrin, 1991
952. Schubert F: *Auf dem Wasser zu singen*, op. 72 (1823). Bärenreiter/Henle München/Kassel 1982
953. Schubert H: Kategorien I, II. Springer, Berlin et al. 1970
954. Schumann R: Kinderszenen, op. 15 (1839). Henle, München 1977
955. Schumann R: Dritter Quartett-Morgen. NZ f. Musik 1838
956. Schweitzer A: Johann Sebastian Bach. (1908) Breitkopf & Härtel, Wiesbaden 1979
957. Scriba C J and P Schreiber: 5000 Years of Geometry: Mathematics in History and Culture. Birkhäuser, 2015
958. Seashore C E: Psychology of Music. McGraw-Hill, New York 1938
959. Shepard R N and L A Cooper: Mental Images and Their Transformations. MIT Press 1982
960. Shepp A: The Geneva Concert. DVD, inakustic LC13921, 2006
961. Shepp A and H Parlan: Goin' Home. CD. SteepleChase (D) SCCD-31079, 1985
962. Shepp A and H Parlan: Trouble in Mind. CD. SteepleChase ? SCCD 31139, 1986
963. Shepp A and H Parlan: En Concert: First Set. CD. 52e Rue Est (F) RECD 015, 1987
964. Shepp A and H Parlan: En Concert: Second Set. CD. 52e Rue Est (F) RECD 016, 1987
965. Shepp A and H Parlan: Reunion. CD. L+R CDLR 45003, 1987
966. Shepp A and H Parlan: Goin' Home. CD. SteepleChase (D) SCCD-31079, 2010
967. Shepp A and H Parlan: Trouble in Mind. CD. SteepleChase ? SCCD 31139, 2010
968. Siebers T: Disability Aesthetics (pp. 2-3). University of Michigan Press, Ann Arbor 2010
969. Serre J-P: Faisceaux Algébriques Cohérents. Annals of Mathematics, 61 (2): 197-278, 1955
970. Serre J-P: Trees. Springer, New York 1980
971. Serres M (Ed.): Eléments d'histoire des sciences. Bordas, Paris 1989
972. Sessions R: Questions About Music. W. W. Norton and Company, 1971
973. Shaffer L H: Musical Performance as Interpretation. Psychology of Music, Vol. 23, 1995
974. http://www.fau.edu/bodymindculture
975. Shusterman R: Somaesthetics: A Disiplinary Proposa. The Journal of Aesthetics and Art Criticism. Vol.57, No.3, pp. 299-313, 1999
976. Simondon G: Du mode d'existence des objets techniques. Aubier, Paris 1989
977. Simonton D K: Thematic fame, melodic originality, and musical zeitgeist: A biographical and transhistorical content analysis. Journal of Personality and Social Psychology, Vol. 38, pp. 972-83, 1980
978. Simonton D K: Thesaurus of scales and melodic patterns. Coleman-Ross Company, New York 1947
979. Siron J: La partition intérieure. Outre Mesure, Paris 1992
980. Slawson W: The Musical Control of Sound Color. Canadian Univ. Music Review, No.3, 1982

981. Slawson W: Sound Color. U. Cal. Press, Berkeley 1985
982. Sloan D: From DARMS to SMDL, and back again. In: Haus G and I Pighi (Eds.): X Colloquio di Informatica Musicale. AIMI, LIM-Dsi, Milano 1993
983. Sloboda J: The Musical Mind: An Introduction to the Cognitive Psychology of Music. Calderon Press, Oxford 1985
984. Sloboda, A S and A C Lehmann: Tracking Performance Correlates of Changes in Perceived Intensity of Emotion During Different Interpretations of a Chopin Piano Prelude. Music Perception, Summer 2001, Vol 19, No. 1, pp. 87-120, 2001
985. Sloterdijk P: Selbstexperiment. Hanser, München & Wien 1996
986. Sluming V et al.: Broca's Area Supports Enhanced Visuospatial Cognition in Orchestral Musicians. The Journal of Neuroscience, Vol. 27(14), pp. 3799-3806, April 4, 2007
987. Smart M A: Mimomania: Music and Gesture in Nineteenth-Century Opera. New Ed edition, University of California Press, Berkeley 2005
988. SMDL committee: Standard Music Description Language Draft, Pittsford, NY 1995
989. Smith III J O: Synthesis of bowed strings. In: Strawn J, Blum T, ICMA: Proceedings of the International Computer Music Conference, San Francisco 1982
990. Smith III J O: Physical Modeling of Musical Instruments, part I. Computer Music Journal, 16 (4), 1992
991. Smith III J O: Physical Modeling Synthesis Update. Computer Music Journal, 20 (2), 1996
992. Sodomka A: KUNSTRADIO at documentaX. http://www.culture.net/orfkunstradio/BIOS/sodomkabio.html, Vienna 1997
993. Spanier E H: Algebraic Topology. McGraw-Hill, New York et al. 1966
994. Späth H: Eindimensionale Spline-Interpolations-Algorithmen. Oldenburg, München et al. 1990
995. Späth H: Zweidimensionale Spline-Interpolations-Algorithmen. Oldenburg, München et al. 1991
996. Spellman A B: Four Lives in the Bebop Business. Pantheon 1966
997. Sperry R W: Neurology and the mind-body problem. American Scientist, 40, 291-312, 1952
998. Spicker V: Das Abstraktmotiv bei Cecil Taylor. MusikTexte, 73/74 March 1998
999. Spivak M: Calculus on Manifolds. Benjamin, New York 1965
1000. Squire R L and Butters N (Eds.): Neuropsychology of Memory. D. Guildford Press, NewYork and London 1984
1001. Stahnke M: Struktur und Ästhetik bei Boulez. Wagner, Hamburg 1979
1002. Stange-Elbe J, G Mazzola: Cooking a Canon with RubatoPerformance Aspects of J.S. Bach's "Kunst der Fuge". In: ICMA (Ed.): Proceedings of the ICMC 98, 179-186, San Francisco 1998.
1003. Stange-Elbe J: Analyse- und Interpretationsperspektiven zu J.S. Bachs "Kunst der Fuge" mit Werkzeugen der objektorientierten Informationstechnologie. Habilitationsschrift, Osnabrück 2000
1004. Stange-Elbe J: Instrumentaltechnische Voraussetzungen für eine computergestützte Interpretation. In: Enders B (Ed.): Proceedings of the Klangart congress '01, Osnabrück
1005. Starr D and R Morris: A General Theory of Combinatoriality and the Aggregate. PNM, 16(1), 364-389, 16(2), 50-84, 1977-78
1006. Starr D: Sets, Invariance and Partitions. JMT, 22(1), 1-42, 1978
1007. Stein S and S Sandor: Algebra and Tiling: Homomorphisms in the Service of Geometry.: Math. Assoc. Amer., Washington, DC 1994
1008. Steinberg R (Ed.): Music and the Mind Machine. Springer, Heidelberg 1995
1009. Steinberg R and A Fairweather (Eds.): Blues-Philosophy for Everyone: Thinking Deep About Feeling Low. John Wiley 2012
1010. Sternberg R J and T I Lubart: The Concept of Creativity: Prospects and Paradigms. In: Sternberg R J (Ed.): Handbook of Creativity. Cambridge University Press, New York 1999
1011. Stinson D R: Combinatorial designs: constructions and analysis. Springer, 2004
1012. Stollberg A, J Weissenfeld, F H Besthorn (Eds.): DirigentenBilder—Musikalische Gesten — verkörperte Musik. Schwabe, Basel 2015
1013. Stravinsky I: The Poetics of Music. Harvard University Press 1947
1014. Steele G L: Common Lisp: The Language, 2nd Edition. Digital Press, 1990
1015. Stone P: Symbolic Composer. http:/www.xs4all.nl/~psto/
1016. Stopper B: Gleichstufig temperierte Skalen unter Berücksichtigung von Oktavstreckungen. In: Hosp I (Hrsg.): Bozener Treffen 1991 — Skalen und Harmonien. Südtiroler Kulturinstitut, Bozen 1992
1017. Straub H: Beiträge zur modultheoretischen Klassifikation musikalischer Motive. Diplomarbeit ETH-Zürich, Zürich 1989
1018. Straub H: Kadenzielle Mengen beliebiger Stufendefinitionen. Unpublished manuscript, Zürich 1999/2000
1019. Straus J N: Voice Leading in Atonal Music. In: Baker J et al. [69], 237-274

1020. Stumpf C: Tonpsychologie. Leipzig 1883-1890
1021. Sundberg J, A Askenfelt, L Frydén: Musical Performance: A Synthesis-by-Rule Approach, Computer Music Journal, 7, 37-43, 1983
1022. Sundberg J and V Verillo: On the Anatomy of the Retard: A Study of Timing in Music. J. Acoust. Soc. Am. 68, 772-779, 1980
1023. Sundberg J: Music Performance Research. An Overview. In: Sundberg J, L Nord, R Carlson (Eds.): Music, Language, Speech, and Brain. London 1991
1024. Sundberg J (Ed.): Generative Grammars for Music Performance. KTH, Stockholm 1994
1025. Sundin N-G: Musical Interpretation in Performance. Mirage, Stockholm 1984
1026. Suzuki T et al.: Musical Instrument Database with Multimedia. In: ICMA (Ed.): Proceedings of the ICMC 96, San Francisco 1996
1027. Suppan W: Zur Verwendung der Begriffe Gestalt, Struktur, Modell und Typus in der Musikethnologie. In: Stockmann D and Steszwski J (Eds.): Analyse und Klassifikation von Volksmelodien. PWM Edition Krakau 1973
1028. Swan, R G: Vector Bundles and Projective Modules, *Transactions of the American Mathematical Society* 105 (2): 264-277
1029. Sylvester D: Interviews with Francis Bacon. Pantheon Books, New York 1975
1030. Szczeciniarz J-J: Réflexion sur Riemann, philosophe mathématicien, in [79], p. 195-221
1031. Symmetrie. Katalogband Vol.1 zur Symmetrieausstellung, Mathildenhöhe, Darmstadt, 1986
1032. Takemitsu T: Confronting Silence. Fallen Leaf Press, Berkeley 1995
1033. Takhtajan L A: Quantum Mechanics for Mathematicians. American Mathematical Society, Providence 2008
1034. Teschl G: Mathematical Methods in Quantum Mechanics. American Mathematical Society, Providence 2009
1035. Tanaka A: Interaction, Experience, and the Future of Music. In: O'Hara K and B Brown (Eds.): Consuming Music Together. Springer, Dordrecht 2006
1036. Tarry G: Le problème des 36 officiers. AFAS, Compt 1900–1901
1037. Tatarkiewicz W: A History of Six Ideas: An Essay in Aesthetics. Polish Scientific Publishers, Warszawa 1980
1038. Taylor Y and H Barker: What's It Like to Be Blind? Faking It. `fakingit.typepad.com` 2007
1039. Taylor C: Burning Poles. VHS, Mystic Fire 1991
1040. Taylor C: Silent Tongues. Freedom 1974
1041. Thalmann F: Musical composition with Grid Diagrams of Transformations. Master's Thesis, University of Bern, Bern 2007
1042. Thalmann F: Gestural Composition with Arbitrary Musical Objects and Dynamic Transformation Networks. PhD thesis, University of Minnesota, 2014
1043. Thalmann F and G Mazzola: The Bigbang Rubette: Gestural music composition with Rubato Composer. In: Proceedings of the International Computer Music Conference, International Computer Music Association, Belfast 2008
1044. Thalmann F and G Mazzola: Affine musical transformations using multi-touch gestures. Ninad, 24:58–69, 2010
1045. Thalmann F and G Mazzola: Gestural shaping and transformation in a universal space of structure and sound. In: Proceedings of the International Computer Music Conference, International Computer Music Association, New York City 2010
1046. Thalmann F and G Mazzola: Poietical music scores: Facts, processes, and gestures. In: Proceedings of the Second International Symposium on Music and Sonic Art, MuSA, Baden-Baden 2011
1047. Thalmann F and G Mazzola: Using the creative process for sound design based on generic sound forms. In: MUME 2013 proceedings, AAAI Press, Boston 2013
1048. Thalmann F and G Mazzola: Visualization and transformation in general musical and music-theoretical spaces. In: Proceedings of the Music Encoding Conference 2013, MEI, Mainz 2013
1049. Tormoen D, F Thalmann, G Mazzola: The composing hand: Musical creation with leap motion and the Bigbang Rubette. In: Proceedings of 14th International Conference on New Interfaces for Musical Expression (NIME), London 2014
1050. Terhardt E: Zur Tonhöhenwahrnehmung von Klängen. II. Ein Funktionsschema. Acustica 26, 187-199, 1972
1051. Thomas d'Aquin: Première question disputée. La vérité (De Veritate). (teaching in the years 1256-1259), Paris, Vrin, 2002
1052. Thom R: Stabilité structurelle et morphogénèse. Benjamin, Reading, MA 1972
1053. Tinctoris J: Opera Theoretica, Vol. 2: Liber de arte contrapuncti. Seay A (Ed.): Amer. Institute of Musicology 1975. gopher:`//IUBVM.UCS.INDIANA.EDU/00/tml/15th/tincon2.text`
1054. Tittel E: Der neue Gradus. Doblinger, Wien-München 1959

1055. Titze I R and D T Talkin: A theoretical study of the effects of various laryngeal configurations on the acoustics of phonation. Acoustical Society of America Journal, Vol. 66, No. 60, 1979
1056. Titze I R: Speed, accuracy, and stability of laryngeal movement in singing. Acoustical Society of America Journal, Vol. 115, p. 2590 ff., 2004
1057. Todd N P M: A Model of Expressive Timing in Tonal Music. Music Percep. **3**, pp. 33-58, 1985
1058. Todd N P M: Towards a Cognitive Theory of Expression: The Performance and Perception of Rubato. Contemporary Music Review **4**, 1989
1059. Todd N P M: A Computational Model of Rubato. Contemporary Music Review **3**, 1989
1060. Todd N P M: The Dynamics of Dynamics: A Model of Musical Expression. J. Acoustic Soc. Am. **91**, pp. 3540-3550, 1992
1061. Todoroff T: A Real-Time Analysis and Resynthesis Instrument for Transformation of Sounds in the Frequency Domain. In: ICMA (Ed.): Proceedings of the ICMC 96, San Francisco 1996
1062. Tomasello M: Why Don't Apes Point? In: Enfield N J & S C Levinson (Eds.): Roots of Human Sociality: Culture, Cognition and Interaction. Berg, pp. 506-524, Oxford & New York
1063. Tomasello M, M Carpenter, U Liszkowski: A New Look at Infant Pointing. Child Development, Vol. 78, No. 3. pp. 705-722, May/June 207
1064. Tymoczko D: The Geometry of Musical Chords. Science, Vol 313, 2006
1065. Tymoczko D: Mazzola's Model of Fuxian Counterpoint. In: Agon, C, G. Assayag, E. Amiot, J. Bresson, J. Mandereau (Eds.): Mathematics and Computation in Music. LNAI 6726, Springer, Heidelberg 2011
1066. Uhde J: Beethovens Klaviermusik III. Reclam, Stuttgart 1974
1067. Uhde J and R Wieland: Forschendes Üben. Bärenreiter 2002
1068. Uhde J and R Wieland: Denken und Spielen. Bärenreiter, Kassel et al. 1988
1069. Ungvary T and R Vertegaal: The Sentograph: Input Devices and the Communication of Bodily Expression. ICMA 1995
1070. Utgoff P E and P B Kirlin: Detecting Motives and Recurring Patterns in Polyphonic Music. Proceedings of the International Computer Music Conference, ICMA 2006
1071. van Dalen D: Logic and Structure. Springer, Berlin et al. 1997
1072. Valéry P: Leçon Inaugurale du Cours de Poétique au Collège de France. Gallimard, Paris 1945
1073. Valéry P: Cahiers I-IV (1894-1914). Celeyrette-Pietri N and J Robinson-Valéry (Eds.), Gallimard, Paris 1987
1074. Valéry P: Philosophie de la danse. Oeuvres, tome I, Variété, "Théorie poétique et esthétique", pp. 1390-1403, Gallimard, 1956
1075. Valéry P: Analecta (1926), in *Tel quel*, tome II, Collection Idées, no 241, Gallimard, 1971
1076. Vallès J: "Penser c'est oublier" *Images des mathématiques.* CNRS, `http://images.math.cnrs.fr/Penser-c-est-oublier.html`, 10 juillet 2015
1077. Valtieri S: La Scuola d'Athene. Mitteilungen des Konsthistorischen Instituts in Florenz 16, 1972
1078. van der Waerden B L: Algebra I, II. Springer, Berlin et al. 1966
1079. van der Waerden B L: Algebra I. (English tranlsation) Springer, New York 2003
1080. van der Waerden B L: Die Pythagoreer. Artemis, Zürich 1979
1081. Varèse E: Erinnerungen und Gedanken. In: Darmstädter Beiträge III. Schott, Mainz 1960
1082. Varèse E: Rückblick auf die Zukunft. edition text+kritik, München 1978
1083. Vercoe B: The Synthetic Performer in the Context of Live Performance. In: ICMA (Ed.): Proceedings of the ICMC 84, San Francisco 1984
1084. Veyne P: Comment on écrit l'histoire, 1959. nouv. ed., Points/Histoire 226, Paris 1996
1085. Vico J-B: La méthode des études de notre temps. Présentation et traduction de A. Pons, (De nostri temporis studiorum ratione, 1708), online, site du Réseau Intelligence de la Complexité, 1981
1086. Vieru A: The Book of Modes. Editura Muzicală, Bucarest 1993
1087. Vinci A C: Die Notenschrift. Bärenreiter, Kassel 1988
1088. Violi P: Beyond the body: towards a full embodied semiosis. In: Dirven R, and R Frank (Eds.): Body, Language and Mind. Mouton de Gruyter, Berlin 2006
1089. Vogel M: Die Lehre von den Tonbeziehungen. Verlag für systematische Musikwissenschaft, Bonn-Bad Godesberg 1975
1090. Vogel M: Arthur v. Oettingen und der harmonische Dualismus. In: Vogel M (Ed.): Beiträge zu Musiktheorie des 19. Jahrhunderts. Bosse, Regensburg 1966
1091. Vogel M: Berechnung emmelischer und ekmelischer Mehrklänge. In: Herf F R (Ed.): Mikrotöne II. Helbling, Innsbruck 1988
1092. Vogt H: Neue Musik seit 1945. Reclam, Stuttgart 1972

1093. Vossen G: Datenbankmodelle, Datenbanksprachen und Datenbank-Management-Systeme. Addison-Wesley, Bonn et al. 1994
1094. Vuza D T: Sur le rythme périodique. Revue Roumaine de Linguistique—Cahiers de Linguistique Théorique et Appliquée 22, no. 1 1985
1095. Vuza D T: Some Mathematical Aspects of David Lewin's Book *Generalized Musical Intervals and Transformations*. Perspectives of New Music, vol. 26, no. 1, 1988
1096. Vuza D T: Supplementary Sets and Regular Complementary Unending Canons (Part One). Perspectives of New Music, vol. 29, no. 2, 1991
1097. Vuza D T: Supplementary Sets and Regular Complementary Unending Canons (Part Two). Perspectives of New Music, vol. 30, no. 1, 1992
1098. Vuza D T: Supplementary Sets and Regular Complementary Unending Canons (Part Three). Perspectives of New Music, vol. 30, no. 2, 1992
1099. Vuza D T: Supplementary Sets and Regular Complementary Unending Canons (Part Four). Perspectives of New Music, vol. 31, no. 1, 1993
1100. Vygotsky L S: Thought and Language (1934). Edited and translated by E Hanfmann and G Vakar, revised and edited by A Kozulin. MIT Press, Cambridge 1986
1101. Wanske H: Musiknotation. Schott, Mainz 1988
1102. Waldvogel M. et al.: *presto* source code. TRIMAX, Zürich 1986-1993
1103. Wallas G: The Art of Thought. Harcourt, New York 1926
1104. Wallis J: Tractatus de Sectionibus Conicis. 1655
1105. Walker J S: Fourier Analysis and Wavelet Analysis. Notices of the AMS, vol.44, No.6, July 1997
1106. Ward J: Gregorian Chant according to the Principles of André Mocquereau of Solesmes. The Catholic Education Press 1923 (Reprint)
1107. Waugh W A O'N: Music, probability, and statistics. In: Encyclopedia of Statistical Sciences, **6**, 134-137, 1985
1108. Webern A: *Variationen op.27* (mit Interpretationsanweisungen). UE 16845, Wien 1980
1109. Wegner P: Interactive Foundations of Object-Oriented Programming. Computer, IEEE Computer Soc., October 1995
1110. Weidman A: Singing the Classical, Voicing the Modern: The Postcolonial Politics of Music in South India. Duke University Press 2006
1111. Weil A: Œuvres scientifiques. Collected papers, Vol. I, Springer, New York, 1979
1112. Weisberg R W: Creativity and Knowledge: A Challenge to Theories. In: R J Sternberg (Ed.): Handbook of Creativity. Cambridge University Press, New York 1999
1113. Weyl H: Symmetrie. Birkhäuser, Basel 1955
1114. Weinberger N M: Musica Research Notes. Music and Science Information Computer Archive, Center for Neurobiology of Learning and Memory, U California Irvine, Vol. III, Issue II, Fall 1996
1115. Whiteside A: Indispensables of Piano Playing. Coleman-Ross, New York 1955
1116. Wicinski A A: Psichologyiceskii analiz processa raboty pianista-ispolnitiela nad muzykalnym proizviedieniem. Izviestia Akademii Piedagogiceskich Nauc Vyp., 25 [Moscow], 171-215, 1950
1117. Widmer G: Learning Expression at Multiple Structural Levels. In: ICMA (Ed.): Proceedings of the ICMC 94, San Francisco 1994
1118. Widmer G: Modeling the Rational Basis of Musical Expression. Computer Music J. 18, 1994
1119. Widmer G and W Goebl: Computational Models of Expressive Music Performance: The State of the Art. Journal of New Music Research 33(3), 203-216, 2004
1120. Wiedemann M and E Zimmermann: Kreativität und Form. Springer Vieweg, Berlin Heidelberg 2012
1121. Wieser H-G: Electroclinical features of the psychomotor seizure. Fischer, Stuttgart and Butterworth, London 1983
1122. Wieser H-G and G Mazzola: Musical consonances and dissonances: are they distinguished independently by the right and left hippocampi? Neuropsychologia 24 (6):805-812, 1986
1123. Wieser H-G and G Mazzola: EEG responses to music in limbic and auditory cortices. In: Engel J Jr, Ojemann G A, Lüders H O, Williamson P D (Eds.): Fundamental mechanisms of human function. Raven, New York 1987
1124. Wieser H-G: Musik und Gehirn. Revue Suisse Méd. 7, 153-162, 1987
1125. Wieser H-G and S Moser: Improved multipolar foramen ovale electrode monitoring. J Epilepsy 1: 13-22, 1986
1126. Wiggins G et al.: A Framework for Evaluation of Music Representation Systems. Computer Music Journal 17:3, 1993
1127. Wiil U K (Ed.): Computer Music Modeling and Retrieval (CMMR 2003). Springer, Heidelberg 2004
1128. Wiil U K (Ed.): Computer Music Modeling and Retrieval (CMMR 2004). Springer, Heidelberg 2005

1129. Wille R: Symmetrie — Versuch einer Begriffsbestimmung. In: Symmetrie, Katalogband Vol.1 zur Symmetrieausstellung, Mathildenhöhe, Darmstadt, 1986
1130. Wille R: Personal Communication. Darmstadt 1982
1131. Wille R: Musiktheorie und Mathematik. In: Götze H and R Wille (Eds.): Musik und Mathematik. Springer, Berlin et al. 1985
1132. Wille R: Bedeutungen von Begriffsverbänden. Preprint Nr. 1058, TH Darmstadt 1987
1133. Wille R: Personal Communication. Darmstadt 1985
1134. Winson I: Brain and Psyche. The Biology of the Unconscious. Anchor Press/Doubleday, Garden City, NY 1985
1135. Wilson F R: The Hand: How Its Use Shapes the Brain, Language, and Human culture. Pantheon Books, New York 1998
1136. Witten E: Magic, Mystery, and Matrix. Notices of the AMS, Vol. 45, Nr. 9, 1998
1137. Wittgenstein L: Tractatus Logico Philosophicus – Logisch-Philosophische Abhandlung, Wilhelm Ostwald (Ed.), Annalen der Naturphilosophie, 14, 1921
1138. Wittgenstein L: Tractatus Logico-Philosophicus (1918). Suhrkamp, Frankfurt/Main 1969
1139. Wolfram S: Software für Mathematik und Naturwissenschaften. In: Chaos und Fraktale. Spektrum der Wissenschaft Verlagsgesellschaft, Heidelberg 1989
1140. Xenakis I: Formalized Music. Indiana Univ. Press, Bloomington 1972
1141. Yaglom A M and I M Yaglom: Wahrscheinlichkeit und Information. Deutscher Verlag der Wissenschaften, Berlin, 1967
1142. Yoneda N: On the homology theory of modules. J. Fac. Sci. Univ. Tokyo, Sct. I,7, 1954
1143. Ystad S et al.: Of Parameters Corresponding to a Propagative Synthesis Model Through the Analysis of Real Sounds. In: ICMA (Ed.): Proceedings of the ICMC 96, San Francisco 1996
1144. Yusuke A et al.: Hyperscore: A Design of a Hypertext Model for Musical Expression and Structure. J. of New Music Research, Vol. 24, 1995
1145. Zahorka O: Versuch einer Charakterisierung des altrömischen Melodiestils. Seminar Lütolf, MWS Univ. Zürich 1993
1146. Zahorka O: From Sign to Sound—Analysis and Performance of Musical Scores on RUBATO. In: Enders B (Ed.): Symposionsband Klangart '95, Schott, Mainz 1997
1147. Zahorka O: PrediBase—Controlling Semantics of Symbolic Structures in Music. In: ICMA (Ed.): Proceedings of the ICMC 95, San Francisco 1995
1148. Zahorka O: RUBATO – Deep Blue in der Musik? Animato 97/3, 9-10, Zürich 1997
1149. Zalamea F: Synthetic Philosophy of Contemporary Mathematics. Urbanomic, Falmouth, UK, & Sequence Press, New York, USA 2011
1150. Zarca B: L'univers des mathématiciens. L'ethos professionnel des plus rigoureux des scientifiques. Presses Univ. Rennes, 2012
1151. Zarlino G: Istitutioni harmoniche. Venezia 1558
1152. Zbikowski Z: Conceptualizing Music: Cognitive Structure, Theory, and Analysis. Oxford University Press 2005
1153. Zbikowski Z: Music, Emotion, Analysis. Music Analysis 29/1-3(2010): 37-60
1154. Zekl G: Topos. Meiner, Hamburg 1990
1155. Zlotin B and A Susman: Creative Pedagogy. Journal of TRIZ. no.4, pp. 9-17 (in Russian) 1991
1156. Zweifel P F: Generalized Diatonic and Pentatonic Scales: A Group-theoretic Approach. PNM, 34(1), 140-161, 1996
1157. Zwicker E and H Fastl: Psychoacoustics. Facts and Models. Springer, Berlin, et al. 1999
1158. Zwicker E and M Zollner: Elektroakustik (2nd edition). Springer, New York et al. 1987
1159. Zwiebach B: A First Course in String Theory. Cambridge University Press, Cambridge et al. 2004

Index

2^{Fu}, 56
$B_\epsilon x$, 229
$Dia(Names/\mathcal{E})$, 1437
$H^i(\Delta/G^I)$, 355
$Mono(\mathcal{E})$, 1437
$R^n\dot{@}B\tilde{@}M$, 169
$T_2(R)$, 186
$\coprod$, 44
$\mathfrak{C}^{(3)}$, 266
$\mathcal{A}_2(U,V;W)$, 1399
$\prod$, 44
$\{\}$, 44
!, 1424, 1432
$!/\alpha$, 331
(G), 1386
$(G:H)$, 1379
$(G^I)^\vee$, 275
(I_1/J_1), 519
(I_7/J_7), 519
(K_1/D_1), 519
(K_7/D_7), 519
$(X/C(X))$, 518
$(X[\varepsilon]/Y[\varepsilon])$, 521
$(X|C(X))$, 518
(f/α), 334
$(r_{i,j})$, 1392
(x,y), 1372
$/N$, 296
0_R, 1385
12-$Temp$, 422
1_R, 1385
2, 1375
3CH, 448
3$Chains$, 260
$< N$, 340
$<_{\mathbf{Mod}}$, 76, 78
$?^s$, 1414
$?_e$, 429
$?_{\beta\sigma\mu}$, 122
$A(f,\zeta,\eta,J(\zeta),U(\eta))$, 1451
AK_ξ-topology, 1237
$ALLSER_n$, 195
$ASCII$, 335
AX, 1434
$A@F$, 1423
$A@MOT$, 384
$A@MOT_F$, 384
$A@MOT_{F,n}$, 384
$A@<_M$, 79
$A \times B$, 1372
A^B, 1375
$A^\star$, 1398
A_q, 1406
A_s, 1406
A_Γ, 1457
$A_{t,y,G}$, 788
$A_{threshold}$, 1348
$AbsDyn$, 68, 641
$AbsDynamicEvents$, 641
$AbsTpo$, 341
$AbsTpoEvt$, 341
$AbsTpoInComp$, 341
$Accent_{U,S}$, 632
Ad, 141
$AddTo(c1,c2,at)$, 807
$Alg_{\mathbb{R}}$, 925
$AllExt_B(m)$, 443
$AllExt_B(n,m)$, 443
$AllSet$, 1372
$Alphabet$, 1376
$Alt(s)$, 1468
$AltMaj(i)$, 475
$AltMaj(i)^{(3)}$, 475
$An(B,M)$, 1400
$ArchaicForm$, 84, 85
Arg, 616
$ArtiSlurs_{S,U}$, 632
Aut, 1378
$Aut(K)$, 144
$Aut(c)$, 1420
$Aut_R(M)$, 1391
$Aut_\Phi(S)$, 1093

G. Mazzola, *The Topos of Music II: Performance*, Computational Music Science,
https://doi.org/10.1007/978-3-319-64444-8

$B(\Delta)$, 1402
$B(f, \zeta, \eta, J(\zeta), U(\eta))$, 1451
$BA \models \alpha$, 1434
$BD_V^{mean}(w)$, 648
$BD_V(w)$, 648
BIT, 44, 60
$B\tilde{@}F$, 168
B_*, 946
B_N, 290
B_n, 966
$B_r(x)$, 1449
$B_{/M}$, 1400
$BarLine$, 68
$Basic$, 343
$BeetMotChordFiber$, 347
$BeetMotChordFiberObject$, 348
$BetterForm$, 84
C, 67, 439, 1349
$C(F)$, 55, 1087
$C(G, F, P, w)$, 191
$C(X)$, 518
$C(i)$, 475
$C+$, 439
$C/OnBeetSon$, 347
$C0$, 439
CD, 342, 676
$CINT(S)$, 208
$CINT(Ser)$, 207
$CINT_m(S)$, 208
$CN(G^I)$, 283, 988
$COM(Cont)$, 206
$CONT_{n,k}(X)$, 206
CP, 391
$CT(D)$, 57
C^*-algebra, 1492
$C^*(K; M)$, 1447
$C^0[a, b]$, 1450
C^5, 420
$C^i(X)$, 518
$C^n(K; M)$, 1447
$C^\star(G^I)$, 307
$C_n(R, X)$, 966
C_{G^I}, 288
Cad, 454
Cc, 518
$CcFourier$, 72
CcM, 518
$CellHierarchy_{BP}$, 596
$Cell_U$, 596
$Cell_{BP}$, 596
$Ch/PianoChord$, 340
$Chain_R$, 1476
$Chrono - Fourier$, 235
$ChronoFourierSound$, 235
Ci, 518
CiM, 518
$Cl_H(U)$, 588
Cl_n, 177
$ClassZChord$, 420
$Closed_X$, 1443
Cm, 439
$Cochain_R$, 1477
$ColimCirc$, 65
$CommaZModule$, 422
$Consonant$, 234
$Conv(D)$, 1462
$Count$, 343
$Covens$, 1445
$Crescendo$, 67
$Ct(GestSem)$, 1183
D, 67, 263, 592, 1347
DEG, 457
DF_0, 1451
DNR, 210
$D^r f$, 1451
$D^\mu_\epsilon(M)$, 400
D_K, 1415
D_f, 1412
$D_x(T_0)$, 660
$D_\epsilon(M)$, 397
$D_{k,n}$, 125
$D_{n,m}$, 1213
$Daughters$, 596
Dc, 518
DcM, 518
$DenOrb(R, n, B, M)$, 170
$Der(L)$, 1408
$Der(X)$, 1469
Df, 1451
Df_x, 1451
Dg, 457
Dg^*, 457
Di, 518
DiM, 518
Dia, 387
$Dia(Names)$, 1437
$Dia\pi$, 392
$Dia^*(Y/\mathcal{E})$, 1089
Dia_k, 387
Dil, 1392
Dir, 656
$Dir(Q)$, 1376
$Dom(X)$, 380
Dp, 457
$Dup(T)$, 335
$Duration$, 67
$DurationValue$, 809
$DynSymb$, 68
E, 67, 1347
$E2M$, 62
$EHLD$, 631, 787
$EHLDGC$, 658

$EMB(A,B)$, 208
ENH, 422
EX, 1432
$E \rightsquigarrow F$, 109
E_N, 1209
$E_{n,m}$, 1213
Ed_{Rg}, 389
$Ed_{\Delta Rg}$, 389
$El(G)$, 403
$El(M)$, 403
$Elast$, 388
$End(A)$, 141
$End(c)$, 1420
$End_R(M)$, 1391
$Envelope$, 71
$EulerChord$, 92
$EulerChordEvent$, 93
$EulerModule$, 62, 90, 92
$EulerModule_q[\varepsilon]$, 508
$EulerPlane$, 93, 96
$EulerZChord$, 420
$Euler\mathbb{R}Module$, 563
$Evts_S(U)$, 630
Ex, 422
$Ex(D, L_{\alpha,\beta})$, 306
$Ex(GestSem)$, 1183
$Ex(L/K)$, 1461
$ExTop_A(F)$, 430
$Ext^n_R(Z,W)$, 945
$Ext_A(M)$, 427
$F \dashv G$, 1423
$F(D)$, 57
$FCM(Alphabet)$, 1376
FD, 676
$FG(Alphabet)$, 1379
$FM(Alphabet)$, 1376
FM-$Object$, 73
$FO(PEX)$, 1435
$FORMS$, 1438
FSH, 343
F_0, 1451
F_x, 1411
F_π, 554
F_{sym}, 554
$Fermata$, 643
$Fermata_{E,S}$, 632
$Fib(e.)$, 758
$Field_U$, 597
$Field_{U,R}$, 597
Fin, 1426
$Fin(Onset)$, 81
$Fin(Pitch)$, 81
$Fin(S)$, 1374
$Flatten$, 75
$Flatten^n$, 75
$Fold$, 1103
$ForSem$, 1092
$FormList(S)$, 60
$Fourier$, 70, 72
$FourierSound$, 71
$Frame_U$, 596
$Func(\mathbf{C},\mathbf{D})$, 1422
$Fund(H)$, 588
$Fushi$, 343
$FushiOrnament$, 343
$FushiPic$, 342
$FushiSTRG$, 343
G, 67, 913, 1350
$G(K)$, 941
$G.x$, 257
G/H, 1379
$G/\Omega \pitchfork \Omega^*$, 633
$GL(n,p)$, 1382
$GL_n(R)$, 1382
$G[f]$, 945
$G \wr H$, 1380
G^I, 252, 987
G^I*, 289
G^{opp}, 1378
G_m, 1378
$G_{v,k,l}$, 185
$Gal(L/K)$, 1460
$Ges^P_t(M)$, 391
$Ges^\mu_t(M)$, 400
$Ges_t(M)$, 391
$GestSem$, 1183
GlC^t, 557
$Glissando$, 67
$GlobPerfScore_{BP.}$, 599
$GlobPerfScore_{BP}$, 598
$Grass_{r,n}$, 1416
H, 67, 588, 1339, 1350
$HA \models \alpha$, 1434
$H\backslash G$, 1379
$H \lhd G$, 1379
H^*-emergent property, 1062
$H^i_\Delta(G^I)$, 355
$H^n(K;M)$, 1447
$H^q(C)$, 1477
$H^\star(G^I)$, 307
H_X, 922
$H_X(x,y)$, 922
$H^\star_l(X^{MaxMet})$, 379
$H_n(K.c.)$, 970
$H_n(R,X)$, 966
$H_q(C)$, 1475
$H_k(f)$, 986
$H^{net}_{n-1}(R,\mathcal{C},\Gamma_*)$, 986
H_{prime}, 1351
$HarMin$, 474
$HarMin^{(3)}$, 474
Hom, 1378, 1385, 1419

Hom^{loc}, 281
I, 573, 592
$I(F)$, 55, 1087
$I(H)$, 577
$ICV(A, B)$, 207
ID, 593
III_X, 262
II_X, 262
$INT(Ser)$, 207
$INT_m(Ser)$, 207
$IV(A, B)$, 208
IV_X, 262
$I \mid i$, 269
I^n, 966
I^q, 1476
I_U, 596
I_X, 262
I_x, 511
Id_A, 141, 1372
Id_e, 1420
$Idempot(M)$, 1376
$Im(f)$, 1375
$Importance$, 342
$InTop(F)$, 429
$InitPerf_U$, 597
$InitSet_U$, 596
$InstruName$, 69
$Int(E)$, 427
$Int(M)$, 427
$IntMod_{12,q}[\varepsilon]$, 511
$IntMod_{12}$, 510
$IntThirds_{3,4,q}[\varepsilon]$, 511
$IntThirds_{3,4}$, 510
$Int^e(M)$, 427
Int_A, 430
Int_g, 1379
$Intonation$, 563
$Item(F)$, 809
J-manifold, 1062
J-shape, 1062
$J(X)$, 1247
$J(\mathfrak{Z}_E, w)$, 654
JCK, 342
$JCKFU$, 344
$J_K(X)$, 1431
J_{Kt}, 266
$K©L$, 939
$K(X)$, 1430
$K.c.$, 967
$KU_{s,t}$, 120
$K_1 \curlywedge K_2$, 178
K_4, 206
K_E, 575
K_b, 708
K_{Tops}, 631
$Ker(f)$, 1379
$Kernel_U$, 597
$Knot$, 73
$Knot_{Basic}$, 74, 270
$Knot_{PerOns}$, 503
Kq, 62
Kt, 62
$K|k$, 1445
L, 67, 1348
$LARM$, 974
LB, 804
LCP, 391
$LPS^{BP,k}_{Instrument}$, 599
$LPS^{BP.}_{Instrument}$, 599
$LPS^{BP}_{Instrument}$, 599
$L_F f$, 1455
$L_\psi f(a, b)$, 236, 1344
$L_{n,1}$, 922
$Label$, 342
$LegatoSlur_U$, 632
$LegatoSlur_{U,S}$, 632
Lev, 269
$LimCirc$, 65
$Limint(F)$, 259
Lin_R, 1391
$Lin_R(G^I, M)$, 289
$List(F)$, 809
$ListEntry_F$, 343
$List_F$, 343
$LocC^t$, 552
$LocPerfScore_{BP}$, 595
Loc^F, 91
$Loudness$, 67
M, 913
M/N, 289
MOT, 384
MOT_F, 384
MOT_n, 384
$MOT_{F,n}$, 384
$M[\varepsilon]$, 107
$M@F$, 1437
$M@$, 1397, 1422
$M \mid f\overset{\iota}{/}Id_A$, 289
$M^\star$, 1376
M_n, 207
M_q, 1406
M_s, 1406
M_t, 449
$M_x(T_0)$, 660
M_{12}, 260
M_{OP}, 384
M_O, 384
$M_{[\varphi]}$, 1392
$M_{\mathbb{C}}$, 994
$M_{d,a}$, 444
$M_{i,j}$, 787
$Maelzel$, 341

Maj, 474
$Makro_{Basic}$, 74, 270
$Makro_{PerOns}$, 503
Man^r, 1466
$Marcato_{U,S}$, 632
$Match_P$, 1432
$MathPitch$, 61
$Max(X)$, 268
$MaxMet(X)$, 268
$Media$, 342
$MelMin$, 474
$MelMin^{(3)}$, 474
$Min(L, S)$, 581
$Min(X, S)$, 581
Mod, 1111
$ModelForm$, 84
$Modulator$, 1115
$MonEnd(F)$, 427
$Mono(\mathcal{E})$, 1089
$Mor(\mathbf{C})$, 1420
Mos^G_n, 310
$Mos^G_{n,\lambda}$, 311
$Mos^G_{n,k}$, 310
$Mother$, 596
$Mt/Motif$, 340
$N(D)$, 57
$N(F)$, 54, 1086
$N(U)$, 1446
$NComp$, 341
NF, 209
$N \rtimes_\phi H$, 1380
N_e, 1376
$N_k(U)$, 1446
N_{red}, 300
$Nat(F, G)$, 1422
$NatMin$, 474
$NatMin^{(3)}$, 474
$NewContainer$, 807
$Note$, 1111
$NoteGroup$, 340
$Number$, 342
OPD, 120
$OPLD_{\mathbb{Z}}$, 437
OS, 552
$Ob(\mathbf{C})$, 1420
$ObExTop_B(F)$, 433
$ObGl$, 987
$ObLoc$, 987
$OnMod_m$, 63
$OnPiMod_{m,n}$, 63
$Onset$, 67, 93
$Onset??_S$, 67
$OnsetProj$, 347
Op, 616
$Open(G)$, 335
$Open(\mathcal{S})$, 1443
$Open_X$, 1411, 1443
$Open_f$, 1411
$Open_{X,x}$, 1411
$Open_{f,x}$, 1411
$OrchSet$, 69
$Overtone$, 1115
$P(E)$, 525
$P(Q)$, 1388
$P(Q)/\sim$, 1420
PEX, 1435
PF, 210
PO, 502
$P \wedge Q$, 346
P^+, 1432
$P_{Träumerei}$, 720
$Para$, 99, 100
$Part(I)$, 310
$Part(a., b.)$, 1462
$Partial$, 72, 1122
$Path(\Gamma)$, 1458
$Pause$, 69
$Per(R)$, 312
$PerCell$, 1254
$PerOns$, 502
$PerOns_{12}$, 503
$Percussion$, 502
$Periods_w$, 1340
$PhysCrescendo$, 68
$PhysDuration$, 68
$PhysGlissando$, 68
$PhysInstr_i$, 555
$PhysLoudness$, 68
$PhysOnset$, 68
$PhysOrchestra$, 555
$PhysPitch$, 68
$PhysicalBruteForceOperator$, 651
$PiMod_{12,(7)}$, 508
$PiMod_n$, 63
$PiThirds_{3,4}$, 510
$PianoSelector$, 47
$Piano\text{-}Note$, 44
$Pitch$, 67
$PitchChange$, 343
$PitchSymb$, 68
$PowerCirc$, 65
$Ptch$, 117
$PythagorasLine$, 460
$QNormalize(c)$, 807
$QReduce(c, [n])$, 807
Q_w, 653
$Q_{\beta\sigma\mu}$, 122
$Q_\theta(E_1, E_2)$, 526
$Q_\theta(K/D)$, 526
$R(Para)$, 267
$R.S$, 106
$R.f$, 143

REd, 389
RT, 809
RTC, 370
$RULES$, 1434
$R[\varepsilon]$, 1387
R-$Gestoid$, 923
$R\langle M\rangle$, 1386
$R\langle Q\rangle$, 1388
RC, 923
R-H_X, 923
R^*, 503
R^*_{Ons}, 503
R^2, 727
R^C, 1392
$R^\times$, 1385
R_R, 1391
R_{G^I}, 289
R_{Vuza}, 312
R_{max}, 596
R_{min}, 596
$Rad(M)$, 1395
$Rad(f)$, 926
$Rad(x)$, 925
Rat, 616
$Rate$, 341
$RelDyn$, 68, 641
$RelDynamicEvents$, 641
$Remainder_{Split} \propto \mu$, 649
$RemoveFrom(c1, c2)$, 807
$Review$, 342
Rg, 386
$Rg_o(M)$, 386
$Rg_p(M)$, 386
$Rhythm(Para)$, 267
$S!$, 344
S, 263
$S(E)$, 525
$S(EX)$, 1432
$S(p, u.)$, 1381
$S/\sim$, 1373
$SEG_k[M^{1/n}]$, 207
$SERM_n$, 125
$SERM_{k,n}$, 125
SER_n, 125
$SER_{k,n}$, 125
$SO_2(71)$, 786
SPE, 234
$S[Alphabet]$, 1386
$S\langle Alphabet\rangle$, 1386
$S\otimes_R?$, 1397
S^1, 921, 950, 971
S^3, 922
$S^{-1}A$, 1406
$S^{-1}M$, 1406
$S^\diamond$, 101
$S_n(K)$, 1447
S_q, 508
$S_q[\varepsilon]$, 508
$S_t^{M_t}$, 449
Sat, 370
Sat_B^F, 434
$Satellite$, 668
$SatelliteLevel$, 1108
$ScalarOperator_w$, 654
$ScoreForm$, 596
$ScoreInstr_i$, 555
$ScoreOrchestra$, 555
Sem, 1092
$SemInTop(F)$, 429
$Sema(\mathcal{E})$, 1439
$SemiEnd(F)$, 427
$Semitone$, 1350
Sg, 457
$Sg(GestSem)$, 1183
$Sh(\mathbf{C}, J)$, 1431
$Simplex_U$, 596
$Simplexes_U$, 596
$Sound$, 121
Sp, 457
$Sp(X)$, 378
$Sp(x)$, 269
$Sp_\mu(x)$, 385
$Spec$, 1412
$Spec(A)$, 1412
$Spec(R)$, 148
$Spec(f)$, 1412
$Special$, 343
$Split \propto \mu$, 649
$Split_{U,\nu}$, 649
$Staccatissimo_{U,S}$, 632
$Staccato_{U,S}$, 632
$StepTune$, 644
$Sub(X)$, 1375
$Sub(BP)$, 588
$SubGal(L/K)$, 1461
$Sub_{M*}(N)$, 395
$Support$, 1336
$SupportForm(E)$, 339
$Switch$, 60
$Syllabic$, 234
$Sym(K)$, 144
$Sym_i(A, B)$, 208
$SymbolicBrueForceOperator$, 650
$SynCirc$, 64
T, 263, 592, 1338
$T(E)$, 575
$T(F)$, 55, 1086
$T(G)$, 1381
$T(p., u.\cdot)$, 1381
TF, 447
$TF_{f,t}$, 447
TG^I, 558

TID, 593
TI_n, 125
TK, 551
TO, 1451
$TRUTH(F)$, 436
$TRUTH(I)$, 335
$TRUTH(h)$, 336
TTO, 207
TX, 1467
T^r_s, 1468
$T^r f$, 1451
$T^t K$, 551
T_0, 660, 1445
T_1, 1445
T_2, 1445
T_S, 927
$T_S(x, y)$, 927
T_f, 927
$T_k K$, 551
$T_{2,\mathbb{R}}$, 391
$T_{X,x}$, 927, 1415
T_Λ, 660
T_{TON}, 484
T_ε, 107
$T_{\zeta,\eta,f}$, 1452
T_{red}, 107
$Tempo$, 562
$TempoOperator_w$, 653
$Tension(\lambda^{\cdot}, \mu_{.}, \omega, \phi_{min})$, 484
$Tenuto_{U,S}$, 632
$Terminal$, 809
Tf, 551, 1451
Tf_k, 552
Tg, 457
$TimeSig$, 69
$TimeSig(p/q)_S$, 632
$Title$, 342
$Top(H)$, 588
Top_S, 631
$TorSeq$, 388
$Toroid^{m,l}_\gamma$, 388
Tp, 457
$Tr(D, T)$, 522
$Trans$, 312
$Trans(D, T)$, 522
$Triv(K)$, 144
$Tune$, 644
$Types$, 1437
$UNICODE$, 335
$U \boxtimes V$, 1400
U^n, 597
U^o, 1443
U_g, 118
U_n, 207
U_s, 118
U_x, 262
U_{n+1}, 597
$U_{x/x+1}$, 262
$Utai$, 344
$V(E)$, 1412
$VF(O)$, 1455
VII_X, 262
VI_X, 262
V^*, 1212
V_X, 262
V_Γ, 1457
$V_{n,x}$, 175
$Val(I)$, 335
$ValCh(p/q = r/s)_S$, 632
$VerbAbsTpo$, 341
$Vowel$, 234
$WP_n(U, \mathbb{T}_\mathbb{R})$, 597
W^*, 498
$W^R_{n,i}$, 631
$Weight(U)$, 597
$WeightList_{BP}$, 597
$Weight_n(U)$, 597
$Weight_{BP}$, 597
$X(G, M)$, 913
$X(R, n, B, M)$, 170
$X(harmo)$, 725
$X(melod)$, 725
$X(metric)$, 725
$X@F$, 1397
X-$chromatic$, 420
X-$harm$, 420
X-$major$, 420
X-mel, 420
X^*, 1488
$X^{(3)}$, 262
X^{MaxMet}, 268
$X^{MetLg[L]}$, 268
$X^{MetPer[P]}$, 268
$X^\star$, 1397
$X^{cyc(g)}_{.}$, 1382
X_6, 459
X^r_n, 175
X_{add}, 725
Y☞Z, 946
$Y_{\mathbf{Aff}}$, 1414
$Y_{\mathbf{ComRings}}$, 1414
$Year$, 342
$Yoneda$, 982
$Z(R)$, 1385
$Z(harmo)$, 725
$Z(metric)$, 725
ZNF, 210
Z_n, 966
$[C/C(X)]$, 518
$[C|C(X)]$, 518
$[K)$, 231
$[K]$, 193

$[L:K]$, 1460
$[M)$, 402
$[M^{1/n}]$, 206
$[R]$, 335
$[S]$, 1422
$[X, A; Y, B]_{X'}$, 1474
$[\frac{n}{i}]$, 311
$[\mathcal{X}_C, \mathcal{X}_F]$, 1246
$[a, b]$, 457
$[b, p, g]$, 267
$[f]$, 943
$[l|x|k]$, 110
$[s]$, 1373
$[x)$, 402
$[x]$, 1454
$[xy]$, 444
&, 1432
$\mathbb{A}^n$, 1415
$\vec{?}$, 944
$\vec{F}$, 942
$\vec{X}$, 914, 1457
$\vec{@}$, 916, 941
$\vec{f}$, 914, 1458
Ab, 1411
Aff, 1412
$\|f\|_1$, 1450
$\|f\|_2$, 1450
$\|f\|_\infty$, 1450
$\|x\|_1$, 1450
$\|x\|_2$, 1450
$\|x\|_\infty$, 1450
$\| \ \|$, 1450
$@M$, 1397, 1422
@, 53, 1396
$@^{red}_R M$, 1401
$@_R$, 1396
$@_{loc}X$, 280
$\mathbb{C}$, 1385
$\mathbb{C}_m$, 1112
CHR, 44
CatMan, 993
Colimit, 1437
$ComGlob_A$, 285
$ComLoClass^{gen,M}_{n+1,End(B)}$, 170
$ComLoClass^{gen,lf,sp}_{n+1,0_R}$, 177
$ComLoClass^{gen}_{n+1,O_R}$, 173
$ComLoClass_{n+1,0_R}$, 177
$ComLoc$, 979
$ComLoc_A$, 142
$ComLoc^{emb}_A$, 143
$ComLoc^{gen}_A$, 143
$ComLoc^{in}_A$, 143
ComMod, 1412
ComRings, 1411
$\mathbf{ComRings}_@$, 1415
ΔRg, 386
$\Delta n\Gamma(G^I)$, 294
Δ, 565, 628, 1432
Δ-formula, 926
$\Delta(G^I)$, 989
Δ_d, 1461
Δ_n, 173, 293
Δ_q, 1476
Δ_{G^I}, 294
$\Delta_{i,j}$, 1403
Den, 331
$Den(x, y)$, 331
$Den_{\mathbf{Colimit}}$, 331
$Den_{\mathbf{Limit}}$, 331
$Den_{\mathbf{Power}}$, 331
$Den_{\mathbf{Simple}}$, 331
$Den_{\mathbf{Syn}}$, 331
Den_∞, 335
Den_∞/sig, 338
ⓓ, 975
Digraph, 914, 938, 1457
$FlatLocNet(\mathcal{C})$, 981
$\mathcal{F}_\psi$, 928
$Formula$, 926
$ForSem$, 1440
ΓB_f, 291
$\Gamma \Lambda^n_f$, 291
Γ, 290, 1364, 1412
Γ-machine, 1038
$\Gamma(G^I)$, 288
$\Gamma(U, F)$, 1412
$\Gamma(f \overset{\iota}{/} Id_A)$, 290
$\Gamma(myFMObject)$, 73
$\Gamma \vec{@} K$, 941
$\Gamma \vec{@} X$, 937
$\Gamma \vec{@}_A \Sigma^I$, 998
$\Gamma \circlearrowright \mathcal{C}$, 984
$\Gamma*$, 1457
$\Gamma^G \vec{@}_A \Sigma^I$, 998
Γ^w, 239
Γ_t, _k386
Γ_t, 385
$\Gamma_t(M, j)$, 309
Γ_{Dir}, 660
$\Gamma_{RedIndia,c}$, 387
$\Gamma_{t,n}$, 385
$Gesture$, 937
$Gesture(F)$, 942
$Gesture(K)$, 941
$_\mu Glob$, 352
$GlobalDigraph$, 996
Grp, 1411, 1421
$Graph$, 1458
$\mathbb{H}$, 1385
$Heyting$, 930
LCTop, 1255
$\Lambda \downarrow$, 628

$\Lambda\uparrow$, 628
$\Lambda\updownarrow$, 628
$\Lambda^s T$, 1468
Λ^∞, 628
$\mathbf{Lie}_R$, 1408
Limit, 1437
LinMod, 1421
$\mathbf{LinMod}_R$, 1421
Loc, 136
$Loc_{End(A)}$, 141
$Loc_{@A}$, 137
$_\mu Loc$, 352
LocNet, 981
$LocNet(\mathcal{C})$, 981
LocRgSpaces, 1411
$\mathbb{M}_R$, 1420
$\mathbb{M}_{m,n}(R)$, 1392
$\mathbf{Mod}^{@}$, 961, 979, 1397
Mod, 1397, 1411, 1421
$\mathbf{Mod}_R$, 1397, 1421
Mon, 1411, 1421
$\mathbb{N}$, 1371, 1420
$_\mu ObGlob$, 352
ObLoc, 132, 979
$_\mu ObLoc$, 352
$ObLoc_{End(A)}$, 141
$ObLoc_{@A}$, 137
$\Omega \propto \mu$, 633
Ω, 598, 628, 720, 1428, 1459
$\Omega(p)$, 483
Ω_{Sh}, 1431
Ω_ω^X, 723
Φ, 1432
Π, 610, 1435
$\Pi_1(X)$, 920
Π_w, 191
Power, 1437
Ψ, 610
$\mathbb{Q}$, 61, 62, 1371, 1385
$\mathbb{Q}_m$, 1112
$\mathbb{R}$, 44, 61, 1371, 1385
$\mathbb{R}\Delta$, 926
$\mathbb{R}_m$, 1112
$\mathbb{R}_{[\mathbb{Q}]}$, 61
RadicalDigraph, 926
$\Rightarrow$, 1342, 1433
$@\mathbb{R}$, 83
Rings, 1411, 1421
$BC(A, W)$, 1453
BP, 588
$B(A, W)$, 1452
$C_1 \times C_2|K$, 587
$GlDiff^t$, 557
Glob, 275
$LocDiff^t$, 552
ObGlob, 275
Tan_R^t, 552
Schémas, 1414
Ens, 1054
$\mathbf{Ens}^\sigma$, 1028
$\mathbf{Ens}_U$, 1420
$\Sigma_I M_i$, 1391
Simpl, 1445
SinLoc, 141
$SinLoc_{End(A)}$, 141
$SinLoc_{@A}$, 141
Sob, 1414
SpaceDigraph, 914
Simple, 1437
denotator, 1112
Syn, 1437
TopCat, 939, 961, 1085
$\overrightarrow{\mathbf{TopCat}}^{©}$, 962, 1086
TON, 447, 448, 483, 485
$\mathbb{T}(h)$, 337
$\mathbb{T}_I$, 335
$\mathbb{T}_I^A$, 335
Tex, 334
$Texig(Den)(h)$, 337
$Texig(Den)_I$, 334
Top2, 1474
Top, 993, 1085, 1421, 1444
©, 939
UNICODE, 1090
Υ, 1435
VAL, 447, 448, 483, 485
VAL_{mode}, 483
VAL_{type}, 483
Ξ, 1432, 1435
Ξ_D, 419
Ξ_K, 419
$\mathfrak{Z}$, 568, 1254
$\mathbb{Z}$, 44, 1371, 1385
$\mathbb{Z}_m$, 1112
$\mathbb{Z}_n$, 1381, 1386
$\mathbb{Z}_2$, 60
$\mathbb{Z}_{71}^4$, 784
$\mathfrak{Z}_{(x,y)}^t$, 1191
$\mathfrak{Z}_{\Lambda,Dir}$, 656
$\mathfrak{Z}_{\Lambda,U}$, 659
$\mathfrak{Z}_\Lambda$, 656
$\mathfrak{Z}_{ex}$, 591
α, 524
α_+, 508, 567
α_-, 508
α_+, 108
α_-, 108
$\bar{X}$, 402
$\bar{z}$, 1385
β, 524
$\beta(x)$, 291
$\bigcap V$, 1372

$\bigcup V$, 1372
$\bigoplus_I M_i$, 1392
$\bigwedge^{\oplus k} \mathcal{L}$, 975
$\perp$, 336, 1433
$\boxtimes$, 1400
$\bullet$, 804
$\frown$, 283
$\check{?}_{@A}$, 137
$\check{S}$, 101
χ, 418, 419
$\chi(Y)$, 1375
$\chi^\star$, 418
χ_σ, 1428, 1459
$\coprod_I M_i$, 1375
δY, 264
δx, 1481
δ, 524, 610
$\delta(X/Y)$, 520
$\delta(Y|X)$, 520
$\delta[X|Y]$, 520
$\delta_{@A}$, 294
δ_{ij}, 1392
$\dot{f}/Id_A$, 295
$\dot{s}$, 295
$\varnothing$, 146, 1371
$\varnothing_R$, 1397
$\exists x$, 1436
$\exists_x P(x,y)$, 346
$\forall x$, 1436
$\forall_x P(x,y)$, 346
$\mathfrak{D}$, 1412
$\mathfrak{H}$, 593
$\mathfrak{MusGen}$, 133
$\mathfrak{M}$, 1436
$\mathfrak{M} \models \alpha[x]$, 1436
$\mathfrak{M}_0$, 459, 952
$\mathfrak{M}_1$, 459
$\mathfrak{M}_2$, 459
$\mathfrak{S}_M$, 1378
$\mathfrak{T}$, 399
$\mathfrak{T}/Ges$, 399
$\mathfrak{T}^{sp}_{Toroid}$, 403
$\mathfrak{T}^{sp}_{Toroid}/Ges$, 403
$\mathfrak{T}_{\mu/Ges}$, 399
$\mathfrak{T}_\mu$, 399, 400
$\mathfrak{T}_\mu/Ges$, 400
$\mathfrak{T}_{t,P,d}$, 397
$\mathfrak{T}_{t,P,d}/Ges_t$, 398
$\mathfrak{W}$, 598
$\mathfrak{gl}(L)$, 1408
$\mathfrak{gl}(n,R)$, 1408
$\mathfrak{h}$, 590, 628
$\mathfrak{h}|U$, 590
$\mathfrak{sl}(L)$, 1408
$\mathfrak{sl}(n,R)$, 1408
γ, 610
$\gamma_i(t)$, 911, 915
$\overrightarrow{GL}(n,p)$, 1382
$\hat{?}$, 56, 131
$\hat{b}$, 721
$\hat{f}$, 131, 1344
->, 1432
$\in$, 1371
$\int_{\mathcal{C}} F$, 1061
$\int_c \omega$, 975
$\int_{\mathbf{C}} F$, 1424
$\int_x f$, 1454
ι, 628
ι_j, 1375, 1392
κ, 454
$\kappa(x)$, 1412
κ_J, 455
κ_{RelDyn}, 642
κ_{orb}, 460
$\langle S \rangle_e$, 1376
$\langle S \rangle$, 423, 1376
$\langle S \rangle$, 1422
$\vdash_{CL} \alpha$, 1434
$\mathbb{D}$, 1021
$\mathbb{L}_4$, 1026
$\mathbb{T}^2$, 966
globfact, 992
Frame$_n$, 983
Frame$_{X.}$, 1254
Sel, 989
$\mathcal{B}$, 1067, 1070, 1443
$\mathcal{B}^n$, 1488
$\mathcal{C}^r$, 1451
$\mathcal{C}^{☞}$, 946
$\mathcal{C}^\tau$, 982
$\mathcal{C}_{Im}$, 1289
$\mathcal{C}_{Ph}$, 1290
$\mathcal{D}(\mathcal{C})$, 1031
$\mathcal{D}_\Gamma \mathcal{X}$, 1037
$\mathcal{E}$, 1437
$\mathcal{E}(x,y)$, 1186
$\mathcal{F}(O)$, 1455
$\mathcal{F}X$, 1469
$\mathcal{G}$, 1067, 1071
$\mathcal{L}$, 1482
$\mathcal{L}(X,Y)$, 1487
$\mathcal{L}(x,y)$, 1185
$\mathcal{L}^p(X, d\mu)$, 1489
$\mathcal{M}$, 172
$\mathcal{M}_r$, 175
$\mathcal{P}$, 1290
$\mathcal{R}$, 1437
$\mathcal{T}^r_s(X)$, 1468
$\mathcal{V}(X)$, 1468
μ, 628
∇, 939
$\neg$, 1433

ω, 349
$\overline{Y}$, 1443
$\overline{f}$, 1447
$\partial Dynamics$, 567
$\partial Intonation$, 567
$\partial Tempo$, 567
∂U, 590, 1443
∂Z, 568
$\partial \text{צ}_{\lambda,ex}$, 591
$\partial f_i/\partial x_j$, 1451
$\partial^Y R$, 580
$\partial_Y R$, 580
∂_q, 1475
∂_n, 966
$\phi(n)$, 1381
πY, 264
$\pi_1(X)$, 920
$\pi_1(X,x)$, 921
π_j, 1392
$\pi_{m,l}$, 393
$\prod_I M_i$, 1374
$\psi_{a,b}$, 236
ρ_D, 1337
$\searrow (g)$, 944
σ-algebra, 1488
$\sigma(X/Y)$, 520
$\sigma(X|Y)$, 520
$\sigma[X|Y]$, 520
$\sim$, 1373
$\sim_P s$, 1362
$\sqrt{x}$, 1376
$\sqsubset$, 398, 1388
$\sqsubset_i^{\lambda^i}$, 482
$\sqsubset_i^{\lambda^i}$, 1388
$\square$, 1400
$\subset$, 1371
$\subseteq$, 1371
$\tau(X,D,S)$, 581
$|Ser|_1$, 229
$|Ser|_2$, 229
θ, 524
$\tilde{A}$, 1412
$\tilde{X}_n^r$, 175
$\times$, 177
$\top$, 336, 1433
$\vec{T}_2(R)$, 185
$\underline{U}$, 591
$\underline{\varDelta}$, 628
צ, 591
$\uparrow$, 915, 1457
$\uparrow^n$, 1457
$\models \alpha$, 1433
φ_x, 661
$\vee$, 1433
$\wedge$, 1433
$\wedge^r$, 1407
$\widehat{M}$, 434
$\wp_{score}$, 912
$]x[$, 265
${}^+\tau_x$, 263
${}^0\tau_x$, 263
${}^G X$, 965
${}^R Glob$, 354
${}^R Glob_A$, 354
${}^{loc}Loc^{@}$, 281
${}_0\mathbf{C}$, 1420
${}_1\mathbf{C}$, 1420
${}_A\varDelta_n$, 293
${}_R R$, 1391
${}_R Sat_A^F$, 434
${}_\mu^R Glob$, 354
${}_\mu^R Glob_A$, 354
$|m|_\alpha$433
$a \sqcup_c b$, 1424
$a \times_c b$, 1424
$abcard_t(M)$, 386
$abcard_t(m)$, 386
$ad(x)$, 1408
ad, 1375
add, 335
at, 448
$b \diamond f$, 721
b_*, 945, 946
$book$, 65, 72
$bottom(x)$, 1390
c, 1348
$c\text{-}space_n(X)$, 205
c^5, 420
c^m, 766
$c^{(3)}$, 461
$c_h^{(3)}$, 461
$c_m^{(3)}$, 461
c_{har}, 474
c_{mel}, 474
c_{nat}, 474
$card(A)$, 1373
$causalEnd$, 767
$causalStart$, 767
$char(F)$, 1386
$codom(f)$, 1419
$colim(\varDelta)$, 1424
$coord(F)$, 1438
$ct_{\mu,\epsilon}(M)$, 409
$cyc(g)$, 1382
d, 389, 1347
dB, 1348
$d\omega$, 973
d^*, 1449
$d_E^2 x$, 721
d_P^*, 394
d_n, 1447
d_t, 389

$d_{1,c}$, 390
$d_{2,c}$, 390
$d_E x$, 721
$d^*_{P,n}$, 394
$d_{\infty,c}$, 390
$deformLiGr$, 766
$den(E)$, 339
df, 1455, 1469
$dim(K)$, 1445
$dim(M)$, 1392
$dom(f)$, 1419
$drap(k)$, 170
dw, 1470
e, 1347
e^M, 1396
e^m, 1396
e_z, 756
enh, 425
ev_p, 148
ex_z, 431
$exp(F,G)$, 334
$exp(ad(x))$, 1409
$ext(E)$, 337
f/α, 129
$f : a \rightarrow b$, 1419
$f@S$, 102
$f \circ g$, 1419
$f \overset{\iota}{/} \alpha$, 274
$f \overset{\iota}{/} \alpha \star M$, 288
$f^{-1}(C)$, 1375
$f_* F$, 1411
$f_{S,T}$, 1406
$finalEnd$, 767
$finalStart$, 767
$frame(F)$, 1438
$fun(F)$, 1438
g, 457, 1350
$g \circ f$, 1372
g_{U,w_s,w_p}, 1225
glb, 347
$grad(f)$, 1455
$groundclass(C^I)$, 313
h, 1349, 1350
h_x, 1411
i, 1432
$i < x, y >$, 207
$i\{x, y\}$, 207
$ic < k >$, 207
$ic\{k\}$, 207
$id(F)$, 1438
$intex_{z,B,G}(Ch)$, 459
$intex_{z,B}$, 432
$ip < a, b >$, 207
$ip\{a, b\}$, 207
is_X, 378
$iso\cap$, 283
j, 1414
jak, 213
jak_π, 218
k-spectroid, 925
k-*Spectroid*, 925
k-*Contra*, 387
k_s, 119
key, 1350
l, 1375
$l(Cont)$, 206
$l(M)$, 268, 1395
$l^2(\mathbb{N})$, 1486
$lambda_w(x)$, 259
lev, 1374
$lev(x)$, 269
lev_i, 269
$lim(\Delta)$, 1424
$lim(\mathbf{D})$, 1387
loc, 280
lub, 347
m^{-1}, 1378
m_x, 1411
m_{12}, 260
mod_o, 420
mod_p, 95
$monex_z$, 431
$myFM^w$, 239
$myFM_w$, 239
$n(U)$, 1445
nW, 409
$nW_\epsilon(M)$, 409
$n\Gamma(\sigma)$, 288
$n_k(U)$, 1445
$n_\infty(K^I)$, 259
$newset$, 146
o, 62
o-*TempClass*, 93
$obintex_{z,B}$, 433
p, 457
$p(M)$, 268
p-*ClassChord*, 93
p-*EulerClass*, 93
p-*Scale*, 95
p^Q_x, 1335
$p^{BP}_{Instrument}$, 599
p_B, 567
p_X, 1467
p_j, 1374
p_{Onset}, 100
$p_{U,V}$, 588
p^x_Δ, 523
$p_{\beta\sigma\mu}$, 121
p_{cf}, 508
p_{int}, 508
p_{meter}, 267

pct, 642
$pr_{\mu,\epsilon}(M)$, 409
$prof$, 379
pv, 62
q, 62
q_X, 1414
$r(\mathfrak{S}_E, w)$, 654
r_H, 1337
r_t, 207
$rel(x, a_i)$, 648
res_{G^I}/Id_A, 294
$res_{@A}$, 294
res^{ι}_{f/Id_A}, 294
$resclass(C^I)$, 313
ret, 298
rev_k, 125
$round(x)$, 1390
$s' \leqslant s$, 1445
$s(\mathfrak{S}_E, w, t)$, 654
$s(x)$, 1412
$s(x, y)$, 1185
sG, 457
sG^*, 457
sP, 457
$scalemod_p$, 95
sem, 1438
set, 146
sig_{Den}, 334
$sing$, 1426
$sp(x)$, 269
$span$, 384
$st^i(x)$, 259
$supp(\chi)$, 337
$sym(A)$, 208
t, 62, 385
$t(F)$, 1438
$t(x, y)$, 1185
t_n, 385
$top(x)$, 1390
$true$, 1428, 1459
$v \wedge w$, 1468
w, 628
$w\text{-}Pitch$, 133
$w\text{-}PitchClass$, 94
$w\text{-}TemperedScale$, 96
w^S_{harmo}, 647
w^S_{metro}, 646
w^S_{motif}, 647
$w_{Evt.,RelEvt..}$, 642
$w_{GrpArti}$, 645
w_{loc}, 763
w_{orth}, 763
x_{hmax}, 720
x_{hmean}, 720
$x_{melodic}$, 720
x_{metric}, 720
x/E, 338
$x < y$, 1373
$x > y$, 1413
$x@\hat{G}$, 257
$x^{\sim}$, 332
$x \succ z$, 227
$x \twoheadrightarrow y$, 402
x_h, 261
x_m, 261
$x_{U,V}$, 445
x_{alt}, 108
$x_{j,s}$, 710
x_{red}, 108
$y(t_i, j)$, 724
$\mathbf{C}/^{opp}b$, 1424
$\mathbf{C}/b$, 1424
$\mathbf{C} \models \alpha$, 1434
$\mathbf{C}^{@}$, 1422
$\mathbf{C}^{opp}$, 1421
$\mathbf{C}^{spaces}$, 1411
$\mathbf{C}^{spaces}_{?}$, 1411
$\mathbf{C}^{spaces}_{X}$, 1411
$\mathbf{a}P$, 1432
$|$, 1432
$|?|$, 1446
$|?|_d$, 1446
$|K|$, 1446
$|f|$, 1446
$|s|$, 1446
ébauche, 1038
Colimit, 59
  denotator, 1113
Limit, 58
  denotator, 1113
List
  denotator, 1113
Loc, 89
Mod$^{@}$, 54
Power, 58, 90
Simple, 58
Syn, 58
T^2, 917
head, 1375
tail, 1375
2-category, 939, 946
2D Fourier series, 1212

A

A-addressed
  function, 288
  gesture, 961
A-parametrized gesture, 1317
A2C, 870
abbreviation, 1053
abelian, 1378
abelian group, finitely generated -, 306

Ableton Live, 1158
absolute
  dynamical sign, 640
  dynamics, 685
  logic, 145
  music, 774
  novelty, 1062
  pitch, 577
  symbolic dynamics, 68
  tempo, 561, 643
absolutely relative, 1028
absorbing point, 432
absorption, 1337
  coefficient, 1337
abstract
  cardinality, 386
  complement, 208
  gestalt, 391
    specialization, 402
  identity, 14
  inclusion, 208
  motif, 386
  onset, 125
  specialization, 402
abstraction, 405, 407, 425, 1026
  concept framework, 385
  textual -, 362
Abstraktmotiv, 909
accelerando, 608, 631, 644
accelerated motion, 607
accentuation, 592
accessory parameter, 831
accumulation point, 1443
acoustical wave, 963
acoustics, virtual -, 701
action
  bodily -, 1305
  complement -, 518
  faithful -, 1378
  free -, 1378
  gestural -, 912
  Lagrangian -, 1188, 1481
  left -, 1378
  local -, 1038
  motor -, 608, 897
  Nambu-Goto -, 1185, 1484
  right -, 1378
  transitive -, 1378
action-homunculus, 869
action-to-cognition layer, 870
activities, fundamental -, 3, 6
activity
  combinatorial -, 198
  construction -, 162
  instinctive -, 623
  interpretative -, 246, 251, 252
actology, 1001
acuteness, 237
Aczel, Peter, 1094
ad-hoc polymorphism, 802
adapted tempo curve, 576
Add-Element, 814
`AddObjects`, 1132
address, 53, 54, 1179, 1397, 1423
  change, 54, 70
    technique, 70
  faithful, 431
  fixed vs. variable, 89
  for a chord, 94
  full -, 431
  fully faithful -, 431
  functor, 141
  killing, 168
  navigation, 140
  subcategory, 1089
  variable, 53
  zero -, 53
addressed
  adjointness, 138
  comma category, 137
adic representation, 1389
adjoint
  functors, 907
  left -, 1423
  right -, 1423
adjointness, addressed, 138
adjunction, 1375
admissible parallel hypergesture, 1250
admitted
  successor, 533
  tonalities, 465
Adorno, Theodor Wiesengrund, v, 154, 245, 248, 548, 569, 574, 609, 623, 652, 889–891, 893, 894, 1182
Adrien, Jean-Marie, 1346
affect, 890, 893
  categorical -, 1266
  vitality -, 1266
affine
  counterpoint group, 391
  dual, 1398
  Lie bracket, 444
  simplex, 1461
  tensor product, 1400
  transformation, modular -, 786
affine functions
  complex of -, 288
  on functorial global compositions, 354
after qualifier, 815
Agawu, Kofi, 329
Age of Enlightenment, 36
aggregate, 207
Agmon, Eytan, 203, 205

agogical
architecture, 799
operator, 720
agogics, 249, 643
global -, 629
primavista -, 629
AgoLogic, 576, 623, 789, 799
Agon, Carlos, xi, 210, 314
Air, 1283
Alain, 145
Alberti, Leon Battista, 856
aleatoric component, 198
aleatorics, 59
Alexander Technique, 868
Alfohrs, Lars V., 1465
algebra, 1385
Boolean -, 103, 1433
general linear -, 1408
group -, 925
Heyting -, 103, 930, 934, 1433
Lie -, 1408
logical -, 1433
monoid -, 1386
quiver -, 1388
Riemann -, 481
tensor -, 1469
algebra of gestures, 1002
algebraic
element, 1460
field extension, 1460
geometry, 550, 1009
topology, 164, 919
universe, 1029
algorithm, 1342
Euclidean -, 1386, 1389
in FM synthesis, 73
off-line -, 761
real-time -, 761
TX802, 236
algorithmic extraction of performance fields, 759
Alhaiya Bilwal, 1306
Alighieri, Dante, 116, 161, 1218
aliquid pro aliquo, 15
all-interval
n-phonic series, 195
series, 200
allomorph, 443
allomorphic extension, 443
allowed successor pairing, 531
almost regular manifold, 974
Alpaydin, Ruhan, 678, 1098, 1099
α-restriction, 433
alphabet
of creativity, 198
of music, 90
alphabetic ordering, 36, 38, 50
alterated note, 107
alteration, 107, 108, 162, 226, 466, 507, 1148
as tangent, 108
direction of -, 788
elementary -, 108
force field, 788
in pitch, 53
pitch -, 788
successively increased -, 788
two-dimensional -, 787
`Alteration`, 1135
altered scale, 480
alternating tensor, 1468
Alunni, Charles, 859, 861, 910, 1036, 1059, 1060, 1177
ambient
space, 90
coproduct -, 105
product -, 104
ambient space
dual -, 107
ambiguity, 245, 251
theory of -, 245
tonal -, 493
ambitus, 261
American
(musical) set theory, 116, 180, 202–211, 979
contour
theory, 385
jazz, 442
notation, 438
theory, 439
amplitude, 238, 1340
modulation, 236, 833
spectrum, 1340
AMS, 1034
Amuedo's decimal normal rotation, 210
Amuedo, John, 209, 211, 439
amusia, 872
analyse créatrice, 1139
analysis, 1338
-by-synthesis, 609, 621
chord -, 438
coherent -, 634
comparative -, 273
complex -, 1465
FM -, 236
gestural -, 1274
immanent -, 378, 383
metrical -, 688
motivic -, 214, 404
musical -, 612
neutral -, 222, 250
normative -, 377
notic -, 1306
principal component -, 739
regression -, 711, 724, 726

situs, 164, 227
sonic -, 694
spectral -, 524
text -, 609
analytic, 1465
analytical
discourse, 12
method, 1012
vector, 723
weight, 549, 553, 646
Anaxagoras, 855
anchor note, 625
Andersen, Anne Sophie, 1277
Andreatta, Moreno, vi, xi, 210, 314, 1036, 1173
Andrieu, Bernard, 868
ANSI-C, 783
anthropic principle, 465, 466, 543
anthropocentricity, 1177
anthropology, computer-aided -, 765
anti homomorphism, ring -, 1385
antisymmetric, 1373
antiworld, 460, 497, 498, 773
anvil, 1354
apophatic theology, 1004
Applebaum, Mark, 1264
application framework, 665
apposition, 16
approach
bigeneric -, 444
categorical -, 801
historical -, 465, 472
nonparametric -, 708
statistical -, 707
systematic -, 472
transformational -, 204
approximate, 683
Aprea, Bruno, 1292
arbitrary, 16
archetypal gesture, 1269
archicortex, 528
Archimedes, 864
architectural principle, 718
architecture
agogical -, 799
modulatory -, 495
Arduino, 1277
area
Broca's -, 872
Wernicke's -, 872
Argerich, Martha, 735, 736, 768
argument, 616
Aristotle, ix, 28, 38, 43, 773, 885
arm-based gesture, 1274
Arnold, Troy, 1276
aroh, 1306
arpeggio, 74, 134, 575, 592, 625
field, 575
arrow, 508, 898, 908, 1037, 1046, 1050, 1375, 1457
self-addressed -, 514
arsis, 1192, 1287, 1325
art, scientific -, 1045
articulated listening, 249
articulating notes, 1303
articulation, 249, 562, 632, 645, 685
double -, 17
field, 565–568
initial -, 578
operator, 591
Artin, Emil, 1087
artistic
difficulty, 1218
fantasy, 570
operator, 1204
artistry, combinatorial -, 199
arts, 5
Ashkenazy, Vladimir, 730, 735, 737
Assayag, Gérard, xi, 791
Assisi, Francis of, 847
associated
metric, 1450
metrical rhythm, 267
topology, 1450
AST, 202, 271, 387, 409, 439
global -, 314–316
software for -, 209
asymmetries of communication, 754
Atari®, 624
Mega ST4, 791
atlas, 252, 557, 987, 991, 1466
A-addressed, 252
projective -, 296
standard -, 293
atlases, equivalent -, 257
atom, semantic -, 442
atomic formula, 1435
atomism, ontological -, 24
atonal music, 204
attack, 1339
attarantate, 874
audience, 1278
auditory
cortex, 524, 527
gestalt, 396
nerve, 1355
representation, 197
augmentation, 134
augmented, 444
Augustinus, 463, 502, 1322
Auroux, Sylvain, 36
auto, 1420
autocomplementarity, 181, 427
function, 419, 519

autocomplementary, marked - dichotomy, 518
autocorrelation, 766
autocorrelogram, 715
autograph, free -, 1052
automorphism, 1378, 1385, 1420
  group, 144, 1391
    of interpretable compositions, 306
  relative -, 1460
autonomy, 6
autoreferential, 849
avaroh, 1306
Avison, Charles, 248
awareness, kinesthetic -, 1281
axiom, 1434
  of choice, 1374
axioms for QM, 1492
axis
  third -, 96
  diachronic -, 328, 472
  fifth -, 96
  of combination, 116, 212
  of selection, 116, 212
  paradigmatic -, 160
  synchronic -, 328, 472
  syntagmatic -, 16, 160
Ayler, Albert, 933

B

Bätschmann's Bezugssystem, 11
Bätschmann, Oskar, 11, 155
Bénabou, Jean, 1038
Béziau, Jean-Yves, 1028
Békésy, Georg von, 1358
Babbitt, Milton, 203, 204
Bach, Johann Sebastian, 114, 115, 118, 121, 161, 189, 199, 203, 248, 249, 325, 488, 571, 608, 688, 709, 711, 752, 753, 890
Bachelard, Gaston, 1059, 1060
background, 415
Bacon, Francis, 5, 859, 860, 1005
Badiou, Alain, 1007, 1016
Bagatelle op.126,2, Beethoven, 894
Bageshri, 1305
Bakhtin, Michail, 856
ball, open -, 1449
Ballade op. 23, 1161
Banach space, 1450, 1486
Banach, Stefan, 1450
band, frequency -, 526
bandwidth, 708, 721
bankruptcy, scientific -, 22
bar grouping, 713
bar-line, 592, 631
bar-lines, 68
Barbin, Évelyne, 1040
Barker, Hugh, 886
barline meter, 97
Barlow, Klarenz, 1368
Barthes, Roland, 15, 160
barycentric coordinate, 1446
base, 1393
  for a topology, 1443
  sheaf on -, 1412
Basic, 74
basic
  extension, 427
  intension, 427
  series, 125
  theme, 202
basilar membrane, 1356
basis, 1430
  calculation, 761
  coordinate, 1347
  deformation, 656
  of a tangent composition, 551
  of disciplines, 5
  parameter, 67, 654
  Schauder -, 1209, 1486
  space, 567, 588, 628
  specialization, 656
basis-pianola operator, 655
Batt-Rawden, Andrew, 1277
Baudelaire, Charles, 218, 493, 799
Bauer, Moritz, 215
Baumgarten, Alexander, 868
beat, 377, 1365
  extraction, 1292
  frequency, 1365
  meter, 97
  strong -, 377
  weak -, 377
beauty, 345
becoming, 1069
Beethoven, Ludwig van, 100, 120, 134, 201, 248, 267, 276, 325, 347, 405, 407, 459, 460, 463, 466, 488, 495, 571, 774, 780, 937, 1178, 1353
before qualifier, 815
behave well, 398
being, 1067, 1068
Bell, John L., 1058
Benjamin, Walter, 548, 569, 652
Benveniste, Émile, 856, 903
Beran operator, 723, 744
Beran, Jan, vi, xi, 201, 202, 488, 612, 707, 723
Berg, Alban, 125, 204, 247, 1260
Berger, Hans, 525
Bernard, Jonathan W., 204
Bernoulli, Jakob, 1007, 1048
Bernoulli, Johann, 1007
Beschler, Edwin, xii
Bessel function, 1344
Bevilacqua, Frédéric, 902

Bezugssystem, Bätschmann's, 11
Bezuoli, Giuseppe, 28
biaffine, 1399
bicategory, 946
  gesture -, 937, 946
bidirectional dialog, 31
big bang, 329
big science, 196
BigBang, 1103, 1187
  rubette, 947, 1097, 1127, 1157
BigBangObject, 1103, 1114
bigeneric
  approach, 444
  major tonality, 449
bigeneric morphemes, construction of -, 445
bijective, 1372
bilinear form, 291
Binnenstruktur, 247
biological inheritance principle, 627
bipolar recording, 524
Bissonanz, 424
Blake, William, 417, 755
block, 589, 784
  type, 311
blues, 885
Bobillier, Etienne, 1027
Boccherini, Luigi, 241, 826, 1264
Boden, Margaret A., 1034
bodily
  action, 1305
  technique, 1280
body, 870, 910, 1271
  Hegel -, 1066, 1070
  performance -, 586
  performative -, 1283
  philosophy of the -, 868
Boesendorfer, 577, 629, 687, 700
boiling down method, 648
book concept, 48
BOOLE, 44
Boole, George, 1433
Boolean
  algebra, 103, 1433
  combination
    of (class) chords, 94
  operation, 784
  topos, 1434
Boquiren, Sidney, 1277
Borel
  function, 1488
  set, 1488
boson, 1184, 1485
bottle, Möbius -, 558
bottom wall, 631
Boulez, Pierre, 30, 35, 36, 40, 89, 90, 114, 127, 245, 246, 253, 287, 303, 779, 781, 831, 839, 859, 1179, 1181, 1363
bound
  Chopin rubato, 625
  variable -, 1436
boundary, 1443
  homomorphism, 966
  manifold with -, 1467
  operator, 969
bounded operator, 1487
Bourbaki, Nicolas, 1028, 1041
bow
  angle, 235
  application, 235
  parameter, 833
  pressure, 235, 833
  velocity, 235, 833
box, 802, 1066
  factory -, 805
  flow -, 1455
  Hegel's -, 1072
  temporal -, 811
Box-Value, 814
bracket, Lie -, 1246, 1408, 1455
Brahms, Johannes, 1264
brain, emotional -, 527
brane, 1184
breaking, symmetry -, 775
breath, 1286
breathe together, 1317
breathing, 1315, 1316
Brendel, Alfred, 729, 730
Bresin, Roberto, 902
brilliance, 237
Brinkman, Alexander, 209
Broca's area, 872
Brofeldt, Hans, 883
Bruhat, standardized - transformation, 1154
Bruijn, Nicolass Govert de, 190, 194, 311
Bruter, Claude-Paul, 1035
Bulwer, John, 855
bundle
  of ontologies, 142
  tangent -, 1451
Bunin, Stanislav, 735
Buteau, Chantal, xi, 376, 403

C

C-major, inner symmetry of -, 123
C-major, 457
c-motif, 100
C-scale frame, 475
C-scheme, 1415
C-space, 1411
c-space, 205

CAC, 801
cadence, 414, 453–462, 465, 817, 840
parameter, 454
Rameau's -, 455
cadential, 454, 465
family, minimal -, 455
formula, 453
set, 455
Cadoz, Claude, 901, 1182
Cage, John, 59, 250, 571
Calbris, Genevieve, 1301
calculation, 1002, 1029
basis -, 761
field -, 761
precision, 639
calculus, 570
of variations, 1471
camera obscura, 603
camino, 1005
Campanella, Tommaso, 849, 856, 1037, 1177
Campos, Daniel G., 1053
Camurri, Antonio, 902
canon, 117, 159, 268
canonical
curve, 720, 733, 739
operator, 207
Para-meter, 268
program, 325
canons, classification of -, 312
Cantor, Georg, 1006
cantus
durus, 261
firmus, 509, 931
mollis, 261
Capova, Sylvia, 735
cardinality, 672, 1373
abstract, 386
of a gestalt, 391
of a local composition, 90
Carey, Mariah, 1303
Carnatic vocalist, 1301
Carnot, Lazare, 1049
Carpitella, Diego, 874
carpus, 912
carrier, 73, 1342
Cartan, Élie, 1055
cartesian, 938
closed, 1429
coordinates, 1007
doubt, 1012
product, 1372, 1424
thinking, 898, 1182
case
linear -, 205
cyclic -, 205
Casella, Alfredo, 183
Castine, Peter, 209
catastrophe, 497–499, 501, 957
theory, 226, 497
categorical
affect, 1266
approach, 801
digraph, 940
manifold, 993
manifolds, category of -, 993
modeling, 1001, 1003, 1045, 1046
categorically natural gesture, 918
categories, equivalent -, 1422
categorization, 1004
category
cocomma -, 1424
cocomplete -, 1425
comma -, 1424
complete -, 1425
finitely
cocomplete -, 1425
complete -, 1425
graph -, 939
homotopy -, 1474
internal -, 1464
isomorphic -, 1421
linear -, 923
matrix -, 1420
mimetic -, 892
of categorical manifolds, 993
of cellular hierarchies, 593
of commutative global composition, 285
of coverings of sets, 293
of denotators, 331–334
of elements, 132, 1061, 1424
of factorizations, 943
of forms, 57
of formulas, 926
of functorial global compositions, 275
of gestures, 1289
of Heyting algebras, 930
of local compositions, 89
of locally almost regular manifolds, 974
of objective global compositions, 275
of performance cells, 586
of points, 981, 1255
of sheaves, 1431
of sketches, 1029
of textual semioses, 337
opposite -, 1421
path -, 1388, 1458
product -, 1422
quotient -, 1420
simplex -, 939
skeleton -, 1421
tangent -, 927
theory, 908, 1002, 1028

topological -, 939, 961, 1085, 1464
category of form semiotics, 1092
category of known simple models, 1062
Cauchy
problem, 1456
sequence, 1450
Cauchy, Augustin, 1450, 1456
Cauchy-Riemann equation, 1465
causal
coherence, 769
depth, 676
relation, 816
causal-final variable, 767
causality, 766
Cavaillès, Jean, v, 859, 861, 873, 909, 924, 1035, 1177, 1181, 1300
Cavalieri, Bonaventura, 1048
CDC Cyber, 525
Čech cohomology, 354, 2131
Celibidache, Sergiu, 117, 574, 1285
cell, 306
complex, 325
Deiters' -, 1356
hair -, 1356
outer
hair -, 1356
pillar -, 1356
performance -, 585, 1254
cellular
hierarchies
category of, 593
classification of -, 591
hierarchy, 590, 596, 1254
product -, 590
restriction -, 590
type of a -, 590
organism, 325
Cent, 1349
center, 1385
central
pitch detector, 1361
CERN, 196
Certeau, Michel de, 1040
Châtelet, Gilles, 859, 862, 872, 894, 898, 1036, 1060, 1177, 1180
chain, 1395
complex, 1475
proof -, 1434
third -, 675
chamber pitch, 563, 1349
change
of frame, 1058
of material, 813
of orientation, 509, 515, 531
of perspective, 324
program -, 784
value -, 632
CHANT, 238, 1314
chant, Gregorian -, 509
character string, 60
characteristic, 1386
function, 335, 1375
gesture, 956
map, 1428
characteristics, method of -, 1456
charge, semantic -, 403
chart, 251, 252
of level j, 269
cheap design, 873
checkboxes
matrix of -, 1110
Cherlin, Michael, 934
chest voice, 1315
Chin, Unsuk, 1295
Chinese philosophy, 1278
chironomic
game, 1327
Choi, Insook, 901
choice, axiom of -, 1374
Chomsky, Noam, 234
Chopin rubato, 550, 562, 576, 624, 765
bound -, 625
free -, 625
Chopin, Frédéric, 82, 625, 743, 1161
chord, 92, 180, 187, 249, 413
structural -, 415
a scale's -, 95
addresses, 94
analysis, 438
circle -, 424
class -, 94
classification, 159, 414
closure, 431
complement, 94
core -, 485
dictionary, 180
difference, 94
diminished seventh -, 463, 497, 499, 501
event, 93
foundation -, 439
fundamental -, 439
inspector, 675
intersection, 94
inversion, 420
isomorphism classes, 180
just - class, 94
n-, 93
pivotal -, 463
progression, 413, 414
prolongational -, 415
self-addressed -, 185
sequence, 486

coherent -, 486
standard -, 436
symbol, 438, 439
tempered - class, 94
tesselating -, 309
union, 94
CHORD-CLASSIFIER, 210, 441
Chowning, John, 1342
Christian religion, 847
christianization, 889
of music, 891
chromatic
(tempered) class chord, 94
Michel -, 478
octave, just -, 96
Roederer -, 478
scale, 418
Vogel -, 478
chronocentricity, 1175, 1177
chronospectrum, 235
Cicero, 856
circle
chord, 424
of fifths, 423
of fourths, 423
circular
colimit, 65
definition, 48, 146
denotator, 72–75
denotators
folding - denotators, 369
form, 64
gesture, 1269
limit, 65
set, 66
synonymy, 64
circularity, 849, 851
conceptual -, 146
of forms, 48
CL, 1434
class, 156, 802
contour -, 206
resultant -, 313
chord, 93
just -, 94
tempered -, 94
contiguity -, 1447
counterpoint dichotomy -, 518
dichotomy -, 518
equivalence -, 1373
ground -, 313
marked
counterpoint dichotomy -, 518
dichotomy -, 518
meta-object -, 814
nerve, 283, 309, 322, 988
number, 180
precedence
list, 804
segment -, 206, 208
set -, 207, 208
third comma -, 265
Vuza -, 312
weight, 188, 283
classical logic, 1434
classification, 828
epistemological -, 5
semiotics of sound -, 240
chord -, 414
geometric -, 177, 989
in musicology, 157
local musical interpretation of -, 173
local theory of -, 157
of canons, 312
of cellular hierarchies, 591
of chords, 159
of motives, 187–190
of music-related activities, 4
of rhythms, 312
of sounds, 10
recursive -, 177
sound -, 232
technique, 169
theory, 831
classifier, subobject -, 1428, 1459
CLOS, 210, 801, 802
closed
cartesian -, 1429
locally -, 1445
path, 1375
point, 228
set, 1443
sieve, 1431
simplex, 1446
closure, 1443
hierarchy -, 588
objective -, 432
closure operator, Kuratowski -, 1237, 1443
Clough, John, 203
cluster
Cortot -, 735, 739
Horowitz -, 735, 739
Clynes, Manfred, 604, 607, 889, 891, 895, 1182
CMAP, 209
coarser, 1443
coboundary map, 1447
cochain, 1477
complex, 1477
of a global composition, 307
singular -, 1447
cochlear Fourier analysis, 1365
cockpit, 1104

cocomma category, 1424
cocomplete category, 1425
cocone, 1423
coconut parable, 1089
coda, 249, 496
code, 212
codification of a symmetry, 128
codomain, 1372, 1419
coefficient
  absorption -, 1337
  largest -, 733
  system, 1447
coefficients, signs of -, 731
cognitive
  dimension, 179
  effort, 179
  embryology, 870
  independence, 179
  layer, 869
  musicology, 21
  psychology, 179, 226
  science, 611
  stratum, 869
coherence, 415, 550, 634, 688
  causal -, 769
  final -, 769
  harmonic -, 447
  inter-period -, 769
coherent
  analysis, 634
  chord sequence, 486
  topology, 1444
Cohn, Richard, 315
cohomological technique, 1061
cohomology, 307, 1477
  Čech -, 354, 1231
  de Rham -, 973
  group, 1447
  l-adic -, 379
  module, 1477
    of a global composition, 307
  resolution -, 355
coinduced
  topology, 1444
Coleman, Ornette, 794, 900, 1183
Coleman, Steve, 377
colimit, 252, 1290, 1424
  circular -, 65
  form, 57
  topology, 1444
colinear, 230
collaborative environment, 197
collaboratory, 32, 667
collective responsibility, 633
color, 1108
  coordinate, 1341
  encoding, 765
  parameter, 835
  sound -, 159
  space, 832
coloring, 784
Coltrane, John, 571, 603, 1295
COM matrix, 387
combination
  axis of -, 116, 212
  linear -, 1391
  weight -, 681
combinatorial
  activity, 198
  artistry, 199
  design, 1025
  topology, 254
combinatoriality, 210
combinatorics, creative -, 246
comes, 159, 199, 688
comma
  category, 1424
    addressed -, 137
  fifth -, 62
  Pythagorean -, 62
  syntonic -, 62, 97
  third -, 62, 265
common
  language, 22
  taste, 752
common-note function, 204
communication, 3, 4, 9, 11, 24
  asymmetries of -, 754
  coordinates, 14
  process, 14
communicative dimension, 22
commutative, 1376
  local composition
    module of a -, 105
  diagram, 1421
  global compositions, 285
  homotopy -, 1474
  local composition, 105
  polynomials, 1386
commutativity relation, 1420
compact, 1445
  locally -, 1464
compact-open topology, 914, 937, 1464
comparative
  analysis, 273
  discourse, 493
comparative criticism, 756
comparison matrix, 206
competence, 331
  historical, 349
  stylistic, 349
complement, 1371

abstract -, 208, 210
action, 518
literal -, 210
theorem, 208
complete
category, 1425
harmony, 827
quiver, 1375
uniform space, 1450
completeness, 36, 42, 49
finite -, 138
completion, semantic -, 49
complex
analysis, 1465
cell -, 325
chain -, 1475
cochain -, 1477
derivative, 1465
gesture, 1042
module -, 288
number, 948
numbers, 1385
of affine functions, 288
quotient -, 289
set -, 314
simplicial -, 779, 1445
simplicial cochain -, 1447
time, 1186, 1218
complexity, 165
degree of -, 162
formal -, 383
measure, 254
of performance, 548
component
aleatoric -, 198
alteration, 107
idempotent -, 1377
irreducible -, 269
metrical -, 267
reduced -, 107
semiotic -, 854
composed frame, 802
composer, 12, 895, 1263
perspective of the -, 246
composition, 162, 1372
t-fold tangent -, 551
global -
standard -, 293
commutative
local -, 105
computer
assisted -, 801
computer-aided -, 774, 791
concept, 571
dodecaphonic -, 124
functorial -, 1180
functorial local -, 101
generic, 174
global -, 41, 140, 831, 987, 1466
N-formed -, 291
oriented -, 291
resolution of a -, 323
global functorial -, 256
global objective -, 252
interpretable -, 304
local -, 41, 75, 89, 90
embedded -, 106
dimension of a -, 178
generating -, 106
projecting -, 178
standard -, 293
local objective -, 90
locally free local -, 175
modular -, 251
musical -, 30
non-interpretable -, 305, 308
non-interpretable global -, 990
tangent -, 551
tools, fractal -, 115
compositional
design, 209
idea, 322
space, 204
compositions
commutative global -, 285
local -
category of -, 89
computation
symbolic -, 801
computational musicology, 21
computer assisted composition, 801
science, 6, 156
computer-aided
anthropology, 765
composition, 774, 791
computer-assisted performance research, 629, 701
conativity, 212
concatenation, 132, 1372
principle, 133, 512
concept, 9, 35
architecture, 152
composition -, 571
construction history, 47
critical -, 1072
denotator -, 665
form -, 665
format, 42
framework, 3, 5, 9
abstraction -, 385
dynamic -, 329
fuzzy -, 164
grouping -, 250

human - construction, 48
leafing, 50, 52
of a book, 48
of instantaneous velocity, 27
of music, 21
paradigmatic -, 229
poietical -, 14
point -, 145
RUBATO®-, 665
score -, 250, 571, 754, 811
set -, 146
space, 21, 31, 32, 1034
spatial -, 894
stable -, 226
surgery, 83–85, 633
concepts
standard of basic musicological -, 92
void pointer -, 32
conceptual
circularity, 146
explicitness, 21
failure, 23
genealogy, 64
identification, 229
laboratory, 30
navigation, 35
precision, 32
profoundness, 92
universality, 92
zoom-in, 19
conceptualization
dynamic -, 67
fuzzy -, 375
human -, 145
precise -, 211
process, 201
concert
form, 793
master, 626
pitch, 576
condition
Dirichlet -, 1483
initial -, 1452
instrumental -, 701
Neumann -, 1483
conducting, mechanism of -, 1291
conductor, 551, 562, 626
gestures of the -, 1285
primitive gesture of -, 1287
role of -, 1288
cone, 1423
configuration
counting series, 191
view -, 1108
conformal, 1466
conjecture, diamond -, 929, 937
conjugation, 1379, 1385
conjugation class
of endomorphisms, 180
of symmetry group, 180
conjunction, 346, 1432
connective, predicate -, 1435
connotation, 16
connotator, 328, 1440
conservation, energy -, 973
consonance, 463, 469, 509, 1364
deformed -, 532, 1236
imperfect -, 522, 532, 541
perfect -, 522, 532, 541
consonance-dissonance, 1353
dichotomy, 419, 519, 543
consonant, 234
interval, 414, 527
mode, 450
constant
functor, 1422
module complex, 288
part, 433
shift, 109
structural -, 1408
constraint
geometric -, 913
gestural -, 617
musical -, 913
physical -, 913
programming, 774, 801
constraints
geometric -, 912
mechanical -, 912
semiotic -, 232
construction
activity, 162
of bigeneric morphemes, 445
recursive -, 42
construction history of concept, 47
construction using compass and straightedge, 1461
contact, 212
point, 235
container, 806, 811
content, 15, 337, 409, 647, 673
interval -, 207
mathematical -, 15
maximal structure -, 1362
musical -, 15, 1286
textual -, 1286
context, 212, 736, 1072
problem, 674
real-time -, 760
contiguity, 16
class, 1447
contiguous simplicial maps, 1447
continuous, 1444

gesture, 817
method, 639
presheaf, 982
stemma, 661
weight, 639
continuum, Kendon -, 854
contour, 159, 387
class, 206
space, 205
theory, 271
American -, 385
contra, 532
contractible, 1474
contraction, 1452, 1474
contrapuntal
form, 249
group, 115
interval, oriented -, 509
meaning of $\mathbb{Z}$-addressed motives, 100
motion shape type, 387
sequence, 531
symmetry, 532, 1235
local character of a -, 533
technique, 159
tension, 531
tradition, 199, 1367
contravariant functor, 1421
contravariant-covariant
rule, 805
control
group, 775
interactive -, 813
of transformation, 200
conversation, topos of -, 826
convex, 1462
hull, 1462
Cooper, Lynn, 872
coordinate
barycentric -, 1446
basis -, 1347
color -, 1341
fifth -, 1350
function, 174
geometric -, 1341
gestural -, 893
octave -, 1350
ontological -, 10
pianola -, 1347
third -, 1350
coordinates, 45
cartesian -, 1007
of existence, 578
coordinator, 43, 1438
form -, 55
gestural form -, 1087
of a form, 55
coproduct, 1375
ambient space, 105
of local compositions, 105
type, 47
Corballis, Michael C., 871
Corbin, Alain, 874
core chord, 485
corporeal, 1264
correlate
electrophysiological -, 524
physiological -, 897
correspondence
exponential -, 1464
Galois -, 1093
cortex, auditory -, 524, 527
Corti, organ of -, 1356
Cortot cluster, 735, 739
Cortot, Alfred, 734, 735, 737
coset
left -, 1379
right -, 1379
cosmology, 465
counterpoint, 134, 419, 508, 523, 826, 966, 1235
dichotomy, 518
class, 518
double -, 513
rubette for -, 1097
theorem, 534, 538
theory, 775, 839
counterpoint dichotomy
class, marked -, 518
countersubject, 689
counting series, configuration -, 191
coupling
monogamic -, 632
polygamic -, 632
Courtine, Jean-Jacques, 874
covariant functor, 1421
covering, 252
equivalent -, 252
family, 1430
motif, 384
sieve, 1430
Cox, Arnie, 1264
cp, 205
CPL, 804
cpset, 206
creation, melodic -, 1077
creative
combinatorics, 246
extension, 201
mathematics, 1033
process, 1066, 1072
space, 1034
creativity, 198, 329, 1001, 1031, 1035, 1038, 1044
alphabet of -, 198

essence of -, 1287
gestural -, 1183
mathematical -, 1002
musical -, 1227
creator, 11, 12
crescendo, 67, 551, 594, 607, 1348
wedge, 641
critical
concept, 1072
distance, 1337
fiber, 755, 758
criticism
comparative -, 756
journalistic -, 730
music -, 634
critique, 755
music -, 751
cross-correlation, stemmatic -, 634
cross-modal, 1274
cross-semantical relation, 612
cube, 1476
n-dimensional -, 966
topographic -, 17, 32
cul-de-sac, 541
interval, 538
cultural mimesis, 874
culture of performance, 623
curve, 230, 1451
canonical -, 720, 733, 739
exact -, 945
integral -, 1245, 1453
intonation -, 563
tempo -, 562, 607, 623, 723, 784
curvilinear reduction, 776
Cusick, Susan, 1264
cutting, 1032
CX5M, Yamaha -, 525
cycle, 1375, 1382, 1454
index, 191, 1382
of variations, 791
pitch -, 207
cyclic
case, 205
extension, 207
group, 1381
interval succession, 207, 208
Czerny, Carl, 624, 764, 911, 914

D

d'Alembert, Jean Le Rond, 5, 36, 50
D-brane, 1482, 1483
da capo, 117
dactylus, 213
grid, 216, 217
Dahlhaus, Carl, 23, 123, 245, 264, 265, 447, 472, 488, 674, 826, 1293, 1367
Damasio, Antonio, 854
dance, 604, 874, 951, 1301
Dannenberg, Roger, 761
darkening, 1315
data, ethnomusicological -, 83
dataglove, 607
daughter, 619
tempo, 562
daughter performance, 984
daughters, 596
Davin, Patrick, 817
Davis, CeDell, 876
DBMS, 665
de la Motte, Helga, 22, 571
De Martino, Ernesto, 874
De Morgan, Augustus, 1002
De Novellis, Romina, 874
de Rham cohomology, 973
Deacon, the Deacon, 1321
Debussy, Claude, 183, 492, 622, 1279
decay, 1339
decomposition
hierarchical -, 710, 720
natural -, 707
orthonormal -, 10
spectral -, 708
Sylow -, 445, 510
Deep Purple, 189
deep-frozen gesture, 912
default weight function, 482
definition
circular, 48, 146
of music, 5
of musical concepts, 97
definitive performance, 895
deformation, 226, 591
basis -, 656
degree of -, 787
hierarchy -, 658
non-linear, 681
non-linear -, 640, 733
of a tempo curve, 576
pianola -, 591, 656
deformed
consonance, 532, 1236
dichotomy, 532
dissonance, 532, 1236
degree, 249, 262, 441, 442, 465
different -, 264
modulation -, 465
of complexity, 162
of deformation, 787
of freedom, 179
of organization, 718
of symmetry, 208
parallel -, 264

system, irreducible -, 457
theory, 436
triadic -, 1245
Deiters' cell, 1356
Delalande, Francois, 607, 608
delay, 833
relative -, 236
Deleuze, Gilles, 859, 860
Deligne, Pierre, 351, 1022, 1179
delta, Kronecker -, 1392
denaturation, sacred -, 891
Dennett, Daniel, 149
denotator, 41, 57–59, 1074
Simple -, 1112
$\mathbb{Z}$-addressed -, 54
Colimit-, 1113
Limit-, 1113
List-, 1113
attributes, 42
circular -, 72–75
concept, 665
genealogy, 41
flow chart, 42
image, 59
language, 595
morphism, 91
name, 45
non-zero-addressed -, 69
ontology, 328
orchestra instrumentation -, 69
philosophy, 153
Pow -, 1113
reference -, 332
regular -, 66–71
self-addressed -, 69
truth -, 335
denotators
circular - folding, 369
linear ordering among -, 50
ordering on -, 75–83
ordering principle on -, 49
relations among -, 89
Denotex, 668, 1441
DenotexRUBETTE®, 668
dense, 229
densification, 817
density
Lagrange -, 1482
Lagrangian -, 1185
deocentricity, 1175
deontology, 848
depiction, 1308
depth, 21, 23, 198
causal -, 676
EEG, stereotactic -, 524
electrode, 525
final -, 676
in mathematics, 23
in musicology, 23
in the humanities, 485
semantic -, 383
derivation, 1408, 1455, 1469
inner -, 1408
outer -, 1408
derivative, 1451
complex -, 1465
exterior -, 973
Lie -, 1455, 1469
derived serial motif, 194
Desain, Peter, 547
Descartes' duality, 1187
Descartes, René, 11, 148, 1005, 1015, 1031, 1364
description
object -, 200
verbal -, 622
design
combinatorial -, 1025
compositional -, 209
matrix, 724
sound -, 1121
Desmond, Paul, 179
development, 249, 496
history, 612
software -, 595
syn- and diachronic of music, 199
Dezibel, 1348
Di, 1103
di-alteration, 108
diachronic, 16
axis, 328, 472
index, 223
normalization, 754
diaffine homomorphism, 1396
diagonal
embedding, 1403
field, 565
diagram, 898, 907, 1017, 1032, 1053, 1421
scheme, 1420
commutative -, 1421
filtered -, 1411
gestural -, 1036
Hasse -, 218, 1374
of forms, 57
diagrammatician, 859
dialog, 827
bidirectional, 31
experimental navigation -, 32
dialogical principle, 828
diameter, 520
diamond conjecture, 929, 937
diastematic, 672
index shape type, 387

notation, 1322
shape type, 387
diatonic scale, 543
dichotomy class, 518
marked -, 518
consonance-dissonance -, 419, 519, 543
counterpoint -, 518
deformed -, 532
interval -, 518
major -, 519, 543
marked counterpoint -, 518
marked interval -, 518
Riemann -, 522
Saussurean -, 15
dictionary of expressive rules, 615
Diderot, Denis, 5, 36, 50
Dieudonné, Jean, 1044
difference, 1371
genealogical -, 756
phenomenological -, 756
different degree, 264
differentiable, 1451
n-chain, 975
manifold, 1466
differential, 1455, 1469
equation, 1452
form, 975
operator, 1470
semantic -, 163
differentiation rules, 610
difficulty, artistic -, 1218
digital age, 35
digraph, 914
categorical -, 940
dual -, 1457
global -, 996
radical -, 926
spatial -, 914, 940, 1457
topological -, 993
digraph (directed graph), 1457
diinjective, 1405
dilatation, 133, 1402
time -, 70
dilinear
homomorphism, 1392
part, 1397
dimension, 1392, 1445
cognitive -, 179
communicative -, 22
of a local composition, 178
of a simplex, 1445
ontological -, 17
diminished, 444
diminished seventh chord, 463, 497, 499, 501
Ding an sich, 21
direct
image, 1411
sum module, 1392
directed graph (digraph), 907, 914, 1375, 1457
direction of alteration, 788
directional endomorphism, 656
Dirichlet condition, 1483
disabled gesture, 875
discantus, 509, 931
disciplinarity, dynamic -, 666
discipline, basic -, 5
discourse
analytical -, 12
comparative -, 493
esthesic -, 14
discoursivity, 36, 42, 49
discrete, 638
interpretation, 254
field, 760
gesture, 817, 944
functor, 944
nerve, 254
topology, 1443
disease, psychosomatic -, 1288
disjoint, 1372
sum, 1424
disjunction, 346, 1432
dissonance, 463, 469, 509, 1364
deformed -, 532, 1236
emancipation of -, 31
dissonant interval, 527
mode, 450
distance, 226, 228
critical -, 1337
Euclidean - for diastematic types, 389
Euclidean - for rigid types, 389
for toroidal types, 390
function, 389
natural -, 362
on toroidal sequences, 390
relative Euclidean - for rigid types, 389
third -, 511
to an initial set, 580
Distributed RUBATO®, 764
distributed laboratory, 32
distributive, 1433
distributor, 689
division
of pitch distances, 61
of time
regular -, 376
divisor, resulting -, 313
Dockery, Wayne, 877
documentation, 3–6
dodecaphonic
composition, 124
composition principle, 115

method, 775, 780
paradigm, 125
series, 125, 162, 194, 247, 253, 325
vocabulary, 199
dodecaphonism, 134, 206
communicative problem of -, 135
esthetic principles of -, 134
domain, 1372, 1419
fundamental scientific -, 5
modulation -, 476
dominance, 218, 269
topology, 231, 402
dominant, 263, 414, 445, 448
role of major scale, 541
seventh, 419
dominate, 227, 1413
Donald, Merlin, 874
double
articulation, 17
counterpoint, 513
doubt
cartesian -, 1012
sceptical -, 1012
drama, musical -, 753
dramaturgy, emotional -, 893
drawing, 1147
Dreiding, André, 291
Dress, Andreas, 291
drill, 1305
driving grid, 787
drum, ear -, 1354
dual
affine -, 1398
ambient space, 107
digraph, 1457
gesture, 940
linear -, 1397
numbers, 508, 1387
space, 1488
dual numbers, 107
dualism between major and minor, 122
duality, Descartes' -, 1187
Dufourt, Hugues, 801
duration, 44, 67
period, 97
dux, 159, 199
DX7, 1342
Dylan, Bob, 885
dynamic
concept framework, 329
conceptualization, 67
disciplinarity, 666
navigation, 40
dynamical
initialization, 578
knowledge management, 329
modularity -, 666
sign
absolute -, 640
relative local -, 640
relative punctual -, 640
dynamically loadable module, 665
dynamics, 248, 249, 562, 564
absolute -, 685
historical -, 221, 223
mechanical -, 608
of performance, 658
primavista -, 629
relative -, 685
symbolic
absolute, 68
relative, 68

E

ϵ-ball, 229
E-MU Xboard, 1151
ϵ-neighborhood, 397
ϵ-paradigm, 229
ear
drum, 1354
inner -, 1355
middle -, 1354
outer -, 1354
ecclesiastical mode, 261, 540, 541
Eco, Umberto, 1050
edge, 1458
editing, geometric -, 783
editor, 802
EEG
depth -, 524
response, 524
semantic charge of -, 524, 525
test, 524
effect, groove -, 788
effective, 1378
Eggebrecht, Hans Heinrich, 21–23, 743
Ego, poetic -, 214, 219
Ehrenfels' transpositional invariance criterion, 91
Ehrenfels, Christian von, 91, 167, 226, 246, 271, 273, 383
Ehresmann, Andrée, 1013, 1029, 1060
Ehresmann, Charles, 934
eigenvector, 899
Eilenberg, Samuel, 1178
Eimert, Herbert, 127, 211
Eitz, Carl, 1350
elastic, 672
shape type, 388
electrode, depth -, 525
electrophysiological correlate, 524
element, 1371
algebraic -, 1460
neutral -, 1376

primitive -, 1460
transcendental -, 1460
varying -, 1023
elementary
alteration, 108
gesture, 817
neighborhood, 403
shift, 109
elementary gesture of the pianist, 1192
elements, category of -, 132, 1061, 1424
Ellington, Duke, 1297
emancipation of dissonance, 31
embedded local composition, 106
embedding, 1422
diagonal -, 1403
number, 208
Yoneda -, 1397, 1423
embodied
gesture, 1274, 1279
musical gesture, 1263, 1264
embodiment, 846, 864, 909, 952, 1176
musical -, 1314
science, 868
embryology, cognitive -, 870
emergence, 1061, 1063
emotion, 528, 604–606, 894
emotional
brain, 527
dramaturgy, 893
function of music, 528
landscape, 241
emotive state, 1266
emotivity, 212
empty
form name, 47
set, 146
string, 45
encapsulated history, 556
encapsulation, 23, 156, 806
speculative -, 28
encoding
color -, 765
formula, rubato -, 618
Encyclopédie, 5, 36, 38, 50
encyclopedia, 36, 362
encyclopedic ordering, 50
encyclopedism, 49
encyclospace, 36, 38, 50
endo, 1420
endolymph, 1355
endomorphism, 1420
directional -, 656
enharmonic -, 426
right-absorbing -, 432
ring, 1391
energy, 608
conservation, 973
spectrum, 1340
enharmonic, 425
endomorphism, 426
group, 426
identification, 425
ensemble rules, 610
enumeration
musical - theory, 190
of motives, 195
theory, global -, 309
envelope, 71, 1339
environment
collaborative -, 197
experimental -, 681
epi, 1420
epilepsy therapy, surgical -, 524
epileptiform potential, 524
epimorphism, 1420
epistemology
of musicology, 27
transitive -, 1059
epsilon
gestalt topology, 398
topology, 397
Epstein, David, 607
equation, 898, 907
Cauchy-Riemann -, 1465
differential -, 1452
Euler-Lagrange -, 1188, 1189, 1227, 1482
Poisson -, 1189, 1197, 1471
Schrödinger -, 1493
spring -, 1341
equations, Maxwell's -, 973
equivalence
phonological -, 215
class, 1373
homotopy -, 1474
paradigmatic transformation -, 212
perceptual -, 229
relation, 250, 1373
syntagmatic -, 215
equivalent
atlases, 257
categories, 1422
covering, 252
norms, 1450
equivariant, 1378
Erlewine, Stephen Thomas, 877
Escher Theorem, 916, 931, 941, 961, 965, 966, 985, 995, 998, 1184
Escher, Maurits Cornelis, 161, 916
EspressoRUBETTE®, 759, 764
essence of creativity, 1287
essential parameter, 831
essentic form, 895

esthesic, 11, 1342
identification, 248
esthesis, 12, 13, 211
esthetic, 211
esthetics, 13, 211
of music, 323
estrangement, 1003
ethnological form, 49
ethnology
inverse -, 754
ethnomusicological data, 83
ethnomusicology, 754
Euclid, 147, 148, 507, 1021, 1047, 1178
Euclidean
algorithm, 1386, 1389
geometry, 290
metric, 229
Euler
function, 1381
module, 62, 179
plane, 93
point, 62, 1349
space, 1349
Euler's identity, 1341
Euler, Leonhard, 62, 478, 509, 1350, 1364, 1497
Euler-Lagrange equation, 1188, 1189, 1227, 1482
Eulerian square, 1025
European score notation, 67
Eustachian tube, 1355
evaluation, 295, 1433
Evans, Bill, 880
event
percussion -, 503
time -, 556
evolution, 628
ex movere, 894
exact
curve, 945
sequence, split -, 1380
exactness, 1002
exchange
of pitch and onset, 127
parameter -, 133, 134
existence, 57, 327
mathematical -, 145, 328, 340
musical -, 340
experiment
mental -, 550
musicological -, 30, 31
physical -, 29
experimental
environment, 681
humanities, 27
material, 330
natural sciences, 27
strategy, 694, 701
experimentation, 813
experiments of the mind, 31
explanatory variable, 724
explicitness conceptual -, 21
exponentiable, 1429
exponential
correspondence, 1464
law, 1464
exposition, 249, 496, 794
expression, 335, 603, 759
human -, 570
instrumental -, 825
of expression, 892
rhetorical -, 570
expressive rules, dictionary of -, 615
expressivity
pure -, 606
rhetorical -, 556
extension, 306, 330, 337, 552
allomorphic -, 443
basic -, 427
creative -, 201
cyclic -, 207
Galois -, 1460
separable -, 1460
strict -, 443
topology, 430
exterior
derivative, 973
score, 571
extraction
beat -, 1292
monotone -, 1019
extraterritorial part, 591
extroversive semiosis, 329

F

f-morphism, 1411
F-to-enter level, 727
face, 969, 1445
operator, 968
fact, 1067
facticity, 327, 345, 464
finite - support, 338
full -, 338
factor
pressure decrease -, 1337
strength -, 610
factory box, 805
faithful
action, 1378
address, 431
functor, 1421
local network, 980
point, 431
False, 1433

'false' voice, 1315
family, 239
  covering -, 1430
  minimal cadential -, 455
  of violins, 828
  violin -, 241, 840
fanfare, 958, 1075
fantasy, artistic -, 570
faster uphill, 610
father, 619
Fauré, Gabriel, 1285
Feldman, Jacob, 607
Feldman, Morton, 250
Felver, Christopher, 1182
Fermat, Pierre de, 23, 1007
fermata, 550, 630, 632, 643
fermion, 1485
Ferrara, Franco, 1292
Ferretti, Roberto, xi
feuilleton, 635
feuilletonism, 751
FFT, 524
fiber, 611, 1375
  critical -, 755, 758
  group, 775
  product, 1387, 1424
    of local compositions, 138
  structure, 756
  sum, 1424
    of local compositions, 140
Fibonacci sequence, 340
Fibonacci, Leonardo, 59, 340
fibration, linear -, 758
fiction, 327, 465
fictitious performance history, 627
field, 597
  arpeggio -, 575
  calculation, 761
  diagonal -, 565
  discrete -, 760
  finite -, 786
  fundamental -, 591
  interpolation, 761, 764
  intonation -, 563
  of equivalence, 157
  of fractions, 1406
  operator, 652
  paradigmatic -, 125
  parallel articulation -, 567
  parallel crescendo -, 567
  parallel glissando -, 567
  performance -, 564, 568, 585, 983
  prime -, 1386
  selection, 803
  skew -, 1385
  tempo -, 562
  tempo-intonation -, 565
  tensor -, 1468
  vector -, 1245, 1451, 1468
  writing, 803
field extension, algebraic -, 1460
fifth, 62, 1350
  axis, 96
  coordinate, 1350
  sequence, 262
figuratio, 847
film music, 603
filtered diagram, 1411
filtering, input -, 760
final
  coherence, 769
  depth, 676
  retard, 607
  ritard, 896
  vertex, 660
finale, 791
finalis, 261
finality, 766
fine arts, 13, 153
finer, 1443
finger, 912
  space, 915
finger-based gesture, 1280
fingering, 248, 607, 622
finite, 1371
  completeness, 138
  cover topology, 353
  field, 786
  locally -, 1446
  monoid, 1376
  multigraph, 1375
finitely
  cocomplete category, 1425
  complete category, 1425
  generated, 1381, 1391
    abelian group, 306
Finscher, Ludwig, 825, 826
Finsler's principle, 145
Finsler, Paul, x, 145, 1094
Fiordilino, Emilio, 1471
first representative, 180
FIS, 204
Fisher, George, 1264
Fitting's lemma, 1396
Fitting, Hans, 1396
fixpoint, 1452
  group, 1378
flasque module complex, 304
flat, 109
  local network, 981
flatten, 74, 1111
flattening operation, 75, 270

Fleischer, Anja, xi, 484
Fleisher, Leon, 883
FLOAT, 44
Floris, Ennio, 1012
flow, 886, 1299, 1300
  box, 1455
  interpolation, 582
Flusser, Vilém, 848
flying carpet, 767
FM, 236, 1342
  -object, generalized, 239
  analysis, 236
  synthesis, 73
folding, 363
  circular denotators, 369
  colimit denotators, 367
  limit denotators, 366
foramen ovale recording, 524
force
  field, alteration -, 788
  modulation -, 466, 469
forces in physics, 534
foreground, 415
form, 43, 52–57, 1004, 1074, 1086
  bilinear -, 291
  circular -, 64
  circularity, 48
  colimit -, 57
  concept, 665
  concert -, 793
  contrapuntal -, 249
  coordinator, 55
  differential -, 975
  essentic -, 895
  ethnological -, 49
  Forte's prime -, 210
  functor, 55
  gestural -, 1086
  identifier, 55, 56
  large -, 1298
  limit -, 57
  list -, 809
  morphisms, wrap -, 331
  musical -, 6
  name, 43, 44
    empty -, 47
  normal -, 209
  of a symmetry, 113
  pointer character, 48
  powerset -, 56
  prime -, 210
  Rahn's normal -, 209
  regular -, 64
  semiotic, global -, 1440
  simple -, 56
    simplify to a -, 63
  sonata -, 249, 496, 791
  space, 55
  Straus' zero normal -, 210
  synonym -, 56
  type, 55
  typology, 55
form semiotics
  morphism of -, 1439
  category of -, 1092
formal
  complexity, 383
  structure, 801
formalism, Lie -, 658
formant, 238
  manifold, 238
  open - set, 238
formoid, radical -, 928
forms
  category of -, 57
  diagram of -, 57
  ordering on -, 75–83
formula, 907, 1435
  Δ- -, 926
  atomic -, 1435
  cadential -, 453
  generalized -, 927
  propositional -, 1435
  quantifier -, 1435
formulas, category of -, 926
Forte's prime form, 210
Forte, Allen, 203, 204, 209, 314
foundation chord, 439
foundations, 1028
four part texture, 826
Fourier analysis
  cochlear -, 1365
  decomposition, 70
  ideology, 233
  paradigm, 232
  representation, 740, 832
  theorem, 10
  theory, 925
  transform, 1344
Fourier series, 2D -, 1212
Fourier's theorem, 1340
Fourier, Jean Baptiste Joseph, 422
fractal, 59, 162, 782
  composition tools, 115
  principle, 799
fractions, field of -, 1406
frame, 585, 596, 802, 1346, 1438
  change of -, 1058
  composed -, 802
  simple -, 802
  space, 55, 1087
  structure, 591

wavelet -, 236
framework, 806
application, 665
concept -, 3, 5, 9
hermeneutical -, 11
free
action, 1378
autograph, 1052
Chopin rubato, 625
commutative monoid, 1376
group, 1379
jazz, 548, 1177
locally -, 1413
module, 1392
monoid, 1376
variable -, 1436
free jazz, v, 13, 900
freedom of choice, 543
Frege, Gottlob, 849, 859, 867, 898, 937, 1180
French gesture theory, 894
frequency, 61, 70, 1340, 1341
band, 526
beat -, 1365
fundamental -, 1340
modulation, 236, 1111, 1342
modulation -, 236, 833
of variable inclusion, 732
Freud, Sigmund, 528
Friberg, Anders, 610
Fripertinger, Harald, xi, 167, 190, 211, 309, 310, 1382
Frost, Robert, 273
Frydén, Lars, 610, 621
Frye, Roger, 1024
fugue, 159, 199
full
address, 431
functor, 1421
model, 726
point, 431
subcategory, 1422
subcomplex, 1445
fully faithful
address, 431
functor, 1421
point, 431
function, 859, 898, 1181, 1372
A-addressed -, 288
autocomplementarity -, 419, 519
Bessel -, 1344
Borel -, 1488
characteristic -, 335, 1375
common-note -, 204
Euler -, 1381
generic -, 804
gradus suavitatis -, 1364, 1497
Green -, 1189, 1471
horizontal poetical -, 781
index -, 1390
interval -, 204
inverse -, 1373
level -, 269, 1374
of a symmetry, 113
poetical -, 16, 116, 212, 241, 248, 774, 781
theory, 265, 437
tonal -, 249, 263, 447
value, tonal -, 447
vertical poetical -, 782
weight -, 1253
function harmony, 32
functional, 1372
gesture, 1034
object, 1036
programming, 801
relation, 1196
semantics, 445
functor, 1179, 1421
address -, 141
constant -, 1422
contravariant -, 1421
covariant -, 1421
faithful -, 1421
form -, 55
full -, 1421
fully faithful -, 1421
gestural form -, 1087
global section -, 1412
God -, 1179
homology -, 1476
limiting -, 962
local -, 926
logical -, 946
module -, 143
nerve -, 988, 1445
of orbits, 1417
open -, 1416
open covering of -, 1416
representable -, 1423
resolution -, 294
simplicial -, 988
support -, 257
topological -, 939
functorial
composition, 1180
global composition, 256
local composition, 101
functors, adjoint -, 907
fundamental
activities, 3, 6
chord, 439
field, 591
group, 920, 1475
groupoid, 920, 1475

note, 441
period, 1340
pitch, 437
scientific domain, 5
series, 115
space, 588
Fundamental Lemma of Calculus of Variations, 1471
fushi, 13, 342
Fux rules, 541, 1235
Fux, Johann Joseph, 523, 541, 840, 1367
fuzziness, 437
fuzzy
concept, 164
conceptualization, 375
logic, 336
set, 163, 1029
theory, 159

G

G-prime form, 208
Gabriel, Peter, 153, 925
Gabrielsson, Alf, 604, 607
gait, 1305
Galilei, Galileo, 27, 29, 32, 548
Galois
correspondence, 1093
extension, 1460
group, 1460
theory, 1087, 1460
Galois, Evariste, x
game, chironomic -, 1327
Garbers, Jörg, xi, 665, 1088, 1441
Garbusow, Nikolai, 181
Gast, Peter, 1033
gate function, hippocampal -, 528
Geary, James, 1022
Gegenklang, 265, 457
Geisser, Heinz, 1176, 1295
Gell-Mann, Murray, 145
genealogical difference, 756
genealogy
conceptual, 64
of denotator concept, 41
poietic, 128
general linear algebra, 1408
pause, 644
position, 322
General Midi, 235
general position, 106, 174
musical meaning of -, 106
generalization, 1026
generalized formula, 927
generated, finitely -, 1381, 1391
generating local composition, 106
Generative Theory of Tonal Music (=GTTM), 255, 376
generator
sound -, 700
time -, 775
generators for a topology, 1463
generic
composition, 174
function, 804
linear visualization, 362
point, 228, 269, 1413
score, 548
genotype, 782
Gentilucci, Maurizio, 871
geocentricity, 1175, 1177
geodesic, 241
geographic
information system (GIS), 667
orientation, 38
geometric
classification, 177, 989
constraint, 913
constraints, 912
coordinate, 1341
editing, 783
intuition, 1036
language, 893
morphism, 946
parameter, 832
realization, 988, 1446
representation, 783, 892
geometry
analytical -, 148
algebraic -, 148, 165, 550, 1009
Euclidean -, 290
germ, 267, 1451
rhythmic -, 127, 266
Germain, Sophie, 1017, 1033
germinal melody, 220, 221, 792
gestalt, 90, 167, 271, 383, 405
abstract -, 391
auditory -, 396
cardinality of a -, 391
global -, 251
musical -, 90, 127
paradigm, 672
psychology, 90
small -, 398
specialization, 402
category, 403
stability, 226
gestoid, 923
gestural
action, 912
analysis, 1274
constraint, 617
coordinate, 893
creativity, 1183
diagram, 1036

form, 1086
groove, 1303
morphism, 1287
performance, 893
philosophy, 893
presheaf, 962, 1086
primitive, 901
rationale, 753
semantics, 753
similarity, 1289
strategy, 1270
symbol, 912
theory, 1182
transformation, 1272
vibration, 893
gestural form
coordinator, 1087
functor, 1087
identifier, 1087
space, 1087
type, 1086
gestural interaction of both hands, 1299
gesturalize, 1154, 1162
gesturally representable, 962
gesture, 604, 607–609, 891, 895, 898, 909, 914, 1001, 1005, 1032, 1036, 1067, 1176, 1180, 1245, 1300
A-addressed -, 961
A-parametrized -, 1317
archetypal -, 1269
arm-based -, 1274
bicategory, 937, 946
categorically natural -, 918
characteristic -, 956
circular -, 1269
complex -, 1042
continuous -, 817
deep-frozen -, 912
disabled -, 875
discrete -, 817, 944
dual -, 940
elementary -, 817
embodied -, 1274, 1279
embodied musical -, 1263, 1264
finger-based -, 1280
functional -, 1034
functor, discrete -, 944
global -, 993
high level -, 1041
human -, 1314
imaginary -, 1042, 1314, 1315
instrumental -, 817, 902, 1318
invisible -, 1302
mathematical -, 1002, 1036, 1037, 1040
melodic -, 1302
mimic -, 893
mirror -, 1270
normative -, 1305
objective -, 1042
of gestures, 894
of locally compact points, 983
of silence, 1327
of stemmata, 1256
orchestral -, 817
performative -, 1283
phenomenology, 848
physical -, 911, 1186, 1229, 1289
physical - curve, 912
semiotic, 1183
sketch of -, 1030, 1038
smartphone -, 1279
spectral -, 1318
speech -, 1311
subjective -, 1042
symbolic -, 911, 912, 1186, 1228, 1229, 1289
symbolic - curve, 912
technology, 1283
theory, 893, 1173
French -, 894
topos, 945
vocal -, 1301, 1313, 1315
gesture disabled, 875
gesture-sound mapping, 1279
gestures
of the conductor, 1285
algebra of -, 1002
category of -, 1289
types of -, 1286
Get-Editor, 814
Get-View, 814
Giannitrapani, Duilio, 525
Gianoli, Reine, 735
Gilels, Emil, 622
Gilson, Etienne, 12
Ginsburg, Carlo, 1040
GIS, 204, 315
structure, 203
Glarean, 261
glide reflection, 1402
glissando, 67, 551, 567, 594, 817, 1350
global, 245
affine functions, module of -, 355
agogics, 629
AST, 314–316
composition, 140, 831, 987, 1466
cochain complex of a -, 307
digraph, 996
enumeration theory, 309
form semiotic, 1440
functorial composition, 256
morphism, 274
gestalt, 251
gesture, 993

hypergesture, 1223
molecule, 292
molecules, morphism of -, 292
morphisms, 246
network, 990
object, 245
objective composition, 252
morphism, 274
performance score, 598
predicate, 454
score, 250, 783
section, 288, 1424
functor, 1412
skeleton, 996
slope, 677
solution, 1454
standard composition, 293
tangent composition, 557
technical parameter, 835
tension, 677
theory, 220
threshold, 675
variational principle, 1231
world-sheet, 1224
Zarlino network, 992
globalization
metrical -, 99
orchestral -, 555
gluing, 1032
God, 1004, 1016, 1177, 1186
functor, 1179
Godøy, Rolf Inge, 902
Goethe, Johann Wolfgang von, 123, 163, 325, 748, 828
Goffman, Ervin, 853
Goldbach conjecture, 30
Goldbach, Christian, 30
golden section, 59
Goldin-Meadow, Susan, 899
Goldstein, Julius, 1361
Göller, Stefan, xi, 362, 1103, 1088, 1441
Goswitz, Robert, vi
Gottschewski, Hermann, 29
Gould, Glenn, 550, 608, 694, 702, 751, 752, 899, 909
GPL, xi
GPS, 598
gradus suavitatis function, 1364, 1497
Graeser, Wolfgang, 113, 115, 203, 249, 890, 1182
Graffman, Gary, 883
Gram identity, 292
grammar
locally linear -, 660
performance -, 615
rule-based -, 615
grand unification, 463
granddaughter, 660
grandmother, 660
graph, 1372, 1375, 1458
category, 939
directed -, 907, 914, 1375
of a FM-denotator, 73
operation -, 1138
Riemann -, 677
unordered -, 1458
weighted -, 239
graphical
interface design, 361
MOP, 813
Grassmann scheme, 1416
Greek prosody, 1324
greeking, 363, 381, 1103
Green function, 1189, 1471
Gregorian chant, 509 1321
Gregorian musical figures, 1328
Greimas, Algirdas Julien, 774–776
grid
dactylus -, 216, 217
driving -, 787
vector
horizontal -, 788
vertical -, 788
Grondin, Jean, 1050
groove
effect, 788
gestural -, 1303
Grothendieck
topology, 149, 353, 1430
topos, 1431
Grothendieck, Alexander, x, xii, 145, 149, 153, 351, 353, 358, 850, 908, 1031, 1038, 1173, 1177, 1179, 1430
ground class, 313
group, 1378
affine counterpoint -, 391
algebra, 925
automorphism -, 144, 1391
cohomology -, 1447
contrapuntal -, 115
control -, 775
cyclic -, 1381
enharmonic -, 426
fiber -, 775
fixpoint -, 1378
free -, 1379
fundamental -, 920, 1475
Galois -, 1460
Hegel -, 1066
homomorphism, 1378
isomorphism, 1378
isotropy -, 1378
Klein -, 206, 451
linear counterpoint -, 391
opposite -, 1378
p-Sylow -, 1380

paradigmatic -, 390
of isometries, 394
product -, 1380
quotient -, 1379
rhythmical -, 810
simple -, 1379
Sylow -, 179
symmetric -, 1378
symmetry -, 144, 180, 470, 672
theory, 157, 212
torsion -, 1381
group-theoretical method, 198, 205
grouping, 376, 415, 608, 629
bar -, 713
concept, 250
hierarchical -, 611
instrumental -, 599
metrical -, 248
of sounds, 74
rules, 610
stemmatic -, 633
structure (=G), 377
time -, 99
groupoid, 922
fundamental -, 920 1475
Growth Point, 854
GTTM, 255, 376, 619
Guérin, Michel, 850
GUI, network -, 1096
Guitart lemma, 1006
Guitart, René, 1006
gyri, Heschl's -, 1360

H

Hölderlin, Friedrich, 9, 116
Hüllakkord, 431
Haegi, Hans, 291
hair cell, 1356
Hajós group, 310, 314
Hajós, Gyorgy, 310
Halle, Morris, 234
Halsey, George, 309, 313
Hameer, 1305
Hamilton operator, 1493
Hamilton's variational principle, 1243, 1481
Hamilton, William, 1385
Hamiltonian, 658
hammer, 1354
Han singing, 1321
hand, 855, 912, 915
impairment, 881
left -, 1289
pianist's -, 911, 912
hand's position, 1192
hands
gestural interaction of both -, 1299
shapes of the -, 1310
handshape, 1311
hanging orientation, 108, 509
Hanslick, Eduard, 15, 248, 251, 775, 828, 1181
harkat, 1306
HarmoRUBETTE®, 744
harmolodic, 794
harmonic
coherence, 447
knowledge, 485
logic, 449
minor, 479
morpheme, 449
morphology, 908
motion, 415
path, 482
progression, 128
semantics, 436
strip, 253, 262, 442
tension, 481, 482
topology, 443
weight, 482, 647
harmonic minor
scale, 472, 474
tonality, 460
harmonical-rhythmical scale, 794
harmony, 90, 181, 523
complete -, 827
jazz -, 276
Keplerian -, 30
Riemann -, 263
rubette for -, 1098
HarmoRUBETTE®, 448, 481, 648, 674, 714
Harnoncourt, Nicolas, 751, 754
Harris, Craig, 209
Hashimoto, Shuji, 607, 613
Hasse diagram, 218, 1374
specialization -, 220
hat, Mexican -, 1345
Hatten, Robert S., 871, 889, 898, 899, 909, 950, 979, 1265
Hauptmann, Moritz, 436
Hausdorff
locally compact -, 983
topology, 1445
Hausdorff, Felix, 1445
Hausegger, Friedrich von, 1293
Hawking, Stephen, 1187
hayashi, 13
Haydn, Joseph, 241, 826, 827
Hazlitt, William, 685
head, 914, 1457
head voice, 1315
heartbeat, 607
Hebb, Donald O, 719
Hegel

body, 1066, 1070
box, 1072
group, 1066
Hegel, Georg Wilhelm Friedrich, 35, 748, 781, 1040, 1066, 1174
Heidegger, Martin, 854
Heijink, Hank, 761
helicotrema, 1355
Helmholtz, Hermann von, 509, 1364, 1365
Hemmert, Werner, xi
Hentoff, Nat, 603
Hermann, Jakob, 1048
hermeneutics, 1051
unicorn of -, 13
Hertz, 1340
Hervé, Jean-Luc, 817
Heschl's gyrus, 524, 1360
Hess system, 524
Hesse, Hermann, v, 164, 573, 1177
Hewitt, Edwin, 309, 313
hexagram of Pascal, 1023
hexameter, 213
Heyting
algebra, 103, 335, 930, 934, 1433
category of - algebras, 930
logic, 436, 1460
Heyting, Arend, 335, 436, 1433
Hichert, Jens, 517, 534, 536, 538, 543
hidden symmetry, 114
hierarchical
decomposition, 710, 720
grouping, 611
organism, 249
smoothing, 708
hierarchy, 250, 555, 631
cellular -, 590, 596, 1254
closure, 588
deformation, 658
metrical -, 375
of performance development, 622
parallel -, 591
performance -, 556
phrasal -, 1311
piano -, 593
space -, 588
standard -, 590
tempo -, 624
violin -, 593
high level gesture, 1041
Hilbert space, 1486
Hindemith, Paul, 123, 414, 422
Hindustani
musician, 1305
vocalist, 1301
Hintergrund, 415
hippocampal
gate function, 528
memory function, 528
hippocampus, 524, 528
histogram, 716
historical
approach, 465, 472
dimension of music, 91
dynamics, 221, 223
instrumentation, 324
localization, 223
musicology, 328
process, 627, 634
rationale, 826
reality, 487
historicity in music, 220
history, 5
development -, 612
encapsulated -, 556
of mathematics, 1040
of music, 6
hit point, 582
problem, 580
Hjelmslev, Louis, 14, 16, 328, 849, 1183
Ho, Jocelyn, vi, vii
Hofmann, Ernst Theodor Amadeus, 248
Hofstadter, Douglas, 1022
holomorphic, 1465
homeomorphism, 1444
homology, 1475
functor, 1476
module, 966, 1475
network -, 986
singular -, 965, 1476
theory, 967
homology module, hypergestural -, 970
homomorphism
boundary -, 966
diaffine -, 1396
dilinear -, 1392
group -, 1378
Lie algebra -, 1408
linear module -, 1391
monoid -, 1376
ring -, 1385
structural -, 1385
homotopy, 916, 920, 957, 1181, 1185, 1289, 1447, 1474
relative -, 1447
category, 1474
commutative, 1474
equivalence, 1474
inverse, 1474
theory, 1474
type, 1474
Honing, Henkian, 547
Horace, 621
horizontal

grid vector, 788
poetical function, 781
poeticity, 214
Horowitz cluster, 735, 739
Horowitz, Vladimir, 730, 735, 736, 738, 768, 899, 909
hue, 1115
Hul, Bopha, 1276
human
conceptualization, 145
expression, 570
gesture, 1314
precision, 623
human phonatory system, 1314
humanism, 828
humanities, 164, 225
experience in the -, 31
experimental -, 27
Husmann, Heinrich, 1366
hypergestural
homology module, 970
singular homology, 1236
hypergesture, 870, 894, 915, 937, 959, 967, 985, 1036, 1184, 1185
global -, 1223
non-singular -, 1247
parallel -, 1245
spatial -, 917
vocal -, 1316
hypermedia, 38
hypernetwork, 985
hyperouranios topos, 21, 38
hyperset theory, 1094
hypothesis, mimetic -, 1264, 1276

I

I, 22
I, the, 892
ICMC, 197, 612
icon, instrumental -, 784
Ide, Takefumi, 1274
idea
compositional -, 322
musical -, 774
ideal, 1386
left -, 1386
right -, 1386
ideas, intermediate -, 1010
idempotent, 1376
component, 1377
identical reproduction, 891
identification
conceptual -, 229
enharmonic -, 425
esthesic -, 248
identifier, 1438
form -, 55, 56
gestural form -, 1087
identity, 405, 408, 1419
of a point, 147
abstract, 14
Euler's -, 1341
Jacobi -, 1408
of a work, 14
resolution of the -, 1494
slice, 275
ideology, Fourier -, 233
IL, 1434
image, 1375
denotator -, 59
direct -, 1411
inverse -, 1375
imaginary
gesture, 1042, 1314, 1315
time, 957, 1189, 1297, 1300
world, 1058
imagination, 5, 1032
imitation, 405, 407
immanent analysis, 383
impairment, hand -, 881
imperfect, consonance, 522, 532, 541
implementation, 627
implication, 346, 1432, 1433
importance, relative -, 648
improvisation, 40, 1278, 1295
jazz -, 179
melismatic -, 1303
improvised music, listen to -, 1295
in absentia, 16, 789
in praesentia, 16, 789
in-time music, 817
inbuilt performance grammar, 753
incarnation, 847
included, literally -, 208
inclusion
abstract -, 208, 210
literal -, 210
incomplete semiosis, 330
incorrect, politically -, 752
indecomposable, 1395
space, 588
independence, cognitive -, 179
index
cycle -, 1382
diachronic -, 223
function, 1390
set, 1373
indiscrete
interpretation, 254
topology, 1443
individual variable, 1435
ineffability, 22, 570
ineffable, 1285

infinite, 1371
interpretation, 259
message, 751
performance, 549
infinitely small, 569
infinitesimal, 638
information, 35
paratextual -, 685
system, geographic -, 667
InfoRUBETTE®, 668
ingenium, 1012
ingenuity, 1015
inharmonic, 675
inharmonicity, 237
inheritance, 628, 802
principle, biological -, 627
property, 395
initial, 1424
articulation, 578
condition, 1452
design matrix, 724
moment, 574
performance, 586, 597
set, 573, 586, 596, 1254
polyhedral -, 580
value, 562
initial set, distance to an -, 580
initialization, dynamical -, 578
injective, 1372
module, 1407
inlet, 805
inner
derivation, 1408
ear, 1355
logic, 623
score, 548, 571, 1295
innervation, mimic -, 891
input
filtering, 760
real-time -, 783
inspector, chord -, 675
instance, 802
instantiation, 802
instinctive activity, 623
instrument
makeshift -, 1279
name, 69
space, 597
instrumental
condition, 701
expression, 825
gesture, 817, 902, 1318
grouping, 599
icon, 784
parameter, 832, 835
technique, 832
variety, 555
vector, 835
voice, 220
instrumentation
historical -, 324
orchestra - denotator, 69
instrumentum, 38
INTEGER, 44
integer, 1385
integral
curve, 1245, 1453
of perspectives, 325
surface, 1456
integrated serial motif, 195
integration
Lebesque -, 1488
method, 683
intellectuality
mathematical -, 1006
musical -, 1006
intelligence, 873
intensification, 799
intension, 330, 427
basic -, 427
topology, 429
intensity, 608
inter-corporeal, 1274
inter-period coherence, 769
inter-sensoriality, 1274
inter-sensory, 1264
interaction
interpretative -, 720
matrix, 709
interactive control, 813
interface design, graphical -, 361
interictal period, 524
interior, 1443
interiorization, 893
interlude, 689
intermediate
ideas, 1010
performance, 622
internal
category, 1464
movement, 897
structure, 254
interpolation, 570, 760
field -, 761, 764
flow -, 582
interpretable, 989
composition, 304
automorphism group of -, 306
molecule, 292, 316
interpretation, 220, 222, 549, 989
discrete -, 254
indiscrete -, 254

infinite -, 259
iterated -, 258
just triadic degree -, 265
metrical, 268
motivic -, 271, 384
of a local composition, 258
of weights, 659
rhythmical -, 268
semantic -, 490
silly -, 254
singleton -, 275
sketchy -, 622
tangent -, 558
tetradic -, 276
third chain -, 260
triadic -, 276, 450, 455, 466
triadic degree -, 261
interpretative
activity, 246, 251, 252
interaction, 720
interpreter, 892, 893
interspace, 569
sequence, 192
structure, 192
interval
unordered p-space -, 207
unordered pc -, 207
class content vector, 207
consonant -, 414, 527
content, 207
contrapuntal -, oriented -, 509
cul-de-sac -, 538
cyclic - succession, 207
dichotomy, 518
dissonant -, 527
function, 204
multiplication, 512
ordered p-space -, 207
ordered pc -, 207
succession, 207
cyclic -, 208
m^{th} -, 208
successive -, 526
time -, 70
vector, 208, 210
interval-class vector, 204
intonation, 562, 563
curve, 563
field, 563
intratextual, 329
introversive semiosis, 329
intuition
geometric -, 1036
mathematical, 1027
musical -, 202
intuitionistic logic, 443, 934, 1434
invariance
transformational -, 226, 271
vector, 208
invariant pcset, 208
invent problems, 1042
invention, 1001
method of -, 1039, 1044
inversa, 692
inverse, 1378
ethnology, 754
function, 1373
homotopy -, 1474
image, 1375
left -, 1378
performance theory, 611, 651, 756
right -, 1378
inversion, 117, 247, 262
chord -, 420
real, 124
retrograde -, 61, 120
tonal, 124
tonal -, 788
inverted weight, 681
invertible, 1385
invisible gesture, 1302
IRCAM, v, 238, 801, 1314
irreducible, 589
component, 269
degree system, 457
topological space, 1413
iso, 1420
isometry, 250, 1449
isomorphic, 1378
category, 1421
isomorphism, 1420
group -, 1378
monoid -, 1376
ring -, 1385
isomorphism classes
of local rhythms, 181
of chords, 180
isotropy group, 1378
isotypic tesselation, 309
ISPW, 761
Issigonis, Sir Alec, 1044
istesso tempo, 555
Italian conducting school, 1292
iterated interpretation, 258

J

Jackendoff, Ray, 250, 254, 376, 379, 619, 721
Jacobi identity, 1408
Jacobi, Carl, 1408
Jacobian, 1190, 1247, 1451
Jacobson, Nathan, 1395

Jakobson, Roman, 16, 116, 157, 212, 222, 234, 241, 250, 329, 774, 781, 1050
James, William, 873
Jandl, Ernst, 799
Japanese philosophy, 1266
Jaques-Dalcroze, Émile, 875
Jauss, Hans Robert, 155
Java, 666
Java2D, 764
jazz, 12, 13, 40, 179, 571, 909
- American -, 442
- CD review, 342
- free -, 13, 548, 900, 1177
- harmony, 276
- improvisation, 179
- lead-sheet notation, 438

JCK, 342
Jefferson, Blind Lemon, 885
jnd, 229
Johnson, Tom, 488, 789
join, 1433
journalistic criticism, 730
joystick, 895
Józef Marja Hoene-Wroński, 319
`JSynNote`, 1111
Julia set, 162
Julia, Gaston, 162
Junod, Julien, 543, 1097
Jurek, Thom, 877
just, 93
- chromatic octave, 96
- class chord, 94
- modulation, 475
- scale, 96
- triadic degree interpretation, 265
- tuning, 1350

just-tempered tuning, 1351
justest
- scale, 266
- tuning, 460

juxtaposition, 61

K

K-net, 898, 924, 1182
k-partition, 310
Köhler, Egmont, 189
Körpersinn, 1182
Kagel, Maurizio, 127, 325
kairos, 826
Kaiser, Joachim, 248, 495, 752
Kan, Daniel, 907
kansei, 607, 613
Kant, Immanuel, ix, 9, 21, 29, 38, 145, 1043
Karajan, Herbert von, ix, 577, 608, 783, 1292
Karg-Elert, Sigfrid, 115, 415
Kashalkar, Ulhas, 1306
Katsman, Roman, 847
Kayser, Hans, 1177
Kendon continuum, 854
Kendon, Adam, 853, 899, 909, 1180, 1301
Kepler, Johannes, 114
Keplerian harmony, 30
kernel, 597, 1379
- Naradaya-Watson -, 709
- Priestley-Chao -, 744
- smoothing, 708, 721
- smoothing -, 708
- symbolic -, 585
- view, 680

kernel (of a performance cell), 1254
key, 784
- function of music, 529
- musicogenic -, 529
- signature, 631

keyboard, 915
- piano -, 912

Keyser, Cassius, 1052
khyal, 1309
killing, address -, 168
kindred, 240
kinesthetic awareness, 1281
kinetic model, 1306
Kircher, Athanasius, 198
Klangrede, 17, 827
Klein group, 206, 391, 451
Klumpenhouwer, Henry, 898, 980, 1182
Klumpenhouwer network, 924
knot in FM synthesis, 73
knowledge, 35, 361
- crash, 339
- harmonic -, 485
- hiding, 198, 362
- human -, 5
- management, dynamical -, 329
- ontology, 345
- private -, 27
- space, 9, 27
- theory of -, 1005

Koenig, Thomas, 687
Kollmann, August, 827
Kopiez, Reinhard, 237, 604, 605
KORG, 1346
Kronecker delta, 1392
Kronecker, Leopold, 1392
Kronman, Ulf, 607, 896
Krull, Wolfgang, 1396
KTH school, 617, 621
Kubalek, Antonin, 735
Kunst der Fuge, 115, 203, 249, 608, 688, 699
Kuratowski closure operator, 1237, 1443
Kuriose Geschichte, 628, 629, 634, 701, 711
Kurth, Ernst, 1293

Kurzweil, 701

L

l, 1348
λ-*abstraction*, 803
l-adic cohomology, 379
L-system, 782
Lüdi, Werner, 548, 791
Lévi-Strauss, Claude, 487
Laban, Rudolf, 874
laboratory
  conceptual -, 30
  distributed -, 32
Lagrange
  density, 1482
  potential, 1218
Lagrangian, 658, 1481
  action, 1188, 1481
  density, 1185
Lakoff, George, 859, 1181
Lamé, Gabriel, 1027
Landry, Elaine, 908
landscape, emotional -, 241
Langer, Jörg, 605
Langer, Susan, 604
language, 17, 853, 893
  common -, 22
  denotator -, 595
  geometric -, 893
  natural -, 1034
langue, 17
Laplace operator, 1471
large
  form, 1298
  performance of a - orchestra, 626
largest coefficient, 733
laryngeal movements, 1316
larynx, 1316
Lasker, Emanuel, 1027
Latin square, 1024
lattice, 1433
law
  exponential -, 1464
  Weber-Fechner -, 1348
Lawrence, David Herbert, 751
Lawvere, William, 149, 357, 934
layer
  action-to-cognition -, 870
  cognitive -, 869
  RUBATO®-, 668
layers
  of reality, 10
  time -, 1299
Lazier, Rebecca, 933
lazy path, 1375
LCA, 525
Le Guin, Elizabeth, 1264
lead-sheet notation, 441, 571
leaf, 949
Leap Motion, 1143
learning
  by doing, 32
  process, 556
leaves of a stemma, 628
Lebesque integration, 1488
left
  action, 1378
  adjoint, 1423
  coset, 1379
  hand, 1289
  ideal, 1386
  inverse, 1378
legato, 645
LEGO, 782
Lehmann, Andreas, 743
Leibnitz, Gottfried Wilhelm, 465, 743, 869, 1178
Leitfaden, vii
λεκτον, 16
Leman, Marc, 902, 1182
lemma
  Fitting's -, 1396
  Guitart -, 1006
  Yoneda's -, 938
length, 268, 1395
  minimal -, 689
  of a local meter, 97
  path -, 1375
lens space, 922
LEP, 196
Lerdahl, Fred, 237, 238, 240, 250, 254, 376, 379, 619, 721
Les fleurs du mal, 219, 799
level, 269
  connotative -, 16
  denotative -, 16
  F-to-enter -, 727
  function, 269, 1374
  meta -, 17
  metrical -, 377
  neutral -, 211, 956
  object-, 17
  sound pressure -, 1348
levels of reality, 10
Levelt, Wilhelm, 1364, 1367
Levi, Beppo, 1040
Lewin, David, 70, 203, 204, 309, 315, 409, 889, 898, 908, 937, 938, 951, 956, 980, 1177, 1182, 1236
Lewis, Clarence Irving, 570
lexical, 344
lexicographic ordering, 50, 76, 1373
Leyton, Michael, xi, 773, 775
LH, 629
library, 51, 367

Lidov, David, 1265
Lie algebra, 1408
  homomorphism, 1408
  linear -, 1408
  bracket, 1246, 1408, 1455
    affine -, 444
  derivative, 1455, 1469
  formalism, 658
  operator, 638
Lie type operator, 1255
Lie, Sophus, 1408, 1455
Ligeti, György, 30, 1179, 1295
limbic
  structure, 524
  system, 528, 606, 1361
limit, 1424
  circular -, 65
  form, 57
  ring, 1387
  topology, 1444
limited
  modulations, 481
  transposition, 126
    mode with -, 126
limiting functor, 962
line, 1375
linear
  (in)dependence, 1393
  algebra, special -, 1408
  case, 205
  category, 923
  combination, 1391
  counterpoint group, 391
  dual, 1397
  fibration, 758
  Lie algebra, 1408
  module homomorphism, 1391
  ordering, 1373
    on a colimit, 77
    on a limit, 77
    on finite subsets, 77
  representation, 1393
  visualization
    generic -, 362
    metrical -, 362
linear ordering among denotators, 50
linearization, 923
linguistics, 160
  structuralist -, 250
Lipschitz, locally -, 1452
Lipschitz, Rudolf, 1452
LISP, 439
list form, 809
listen to improvised music, 1295
listener, 11, 13
listening
  articulated -, 249
  modes of -, 1281
  music -, 1353
  procedure, 610
Liszt, Franz, 16, 18, 495, 1076
literally included, 208
Lloyd, Sam, 1025
Lluis Puebla, Emilio, xi
local, 245
  action, 1038
  character of a contrapuntal symmetry, 533
  composition, 75, 89, 90
    commutative -, 105
    embedded -, 106
    functorial -, 101
    generating -, 106
    morphism, 105
    objective -, 90, 979
    sequence of a -, 192
    wrapped as -, 91
  compositions
    coproduct of -, 105
    fiber sum of -, 140
    product of -, 104
  functor, 926
  meter, 97
    length of a -, 97
    period of a -, 97
  meters, simultaneous -, 500
  morphism, 1411
  network, 980, 1253
    faithful -, 980
    flat -, 981
  optimization, 676
  orientation, 264
  Para-meter, 267
  performance score, 595, 1254
  rhythm, 99, 106
  ring, 1395
  score, 250, 783
  solution, 1452
  standard composition, 293
  symmetry, 533, 534
  technical parameter, 835
  threshold, 675
local topography, 17
local-global patchwork, 250
locality principle, 762
localization, 1406
  historical -, 223
  of epilepsy focus, 524
  of musical existence, 21
locally
  closed, 1445
  compact, 1464
  finite, 1446

free, 1413
linear grammar, 660
Lipschitz, 1452
ringed space, 1411
trivial structure, 252
locally almost regular manifolds, category of -, 974
locally compact Hausdorff, 983
Lochhead, Judith, 1264
Locke, John, 1010
locus, Riemann -, 676
logarithmic perception, 550
LoGeoRUBETTE®, 668
logic, 345, 1039, 1067
absolute, 145
classical -, 1434
fuzzy, 336
harmonic -, 449
Heyting -, 436, 1460
inner -, 623
intuitionistic -, 443, 934, 1434
musical -, 264
of orbits, 200
of toposes, 227
performance -, 556
performing -, 774
predicate -, 436
topos -, 930
logical, 1430
algebra, 1433
connective symbol, 1432
functor, 946
motivation, 639
switch operator, 60
time, 502
loop, 1375
Lord, John, 189
Loria, Gino, 1048
loudness, 44, 67, 608, 1348
LPS, 595, 621, 1254
Lubet, Alex, vi
Luening, Otto, 250
Lussy, Mathis, 615

M

m^{th} interval succession, 208
M-theory, 1485
M.M., 552, 562
Mälzel, Johannn Nepomuk, 341
Möbius
bottle, 558
strip, 476
Möbius strip, 252, 263, 442
Müller, Stefan, v, 1088
Mälzel, Johannn Nepomuk, 552, 562, 571
Mälzel metronome, 28, 1347
Möbius strip, 451, 781
Müller, Stefan, xi, 910
Mac Lane, Saunders, 908, 1035, 1178
Mac OS X, 665, 669
Machado, Antonio, 1004
machine, 1037
Γ-, 1038
performance -, 703
precision, 623
Turing -, 553
MacLean, Paul, 528
macro, 270
-event, 74
germ, 270
macrogesture, 1311
MacroScore, 1106
Mahler, Gustav, 1287
Maiguashca, Mesias, 59, 115
Majithia, Roopen, 886
major, 122, 541
dichotomy, 519, 543
mode, 448
scale, 473, 474
dominant role of -, 541
third, 1350
tonality, 460, 478
bigeneric -, 449
major-minor problem, 122
makeshift instrument, 1279
making music, 22
Mallarmé, Stéphane, 850
Malt, Mikhail, 811
manifold, 251, 913
J-, 1062
almost regular -, 974
categorical -, 993
differentiable, 1466
formant -, 238
musical -, 241
of opinions, 828
semantic -, 241
tangent -, 1467
with boundary, 1467
Manin, Yuri, 1177, 1181
Mannone, Maria, vi, vii, 1219
Mansouri, Baya, 1060
map, 1372
characteristic -, 1428
coboundary -, 1447
performance -, 586
refinement -, 275
simplicial -, 1445
mapping, gesture-sound -, 1279
maquette, 811
Marceau, Marcel, 865, 875
Marek, Ceslav, 248, 910
marked

counterpoint dichotomy, 518
class, 518
dichotomy
autocomplementary -, 518
class, 518
rigid -, 518
strong -, 518
interval dichotomy, 518
Marquis, Jean-Pierre, 908
Marrou, Henri-Irénée, 1040
Marx, Adolf Bernhard, 496
Maschke, Heinrich, 1395
Mason, Robert, 108, 466
Mason-Mazzola theorem, 109
Mason's theorem, 109
mass-spring, 1346
Massinger, Philip, 679
master, concert -, 626
matching, 761
of structures, 718
score-performance -, 761
material
change of -, 813
experimental -, 330
musical -, 811
of music, 90
time, 502
Math-motif, 406
mathêmata, 1015
Mathematica®, 769
mathematical
creativity, 1002
existence, 145, 328
gesture, 1002, 1036, 1037, 1040
intellectuality, 1006
intuition -, 1027
model, 464
morphism, 282
object, 1021
overhead, 512
pulsation, 1002, 1007, 1026, 1033, 1041, 1042, 1058, 1059, 1064
mathematically equivalent morphisms, 282
mathematics, 5, 160
creative -, 1033
history of -, 1040
matrilineal, 628
scheme, 626
matrix, 898, 1392
category, 1420
comparison -, 206
design -, 724
initial design -, 724
interaction -, 709
of checkboxes, 1110
product, 1392
Riemann -, 447, 481, 675
value -, 766
verse -, 213
matrix-like sketch, 1057
Matterhorn, 151
Mattheson, Johann, 248, 827, 1365
MAX, 115, 210, 441, 789
Max/MSP, 1127
maximal, 313
meter
nerve topology, 379
topology, 268, 378
structure content, 1362
Maxwell's equations, 973, 1484
mayamalavagaula, 543
Mayer, Günther, 221
Mazur, Barry, 1032
Mazzola, Christina, xii
Mazzola, Guerino, vii, 218, 502, 503, 612, 721, 783, 890, 900, 1035–1038, 1045, 1163, 1173, 1176, 1295
Mazzola, Guerino, 1181
Mazzola, Patrizio, vi
Mazzola, Silvio, xii
McCullogh, Karl-Erik, 854
McNeill, David, 853, 899, 909, 1180
McPartland, Marian, 883
MDZ71, 239
mean
performance, 727
tempo, 727
meaning, 849, 853, 909
of sound, 241
paratextual -, 330
textual -, 330
topological -, 158
transformational -, 158
measurable, 1488
measure for complexity, 254
measurement, 28
mechanical
constraints, 912
dynamics, 608
model, 896
ritard, 896
mechanism
modulation -, 465
of conducting, 1291
mediante tuning, 1351
mediation, 774
meet, 1433
mela, 543
melakarta, 543
melismatic improvisation, 1303
MeloRUBETTE®, 744
melodic
charge, 610

creation, 1077
gesture, 1302
minor, 480, 541
motion, 1305
movement, 1302
variation, 794
melodic minor
scale, 473, 474
tonality, 460
melody, 226, 271
germinal -, 220, 221, 792
retrograde of a -, 114
melody-as-motion, 1307
MeloRUBETTE®, 384, 408, 647, 672
membrane
basilar -, 1356
Reissner's -, 1355
tectorial -, 1357
memory, 527, 528
function
hippocampal -, 528
mental
experiment, 550
organization, 35
time, 547
tone parameters, 67
mental 3D rotation, 872
Merleau-Ponty, Maurice, 854, 860, 867, 871, 1031, 1264
Mersenne, Marin, 1364
MES (Memory Evolutive System), 1060
message, 12, 24
infinite -, 751
passing, 802
messaging, 156
Messiaen
mode, 126
scale, 794
Messiaen, Olivier, 126, 127, 134, 794, 1266
meta-gesture, 1286
meta-object, 802, 811, 814
class, 814
protocol, 813
meta-programming, 801, 813
meta-vocabulary, 199
Metal, 1281
metal in the voice, 1315
metalanguage, 212
metalevel, 17
metamere, 1362
metaphor, 24
physical -, 1288
metasystem, 17
meter, 97, 375–381, 1347
beat -, 97
barline -, 97
local -, 97
method, 156, 802, 1005, 1007
analytical -, 1012
boiling down -, 648
continuous -, 639
dodecaphonic -, 775, 780
group-theoretical -, 198, 205
integration -, 683
of characteristics, 1456
of invention, 1039, 1044
operational -, 28
selection, 805
statistical -, 612, 673
metric, 1449
associated -, 1450
Euclidean -, 229
Minkowski -, 1484
metrical
analysis, 688
component, 267
globalization, 99
grouping, 248
hierarchy, 375
level, 377
linear visualization, 362
profile, 689
quality, 376
rhythm, associated -, 267
similarity, 163, 388
structure (=M), 377
weight, 375, 376, 646
MetroRUBETTE®, 744
metronome, 99
Mälzel -, 28, 1347
MetroRUBETTE®, 376, 670
Mexican hat, 236, 1345
Meyer wavelet, 1346
Meyer-Eppler, Werner, 1353, 1362
mezzoforte, 1349
Michel chromatic, 478
Michel-Angelo, 857
micro
-motif, 646
timing, 220
micro-gesture, 1223
micrologic, 569
microstructure
timing -, 719
microtiming, 1303
middle ear, 1354
middleground, 415
MIDI, 235, 783, 899, 912, 964, 1096, 1116, 1151, 1164
velocity, 913
Mikaleszewski, Kacper, 622
Milmeister, Gérard, 1035, 1088, 1096, 1103
mimesis, 891
cultural -, 874

mimetic
category, 892
hypothesis, 1264, 1276
mimic
gesture, 893
innervation, 891
mind/body problem, 1264
minimal
cadential set, 455
length, 689
Minkowski, Hermann, 310, 1382
Minkowski metric, 1484
minor, 122
harmonic -, 479
melodic -, 480, 541
mode, 448
natural -, 478
tonality, 478
mirror, 947, 956, 1268
gesture, 1270
neuron, 871, 885, 1264
Mittelgrund, 415
Mitzler, Laurentz, 1365
mixed
sketch, 1028
weight, 671
modal
structure, 314
synthesis, 1346
mode, 126
aeolian -, 261
authentic -, 261
consonant -, 450
dissonant -, 450
dorian -, 261
ecclesiastical -, 261, 540, 541
hypoaeolian -, 261
hypodorian -, 261
hypoionian -, 261
hypolocrian -, 261
hypolydian -, 261
hypomixolydian -, 261
hypophrygian -, 261
ionian -, 261
locrian -, 261
lydian -, 261
Messiaen -, 126
mixolydian -, 261
phrygian -, 261
plagal -, 261
rhythmic -, 502
with limited transpositions, 126
model, 1436
kinetic -, 1306
mathematical, 464
mechanical -, 896
notic -, 1306
physical -, 27
template fitting -, 1361
modelage, 1003
modeling
categorical -, 1001, 1003, 1045, 1046
physical -, 236, 701
models, category of known simple -, 1062
modes of listening, 1281
modification, 1004
of functional relations, 813
syntax -, 813
modular
affine transformation, 786
composition, 251
modularity dynamical, 666
modulatio, 463
modulation, 459, 463–486, 840, 951, 952
amplitude -, 236, 833
degree, 465
domain, 476
force, 466, 469
frequency, 236, 833
frequency -, 1111, 1342
just -, 475
mechanism, 465
path, 491
pedal -, 499
pitch -, 236, 833
plan, 498, 499
quantized -, 470
quantum, 466, 467, 470, 471
rhythmical -, 473, 501, 503, 794
theorem, 470
theory, 1243
topos-theoretic background of -, 467
well-tempered -, 469
modulations, limited -, 481
modulator, 73, 236, 470, 490, 491, 494, 1342
modulatory
architecture, 495
region, 486
module, 1391
as basic space type, 59
cohomology -, 1477
complex, 288
constant -, 288
flasque -, 304
of A-addressed forms, 289
representative -, 298
retracted -, 289
direct sum -, 1392
dynamically loadable -, 665
free -, 1392
functor, 143
homology -, 966, 1475

injective -, 1407
of a commutative local composition, 105
of global affine functions, 355
product -, 1392
projective -, 1406
semi-simple -, 1394
shaping -, 665
simple -, 1394
structuring -, 665
modules in music, 60
modus ponens, 1434
molecule, 291
global -, 292
interpretable -, 292, 316
Molino, Jean, 11, 13, 574
MOLS, 1025
moment, initial -, 574
Monk, Thelonious, 880, 1296
mono, 1420
monochord, 22
monogamic coupling, 632
monoid, 1376
algebra, 60, 1386
finite -, 1376
free -, 1376
free commutative -, 1376
homomorphism, 1376
isomorphism, 1376
morpheme -, 444
multigeneric -, 446
trigeneric -, 444
word -, 1376
monomorphism, 1420
monotone extraction, 1019
Montaigne, Michel de, 1040
Monteverdi, Claudio, 754
Monti, Georges, 1060
Montiel Hernandez, Mariana, xii, 274
mood, 604
MOP, 813
graphical -, 813
Morlet wavelet, 1345
morpheme
harmonic -, 449
monoid, 444
Morphemfeld, 417
morphic, 617, 1439
morphing, 788
morphism, 162, 908, 941, 1419
t-fold differentiable tangent -, 552
t-fold tangent -, 552
geometric -, 946
gestural -, 1287
global -, 246
local -, 1411
mathematical -, 282
mathematically equivalent -, 282
of denotators, 91
of form semiotics, 1439
of formed compositions, 291
of functorial global compositions, 274
of functorial local compositions, 130
of gestures, 915
of global molecules, 292
of local compositions, 105, 128–132
of objective global compositions, 274
of objective local compositions, 129
of performance cells, 586
spatial digraph -, 942
tangent -, 551, 557
morphology, harmonic -, 908
Morris, Robert, 203–205, 211, 314, 316, 409
Morrison, Joseph, 1346
MOSAIC, 1346
mosaic, 310
mother, 596, 619
primary -, 628, 629
prime -, 629
tempo, 562
mother performance, 984
motif, 100, 159, 228, 271
abstract -, 386
classification, 187–190
covering, 384
Reti's definition of a -, 404
rhythmic -, 504
serial, 125
space, 384
$\mathbb{Z}$-addressed -, 100
motif i, 1267
motif ii, 1269
motion, 604, 607, 897
accelerated -, 607
harmonic -, 415
melodic -, 1305
neurophysiological -, 897
sense of -, 608
tracking, 1290
trigger, 607
motivated, 16
motivation, 345
geometric -, 346, 347
logical -, 346, 639
motives, enumeration of -, 195
motivic
analysis, 214, 404
interpretation, 271, 384
nerve, 384
simplex, 385
weight, 408, 647
work, 276
zig-zag, 277, 781

motor action, 608, 897
Motte-Haber, Helga de la, 863, 933
movement, 893
internal -, 897
melodic -, 1302
tensed -, 532
movements, laryngeal -, 1316
Mozart, Wolfgang Amadeus, 189, 377, 490
MSC (Mathematical Subject Classification), 1034
Müller, Stefan, 756, 1441
multigeneric monoid, 446
multigraph, 1375
finite -, 1375
multimedia object, 362, 370
multiple, 1016
multiple-dispatching, 802
multiplication
interval -, 512
scalar -, 1391
multiplicity, 208
principle, 1013, 1064
multiverse, 1175
Mumford, David, 300
Murenzi wavelet, 1345
music, 3, 7, 9, 13, 22
absolute -, 774
alphabet of -, 90
atonal -, 204
christianization of -, 891
composition technology, 464
concept of -, 21
critic, 249
role of -, 752
criticism, 248, 634
critique, 751
definition of -, 5
deixis, 16
emotional function of -, 528
esthetics of -, 323
fact of -, 10
film -, 603
historical dimension of -, 91
history, 6
in-time -, 817
key function of -, 529
listening, 1353
material of -, 90
North Indian raga -, 1305
psychology, 237, 250
research, 7
semiotic perspective of -, 15
software, 250
syn- and diachronic development of -, 199
tape -, 250
theory, 669
thinking -, 22
music data, spectral -, 964
music theory
professional -, 202
topological -, 1236
musical
concepts, definition of -, 97
analysis, 612
composition, 30
constraint, 913
content, 1286
creativity, 1227
drama, 753
embodiment, 1314
gestalt, 90, 127
idea, 774
intellectuality, 1006
intuition, 202
logic, 264
manifold, 241
material, 811
onset, 1347
ontolog, 21
ontology, 845
process, 811
prosody, 220
reality, 142
semantics, 134
sign, 895
string theory, 910
taste, 529
tempo, 28
time, 892
topography, 17
unit, 90
musician
Hindustani -, 1305
performing -, 912
musicking, 1301
musicological
experiment, 30, 31
ontology, 328
musicology, 3, 13, 669, 719
cognitive -, 21
computational -, 21
historical -, 328
systematic -, 328
traditional -, 22, 28
Musikalisches Opfer, 121
Musin, Ilya, 1289
musique concrète, 250
Muzzulini, Daniel, 442, 456, 469, 481

N

n-chain, differentiable -, 975
n-circle, 423
n-cube, 553

singular -, 966
n-dimensional cube, 966
N-formed global composition, 291
n-modular pitch, 205
n-phonic series, all-interval -, 195
N-quotient, 296
Nagasawa, Nobuho, 1276
Nambu-Goto action, 1185, 1484
name, 64
instrument, 69
of a denotator, 45
of a form, 44
names, ordering on -, 76
naming policy, 44, 45, 58
Naradaya-Watson kernel, 709
narration, 773
narrativity, theory of -, 774
Nattiez, Jean-Jacques, 221, 249, 390, 779, 1178
natural, 1422
decomposition, 707
distance, 362
language, 1034
minor, 478
transformation, 1422
natural minor tonality, 460
natural sciences, experience in the -, 31
nature
exterior -, 29
interior -, 29
nature's performance, 765
navigation, 31, 38
address -, 140
conceptual -, 35
dynamic -, 40
productive -, 39
receptive -, 38, 39, 75
topographical -, 19
trajectory, 31
visual -, 361
negation, 346, 1432, 1433
neigborhood, 163
neighborhood, 226, 672, 1443
elementary -, 403
neo-Riemannian theory, 979, 988, 990
nerve, 776, 1445
auditory -, 1355
class -, 283, 309, 322, 988
discrete -, 254
functor, 988, 1445
motivic -, 384
of a global functorial composition, 282
of a global objective composition, 253
weight, 379
induced -, 379
network, 980
global -, 990
global Zarlino -, 992
GUI, 1096
homology, 986
Klumpenhouwer -, 924
local -, 980, 1253
non-interpretable global -, 991
Neuhaus, Harry, 622
νεῦμα, 159
Neumann condition, 1483
Neumann, John von, 1494
neumatic notation, 892
neume, 15, 158, 571, 890, 912, 1182, 1293, 1322, 1327
neural pitch processing, 1361
neuron, mirror -, 871, 885, 1264
neuronal oscillator, 606
neurophysiological motion, 897
neurosis semiotic -, 849
neutral, 11, 1342
analysis, 222, 250
element, 1376
level, 211, 956
neutral level, 12, 13
neutralization, 465
Newton, Isaac, 329, 1048
NeXT, 665, 687
NEXTSTEP, 665, 669
nexus, 314
Nicolas, François, 1006
Nietzsche, Friedrich, 850, 1033
nihil ex nihilo, 25
nilpotent, 1396
Node, 1106
Noether, Emmy, 115, 889, 1027, 1087
Noether, Max, 1027
Noh, 13, 342, 632
Noland, Carrie, 1265
Noll, Thomas, xii, 69, 181, 417, 421, 422, 425, 427, 432, 435, 443, 444, 449, 463, 469, 520, 523, 612, 675, 908, 980, 1088, 1377, 1441
non-commutative polynomials, 1386
non-interpretable
composition, 305, 308
global
composition, 990
network, 991
non-invertible symmetry, 127
non-lexical, 344
non-linear deformation, 640, 681, 733
non-linearity, 1359
non-singular hypergesture, 1247
nonparametric approach, 708
norealworld, 1053
norm, 17, 1450
normal, 1492
form, 209
Rahn's -, 209

order, 210
subgroup, 1379
normalization
diachronic -, 754
synchronic -, 754
normative
analysis, 377
gesture, 1305
norms equivalent -, 1450
North Indian raga music, 1305
not parallel, 654
notation, 891
American jazz -, 438
diastematic -, 1322
European score -, 67
lead-sheet -, 438, 441, 571
neumatic -, 892
square -, 1329
notched tone space, 1363
note
alterated, 107
anchor -, 625
satellite -, 625
symbol, 912
note-against-note, 509, 531
nothing, 1067
nothingness, 1068
notic
analysis, 1306
model, 1306
notional scenery, 1034, 1046
novelty, absolute -, 1062
number
complex -, 948
embedding -, 208
prime -, 227
ring, 1112
numbers
complex -, 1385
dual -, 508, 1387
rational -, 1385
real -, 1385
Núñez, Rafael, 859, 1181

O

object, 156, 802, 1005, 1419
description, 200
functional -, 1036
global -, 245
mathematical -, 1021
multimedia -, 362, 370
prototypical -, 229
visualization principle, 362
object-oriented programming, 48, 595, 627, 630, 633, 801, 802
objective
closure, 432
gesture, 1042
global - composition, 252
local - composition, 90
local composition, 979
trace, 101
Objective C, 665, 679
objectlevel, 17
objectystem, 17
observation, 28
OCR, 630
octave, 62, 1350
coordinate, 1350
period, 93
octave class, 117
ODE, 652, 683, 1451
Oe, Kenzaburo, 1266
Oettingen, Arthur von, 115, 123, 415, 423, 424, 1350
OFF, 913
off-line algorithm, 761
ON, 912
ON-OFF, 60, 964
ondeggiando, 592
oniontology, 845
onomatopoiesis, 16, 776
onset, 44, 67
abstract -, 125
musical -, 1347
origin, 97
physical -, 1347
self-addressed -, 70
time, 1335
weight, 99
ontological
atomism, 24
coordinate, 10
dimension, 17
perspective, 5
shift, 142
ontology, 9, 142, 149, 152, 328, 1001, 1018
musical -, 845
musicological -, 328
denotator -, 328
knowledge -, 345
musical -, 21
time -, 775
Onuma, Shiro, 1295
opacity, 1108
open
ball, 1449
formant set, 238
functor, 1416
semiosis, 330
set, 227, 1443
source, 666
covering of a functor, 1416

Open-Editor, 814
OpenMusic, 210, 315, 774, 782, 801–821
openness, 237
operation, 1132
  Boolean -, 784
  flattening -, 75, 270
  graph, 1138
operationalization, 201
operationalized thinking, 161
operator, 598, 616, 619
  Beran -, 744
  agogical -, 720
  articulation -, 591
  artistic -, 1204
  basis-pianola -, 655
  Beran -, 723
  boundary -, 969
  bounded -, 1487
  canonical -, 207
  differential -, 1470
  face -, 968
  field -, 652
  Hamilton -, 1493
  Laplace -, 1471
  Lie -, 638
  Lie type -, 1255
  performance -, 598, 612, 637–661, 1253
  physical -, 616, 651
  pianola -, 659
  prima vista -, 616
  smoothing -, 721
  splitting -, 649
  sub-path -, 483, 1388
  support -, 1336
  symbolic -, 616, 650
  tempo -, 653
  test -, 652
  Todd -, 619
  TTO -, 207
  validation -, 349
opinions, manifold of -, 828
opposite
  category, 1421
  group -, 1378
opposition, 16
optimal path, 675
optimization, local -, 676
Oram, Celeste, 1264, 1274, 1277
orbit, 1378
  set-theoretic -, 1417
  space, 1378
orbits, functor of -, 1417
Orchestervariationen, 115
orchestra instrumentation denotator, 69
orchestral
  gesture, 817
  globalization, 555
orchestration, 785
order, 1379, 1381
  normal -, 210
  of a PDE, 1455
ordered
  p-space interval, 207
  pair, 1372
  pc interval, 207
ordering, 362
  alphabetic, 50
  alphabetic -, 36, 38
  encyclopedic, 50
  lexicographic -, 50, 76, 1373
  linear -, 1373
    on a colimit, 77
    on a limit, 77
    on finite subsets, 77
  on
    coefficient rings, 79
    compound (naive) denotators, 51
    compound (naive) forms, 51
    coordinators, 76
    denotators, 75–83
    diagrams, 76
    direct sums, 79
    forms, 75–83
    identifiers, 76
    matrix modules, 79
    Mod, 78–80
    morphisms, 80
    names, 76
    simple forms, 76
    types, 76
    universal construction functors, 77
    $\mathbb{Z}\langle ASCII\rangle$, 79
  partial -, 1373
  powerset -, 51
  principle, 361
    on denotators, 49
  universal -, 39
ordinal, 1420
Oresme, Nicholas, 27, 548
organ of Corti, 1356
organic
  composition principle, 717
  principle, 163
organism
  cellular, 325
  hierarchical -, 249
organization
  degree of -, 718
  mental -, 35
orientation, 7, 262, 509, 1468
  hanging -, 108
  change of -, 509, 515, 531

geographic -, 38
hanging -, 509
local -, 264
ontological -, 9
recursive -, 19
sweeping -, 108, 509
oriented
contrapuntal interval, 509
global composition, 291
origin, 267
of onset, 97
original, 892
OrnaMagic, 780, 787–789
ornament, 592, 786
pattern, 202
OrnamentOperator, 646
Orsini, Marco, 1293
orthogonality principle, 762
orthonormal decomposition, 10
Orthonormalization, 725
oscillator, 605
neuronal -, 606
oscillogram, 606
Osgood, Charles, 163
ostinato, 811
ottava battuta, 541
outer
derivation, 1408
ear, 1354
hair
cell, 1356
pillar
cell, 1356
outlet, 805
output, *presto*®-, 784
oval window, 1354, 1358
overhead, mathematical -, 512
overloading, 802
Oxley, Tony, 863

P

p-group, 1380
p-pitch, 206
p-scale, 95
p-space, 206
p-Sylow group, 1380
Pólya, George, 1013, 1039
painting, 151, 783
pair
ordered -, 1372
polarized -, 532
simplicial -, 1447
topological -, 1474
Yoneda -, 1437
Palestrina–Fux theory, 540
Pallasmaa, Juhani, 855, 875
Palmer, Richard, 884
paper science, 145
Papez, James, 528
Paré, Ambroise, 29
Para-rhythm, 267
paradigm, 16
dodecaphonic -, 125
Fourier -, 232
general affine -, 134
gestalt -, 672
phonological -, 220
$\pi\alpha\rho\grave{\alpha}\delta\varepsilon\iota\gamma\mu\alpha$, 158
paradigmatic
concept, 229
field, 125
group, 390
strategy, 780
theme, 221, 222, 390
tool, 789
transformation equivalence, 212
paradigmatics, uncontrolled -, 165
parallel, 654
articulation
field, 567
crescendo field, 567
degree, 264
glissando field, 567
hierarchy, 591
hypergesture, 1245
admissible -, 1250
not -, 654
performance
field, 568
map, 567
process, 1139
space, 590
Parallelklang, 457
parameter
accessory -, 831
basis -, 67, 654
bow -, 833
cadence -, 454
color -, 835
essential -, 831
exchange, 133
geometric -, 832
global technical -, 835
instrumental -, 832, 835
local technical -, 835
pianola -, 67, 654
primavista -, 594
space, 357
stemma, 1257
system -, 473
technical -, 236
vibrato -, 833

parameter exchange, 134
parametric polymorphism, 802
paratextual, 632
  information, 685
  meaning, 330
paratextuality, 349
Parker, William, 863
Parlan, Horace, 875
Parncutt, Richard, 607
parole, 17
part, 246, 250, 273
  dilinear -, 1397
  extraterritorial -, 591
  translation -, 1397
partial, 72, 423
  ordering, 1373
partial differential equation (PDE), 1471
partials, 10
participation value, 525
particle physics, 466
partition, 1372, 1462
  intérieure, 13
partitioning, 210
Pascal, hexagram of -, 1023
Pascal, Pascal, 1023
passing, message -, 802
patch, 802, 811
patchwork, local-global -, 250
path, 1375
  category, 1388, 1458
  closed -, 1375
  harmonic -, 482
  lazy -, 1375
  length, 1375
  modulation -, 491
  optimal -, 675
patrilineal, 628
pattern, 202
pause, 69, 631
  general -, 644
pc, 207
pc-space, 207
pcseg, 207
pcset, 207
  invariant -, 208
PDE, 657, 1455
  quasi-linear -, 1455
PDE (partial differential equation), 1471
Peano axioms, 30
Peck, Robert, 990
pedal
  modulation, 499
  voice, 499
Pederson, Jimmi, 879
peer, 668
Peirce, Benjamin, 1053
Peirce, Charles Saunders, 1053
percept of self-motion, 897
perception, logarithmic -, 550
perceptional pitch concept, 1363
perceptual equivalence, 229
percussion, 220
  event, 503
perfect consonance, 522, 532, 541
performance, 1253
  body, 586
  cell, 585, 1254
  cells
    category of -, 586
    morphism of -, 586
  complexity of -, 548
  culture of -, 623
  daughter -, 984
  definitive -, 895
  development, hierarchy of -, 622
  dynamics of -, 658
  field, 564, 568, 585, 983
    parallel -, 568
    prime mother -, 631
  fields, algorithmic extraction of -, 759
  gestural -, 893
  grammar, 615
    inbuilt -, 753
  hierarchy, 556
  history
    fictitious -, 627
    real -, 627
  infinite -, 549
  initial -, 586, 597
  intermediate -, 622
  logic, 556
  machine, 703
  map, 586
    parallel -, 567
  mean -, 727
  mother -, 984
  nature's -, 765
  of a large orchestra, 626
  operator, 598, 612, 637–661, 1253
  plan, 622
  primavista -, 630
  procedure, 610
  real-time -, 607
  research, computer-assisted -, 629, 701
  score
    global -, 598
    local -, 595, 1254
  structural rationale of -, 325
  synthetic -, 610
  theory, 319, 324, 909, 912
    inverse -, 611, 651, 756
  tradition, 753

vector field, 1253
performance field, time -, 1191
PerformanceRUBETTE®, 583, 652, 654, 679, 733
performative
body, 1283
gesture, 1283
performer, 24, 895, 1263
performing
logic, 774
musician, 912
perilymph, 1355
period, 415
fundamental -, 1340
in the Euler module, 93, 94
interictal -, 524
octave -, 93
of a local meter, 97
of a Vuza canon, 313
of a Vuza rhythm, 312
of duration, 97
temporal -, 376
periodicity, 708
higher level -, 99
Perle, George, 203
permutation, 1373
perspective, 24, 150, 152, 324, 466
change of -, 324
f-, 275
of the composer, 246
ontological -, 5
variation of -, 151
perspectives, integral of -, 325
Peterson, Oscar, 875, 882
Petsche, Hellmuth, 524, 525
Pfeifer, Rolf, 873
phase, 1340
portrait, 1454
spectrum, 238, 1340
phaticity, 212
PHENICX, 1290
phenomenological difference, 756
phenomenology, 1001
gesture -, 848
phenotype, 782
philosophy, 5
Chinese -, 1278
denotator -, 153
gestural -, 893
Japanese -, 1266
of the body, 868
Yoneda -, 828, 1475
phonatory system
first part (diaphragm), 1314
human -, 1314
second part (larynx), 1314
third part (resonant space), 1314
phoneme, 234
phonological
equivalence, 215
paradigm, 220
poeticity, 215
photography, 151
phrasal hierarchy, 1311
phrasing, 248
physical
constraint, 913
gesture, 911, 1186, 1229, 1289
gesture curve, 912
metaphor, 1288
model, 27
modeling, 236, 701
onset, 1347
operator, 616, 651
pitch, 1349
ritard, 896
sound, 71
time, 547
tone parameters, 68
PhysicalOperator, 683
physics, 5
particle -, 466
physiological correlate, 897
p_i-rank, 1381
pianissimo, 607
pianist, 912
elementary gesture of the -, 1192
pianist's hand, 911, 912
piano
hierarchy, 593
keyboard, 912
Piano concert No.1, 201, 202
pianola
coordinate, 1347
deformation, 591, 656
operator, 659
parameter, 67, 654
space, 567, 588, 628
specialization, 659
piecewise smooth, 1340
Pierce, Alexandra, 889
Pinocchio, 370
pitch, 44, 67
-class, self-addressed -, 69
-class set, 203, 207
absolute -, 577
alteration, 53, 788
chamber -, 563, 1349
class, 93, 116
segment, 207
set, 207
concept, perceptional -, 1363
concert -, 576

cycle, 207
detector, central -, 1361
difference, 62
distance, 61
fundamental -, 437
mathematical -, 61
modulation, 236, 833
physical -, 1349
processing, neural -, 1361
segment, 207
spaces, 205
symbolic -, 68
pivot, 470
pivotal chord, 463
pixel, 343
Plücker, Julius, 1027
plan, performance -, 622
plane transformation, 786
Plato, 27, 38, 165
platonic ideas, 21
playing, 22
structural -, 1267
Plomp, Reiner, 1364, 1367
Podrazik, Janusz, 210
Poe, Edgar Allan, 773
poetic Ego, 214, 219
poetical
function, 16, 116, 212, 241, 248, 774, 781
functions
spectrum of -, 218
poeticity, 116, 212
vertical -, 214
horizontal -, 214
phonological, 215
poetics
timbral -, 241
verse -, 248
poetology, 211
poiesis, 12, 211
retrograde -, 14
poietic, 11, 1342
genealogy, 128
Poincaré, Henri, 1004, 1071, 1079, 1180
point, 147–149, 908, 1371
generic -, 228
absorbing -, 432
accumulation -, 1443
closed -, 228
concept, 145
etymology, 147
Euler -, 1349
faithful -, 431
full, 431
fully faithful -, 431
generic -, 269, 1413
identity, 147
turning -, 465
pointed topological space, 1474
pointer, 23, 38, 146, 849, 1183
scheme, 48
points, category of -, 981, 1255
Poisson equation, 1189, 1197, 1471
polarity, 519, 527
at x, 523
in musical cultures, 543
profile, 228
polarized pair, 532
politically incorrect, 752
Pollock, Jackson, 860, 1291
Pólya
enumeration theory, main theorems of -, 191
theory, 191
weight function, 191
Pólya, George, 190, 310
polygamic coupling, 632
polygon, 785
polyhedral initial set, 580
polymorphism
ad-hoc -, 802
parametric -, 802
polynomials
commutative -, 1386
non-commutative -, 1386
polyphony, 826
polyrhythm, 800
polysemy, 108, 164
Popelard, Marie-Dominique, 856
Popper, Karl, 828, 1039
Porphyrean tree, 157
portrait, phase -, 1454
position
general -, 322
hand's -, 1192
privileged -, 520
Posner, Roland, 214, 782
possible, world of the -, 1052
post-serialism, 201
potential
epileptiform -, 524
Lagrange -, 1218
sink -, 608
Pow, denotator, 1113
power
spectral -, 525
window, 524
powerset, 1375
form, 56
ordering, 51
type, 47
PR, 377
practising, 633, 634
pre-Hilbert space, 1341

pre-morphism, 338
pre-object, 338
precise conceptualization, 211
precision, 1298
  calculation -, 639
  conceptual -, 32
  human -, 623
  machine -, 623
PrediBase, 665
predicate, 337
  atomic -, 339
  connective, 1435
  deictic, 346
  global -, 454
  logic, 436
  mathematical -, 340, 345
  morphic -, 338
  objective, 338
  primavista -, 341, 632
    European -, 341
    non-European -, 341
  punctual -, 338
  PV -, 341
  relational -, 338
  shifter -, 344
  textual -, 447, 454
  variable, 1435
preferences, 675
prehistory of the string quartet, 826
Prélude op. 28, No. 4, 743
presemiotic, 848, 859, 861, 910, 1180, 1293
presence, 409, 647, 673
presentation, 1030
presheaf, 1422
  continuous -, 982
  gestural -, 962, 1086
pressure
  bow -, 833
  decrease factor, 1337
  variation, 1335
presto®, 41, 115, 201, 202, 219, 220, 239, 437, 525, 576, 624, 780, 782, 783–789, 791, 1163
  output, 784
Pribram, Karl, 980
Priestley-Chao
  kernel, 744
prima vista operator, 616
primary mother, 628, 629
primavista, 556, 630
  agogics, 629
  dynamics, 629
  parameter, 594
  performance, 617, 630
  predicate, 632
PrimaVista Browser, 1103
PrimavistaOperator, 630
PrimavistaRUBETTE®, 685
prime, 1389
  field, 1386
  form, 210
  mother, 629
    performance field, 631
  number, 227
  spectrum, 227, 239, 925, 1412
  stemma, 628
  vector, 62, 1351
primitive
  element, 1460
  gestural -, 901
  gesture of conductor, 1287
principal component analysis, 739
principal homogeneous G-set, 204
principle
  anthropic -, 465, 466, 543
  architectural -, 718
  concatenation -, 133, 512
  dialogical -, 828
  fractal -, 799
  locality -, 762
  multiplicity -, 1013, 1064
  normative -, 378
  object visualization -, 362
  of relevance, 15
  ordering -, 361
  organic composition -, 717
  organic -, 163
  orthogonality -, 762
  packing -, 362
  sonata -, 135
  variation -, 325
  visualization -, 1107
priority, 725, 734
privileged position, 520
problem, 1032
  Cauchy -, 1456
  context -, 674
  hit point -, 580
  mind/body -, 1264
  wild -, 756
procedure
  listening -, 610
  performance -, 610
  rule based -, 615
  rule learning -, 615
  statistical -, 198
  X-ray -, 892
process, 17, 330
  creative -, 1066, 1072
  historical -, 627, 634
  learning -, 556
  musical -, 811
  of conceptualization, 201

parallel -, 1139
unfreezing -, 912
view, 1138
process I, 1267
process II, 1270
product, 1374
ambient space, 104
Cartesian -, 1372
cartesian -, 1424
category, 1422
cellular hierarchy, 590
fiber -, 1387, 1424
group, 1380
matrix -, 1392
module, 1392
of local compositions, 104
of the cells, 587
ring, 1387
semidirect -, 1380
tensor -, 1388
topology, 1444
type, 46
weight function, 191
wreath -, 1380
production, 3, 4
of a musical work, 12
profile, metrical -, 689
program
canonical -, 325
change, 784
programme narratif, 774
programming
constraint -, 774, 801
functional -, 801
language
visual -, 801
object-oriented -, 48, 155, 595, 627, 630, 633, 801, 802
progression
harmonic, 249
chord -, 414
contrapuntal, 249
harmonic -, 128
projecting local composition, 178
projection, 128, 1374
projective
atlas, 296
functions, 297
module, 1406
Prokofiev, Serge, 183
prolongational reduction (=PR), 377
pronoun, 856
proof, 1014
chain, 1434
propagation, sexual -, 627, 628, 637
property
H^*-emergent -, 1062
inheritance -, 395
propositional
formula, 1435
variable, 1432
prosody
Greek -, 1324
musical -, 220
protocol, meta-object -, 813
prototype, 198
prototypical object, 229
pseg, 207
pseudo-metric, 1449
on abstract gestalt space, 394
psychological reality, 548
psychology, 5, 6
cognitive -, 179, 226
gestalt -, 90
music -, 237, 250
psychometrics, 163, 228
psychosomatic disease, 1288
Puckette, Miller, 761
pullback, 1424
pulsation, 1036
mathematical -, 1002, 1007, 1026, 1033, 1041, 1042, 1058, 1059, 1064
rhythmical -, 1326
punctus contra punctum, 931
pure expressivity, 606
pushout, 1424
PVBrowserRUBETTE®, 668
Pythagoras, 436
Pythagorean, 30
school, 97, 340
tonality, 461
tradition, 22, 154, 1364
tuning, 266, 478, 1350
Pythagoreans, 11

Q

quadrant, 1281
quale, 570
qualifier
after -, 815
before -, 815
quality, metrical -, 376
quantifier
existence -, 346
formula, 1435
universal -, 346
quantization, 689
quantized modulation, 470
quantum, modulation -, 466, 467, 470, 471
quantum mechanics (QM), 417, 425, 1485
quartet, string -, 774, 825
quasi-coherent, 1413
quasi-compact, 1445

quasi-homeomorphism, 1414
quasi-linear PDE, 1455
quaternions, 1385
quatuor dialogué, 827
question, open -, 1072
quid, 1204
Quintilian, 856
quiver, 1375
  algebra, 1388
  complete -, 1375
  Riemann -, 482
  Riemann index -, 483
  stemma -, 660
quotient
  category, 1420
  complex, 289
  dominance topology, 231
  group, 1379
  ring, 1386
  topology, 1444

R

radical, 1376
  digraph, 926
  formoid, 928
Radl, Hildegard, 460, 473, 477, 480
Raffael, 154
Raffman, Diana, 22, 570
raga, 543, 1309
Rahaim, Matt, vi, vii
Rahn, John, 203–205, 409
Ramachandran, Vilayanur S., 873, 885
Rameau's cadence, 455
Rameau, Jean-Philippe, 414, 422, 436, 455, 1365
ramification
  mode, 42
  world-sheet -, 1230
Ramstein, Christophe, 901
random, 14
rank, 1392
  torsion-free -, 1381
Raphael, 165
ratiocentricity, 1175, 1177
rational numbers, 1385
rationale, 616
  gestural -, 753
  historical -, 826
Ratner, Leonard, 329
Ratz, Erwin, 466, 497
Ravel, Maurice, 183
RCA, 525
real, 683
  inversion, 124
  numbers, 1385
  performance history, 627
  time, 1189
  world, 1044
real-time
  algorithm, 761
  context, 760
  input, 783
  performance, 607
reality, 9, 10, 1018
  historical -, 487
  levels of -, 10
  mental -, 11
  musical -, 142
  physical -, 10
  psychological -, 11, 548
realization, geometric -, 988, 1446
reason, 5
recapitulation, 249, 496
receiver, 212
reception, 3, 4
receptive navigation, 75
recitation tone, 261
recombination, 813
  weight -, 640
reconstruction, 406
recording
  bipolar -, 524
  foramen ovale -, 524
recta, 692
recursive
  classification, 177
  construction, 42
  orientation, 19
  typology, 42, 49
reduced
  diastematic shape type, 387
  strict style, 541
reductio ad absurdum, 1018
reduction, 1400
  curvilinear -, 776
reductionism, 6
Reeves, Hubert, 167
reference
  denotator, 332
  tonality, 448
referentiality, 212
refinement, 1462
  map, 275
reflection, 814
  glide -, 1402
reflexive, 1373
reflexivity, 801
Regener, Eric, 204
region, modulatory -, 486
register, 784, 803
regression analysis, 711, 724, 726
regulae, 1016
regular

denotator, 66–71
division of time, 376
form, 64
representation, 1393
structure, 708
regularity, time -, 99
rehearsal, 556, 609, 612, 633, 634
Reichhardt, Johann Friedrich, 827
reification, 814
Reinhardt, Django, 876
Reissner's membrane, 1355
relation
causal -, 816
commutativity -, 1420
cross-semantical -, 612
equivalence -, 250, 1373
functional -, 1196
K -, 314
Kh -, 314
KI -, 314
temporal -, 816
relative
absolutely -, 1028
automorphism, 1460
delay, 236
dynamics, 685
homotopy, 1447
importance, 648
motivic topology, 400
symbolic dynamics, 68
tempo, 561, 685
topology, 1444
relative local
dynamical sign, 640
tempo, 643
relative punctual
dynamical sign, 640
tempo, 643
relevance, principle of -, 15
religion, Christian -, 847
Rellstab, Ludwig, 18
Remak, Robert, 1396
Remove-Element, 814
renaming, 47
reparametrization, 1185
repetition, 117, 1305
sequential -, 1304
replay, 117
Repp, Bruno, 719, 723, 739, 768
representable
functor, 1423
gesturally -, 962
representation, 1030
adic -, 1389
auditory -, 197
Fourier -, 740, 832
geometric -, 783, 892
linear -, 1393
regular -, 1393
score -, 610
textual -, 776
representative
first -, 180
module complex, 298
reprise, 799
reproduction, identical -, 891
res
cogitans, 1186
extensa, 1186
reset, 60
resolution, 357, 831, 840, 989
cohomology, 355
functor, 294
of a global composition, 294, 323
resolution of the identity, 1494
resolvent set, 1493
response, EEG -, 524
responsibility, collective -, 633
restriction
cellular hierarchy -, 590
of modulators, 498
scalar -, 107, 1392
resultant class, 313
resulting divisor, 313
retard, final -, 607
Reti, Rudolph, 165, 225, 376, 383, 403, 672, 721
Reti-motif, 406
retracted module complex, 289
retraction, 1420
retrograde, 14, 23, 119, 127, 134, 208, 247
address involution, 125
inversion, 61, 120, 134
of a melody, 114
retrogression, 207
reverberation time, 1337
reversed order
score played in -, 119
tape played in -, 120
sound, 121
revolution, experimental -, 29
Reye, Theodor, 1023
RGB, 1115
RH, 525, 629
rhetorical
expression, 570
expressivity, 556
shaping, 556
rhetorics, 827
rhythm, 117, 126, 375–381, 807, 1324
local -, 99, 106
throbbing -, 1271
Vuza -, 312

rhythmic
germ, 127, 266
mode, 502
motif, 504
scale, 503
rhythmical
group, 810
modulation, 473, 501, 503, 794
theory, 502
pulsation, 1326
structure, 794
rhythms, 97
classification of -, 312
local -, isomorphism classes of -, 181
Richards, Whitman, 607
richness, semantic -, 570
Richter, Sviatoslav, 622
Riemann
algebra, 481
dichotomy, 522
graph, 677
harmony, 263
index quiver, 483
locus, 676
matrix, 447, 481, 675
quiver, 482
transformation, 315
Riemann Mapping Theorem, 1466
Riemann, Bernhard, 251
Riemann, Hugo, 99, 123, 160, 201, 205, 251, 375, 414, 417, 436, 447, 449, 463, 469, 481, 484, 509, 523, 669, 674, 693, 714, 721
Ries, Ferdinand, 825, 827
right
action, 1378
adjoint, 1423
coset, 1379
ideal, 1386
inverse, 1378
right-absorbing endomorphism, 432
rigid, 262, 279, 466, 470, 473, 672
difference shape type, 386
marked dichotomy, 518
shape type, 386
rigor, 1039
Riguet, Jacques, 1055
Rilke, Rainer Maria, 850
ring, 1385
anti-homomorphism, 1385
endomorphism -, 1391
homomorphism, 1385
isomorphism, 1385
limit -, 1387
local -, 1395
number -, 1112
product -, 1387
quotient -, 1386
self-injective -, 1407
simple -, 1386
ringed space, 1411
ritard
final -, 896
mechanical -, 896
physical -, 896
ritardando, 608, 644
Rizzolatti, Giacomo, 873
RMI, 668
Roederer chromatic, 478
Rogers, Carl, 1032
Roland R-8M, 220, 791
role
exchange, 61
of a music critic, 752
of conductor, 1288
Rostand, François, 1043
Rota, Nino, 1292
rotation, 202, 207, 208, 786, 899
Amuedo's decimal normal -, 210
roughness, 1366
round window, 1358
Rousseau, Jean-Jacques, 248
row-class, 208
Rubato Composer, 947, 1095
RUBATO®, 41, 376, 628, 649, 650, 665–668, 719, 736, 759
concept, 665
Distributed -, 764
layer, 668
rubato
Chopin -, 550, 562, 576, 624, 765
encoding formula, 618
Rubato Composer, 979
rubette, 665, 669–686, 1096
BigBang -, 947, 1097, 1127, 1157
for counterpoint, 1097
for harmony, 1098
ScorePlay -, 1105, 1107
Select2D -, 1105
Rufer, Joseph, 125, 134
rule
based procedure, 615
contravariant-covariant -, 805
Fux -, 541
learning procedure, 615
preference - (=PR), 377
well-formedness - (=WFR), 377
rule-based grammar, 615
rules
differentiation -, 610
ensemble -, 610
grouping -, 610
Runge-Kutta-Fehlberg, 652, 683

Ruwet, Nicolas, 222, 249, 779

S

S-duality, 1184, 1484
Sabine's formula, 1338
Sachs, Klaus-Jürgen, 531, 532
Sacks, Oliver, 885
sacred denaturation, 891
Saint-Victor, Hugues de, 846, 862, 909, 1180
Salzer, Friedrich, 414
Sandall, Robert, 884
Sander, Emmanuel, 1022
Sands' algorithm, 310
Sands, Arthur, 310
sapat, 1306
satellite, 370, 1104, 1114
note, 625
saturation, 434, 1406
sheaf, 434, 444
Saussure, Ferdinand de, 15, 160, 199, 222, 250, 472, 854, 856, 867, 1180
Sawada, Hideyuki, 607
SC, 207
scala
media, 1355
tympani, 1355
vestibuli, 1355
scalar, 1391
multiplication, 61, 1391
restriction, 107, 1392
ScalarOperator, 683
scale, 94, 442
12-tempered -, 260
major -, 473
melodic minor -, 473, 474
altered -, 480
chromatic -, 418
diatonic -, 543
harmonic minor -, 472, 474
harmonical-rhythmical -, 794
just -, 96, 260
justest -, 266
major -, 262, 474
Messiaen -, 794
minor
harmonic -s, 262
melodic -, 262
rhythmic -, 503
whole-tone -, 541
SCALE-FINDER, 210, 441
SCALE-MONITOR, 210, 441
scales, common 12-tempered, 96
Scarlatti, domenico, 1157
scatterplot, 713
scenery, notional -, 1034, 1046
sceptical doubt, 1012
Schönberg, Arnold, 30, 90, 115, 125, 127, 134, 135, 183, 199, 201, 203, 204, 212, 247, 253, 262, 325, 413, 422, 442, 463, 465, 466, 502, 775, 780, 1243
Schäfer, Sabine, 687
Schaeffer, Pierre, 250
Schauder basis, 1209, 1486
scheme, 1414
diagram -, 1420
Grassmann -, 1416
matrilineal -, 626
mental -, 13
Molino's -, 11
sonata -, 503
Schenker, Heinrich, 270, 329, 414
scherzo, 791
Schmidt, Erhard, 1396
Schmitt, Jean-Claude, 846, 903, 909, 1180
Schneider, Albrecht, 902
Schoenberg, Arnold, 933
school
KTH -, 617, 621
Pythagorean -, 97, 340
Zurich -, 612
School of Athens, 154, 165
Schopenhauer, Arthur, 748
Schrödinger equation, 1493
Schubert, Franz, 214, 232, 792
Schumann, Robert, 407, 628, 629, 673, 701, 711, 743, 784, 785, 827, 894, 1260
Schweizer, Albert, 688
science
cognitive -, 611
computer -, 156
contemplative -, 27
doing -, 28
embodiment -, 868
experimental -, 29
paper -, 145
scientific art, 1045
scientific bankruptcy, 22
score, 11, 13, 60, 341, 783, 891, 899, 909, 912, 1105, 1263, 1287
concept, 250, 571, 754, 811
European - notation, 67
exterior -, 571
generic -, 548
global -, 250, 783
inner -, 548, 571, 1295
interior -, 13
local -, 250, 783
played in reversed order, 119
representation, 610
semantics, 574
transformation -, 785
score-following, 761
score-performance matching, 761

ScorePlay rubette, 1105, 1107
Scriabin, Alexander, 182, 482, 799
SEA, 384
section, 1412, 1420
  global -, 288, 1424
segment
  class, 206, 208
  pitch -, 207
Seifert-Van Kampen theorem, 929
Select2D rubette, 1105
selection
  axis of -, 116, 212
  field -, 803
  method -, 805
  stepwise forward -, 727
self-addressed
  chord, 185
  contrapuntal intervals, 514
  denotator, 69
  onset, 70
  pitch-class, 69
self-addressed arrow, 514
self-adjoint, 1492
self-injective ring, 1407
self-modulating, 1342
self-motion, percept of -, 897
self-referential, 19, 146
self-similar time structure, 799
semantic
  atom, 442
  charge, 403
    of EEG, 524, 525
  completion, 49
  depth, 383
  differential, 163
  interpretation, 490
  loading, 42
  manifold, 241
  richness, 570
semantics
  functional -, 445
  gestural -, 753
  harmonic -, 436
  incomplete, 83
  musical -, 134
  of weights, 408
  score -, 574
semi-simple module, 1394
semidirect product, 1380
semigroup, 1376
semiosis, 9
  extroversive -, 329
  incomplete -, 330
  introversive -, 329
  open -, 330
  paratextual -, 349
  textual, 334
  textual -, 348
semiotic, 1265
  component, 854
  constraints, 232
  gesture -, 1183
  marker, visual -, 813
  neurosis, 849
  of $\mathcal{E}$-forms, 1438
semiotical symmetry, 134
semiotics, 5, 14
  of sound classification, 240
semitone, 62
sender, 12, 212
sense of motion, 608
sentence, 1432, 1436
  valid -, 1433
sentic state, 604
sentograph, 895
separable extension, 1460
separating module complex, 296
sequence
  Cauchy -, 1450
  chord -, 486
  contrapuntal -, 531
  Fibonacci -, 340
  interspace -, 192
  of a local composition, 192
sequencer, 789
sequential repetition, 1304
sequentialization, 776
serial
  motif, 125
  technique, 127–128
serial motif
  integrated -, 195
  derived -, 194
serialism, 201
series
  all-interval -, 194, 200
  basic -, 125
  dodecaphonic -, 125, 162, 194, 247, 253, 325
  fundamental -, 115
  (k, n)-, 125, 194
  n-phonic -, 125, 194
  time -, 707
Serre, Jean-Pierre, 1027, 1476
Sessions, Roger, 889, 894, 1182
set, 1371
  Borel -, 1488
  cadential -, 455
  circular -, 66
  class, 207, 208
  closed -, 1443
  complex, 204, 314
    theory, 204

concept, 146
empty -, 146
fuzzy -, 163, 1029
in AST, 203
index -, 1373
initial -, 573, 586, 596, 1254
minimal cadential -, 455
of operations, 209
open -, 227, 1443
pitch-class -, 203
resolvent -, 1493
small -, 1420
source -, 203
support -, 252
theory, 249, 1028
SET-SLAVE, 209, 441
set-theoretic orbit, 1417
SETI (Search for ExtraTerrestrial Intelligence), 1175
seventh
dominant -, 419
natural -, 423
subdominant -, 420
tonic -, 420
sexual propagation, 627, 628, 637
SGC, 208
Shakespeare, William, 637
shape, 405, 1004, 1056
J-, 1062
type, 385
shape type
contrapuntal motion -, 387
diastematic -, 387
diastematic index -, 387
elastic -, 388
reduced diastematic -, 387
rigid difference -, 386
rigid -, 386
toroidal sequence -, 388
toroidal -, 388
shapes of the hands, 1310
shaping, 1148
module, 665
rhetorical -, 556
vector, 723
sharp, 109
sheaf, 1415, 1431
on a base, 1412
saturation -, 434
sheafification, 1432
shearing, 119, 123, 133, 202, 1402
sheaves, category of -, 1431
Shepard, Roger, 872
Shepp, Archie, 877
Sherman, Robert, 179
shift, 109
constant -, 109
elementary -, 109
ontological -, 142
shifter, 574, 578, 856, 892, 895
esthesic -, 344
poietic -, 344
Shusterman, Richard, 868
Siebers, Tobin, 880
sieve, 1428
closed -, 1431
covering -, 1430
sight reading, 556, 722
sign, 14, 850, 1047
deictic -, 16
lexical -, 16
musical -, 895
shifter -, 16
signature, 592
key -, 631
time -, 631
significant, 14
significate, 14
signification, 14, 15, 337
process, 14
signs
of coefficients, 731
system of -, 5
similarity, 159, 163, 226, 228
gestural -, 1289
metrical -, 163, 388
simple
form, 56
simplify to a -, 63
frame, 802
group, 1379
module, 1394
ring, 1386
simple forms, ordering on -, 76
simplex, 596, 988, 1445
affine -, 1461
category, 939
closed -, 1446
dimension of -, 1445
motivic -, 385
singular -, 1447
standard -, 1447, 1461
simplicial
cochain complex, 1447
complex, 779, 1445
functor, 988
map, 1445
metrical weight, 269
pair, 1447
weight, 283, 988
simplify to a simple form, 63
Simula, 802
simultaneous local meters, 500

singing, Han -, 1321
singleton interpretation, 275
singular
  n-cube, 966
  cochain, 1447
  homology, 965, 1476
  simplex, 1447
singular homology, hypergestural -, 1236
sink potential, 608
Siron, Jacques, 571
Sirone (Norris Jones), 1176
sister, 660
site, 1430
  Zariski -, 1415
skeletal space, 918, 974
skeleton, 870, 1445
  category, 1421
  global -, 996
  space-time -, 1226
sketch, 1056, 1059
  matrix-like -, 1057
  mixed -, 1028
  of gesture, 1030, 1038
sketches, category of -, 1029
sketchy interpretation, 622
skew field, 1385
skin, 1274
slave tempo -, 624
Slawson, Wayne, 237
slice, 102
  identity -, 275
f-slice, 275
Sloboda, John, 665, 743
slope, global -, 677
slot, 802
slur, 631, 632
SMAC, 612
small
  gestalt, 398
  infinitely -, 569
  set, 1420
smallness, 237
smartphone, 1274
  gesture, 1279
Smith III, Julius O, 1346
smooth, piecewise -, 1340
smoothing
  hierarchical -, 708
  kernel, 708
  kernel -, 708, 721
  operator, 721
SMPTE, 784
SNSF, 612, 665
sober, 1414
  weight, 379
socle, 1396
software
  development, 595
  engineering, 152
  for AST, 209
  music -, 250
solution
  global -, 1454
  local -, 1452
somaethetics, 868
sonata
  form, 249, 496, 791
  principle, 135
  scheme, 503
  theory, 496
sonification, 1111, 1116
sound, 1313, 1335
  classification, 232
  color, 159
  colors, space of -, 237
  conceptualization of -, 14
  design, 1121
  generator, 700
  grouping, 74
  meaning of -, 241
  natural -, 11
  physical -, 71
  pressure level, 1348
  reversed -, 121
  speech, 827
  transformation, 120
Sound Pattern of English (=SPE), 234
sounding analysis, 694
SoundScore, 1107
source
  open -, 666
  set, 203
space, 892, 893, 1067, 1438
  ambient -, 90
  Banach -, 1450, 1486
  basis -, 567, 588, 628
  color -, 832
  compositional -, 204
  concept -, 1034
  contour -, 205
  creative -, 1034
  dual -, 1488
  Euler -, 1349
  finger -, 915
  form -, 55
  fundamental -, 588
  gestural form -, 1087
  hierarchy, 588
  Hilbert -, 1486
  indecomposable -, 588
  instrument -, 597
  lens -, 922

locally ringed -, 1411
motif -, 384
of sound colors, 237
of spectral values, 1317
of vocal gestures, 1315
orbit -, 1378
parallel -, 590
parameter -, 357
pianola -, 567, 588, 628
pre-Hilbert -, 1341
ringed -, 1411
skeletal -, 918, 974
tangent -, 551, 1467
top -, 588
topological -, 1443
vector -, 1392
space-time, 892
skeleton, 1226
span, 520, 672
time -, 70
spanning tree, 1458
spatial
concept, 894
digraph, 914, 940, 1457
morphism, 942
hypergesture, 917
spatialization, 891
SPE, 234
special linear algebra, 1408
specialization, 162, 218, 231, 402, 591, 1026
abstract -, 402
abstract gestalt -, 402
basis -, 656
co-inherited -, 402
gestalt -, 402
Hasse diagram, 220
inherited -, 402
pianola -, 659
topology, 403
specialize, 227
species, 239
spectral
analysis, 524
decomposition, 708
gesture, 1318
music data, 964
participation vector, 524, 525
power, 525
vector, 832
Spectral Theorem, 1494
spectral values, space of -, 1317
spectroid, 908, 925
spectrum, 721
amplitude -, 1340
energy -, 1340
of poetical functions, 218
phase -, 238, 1340
prime -, 227, 239, 925, 1412
speculum mundi, 36
speech, 17
gesture, 1311
sound -, 827
vocalization, 1302
Sperry, Roger, 873
Spicker, Volker, 909
SPL, 1348
spline, 1461
of type $\mathcal{T}$, 1462
split
exact sequence, 1380
local commutative composition, 176
SplitOperator, 682
splitting, 629
operator, 649
spring equation, 1341
SQL, 668
square
Eulerian -, 1025
Latin -, 1024
notation, 1329
Staatliche Hochschule für Musik, 629
Staatliche Hochschule für Musik, 628, 687
stability, gestalt -, 226
stabilizer, 1378
stable concept, 226
staccato, 645
staff, 1322
stalk, 1411
standard
global - composition, 293
atlas, 293
chord, 436
hierarchy, 590
local - composition, 293
of basic musicological concepts, 92
simplex, 1447, 1461
composition, 173
standardized tempo, 723
Stange-Elbe, Joachim, xii, 628, 687
state
emotive -, 1266
sentic -, 604
stationary voice, 499
statistical
approach, 707
method, 612, 673
procedure, 198
Steibelt, Daniel, 120, 134
Steinway, 791
stemma, 556, 612, 619, 621–635, 1253
continuous -, 661
leaves of a -, 628

parameter -, 1257
prime -, 628
quiver, 660
tempo -, 624
theory, 736, 755
tree, 660
stemmata, gesture of -, 1256
stemmatic
cross-correlation, 634
grouping, 633
Stendhal, 857
Stendhal syndrome, 1288
Stengers, Isabelle, 27, 548
stepwise forward selection, 727
stereocilia, 1356
stereotactic depth EEG, 524
Stern, Daniel, 1266
Stimmigkeit, 848
stirrup, 1354
Stockhausen, Karlheinz, 59, 127, 233, 316
Stokes' theorem, 973, 1244, 1248, 1469, 1470
Stokes' theorem for hypergestures, 976
Stolberg, Leopold, 214, 232, 792
Stone, Peter, 115
Stopper, Bernhard, 93
strategy
experimental -, 694, 701
gestural -, 1270
paradigmatic -, 780
target-driven -, 694, 701
stratum, cognitive -, 869
Straub, Hans, xii, 189, 284, 442, 455, 456, 1517
Straus' zero normal form, 210
Stravinsky, Igor, 1293
strength factor, 610
stretching, 202
time -, 650
strict
extension, 443
style, 541
reduced -, 541
STRING, 44
string
empty, 45
landscape, 1175
of operations, 209
quartet, 69, 241, 774, 825
prehistory of the -, 826
theory, 825
theory, 910, 1173, 1175, 1482
musical -, 910
String Trio op.45, 933
strip
harmonic -, 253, 262, 442
Möbius -, 451, 476, 781
strong marked dichotomy, 518
structural
constant, 1408
homomorphism, 1385
playing, 1267
rationale of performance, 325
structuralist linguistics, 250
structuration, 1026
structure
formal -, 801
frame -, 591
internal -, 254
interspace -, 192
limbic -, 524
local vs. global -, 89
locally trivial -, 252
modal -, 314
of fibers, 756
regular -, 708
rhythmical -, 794
transitional -, 464
structures, matching of -, 718
structures-mères, 1028
structuring module, 665
Stucki, Peter, xii
style, 718
strict -, 541
sub-complex Kh, 210
sub-path operator, 483, 1388
subbase for a topology, 1443
subcategory, 1421
address -, 1089
full -, 1422
Yoneda -, 1437
subclass, 802
subcomplex, 1445
full -, 1445
subconscious, 529
subdivision, 622
subdominant, 263, 414, 445, 448
seventh, 420
subgroup, normal -, 1379
subject, 22
subjective gesture, 1042
subjectivity, 30
subobject, 1428
classifier, 1428, 1459
relation, 76
substance, 43
substitution theory, 1363
subtyping, 805
succession, 774
interval -, 207
successive interval, 526
successively increased alteration, 788
successor
admitted -, 533

pairing
allowed -, 531
suchness, 851, 1179
sukoon, 1306
sum
direct, 63
disjoint -, 1424
fiber -, 1424
SUN, 668
Sundberg, Johan, 553, 607, 609, 615, 743, 896, 1182
super-summativity, 167, 226, 271
superclass, 802
supersensitivity, 688
superstring theory, 1485
support, 337
functor, 257
of a local composition, 90
operator, 1336
set, 252
supporting valence, 1362
surface, integral -, 1456
surgery, concept -, 633
surgical epilepsy therapy, 524
surjective, 1372
suspension, 722
sustain, 1315
Sutera, Salvatore, 1315
svara, 1306
Swan, Richard, 1027
sweeping orientation, 108, 509
Swing, 764
switch, vocabulary -, 240
Sylow
decomposition, 79, 445, 510
group, 179
Sylow, Ludwig, 1380
Sylvester, David, 860
symbol
gestural -, 912
logical connective -, 1432
note -, 912
symbolic
absolute dynamics, 68
computation, 801
gesture, 911, 912, 1186, 1228, 1229, 1289
gesture curve, 912
kernel, 585
operator, 616, 650
pitch, 68
relative dynamics, 68
Symbolic Composer, 115
SymbolicOperator, 683
symmetric, 1373
group -, 1378
SYMMETRICA, 311
symmetries
in music, 14, 114–128
musical meaning of -, 132
semantical paradigm for -, 133
symmetry, 91, 98, 113, 162, 1402
of parameter roles, 127
breaking, 775
codification of a -, 128
contrapuntal -, 532, 1235
degree of -, 208
form of a -, 113
function of a -, 113
group, 144, 180, 470, 672
conjugation class of the -, 180
hidden -, 114
inner - of C-major, 123
local -, 533, 534
non-invertible, 127
semantical function of -, 113
semiotical -, 134
transformation, 249, 250
underlying -, 129
Synaesthesia Playground, 1274
synaesthetic, 885
synchronic, 16
axis, 328, 472
normalization, 754
synonym form, 56
synonymy
circular -, 64
type, 47
syntagm, 16
syntagmatic equivalence, 215
syntax modification, 813
Synthesis, 219, 473, 501, 503, 780, 781, 787, 791–799
synthesis, 1338
modal -, 1346
synthesizer, 1121
synthetic performance, 610
syntonic comma, 97
system
auditory -, 11
coefficient -, 1447
Hess -, 524
limbic -, 528, 606, 1361
meta -, 17
object-, 17
of signs, 5
non-linguistic -, 15
parameter, 473
vestibular -, 608, 897
weight -, 632
systematic
approach, 472
musicology, 328
understanding, 826

T

t-fold
tangent
composition, 551
morphism, 552
t-fold differentiable
tangent morphism, 552
t-gestalt, 391
tönend bewegte Formen, 251, 775
tableau, 1024
tactus, 377
tail, 914, 1457
Takemitsu, Toru, 1263, 1266
Takhtajan, Leon A., 1485
tala, 1309
tangent, 108, 511
bundle, 1451
category, 927
composition, 551
basis of a -, 551
global -, 557
interpretation, 558
manifold, 1467
morphism, 551, 557
space, 551, 1467
Zariski -, 927, 1415
tensor, 1468
torus, 511
tape music, 250
taquin, 1025
target-driven strategy, 694, 701
Tarry, Gaston, 1027
taste
common -, 752
musical -, 529
tautology, 1433
Taylor, Cecil, v, 548, 799, 862, 874, 909, 1182, 1291, 1295
Taylor, Yuval, 886
technical parameter, 236
technique
bodily -, 1280
cohomological -, 1061
instrumental -, 832
technology, gesture -, 1283
tectorial membrane, 1357
teleportation, 1180
telling time, 773
tempered, 93
class chord, 94
scale space, 96
tuning, 1350
template fitting model, 1361
tempo, 547, 550, 552, 561, 896
absolute -, 341, 561, 643
curve, 202, 220, 562, 607, 623, 723, 784
adapted -, 576
deformation of -, 576
daughter -, 562
discrete -, 28
field, 562
hierarchy, 624
istesso -, 555
mean -, 727
mother -, 562
musical -, 27, 28
operator, 653
relative -, 561, 685
relative local -, 643
relative punctual -, 643
slave, 624
standardized -, 723
stemma, 624
weight -, 653
tempo-intonation field, 565
TempoOperator, 683
temporal
box, 811
period, 376
relation, 816
tenor tone, 261
tensed movement, 532
tension, 414, 647
contrapuntal -, 531
global -, 677
harmonic -, 481, 482
tensor, 1468
algebra, 1469
alternating -, 1468
field, 1468
product, 1388
affine -, 1400
tangent -, 1468
Terhardt, Ernst, 1368
terminal, 1424
terminology, 203
territory, 591
Teschl, Gerhard, 1485
tesselating chord, 309
tesselation, isotypic -, 309
test
EEG -, 524
operator, 652
Turing -, 791
Wilcoxon -, 526
tetractys, 30, 1364
tetradic interpretation, 276
tetrahedron, 4, 6
text analysis, 609
textual
abstraction, 362
content, 1286
meaning, 330

predicate, 447, 454
representation, 776
semioses
category of -, 337
textuality, 334–348
texture, four part -, 826
Thalmann, Florian, vi, vii, 979, 1036, 1097
The Topos of Music, 843
theme, 271, 415
basic -, 202
paradigmatic -, 221, 222, 390
Reti's definition of a -, 404
theology, apophatic -, 1004
theorem, 1434
complement -, 208
counterpoint -, 534, 538
Fourier's -, 1340
Mason's -, 109
Mason-Mazzola -, 109
modulation -, 470
Seifert-Van Kampen -, 929
Stokes' -, 973, 1244, 1248
theorist, 1263
theory
American jazz -, 439
American set -, 979
catastrophe -, 226, 497
category -, 908, 1002, 1028
classification -, 831
contour -, 271
counterpoint -, 775, 839
degree -, 436
Fourier -, 925
function -, 437
Galois -, 1087, 1460
gestural -, 1182
gesture -, 893, 1173
global -, 220
group -, 212
homology -, 967
homotopy -, 1474
hyperset -, 1094
landscape -, 1175
modulation -, 1243
music -, 669
neo-Riemannian -, 979, 988, 990
of ambiguity, 245
of knowledge, 1005
of narrativity, 774
Palestrina-Fux -, 540
performance -, 319, 324, 909, 912
rhythmical modulation -, 502
set -, 249, 1028
sonata -, 496
stemma -, 736, 755
string quartet -, 825
string -, 910, 1173, 1175, 1482
substitution -, 1363
superstring -, 1485
transformational -, 889, 898, 908, 938, 947, 965, 979, 1119
valence -, 1353
wavelet -, 1344
thesis, 1192, 1287, 1325
world-antiworld -, 497
Thiele, Bob, 603
thinking, 22
by doing, 28, 30
cartesian -, 898, 1182
music, 22
operationalized -, 161
thinking music, 22
third, 414
axis, 96
chain, 260, 438, 675
closure, 260
interpretation, 260
minimal -, 260
weak -, 260
comma, 265
class, 265
coordinate, 1350
degree tonality, 450
distance, 511
major -, 62, 1350
weight, 676
3D vision, 361
threshold
global -, 675
local -, 675
throbbing rhythm, 1271
tie, 592
Tierny, Miles, 149, 357
tiling lattice, 426
timbral poetics, 241
time, 4, 338, 892, 1067, 1295, 1326, 1347
-slice, 250
-span reduction (=TSR), 377
complex -, 1186, 1218
dilatation, 70
event, 556
generator, 775
grouping, 99
imaginary -, 957, 1189, 1297, 1300
interval, 70
layers, 1299
logical -, 502
material -, 502
mental -, 547
musical -, 892
onset -, 1335
ontology, 775

performance field, 1191
physical -, 547
real -, 1189
regularity, 99
reverberation -, 1337
series, 707
signature, 69, 631
span, 70
reduction, 619
stretching, 650
structure, self-similar -, 799
telling -, 773
told -, 773
unfolding -, 1298
time span, 980
timed co-performance, 1302
timing
micro -, 220
microstructure, 719
Tinctoris, Johannes, 517
Tizol, Juan, 1297
Todd operator, 619
Todd, Neil McAngus, 555, 608, 610, 612, 621, 897, 1182
ToE, 869
told time, 773
tolerance, 642, 681
ToM_CD, vii
Tomasello, Michael, 856
Ton, 507
tonal
ambiguity, 493
function, 249, 263, 447
value, 447
inversion, 124, 788
tonalities, admitted -, 465
tonality, 249, 263, 414, 436, 447, 453
harmonic minor -, 460
major -, 460, 478
melodic minor -, 460
minor -, 478
natural minor -, 460
Pythagorean -, 461
reference -, 448
third degree -, 450
tone
recitation -, 261
space, notched -, 1363
tenor -, 261
tone parameters
mental -, 67
physical -, 68
tonic, 261, 414, 445, 448
seventh, 420
tonical, 263
Tonort, 507
tonotopy, 1361
tool, paradigmatic -, 789
top space, 588
top-down, 622
topic, 38, 329
topographic cube, 17, 32
topographical navigation, 19
topography, 9, 114
local -, 17
local character of -, 24
musical -, 17
topological
category, 939, 961, 1085, 1464
digraph, 993
functor, 939
meaning, 158
music theory, 1236
pair, 1474
space, 1443
irreducible -, 1413
topological space, pointed -, 1474
topology, 38, 157, 163, 225
AK_ξ-, 1237
algebraic -, 164, 919
associated -, 1450
base for a -, 1443
coherent -, 1444
coinduced -, 1444
colimit -, 1444
combinatorial -, 254
compact-open -, 914, 937, 1464
discrete -, 1443
dominance -, 231, 402
epsilon -, 397
epsilon gestalt -, 398
extension -, 430
finite cover -, 353
generators for a -, 1463
Grothendieck -, 149, 353, 1430
harmonic -, 443
Hausdorff -, 1445
indiscrete -, 1443
Lawvere–Tierny -, 357
limit -, 1444
maximal meter -, 268, 378
maximal meter nerve -, 379
on gestalt spaces, 395
on motif spaces, 395
product -, 1444
quotient -, 1444
quotient dominance -, 231
relative -, 1444
relative motivic -, 400
specialization -, 403
subbase for a -, 1443
uniform -, 1444
weak -, 1444

Zariski -, 164, 239
topor, 1438
topos, 3, 9, 21, 914, 979, 1029, 1177, 1429
Boolean -, 1434
gesture -, 945
Grothendieck -, 1431
hyperouranios -, 21
logic, 435, 930
of conversation, 826
Platonic -, 148
topos-theoretic background of modulation, 467
toroidal
sequence shape type, 388
shape type, 388
torsion group, 1381
torsion-free rank, 1381
torus, 917
tangent -, 511
TOS, 605
Toscanini, Arturo, 1292
total, 1372, 1373
touch, 894
Tower, Joan, 1264
Träumerei, 673, 709, 711, 739, 768
trace, objective -, 101
track, 250
tracking, motion -, 1290
tradition, 331
contrapuntal -, 199, 1367
performance -, 753
Pythagorean -, 22, 1364
traditional musicology, 22
transcendence, 21
transcendental element, 1460
transform, Fourier -, 1344
TransforMaster, 789
transformation, 405, 407, 774
control of -, 200
gestural -, 1272
natural -, 1422
of sound, 120
plane -, 786
Riemann -, 315
score, 785
symmetry -, 249, 250
transformational
approach, 204
invariance, 226, 271
meaning, 158
theory, 889, 898, 908, 938, 947, 965, 979, 1119
transit, 1037
transitional structure, 464
transitive, 1373
action, 1378
epistemology, 1059
transitivity, 229
translation, 133, 1396
part, 1397
transposability, 167
transposition, 116, 134, 226, 513
limited -, 126
transvection, 119, 133
tree, 335
spanning -, 1458
stemma -, 660
triad, 90, 414
augmented -, 262
diminished -, 262
major -, 262
minor, 262
triadic
degree, 1245
interpretation, 261
interpretation, 276, 450, 455, 466
trigeneric monoid, 444
trigger, motion -, 607
trill, 74, 625
True, 1433
Truslit, Alexander, 1293
truth, 1002, 1016
denotator, 335
TTO operator, 207
tube, Eustachian -, 1355
Tudor, David, 250
tuning, 249
just -, 1350
just-tempered -, 1351
justest -, 460
mediante -, 1351
Pythagorean -, 266, 478, 1350
tempered -, 1350
12-tempered -, 90
well-tempered -, 1350
turbidity, 123
Turing
machine, 553
test, 791
turning point, 465
12-tempered
scales, common -, 96
tuning, 90
two-dimensional
alteration, 787
TX7, Yamaha -, 525
type, 43, 1438
casting, 333
change, 331
coproduct -, 47
form -, 55
gestural form -, 1086
homotopy -, 1474
of a cellular hierarchy, 590

powerset -, 47
product -, 46
shape -, 385
synonymy -, 47
types
of gestures, 1286
ordering on -, 76
typology
of forms, 55
recursive, 49
recursive -, 42

U

Uhde, Jürgen, 248, 466, 497, 890, 893, 916, 950, 957, 1182
Unbewusstes, 528
uncertainty relation, 245, 425
uncontrolled paradigmatics, 165
underlying symmetry, 129
understanding, 325, 828
musical works, 324
systematic -, 826
Underwater, 1283
unfolding, 776
time, 1298
unfreezing process, 912
unfreezing gestures, 1287
Ungvary, Tamas, 895
unicorned view, 752
uniform topology, 1444
uniformity, 1444
union, 1371
unit, musical -, 90
unitary, 1492
unity, 36, 42, 49
universal ordering, 39
universe, 1175, 1420
algebraic -, 1029
of structure, 329
of topics, 329
unordered
graph, 1458
p-space interval, 207
pc interval, 207
UPIC, 1159
Ursatz, 329
Ussachevsky, Vladimir, 250
Utai, 13, 342

V

vakr, 1306
Valéry, Paul, v, 13, 41, 155, 547, 553, 561, 574, 585, 638, 860, 874, 889, 899, 1042, 1044, 1064, 1177, 1182
valence, 1362
supporting -, 1362
theory, 1353
valid sentence, 1433
validation operator, 349
valuation, interpretative -, 13
value
change, 632
initial -, 562
matrix, 766
participation -, 525
vampire, 947
Vanbremeersch, Jean-Paul, 1060
Varèse, Edgar, 323, 324
variable
bound, 1436
causal-final -, 767
explanatory -, 724
free, 1436
inclusion, frequency of -, 732
individual -, 1435
predicate -, 1435
propositional -, 1432
variable address, 53
variation, 405, 407, 507, 787
melodic -, 794
of the perspective, 151
pressure -, 1335
principle, 325
variational principle
global -, 1231
Hamilton's -, 1243, 1481
variations
calculus of -, 1471
cycle of -, 791
varieties of sounds, 232
variety instrumental -, 555
varying element, 1023
vector, 1391
analytical -, 723
field, 1245, 1451, 1468
instrumental -, 835
interval -, 208, 210
interval-class -, 204
invariance -, 208
prime -, 1351
shaping -, 723
space, 1392
spectral -, 832
spectral participation -, 524, 525
vector field, performance -, 1253
velocity, 608, 784, 1349
instantaneous -, 27
bow -, 833
concept of instantaneous -, 27
MIDI -, 913
physical -, 27
verbal description, 622
Vercoe, Barray, 761

Verdier, Jean-Louis, 354
Verillo, Ronald, 607
verse
  matrix, 213
  poetics, 248
vertex, 1375, 1445, 1457
  final -, 660
vertical
  grid vector, 788
  poetical function, 782
  poeticity, 214
vestibular system, 608, 897
Veyne, Paul, 1040
Viète, François, 1033
vibration, gestural -, 893
vibrato, 235, 237, 607, 833
  parameter, 833
Vico, Giovan Battista, 1012
Vieru, Anatol, 211, 314
view, 802
  configuration, 1108
  kernel -, 680
  process -, 1138
  unicorned -, 752
Vigarello, Georges, 874
Villon, François, 116, 198, 199
Vinci, Leonardo da, 856
viola, 825
Viola, Bill, 1265
Violi, Patrizia, 867
violin, 825
  family, 241, 828, 840
  hierarchy, 593
violoncello, 825
virtual acoustics, 701
visual
  navigation, 361
  programming language, 801
  semiotic marker, 813
visualization, 760, 761
  principle, 1107
vitality affect, 1266
vocabulary
  dodecaphonic -, 199
  extension, 40
  switch, 205, 240
vocal
  gesture, 1301, 1313, 1315
  hypergesture, 1316
vocal folds, 1321
vocal gestures, space of -, 1315
vocalist
  Carnatic -, 1301
  Hindustani -, 1301
vocalization, speech -, 1302
Vogel chromatic, 478
Vogel, Martin, 97, 418, 423, 427, 473, 478, 1350, 1499
voice, 509, 1313
  crossing, 509
  instrumental -, 220
  leading, 249
  pedal -, 499
  stationary -, 499
Voisin, Frédéric, 817
Volkswagen Foundation, 665
volume, 188
Vordergrund, 415
vowel, 234
Vuza
  class, 312
  rhythm, 312
Vuza, Dan Tudor, 70, 211, 268, 309, 312
Vygotsky, Lev Semyonovich, 854

W

W, 498
Wagner, Richard, 212, 670
walking, 607, 1304
wall, 1066
  bottom -, 631
Wall, Anthony, 856
Wallis, John, 1007
wallpaper, 1133, 1163
walls, 1072
  extended -, 1072
  open -, 1072
Wanderley, Marcelo, 901, 1182
Ward, Artemus, 615
wave, 1340
  acoustical -, 963
waveguide, 1346
wavelet, 236, 1344
  frame, 236
  Meyer -, 1346
  Morlet -, 1345
  Murenzi -, 1345
  theory, 1344
wavelet-transformed, 1344
weak topology, 1444
Weber-Fechner law, 1348
Webern, Anton von, 124, 125, 127, 163, 204, 246, 325, 711, 752
Wedderburn, Joesph, 1395
wedge, crescendo -, 641
wedge product, 1468
Wegner, Peter, 27
weight, 597, 610, 612, 619, 638, 673
  analytical -, 549, 553, 646
  class -, 283
  combination, 681
  continuous -, 639
  function, 1253

default -, 482
Pólya -, 191
product -, 191
harmonic -, 482, 647
induced nerve -, 379
inverted -, 681
metrical -, 375, 376, 646
mixed -, 671
motivic -, 408, 647
nerve -, 379
onset -, 99
profile, 218
recombination, 640
simplicial -, 283, 988
simplicial metrical -, 269
sober -, 379
system, 632
tempo, 653
third -, 676
watcher, 681
weighted graph, 239
Weil, André, 1041
Weissmann, Jody, 1035
well-ordered, 1373
well-tempered
modulation, 469
tuning, 1350
Weltgeist, 1174
Wernicke's area, 872
Weyl, Hermann, 161
WFR, 377
whatness, 21, 1179
whereness, 21
White, Andrew, 571
whole, 246, 273
whole-tone scale, 541
Whymper, Edward, 151
Wicinski, A.A., 622
Widmer, Gerhard, 611
Wieland, Renate, 248, 889, 893, 894, 899, 916, 950, 1182
Wieser, Heinz-Gregor, 524
Wilcoxon test, 526
wild problem, 756
Wiles, Andrew, 1179
Wille, Rudolf, 3, 113, 453
window
oval -, 1354, 1358
power -, 524
round -, 1358
Winson, Jonathan, 528
Witten, Edward, 1177, 1485
Wittgenstein, Ludwig, 38, 327, 1178
Wittgenstein, Paul, 883
WLOG, 110
Wolff, Christian, 250
word, 60, 1376
monoid, 1376
work, 11, 13
identity of a -, 14
motivic -, 276
production of a -, 12
world, 460, 497, 498
imaginary -, 1058
of the possible, 1052
real -, 1044
world-antiworld thesis, 497
world-line, 1483
world-sheet, 1185, 1186, 1192, 1471, 1483
global -, 1224
ramification, 1230
wrap form morphisms, 331
wrapped as local composition, 91
wreath product, 1380
writing, field -, 803
Wulf, Bill, 32, 667
Wyschnegradsky, Ivan, 93

X

X-ray procedure, 892
Xenakis, Iannis, 30, 211, 1159
XML, 1096

Y

Yamaha, 687, 700, 1342, 1346
CX5M, 525
RX5, 220, 791
TX7, 525
TX802, 220, 236, 239, 791
Yemenite cantor, 1301
Yoneda
embedding, 1397, 1423
pair, 1437
philosophy, 92, 145, 152, 466, 828, 1475
subcategory, 1437
Yoneda Lemma, 142, 280, 324, 962, 870, 938, 1006, 1022, 1031, 1034, 1037, 1045, 1056, 1061, 1073, 1179, 1255, 1474
Yoneda, Nobuo, 145, 245, 323, 828, 851, 1179

Z

$\mathbb{Z}$-addressed motives, contrapuntal meaning of -, 100
Z-relation, 210
Zahorka, Oliver, 628, 665, 687, 1088
Zalamea, Fernando, 1037, 1060
Zarca, Bernard, 1028, 1043
Zariski, Oskar, 164
Zariski
site, 1415
tangent, 109
space, 927, 1415
topology, 164, 239
Zarlino, Gioseffo, 122
Zbikowski, Lawrence, 1265, 1308

Zermelo, Ernst, 1374
zero address, 53
zig-zag, motivic -, 277, 781
Zurechthören, 1353
Zurich school, 612, 979
Zwiebach, Barton, 1482

The manufacturer's authorised representative in the EU is Springer Nature Customer Service Centre GmbH, Europaplatz 3, 69115 Heidelberg, Germany. If you have any concerns regarding our products, please contact ProductSafety@springernature.com

Printed and bound by CPI Group (UK) Ltd, Croydon, CR0 4YY
01/10/2026
02225992-0002